QUALITY COSTS: IDEAS & APPLICATIONS, VOLUME 1

A Collection of Papers

ASQC Quality Costs Committee

Andrew F. Grimm, Editor

Quality Press
Milwaukee

QUALITY COSTS: IDEAS & APPLICATIONS, VOLUME 1

A Collection of Papers

ASQC Quality Costs Committee

Andrew F. Grimm, Editor

Library of Congress Cataloging-in-Publication Data
Quality costs : ideals and applications / [compiled by] Jack Campanella.
p. cm.
Includes bibliographical references.
ISBN 0-87389-047-7 (v. 1)
1. Quality control — Costs. I. Campanella, Jack
TS156.Q3626 1989
658.5'62 — dc20

2nd Edition, 1987
1st Edition, 1981

1098765432

ISBN 0-87389-046-9

Printed in the United States of America

Quality Press, American Society for Quality Control
310 West Wisconsin Avenue, Milwaukee, Wisconsin 53203

Quality Costs Articles

1970 Transactions **Page**

A Cost-Determined Quality Control Plan for Adjustable Processes *William J. Kennedy* 1

Reducing Manufacturing Costs While Maintaining Reliability and Quality *T. J. Cartin* 5

A Cost Oriented Quality Control System *Joseph Sadowski* 11

Cost-Effective Quality *Michael J. Dwyer* 15

Activities for Reduction of User's Costs *Yohei Koga* 19

1971 Transactions

Prediction of Human Performance in a Quality Control System: A Step Towards Cost Control *A. Raouf* 29

Uncovering the Hidden Costs of Defective Material *C. B. Rogers* 35

1972 Transactions

Methods for Selling Total Quality Cost Systems *A. Q. Whitton* 43

Reducing Quality Costs *W. N. Moore* 53

Honeywell's Cost Effective Defect Control Through Quality Information Systems *Ronald R. Cerzosimo* 61

1973 Transactions

Optimizing Attribute Sampling Costs — A Case Study *Edgar W. Dawes* 71

Quality Cost Measurement and Control *Anthony Agnone* 79

Quality Cost at Work *John T. Hagan* 89

1974 Transactions

Reducing Clerical Quality Costs *William Latzko* 95

Quality Costs: Where are They in the Accounting Process? *A. F. Grimm* 101

Attacking Quality Costs *Robert G. Stenecker* 117

1975 Transactions

Quality Cost Management for Profit *Richard K. Dobbins* 127

Quality Costs Articles (cont.)

1976 Transactions **Page**

Cost Effectiveness of Corrective Action — Quality Costs — A Place for Decision Making and Corrective Action
Richard K. Dobbins 139

Quality Costs — A Place on the Shop Floor *Edgar W. Dawes* 149

Quality Cost — A Place for Financial Impact *O. G. Kolacek* 157

Quality Cost — A Place for Evaluation of Customer Satisfaction *H. James Harrington* 163

Cost Quality Control Chart for Variables *A. D. Oak* 165

Life Cycle Costing — A New Dimension for Reliability Engineering Challenge *T. David Kiang* 169

A Business Performance Measure of Quality Management
R. A. Cawsey 179

1977 Transactions

The Hidden Aspect of Vendor Quality Cost *William O. Winchell* 185

Management Budget Control: Quality Labor Standards
Ronald L. Pollard 191

1978 Transactions

Extending Effectiveness of Quality Cost Programs
Richard K. Dobbins 201

A Method for Predicting Warranty Costs *Andrew F. Grimm* 207

Quality Cost and Profit Performance *F. X. Brown* 215

Quality Costs — What Does Management Expect?
Frank M. Gryna 223

The Philosophy and Usefulness of Quality Costs
William N. Moore 233

Reducing Appraisal Costs *Edgar W. Dawes* 239

Reducing Failure Costs and Measuring Improvement
W. O. Winchell 245

1979 Transactions

Developing a Cost Effective Program — How to Start
Victor J. Goetz 251

Can Quality Cost Principles be Applied to Product Liability
Dorian Shainin 257

Quality Costs Articles (cont.)

Page

Quality Costs — A New Perspective *Robert C. Kroeger* 259

Bayesian Cost Analysis of Rectifying Rejected Lots
David L. Kimble 271

1980 Transactions

Guide for Reducing Quality Costs *James F. Zerfas* 279

Source Surveillance and Vendor Evaluation Plan
Narendra S. Patel 285

Quality Costs and Strategic Planning *F. X. Brown* 297

Cost Reduction Through Quality Management *Frank Scanlon* 303

Guide for Managing Vendor Quality Costs *W. O. Winchell* 309

Quality Costs — A Review and Preview *Clyde W. Brewer* 315

Quality Costs Principles — A Preview *Frank J. Corcoran* 321

Innovations in Quality Costs in the New Decade
Clayton C. Brewer 327

Quality Program Modeling for Cost Effective Tailoring
W. C. Wilhelm 337

Quality/Reliability Challenges for the 1980s *T. Gurunatha* 351

1981 Transactions

Computer Isolation of Significant Quality Costs *J. E. Mayben* 359

How to Effectively Implement a Quality Cost System
Alvin Gunneson 375

Our Only Output is Information *R. W. Stalcup* 387

Cost Improvement Through Quality Improvement *Frank Scanlon* 391

Quality Cost — A Key to Productivity *H. James Harrington* 397

Quality Cost Analysis: A Productivity Measure *L. James Esterby* 413

Selection of MIL-STD-105D Plans Based on Costs
Burton S. Liebesman 425

The Cost of Software Quality Assurance *W. D. Goeller* 437

Guide for Managing Vendor Quality Costs *W. O. Winchell* 447

Quality Optimization via Total Quality Costs *A. H. Zaludova* 455

Quality Costs Can be Sold: Part I *John D. Breeze* 467

Quality Costs Can be Sold: Part II *John D. Breeze* 475

Minimized Cost Sampling Technique *Dennis D. Lee* 483

Quality Costs Articles (cont.)

1982 Transactions **Page**

Guide for Reducing Quality Costs *Ronald J. Wiliams* 491

Improved Productivity Through Quality Management
Frank Scanlon 499

Study Costs and Improve Productivity *William J. Ortwein* 517

A Participatory Approach to Quality *Alvin Gunneson* 527

Mr. CEO, Your Company's Quality Posture is Showing!
Richard K. Dobbins 535

Measuring Quality Costs by Work Sampling *L. James Esterby* 543

Minimizing the Cost of Inspection *William J. Latzko* 549

"Q" Costs/Results: The Petroleum Industry *Clyde W. Brewer* 557

Principles of Quality Costs *Jack Campanella* 563

Quality Costs — Pay *James Demetriou* 581

A COST-DETERMINED QUALITY CONTROL PLAN FOR ADJUSTABLE PROCESSES

William J. Kennedy, Jr., Ph.D. Candidate
Department of Industrial Engineering
Virginia Polytechnic Institute
Blacksburg, Virginia

Simulation can be used to determine the quality control plan which will yield the most net revenue for an adjustable production process. Five variables make up this control plan: the upper and lower mean control limits, the range control limit, the sample size, and the interval between inspections. The simulation technique used determines the optimum values of these five variables for a given quality-based pricing system.

In this model, a product being manufactured has a critical dimension upon which depends the price paid for the product. The dimension comes from a normal probability distribution. At the beginning of each production run, the mean of this distribution is adjusted to the specification mean and the standard deviation adjusted to less than the specification standard deviation. At some random time after the start of a production run, a disruptive event occurs. Immediately, the mean and standard deviation begin to increase linearly with time, and they continue to increase until either an inspection shows the process to be out of control or the production run ends. The process is out of control if the inspection sample mean is too low or too high or if the sample range is too high, where "low" and "high" are determined by the relevant control limits. If the process is out of control at an inspection, production is stopped. The process is then adjusted until the probability distribution of the critical dimension is the same as it was at the beginning of the run. Once an adjustment has been made, the critical dimension behaves as it did after the beginning of the production run.

Certain assumptions are made about the various probability distributions involved. The time between an adjustment and the next disruptive event has the negative exponential distribution, and the measurements of the critical dimension come from a normal population whose mean and standard deviation have been described. Three other assumptions are made: (1) the inspections are equally spaced in time, (2) the number of items inspected is the same for all inspections, and (3) the inspected items come from those items produced immediately before the inspections.

The customer determines the price per item on the basis of a random sample from the production run. He assumes that the percentage of defective items in his sample is the same as the percentage defective in the entire production run, and, on the basis of this assumption, he pays for the fraction of produced items which are not defective. The producer is represented in the customer's inspection facility to verify this fraction defective. The producer and consumer have agreed that an item will be considered defective if the critical dimension is outside of negotiated specification limits.

The quality control cost and the net revenue are easily calculated. Denote the total variable quality control cost by TC_{qc}. Then

$TC_{qc} = N_i \times C_i + N_u \times C_u + N_a \times C_{adj}$, where

N_i = number of inspections

C_i = cost per inspection

N_u = number of units inspected

C_u = cost per unit inspected

N_a = number of adjustments

C_{adj} = cost per adjustment.

Denote the revenue from the sale of the product by R_{in} .

$R_{in} = f \times N_p \times C_b - (1-f) \times N_p \times C_r$, where

f = fraction of items accepted

N_p = number of items produced in a production lot

C_b = price paid per accepted item

$1 - f$ = fraction of items rejected

C_r = cost to producer of a rejected item.

The revenue function to be maximized is: $= R_{in} - TC_{qc}$.

It is the purpose of this simulation to find the values of the quality control variables which together produce the most net revenue. The variables to be determined are: the upper mean control limit, the interval between inspections, and the number of items sampled per inspection. If the control limits are loosened, the number of adjustments will decrease, causing the quality control cost to drop. The number of defective items reaching the customer, however, will increase, causing a decrease in revenue coming in. The overall effect on net revenue will depend on the cost coefficients C_i, C_u, C_{adj}, C_b, and C_r. If the interval between inspections is shortened, more inspections will be made per run, causing the outgoing quality to improve and the incoming revenue to increase. The increased number of inspections will, however, raise quality control cost. The same reasoning holds for increasing the size of the sample taken at each inspection. Hence finding an optimum value of each of the five variables involves making a compromise between increased revenue and increased quality control cost.

The optimization technique used is a variation of the classical gradient technique. All gradient techniques depend on the fact that the maximum rate of increase of a function of several variables is in the direction of the vector of first partial derivatives of the function, the so-called gradient vector. In the classical technique, a starting point is chosen and the gradient evaluated at that point. The function is then calculated along the line from the starting point in the direction of the gradient until a maximum is reached. There the values of the variables giving the maximum become the coordinates of the new starting point. The gradient is then recomputed and a new maximum found. Proceeding in this way, successive maxima are found until the difference between two successive maxima is less than some preset amount.

The technique described above depends implicitly on the assumption that a fixed set of variable values will give one and only one value of the function to be maximized. In a simulation, however, this assumption is not valid. In a simulation, the function to be maximized will be known only as the mean of values generated by a series of simulation runs. This causes problems in the computation of the gradient vector. Where the function f to be maximized is a known differentiable function of the variables $x_1, x_2, \ldots, x_n$, it is possible to compute

$\frac{\partial f}{\partial x_i}$ approximating from

$$(1) \quad \frac{\partial f}{\partial x_i} = \frac{f(x_1, x_2, \ldots, x_i+d_i, x_{i+1}, \ldots, x_n) - f(x_1, \ldots, x_n)}{d_i}$$

for $i = 1, 2, \ldots, n$.

In a simulation, however, $f(x_1, \ldots, x_n)$ is not known. Instead, $\bar{f}(x_1, \ldots, x_n)$ is known where $\bar{f}$ is the mean value of f taken from all values of f obtained from, say, m simulations all using the same values of $x_1, \ldots, x$. Hence instead of having fixed values of f to work with in equation (1) we have only values of $\bar{f}$. With these values of $\bar{f}$ it is possible to obtain an expression analogous to equation (1), namely

$$(2) \quad \frac{\partial f}{\partial x_i} = \frac{\bar{f}(x_1, \ldots, x_{i-1}, x_i + d_i, x_i + {}_1, \ldots, x_n) - \bar{f}(x_1, \ldots, x_n)}{d_i}$$

But the indicated difference may not be statistically significant. If a t-test shows no significant difference between the two values of $\bar{f}$, one of three things can be done. First, increase the value of d_i. Second, increase the number of runs so as to increase the power of the t-test. Third, lower the α-level and pick the d_i large enough that the two values of $\bar{f}$ in equation (2) should be different. Then define some maximum number of runs n_{max} that is acceptable for each calculation of $\frac{\partial f,}{\partial x_i}$ say $n_{max} = 100$. Start the computation using a given number of runs to calculate each of the two values of $\bar{f}$. If their difference is not statistically significant, double the number of runs and repeat. Continue this process until either the difference is significant, in which case calculate $\frac{\partial f}{\partial x_i}$ by equation (2), or the number of runs exceeds n_{max}, in which case set $\frac{\partial f}{\partial x_{i+1}}$. If all of these partial derivatives are found equal to zero, the maximum can be assumed to have been reached. It has been the author's experience, however, that usually some other maximization criterion is met first.

There is a further problem. In the deterministic case, the search for a maximum is discontinued when the difference between two successive maxima is less than some prescribed amount. In the simulation case, however, a better policy seems to be to discontinue the search when the difference between two successive optima is not statisically significant.

The technique presented here seems to have an advantage over most search techniques for optimizing a simulation in that all variables are changed at the same time, rather than one variable being searched each time.

The hypothesis underlying this entire study was that the five variables of the optimum control plan would be influenced by the quality control costs and by the quality-dependent revenue. This hypothesis has been confirmed by the computer runs made on the IBM360/65 at Virginia Polytechnic Institute. The variable that had the most effect on net revenue was the number of inspections per production run. Changes in the upper and lower mean control limits significantly changed the net revenue, as did changes in the range control limit. The variable that had the least effect was the number of items in each inspected sample.

These results indicate that a simulation technique can be used to find the optimum-revenue quality control plan for an adjustable process. In using a qualilty-dependent revenue for the product, and in finding an optimum range control limit, I believe we have provided a useful new tool for the cost-conscious businessman.

REFERENCES

1. Duncan, Acheson J., *Quality Control and Industrial Statistics,* Homewood, Illinois: Richard D. Irwin, Inc., 1965.
2. Duncan, Acheson J., "The Economic Design of $\bar{x}$ Charts Used to Maintain Control of a Process" (1956) *Journal of the American Statistical Association.* 51: 228-242.
3. Gibra, Isaac N., "Optimal Control of Processes Subject to Linear Trends" (1967) *The Journal of Industrial Engineering.* 18: 35-51.
4. Grant, Eugene L., *Statistical Quality Control,* New York: McGraw-Hill, 1946.
5. Hartley, H. O., "The Range in Random Samples" (1942) *Biometrika.* 32: 334-348.
6. Knappenberger, H. Allen, and Grandage, A.H.E., "Minimum Cost Quality Control Tests" (1969) *AIIE Transactions.* 1: 24-32.
7. Lord, E., "The Use of Range in Place of Standard Deviation in the t-test"(1947) *Biometrika,* 34: 41-67.
8. Parzen, Emanuel, *Modern Probability Theory and Its Applications,* New York: John Wiley and Sons, 1960.
9. Pritsker, A. Alan B., "The Setting of Maintenance Tolerance Limits" (1963) *The Journal of Industrial Engineering.* 14: 115-119.
10. Roeloffs, Robert, "Acceptance Sampling Plans With Price Differentials" (1967) *The Journal of Industrial Engineering.* 18: 96-100.
11. Shewhart, Walter A., *Economic Control of Quality of Manufactured Product,* New York: D. Van Nostrand Company, Inc., 1931.

REDUCING MANUFACTURING COSTS WHILE MAINTAINING RELIABILITY AND QUALITY

T. J. Cartin, Operations Manager
Bendix Communications Division, Baltimore, Maryland

It is a common belief that manufacturing costs can be cut, but at the expense of quality and reliability. In fact this is very often what happens. But it doesn't happen because there is some inviable relationship between costs and quality. Most often it happens because there is inadequate cost information, the product quality and reliability level is not known, and the cost relationship between manufacturing, quality and reliability is not measured. How many managers know with any accuracy what their in-process or end item quality level or costs are? How many managers consider quality as distinct manufacturing cost? Unless you know the value of all manufacturing costs you are playing "quality roulette" when you make a manufacturing change to reduce costs when the effect on the quality or its time related cousin, reliability, is not first determined, or will be quickly detected. In fact, you cannot make an intelligent manufacturing process change or cost reduction unless you know your in-process quality costs and level and have some knowledge of product reliability.

Manufacturing costs can be reduced while maintaining product quality and reliability. In fact it has been found that with the proper system, quality (cost and level) is often improved. The key is an accurate cost reporting system combined with the proper quality level reporting.

The terms manufacturing costs, quality level and reliability are used in a practical sense. Manufacturing costs include manufacturing labor reductions through methods improvement. Quality level refers to knowing the reject level of vendors' parts and the in-process acceptance level of your products in inspection and test. The cause of rejections in the order of their frequency and cost must also be known. Reliability is used in reference to what the customer experience is, in terms of returned material or warranty costs.

Fundamental to the success in achieving the goals of this program may be the need to change the common approach of industrial engineering to cost identification. They must learn to consider total costs of manufacture. This may seem elementary and standard procedure but in many industries it is not. The focus on *manufacturing* cost elements often creates tunnel vision. This is a result of the traditional emphasis of industrial engineering education reinforced by common organizational practices, at least in many industries which assemble and test their products. In these industries the industrial engineering department concerns itself with the methods, labor standards and costs of the assembly process. Testing a product is usually not approached the same way because it is more technical in nature and labor standards are much less subject to study because of the variable of trouble finding and subsequent retest. Also, test equipment often involves a significant investment and once applied it "fixes" the test cost and makes cost improvement less practicable. Most products are getting more complex so this test cycle has become a larger factor in cost. The common organizational practices mentioned refer to a common practice of having assembly labor controlled by industrial engineering and test operation by an engineering group in another organization. The education and practices of these different groups tend to create a significant barrier between them resulting in each going his own way. There is an obvious emphasis being placed on quality and reliability, and manufacturing costs can no longer be reduced without consideration, or at the expense, of quality.

QUALITY INFORMATION

Quality information is derived from an in-process defect reporting system. From a good system one can determine which problems are most frequent and which are the more costly. It can indicate whether a problem is with a vendor, in the design, or in the manufacturing process. Such problems *are* in the manufacturing costs but intelligent decisions can't be made on reducing this cost unless the cause is clearly identified and permanently fixed. Too often in manufacturing, "fixes" are made by the foreman or industrial engineer based on their guess. The symptom is treated but the cause is not removed.

Such a complete cost and yield reporting system may appear to be a costly task but a very adequate system can be operated for $0.03 per direct labor dollar. In military contracting in particular, a product must be built exactly to the drawings with no authority to deviate from even a minor dimension and the Quality Specification MIL-Q-9858 requires extensive records and controls through manufacturing inspection and testing. A quality reporting system is mandatory.

LABOR REPORTING

Figures 1, 2 and 3 show the assembly, production test, and inspection labor expenditures for the first year and a half of production of a complex electronics package for military aircraft. The "bid cost" is presented in terms of a learning curve and actual costs are reported in the same form but as cumulative and two week period actuals. The labor improvements in assembly are due almost exclusively to methods improvement in addition to some initial learning. Test performance is one good indicator of vendor and manufacturing process quality. Not unusual in this type of product, test time is very significant, running 40% of assembly labor. A minor process or part quality defect can result in considerable trouble shooting time, and repair and retest. Most of the test time improvement shown is a result of correcting quality problems. In the one year period from the 400th unit to the 5000th unit, assembly labor dropped 50% per unit and test labor 66% per unit. Also, out of 5700 units shipped, there has been only 2% defects reported by the customer. Of those so far verified less than half are actually defective.

The other direct cost is inspection. This, as in test, will reflect not only a quality cost but quality level. The two tend to be inversely related (assuming you hold the same quality standard when in trouble). Inspection costs are reported on a percent of assembly hours or total hours. With a constant reduction in assembly hours per unit product even a constant inspection percent means lower inspection costs per unit. On this program inspection costs are now 18% of assembly. The inspection time dropped from 12 to 9 hours per unit.

QUALITY REPORTING

Cost reports will indicate conformance between actuals and the plan (targets) but are not sufficient to indicate the cause when a deviation occurs. Even where a job is below target, cost reduction efforts continue and as long as they continue they are a potential cause of quality deterioration. Therefore, a sensitive in-process defect reporting system must be maintained. Such a system can provide information as shown in Figures 4 and 5. Figure 4 illustrates the overall product quality experience in test and inspection in terms of percent accepted and defects per unit. This is useful trend information. The basic data reported by each inspector and tester can be easily manipulated in a computer to indicate different information. For example, Figure 5 shows the cause of defects in a printed circuit assembly line. It shows that a significant improvement was made in soldering and parts installation defects when the assembley line was changed to a progressive assembly with shorter cycle times. In another case, an automatic lead cutting machine was introduced to slice off the leads of 350 components inserted onto each printed circuit board. This resulted in a major assembly labor saving. However, shortly after, there was a sudden jump in test failures caused by a diode failure. Quick detection and analysis showed that the lead cutter was overstressing the internal connection in a certain type of diode on one type of board. Manual lead cutting was immediately reinstated until the problem could be solved. Both costs and quality levels were maintained.

The continuous data reporting system made detection and accurate cause removal possible. Without accurate, time based, quantitative data there is no telling what would have been blamed for the failure, (i.e., bad batch of diodes, test overstress, soldering temperature). Also, if the board had simply repaired and shipped, without cause removal, more of those overstressed diodes would have failed for the customer.

Each reporting system should be designed for the product and process. In this case, as the program progressed and the number of significant problems decreased, the charts were dropped and replaced with a short report listing the top five defects in-process for any report period.

A cost and quality reporting system has many side benefits. It is very educational to manufacturing and quality control engineers and supervisors. Management must also have this kind of information to properly evaluate their cost reduction programs. Too often a procurement or manufacturing cost reduction creates a quality problem that results in higher costs but is not traced back to their true cost, a cost reduction. A real supervisor morale booster is also generated through the fact that supervisors see their problems being solved, and the effect of the solution on cost.

This experience, as well as that on other production programs, demonstrates that product quality and reliability can be maintained while reducing manufacturing costs.

OTHER FACTORS

A cost factor not usually given proper and regular attention is the assignment of the minimum manufacturing job classification to each operation. There is a tendency for line foremen to want the most flexibility possible in the assignment of his operators. If not regularly policed this results in operators performing work of a lower classification. A careful periodic review of this factor has resulted in a downgrading of as many as 30% of assembly operators. This does not affect quality.

COST REPORTING

A semi-monthly manufacturing cost report is started at the beginning of every contract and typically reports in the following format with more or less detail than shown.

Assembly Labor	Equivalent Quantity Completed	Unit $ Target	Unit $ Actual	Total $ Target	Total $ Actual	% Actual vs. Target Current	— Prior
Major Subassemblies (Itemized)	17.34	354	279	6,132	4,828	79	81
Set-up Labor		8.6%	10.1%	5,299	6,265	119	115
Rework		13.1%	5.6%	8,045	3,462	43	45
Contract Target				178,809	41%Spent	43% completed	
Text labor							
Major Subassemblies	12.56	1,547	1,471	19,429	18,472	95	93
Inspection Labor							
Total to Date	15.64	669	586	10,463	9,153	87	89
% of Assembly		13.6%	12.4%	26,821	35% Spent	39% Completed	

On smaller jobs the trend charts are not applicable but a cost report is published.

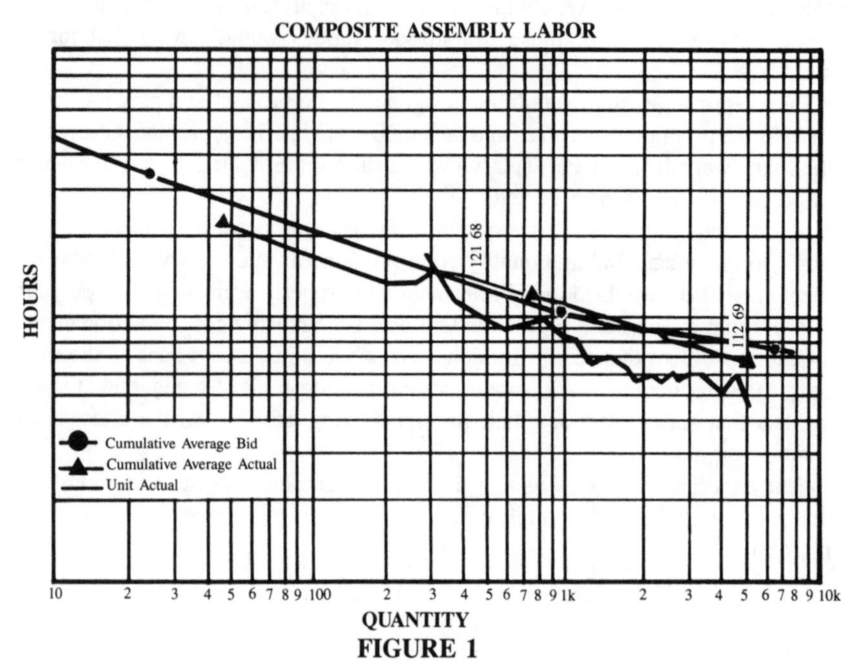

FIGURE 1

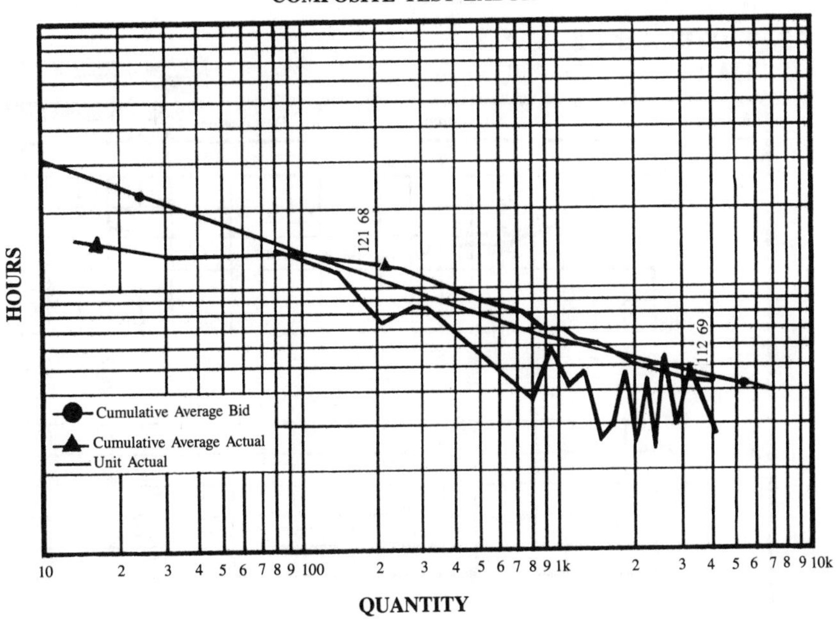

QUANTITY

FIGURE 2

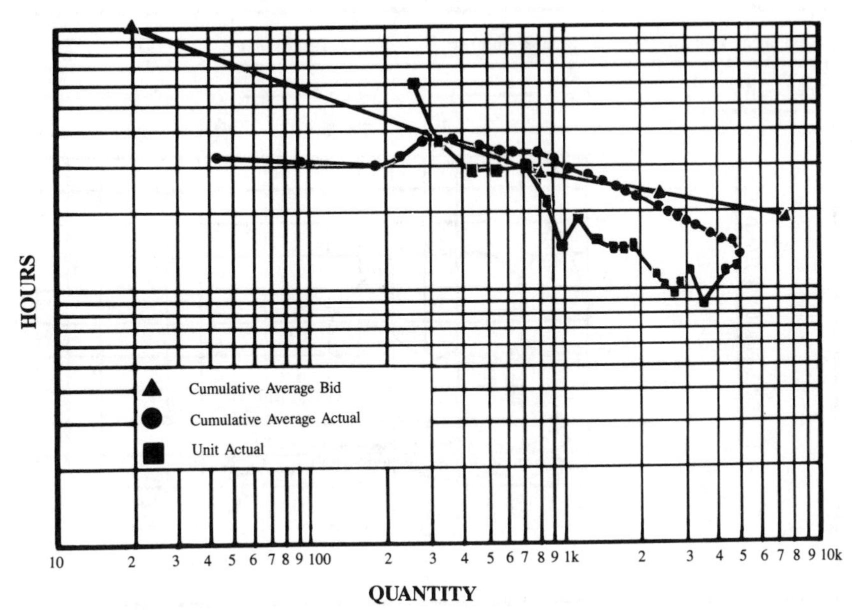

QUANTITY

FIGURE 3

QUALITY PERFORMANCE CHART

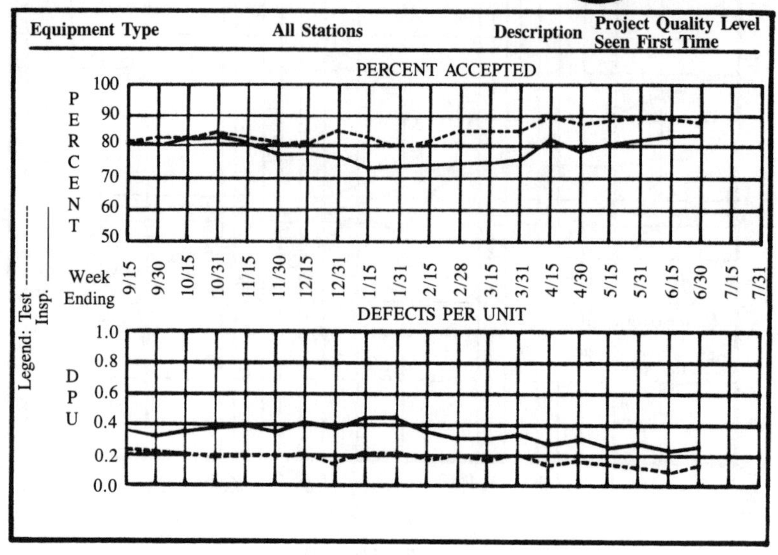

FIGURE 4

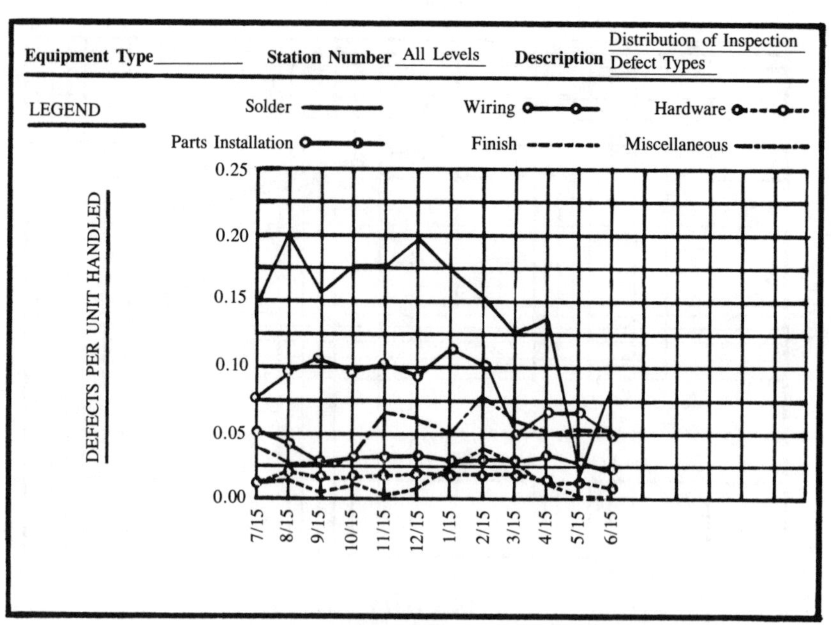

FIGURE 5

A COST ORIENTED QUALITY CONTROL SYSTEM

Joseph R. Sadowski, Supervisor, Statistical Quality Control
Crucible Specialty Metals Div., Colt Industries
Syracuse, New York

Early in 1964 a review of Crucible Specialty Metals' quality reporting procedure showed some serious deficiencies. The basic quality parameter consisted of a monthly report of rejections compiled by the Accounting Department and was issued in the middle of the following month. Rejected items were identified with the order number, grade of steel, amount rejected, and a 3-digit code identifying: (lst digit) the stage of rejection — whether a billet, bar, coil, forging, etc., (2nd digit) the department responsible for the rejection, and (3rd digit) the reason for the rejection.

Some of the primary drawbacks of the procedure were:

1. The bulk of the rejection reporting was done by the Inspection Departments. This lead to bias in assigning responsibility for and in the accuracy* of reporting of rejections.
2. The 3-digit rejection code used was too generalized, making a rejection analysis and identification of problem areas time-consuming and inaccurate. As an illustration, the code 3-2-3 meant that the rejection occurred in the hot-rolled bar stage, one of five rolling mills was responsible for the rejection and the reason for the rejection was some type of surface defect.
3. Except for large "publicized" rejection losses, departments responsible for rejections were not apprised of these losses when they occurred, and the details surrounding the rejections were not "remembered" or available due to the time lag between occurrence and apprisal of responsibility. As an example, January rejections would be reported in the middle of February, thus making a 5 to 6 week occurrence — reporting interval for rejections occurring in the first week in January.
4. The dollar losses of the rejections were not known until the Accounting Department report was issued.
5 Departments responsible for rejections were not charged for the dollar losses incurred.

It became obvious that the procedure would have to be revamped so that rejection information was disseminated as quickly as possible and problem areas accurately pinpointed to enable corrective action to be taken.

The first step taken was the reorganization of the quality control section of the Metallurgical Department. All Inspection Departments were transferred to the quality control section and the entire Metallurgical Department was renamed the Technical Services Division. A statistical section was added to act as liaison among

* Since metal losses at the Division can only be reported as yield losses or rejections, many "non-significant" rejections were reported as yield losses thereby distorting both the yield and rejection performance data.

the quality, operating and accounting groups. The main function of this section is to isolate areas of quality problems and channel this information to the quality and operating groups for corrective action.

The second step was the revision of the rejection reporting procedure as follows:

1. Defining rejections to distinguish these from yield losses.
2. Developing a more descriptive and specific rejection code. The original three-digit code was expanded to six-digits. Contrasted to the example used previously, the current code would be 3-22-304. The first digit identification was not changed; the rejection occurred in the hot-rolled bar stage. The next two digits refer to a specific rolling mill rather than any one of the five; in this case it would be the 14″ mill. The last three digits give the specific reason — scale pits — rather than the generalized information of a "surface defect". The coding of rejections and assignment of responsibility is vested solely with the Technical Services Division.
3. Revising the rejection reporting form to include all data in the format required by the departments involved, i. e., Operating, Accounting and Technical Services. The current form is a five-part form with one copy distributed to each of the following departments:

   Production Planning and Scheduling
   Accounting
   Technical Services
   Responsible Department
   Originating Department

   These rejection notices are distributed within 24 to 48 hours after coding. The benefits of this early reporting to the responsible department are obvious.
4. One of the more significant items of information added to the revised rejection report was the inclusion of the "estimated value" of the rejection by the Technical Services coders. A rather generalized table of rejection costs per pound by grade group and stage is used for this purpose. This provides all parties with a reasonably accurate estimate of rejections as they occur on a daily basis. Additionally, the monthly Accounting Department report with the actual dollar losses was revised and is issued weekly instead of monthly to keep departments informed regarding their performance on a more current basis. Each weekly report also carries the cumulative total for the month so that the final weekly report in a given month is also automatically the monthly report.
5. Perhaps the most significant change in the procedure was the allocation of the rejection losses to the departmental cost control reports. This served as a strong stimulus for the producing areas to actively participate with Technical Services personnel in corrective action.

These changes brought all of the quality-oriented, technically trained personnel into one group so that the inspectors, inspection foremen, metallurgists and metallurgical technicians continuously interact with the operation personnel directly in the manufacturing areas while material is being produced. These daily face-to-face relationships have greatly improved the accuracy of the art of communication.

Since the revised procedure went into effect in April 1964, action taken based on analysis of rejection performance and other data has, in many areas, improved quality and lowered costs. However, favorable results were not obtained in all cases. Some chronic offenders continued to resist all corrective efforts, sometimes despite procurement of additional or more specialized equipment. If after a thorough study by accounting, operating, and the various staff departments a product was still a "loser", abandonment of this market area was given serious consideration. This action was taken on some products and provided two benefits. In addition to shedding a "profit-eater", productive facilities and personnel energies were released for profitable products. In contrast to these cases, quality improvements on some other products has resulted in marketing these items with unconditional guarantees.

COST-EFFECTIVE QUALITY

Michael J. Dwyer, Supervisor, Industrial Quality Engineering
Conductron-Missouri, St. Charles, Missouri

In times of economics stress when management must look around for places to cut manufacturing costs, it is inevitable that their collective eyes come to rest on the Quality organization — and with good cause! The Quality organization produces no hardware to sell, has no direct profit responsibility, and cannot state with any degree of certainty the impact of contemplated changes on profits. A continuing cost-effectiveness study of your Quality organization will give you answers on the impact of decisions affecting Quality as well as initiating profit-consciousness among your Quality people.

Quality costs have been categorized into three groups: Prevention costs, Detection costs, and Failure costs. For a cost-effective Quality study to be meaningful, the following definitions should apply to the data used:

Prevention — Costs incurred in the planning and analysis necessary to prevent defects in the products, either hardware or software;

Detection — Costs incurred in determining compliance of the products with pre-planned standards of acceptance.

Failure — Costs incurred when the products fail to conform to pre-planned standards of acceptance.

Simple economic logic dictates the relationships of the categories to each other and to a minimal Quality cost applicable to each individual company. When the cost of preventing an unfavorable event is higher than the cost of detecting it, economic resources (men, money, etc.) can best be utilized in the area of detection. It follows, then, that costs will be minimal when prevention cost curve crosses the detection cost curve. This is demonstrated by plotting curves as in Figure I.

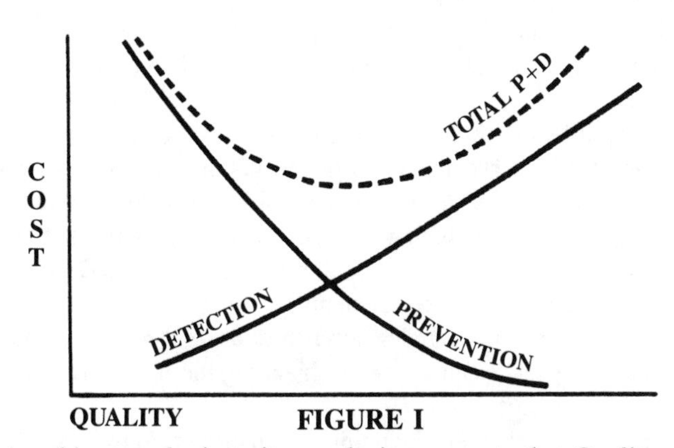

FIGURE I

Pursuing this same logic to its conclusion suggests that Quality costs are optimized when Failure costs equal the sum of Prevention costs and Detection costs. Figure II shows this relationship graphically.

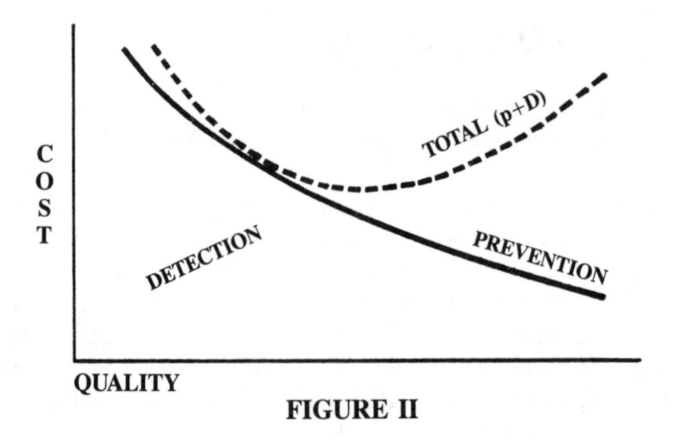

FIGURE II

With the economic rationale above, it is now possible to construct a self-correcting Quality cost system which will seek the optimum cost for any given business situation. It is self-correcting in that if Failure costs increase, Prevention and Detection costs must be increased to maintain the equality, and if Failure costs decline, Prevention and Detection costs should be reduced. Very little experience should be necessary to eliminate over-correcting.

In order to apply the foregoing plan intelligently, it is necessary to sub-divide and define the elements making up the three categories of Quality cost.

1. **Prevention —**
   a. Quality Manager and Staff — In essence, the only purpose of a Quality Manager and his staff is the prevention of failures, therefore, include all salaries and expenses under Preventions.
   b. Quality Systems, Procedures, and Standards — All of the above are aimed at preventing failures, and therefore, all salaries and expenses of this activity are Prevention costs.
   c. Quality Audit and Corrective Action — All salaries and expenses attributable to activities of this nature are geared to prevention of future defects.
   d. Vendor Quality Surveys and Planning — Salaries and expenses including travel expended surveying vendors and planning to survey vendors including time spent with Purchasing and Engineering personnel.
   e. Procurement Quality Review — This includes that portion of Quality Engineering expenses attributable to review of Purchase Orders, Requisitions, etc., for Quality requirements.
   f. Quality Engineering Planning for Incoming Inspection — Includes all wages and expenses caused by pre-planning the incoming inspection of material received from vendors.
   g. Quality Engineering Pre-Planning for In-House Activities — Includes wages and expenses for pre-planning in-house activities by Quality Engineers.

h. Quality Data and Statistics — This encompasses the cost of Quality Data accumulation, analysis, and feed-back, including E.D.P. costs where applicable.
i. Quality Specification for Materials and Processes — This includes the cost of qualifying materials for use and the planning and periodic testing required to control special processes.
j. Quality Training — Includes wages and expenses for Quality Training including posters, etc.

2. **Detection —**
   a. Receiving Inspection — Includes all costs incurred in inspecting material coming from vendors, sub-contractors, etc. This does not include laboratory testing expenses which are covered in 2c, below.
   b. In-Process Inspection and Test — This includes all other inspection costs except:
      1. Handling Scrap and Rework.
      2. Repair, Re-test, and Trouble-shooting.
   c. Laboratory Acceptance Testing — Includes wages and expenses necessitated by special acceptance tests requiring laboratory equipment or procedures.
   d. Calibration and Maintenance of Measuring Equipment — All costs of maintaining the required accuracy of measuring instruments.
   e. Source Inspection at Vendor's Plants — Includes salaries, travel expense, etc., expended to inspect material at a supplier's facility.

3. **Failure —**
   a. Scrap — Includes total losses, including lost labor, resulting from disposal of unusable defective material.
   b. Rework — Includes all material, labor, and overhead expended to return defective material to a usable condition.
   c. Material Review — Includes all expenditures by Quality, Engineering, Purchasing, etc., for evaluation, disposition, repurchasing, and additional handling of defective material.
   d. Trouble-shooting and retest — Includes cost of fault isolation and the resultant retesting of material following a failure.
   e. Failure Analysis, Reporting, and Corrective Action Investigation — Includes all costs associated with the investigation of failures, excluding 3c above.
   f. Customer Service — Includes salaries and expenses required for liason with customers to adjust complaints arising from product failures (either hardware or software).
   g. Returned Material — Includes all additional costs of special handling for returned material, including the additional accounting costs.
   h. Lost Sales — This item, which is at best an estimate, includes the expenditures by the marketing organization and others attempting to overcome

a bad Quality reputation. It also includes an estimate of the amount of profit lost due to poor Quality.

The accumulated cost data, segregated into the above catetgories, will now tell you and other management whether the Quality organization is balanced and proportional to the job to be done. By analyzing the "Significant Few" areas of largest expenditure, it permits the knowledgeable manager to control costs without causing an adverse unbalance. Quality management can forecast changes in any area of the total Quality picture and take appropriate action to reduce the undesired effect by increasing prevention or detection efforts. Being forewarned, a good manager can shift personnel from one activity to another and utilize cost elements for maximum benefit.

When total Quality cost is compared to the cost of sales, direct labor, or any other measure of productive effort, management has an effective, orderly approach to cost reductions in the area of Quality. An additional benefit accrues in the area of bidding new work. With a firm appraisal of Quality costs, it is possible to accurately forecast Quality costs into new work, permitting management to establish realistic goals.

All of the activity thus far can be accomplished without involvement of top management. But for maximum Quality cost reduction, top management must become involved. Top management must set a goal and establish the policy which will attain that goal. The goal, realistically, can be only one thing — to make the best Quality product on the market.

ACTIVITIES FOR REDUCTION OF USER'S COST

Yohei Koga, Director, Quality Assurance Dep't
KOMATSU, LTD., Japan

1. INTRODUCTION

Komatsu, Ltd., a Japanese firm established in 1921, has today the paid-up capital of $48,000,000 and employs a total of 17,000 workers. Annual sales volume for the year 1969 has reached $576,000,000. The Company's main products are construction equipment centering on bulldozers. The Company holds the world's second in bulldozer production. It retains Japan's first in forklift truck production.

"Quality control refers to a set of activities designed to supply to users the world over such products as will meet their satisfaction and, to that end, to economically manufacture, develop technology therefore, and carry out sales and services." This is how Komatsu, Ltd. defines "QC". One of the activities of the Company's QC is gauged by a yardstick called "user's cost." This article describes the thought and processes of the QC maintained by the Company.

2. USER'S COST

2.1 Characteristics of construction machines in service

For the purpose of this article, and in order to properly appreciate the significance of user's cost, it is necessary to note the following facts on construction machines in service:

(1) A construction machine is used to earn profit. The profit is measurable with a fairly high degree of accuracy.

(2) A construction machine is a self-powered vehicle but, unlike automobiles, is not registered with a centralized public agency. Hosts of construction machines at work are hard to trace for statistical purposes. They shift from one place to another after construction jobs.

(3) In the sense of economics, a construction machine is large-size durable goods.

(4) No two identical machines are worked in the same manner. The method of working varies with the type of soil, terrain conditions and other factors of duty.

2.2 Definition of user's cost

User's cost is based on four cost elements.

The first element is *ownership cost* (= depreciation + tax + insurance + interest).

It is a cost arising from the mere fact of owning the machine.

The second element is *operating cost* (= fuel & lubricant cost + expendable supplies cost + repair cost + operator's wages).

The third element is *machine efficiency* determined as follows:

$$\frac{\text{hours worked}}{\text{hours worked} + \text{servicing time (hrs)} + \text{down time (hrs)}} \times \underbrace{(1 - \text{“x”})}_{\text{Depletion for hours worked}} \times \underbrace{(- \text{“y”})}_{\text{Degree of operator's fatigue}} \times \text{factor of machine adaptability to the job}$$

The last element is *production volume* in cubic meters of material handled. With the foregoing elements, user's cost is defined by the following formula:

$$\text{USER'S COST} = \frac{\text{Ownership cost} + \text{operating cost}}{\text{machine effeciency x production volume}}$$

User's cost is expressed in terms of YEN currency.

2.3 **Reduction of user's cost**

Factors of user's cost are indicated in Fig. 1 according to KOMATSU'S own QC technique which combines a cause-effect diagram with a Pareto diagram. For each factor, we set up a target, and assign a person to the task of directing improvement activities for reaching to that target.

Activities are pushed ahead methodically under what we call "flag shape control system," a system devised to reduce the user's cost.

The formula in item 2.2 above tells that the factors must be worked on to bring about changes as follows:

(1) Increase of production volume.
For a given model of machine, the highest possible work capability should be realized without sacrificing the reliability and durability of the machine. Rather, both reliability and durability should be upped.

(2) Increase of machine efficiency.
A high rate of earthmoving production is not possible with a machine that necessitates long or frequent shutdowns (down time) even if its capability were maximized. Thus, reliability and durability must be improved, and the machine must be made more amenable to repair work. Ready availability of replacement parts and smooth supply of these parts will shorten down time. Improved repair techniques will result in speedy repair. Repair work will be made easier if more of the feature of unitized components is incorporated into the design of the machine. A machine easy to service in everyday use is another requirement.

(3) Lowering of operating cost.
This can be accomplished mainly by reducing maintenance cost and overhaul cost. These two costs are dependent on reliability and durability.

(4) Lowering of ownership cost.
Economics of pricing is involved here. Cost elements to become the basis of pricing the machine must be more carefully analyzed and accounted for in order to establish a pricing level that will not only assure a reasonable profit margin but also minimize the share of the machine's selling price in the makeup of the user's cost.

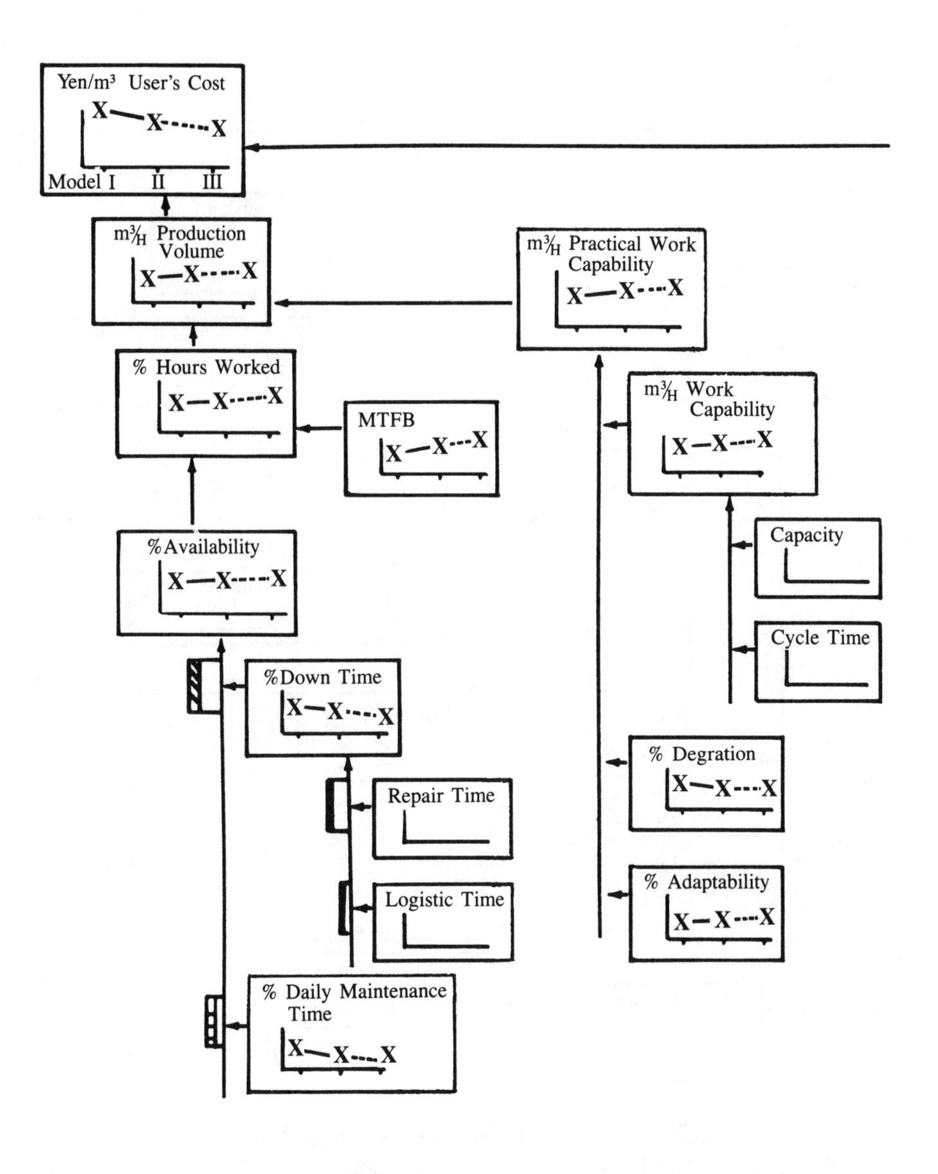

FIGURE 1 — Flag Shape Control System for Reduction of User's Cost

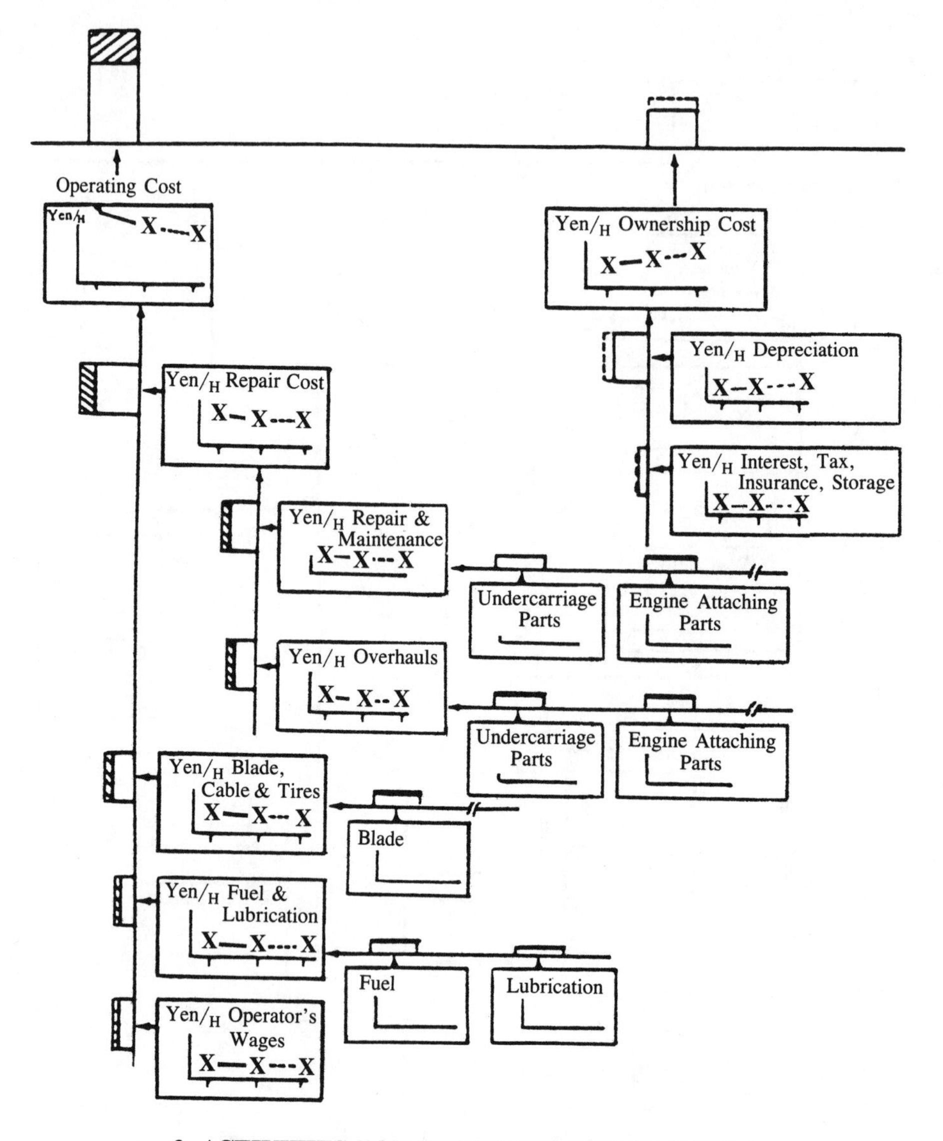

3. ACTIVITIES FOR REDUCING USER'S COST

3.1 Production volume, durability, and reliability

Improvement of these three factors is to be accomplished in the processes of developing a new product machine and also of effecting changes in an existing model.

To that end, we institute a special project. Under this project, we gather information from principle markets, home and overseas. Also, through cooperation of users, we designate some of the machines in operation at job sites as "monitor machines" and collect necessary data on them.

Collected data are then analyzed to set up long-term and short-term quality targets to be attained in the new machine being developed or in the changes to be made in the existing model.

How targets are set up will be illustrated by taking up *production volume* as an example. We designate production volume as parameter V and user's cost as parameter U, and relate these two in this manner:

$$U = \frac{C}{V},$$ where C is a constant given to the model.

Let U_m and V_m stand for the current user's cost and production volume, respectively, and U_n for the target value. From V= C/V, the new production volume required for U_n is determined. We then compare the target V_n against the user's cost U_c and production volume V_c of a competitor's machine.

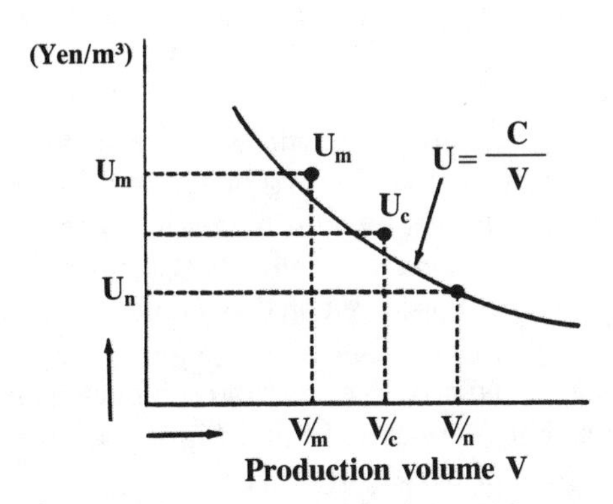

Figure 2 — Relation between Production Volume and User's Cost

We have QC evaluators in our plants, engineering department and sales-service department. They evaluate the quality targets set as above and determine whether or not the targets are consistent with and adequate for the suggestions and wishes of home and overseas users. We call this determination "A" evaluation.

Trial production of the new machine is commenced after the "A" evaluation has been completed to verify the validity of the quality targets. Upon the assembly of the trial machine, we run a performance test on it to check how close we did get to the targets in this machine. We also check on the ease of operation, stability, maintainability and operating cost, under what we term "B" evaluation.

An endurance test is run on the trial machine for different kinds of soil. Major components are examined against the test results reported on them from the factory. These checks come under "C" evaluation. The methods of testing for the above evaluations are continually modified, as necessary, on the basis of the changes noted in the data being received from fields. Machines of the prototype model proved by the trial machine are sold in the test market as initial machines. They are subjected to users' evaluation and suggestion. Data are subsequently collected on them for MTBF, machine efficiency and other factors to check the user's cost. Regular production of the new model machines is commenced only after improvements have been effected as dictated by the information collected from the performances of the initial machines in fields.

Users of the early machines from regular production are consulted with for opinions. Some of the machines manufactured during this period are designated as "monitor machines" from which data are systematically gathered for analytical and evaluative purposes. "D" evaluation refers to examination of these data. After commencement of the regular production, "monitor machines" are designated at regular intervals from the series of machines being turned out from the factory. These "monitor machines" are placed under a centralized system designed to provide information necessary for reduction of user's cost, and the information in the form of data is constantly fed back to the engineering department for use in carrying out our development work for new products and model changes. It will be noted that at the various stages the factors of user's cost are noted, analyzed and evaluated to see if the quality targets are being fully met.

3.2 Machine efficiency

As has been stated, machine efficiency will get better if mechanical troubles are less frequent or less extensive to result in shorter down time or if the machine can be more easily serviced. In other words, higher machine efficiency depends on higher machine reliability, as we have already described.

There is one other aspect to be considered on down time. It is after-service given to the machines. We think that service and sales operation should go hand in hand. For this line of thinking, we hold the manager of each branch office responsible for these two kinds of activity, service and sales. Fig. 3 is a block diagram showing how each branch office of ours is organized.

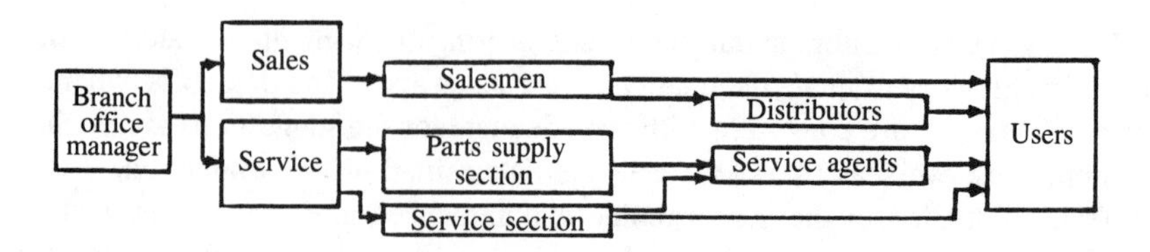

FIGURE 3 — Sales-service Organization

The function of service in respect to reduction of user's cost may be summarized as follows:

1) To make periodic visits to users and give them guidance in the correct use of the machine.
2) To collect data and feed them back, thereby participating in product improvement, new product development and other related activities.
3) To determine a proper method of repair for a failed machine and see to it that the machine is correctly repaired and restored to duty.

Speedy repair work is based on two important conditions. One is ready availability of replacement parts, and the other is advanced techniques of repair. For the former condition, we keep the supply of replacement parts under a system called "availability control"; it is based on the availability ratio:

$$\frac{\text{Availability of parts in stock}}{\text{Number of orders for parts}}$$

For the accomplishment of this object we keep a collective control over the production, inventory, shipment and distributor's inventory of these parts.

On the other hand in order to keep down time in machines to the minimum we need to improve the availability of replacement parts and at the same time to deliver as soon as possible, whenever requested, replacement parts users from our parts depots; accordingly we are setting up a system which will insure delivery to users within 24 hours.

It is well understood that the high quality of service agents and service engineers, who render services to the machines, has an important bearing on user's cost, hence we assist our distributors by extending necessary guidance in the improvement of management and financial assistance in the reinforcement of service facilities and the education of service technicians, inclusive of QC circle activities, in the improvement of their servicing and the shortening of their servicing hours; thus we are constantly making every effort in cutting user's cost.

3.3 Relationship between selling price and user's cost

A substantial portion of ownership cost is the depreciation, an item on which the selling price has direct bearing. A fair profit which the manufacturer of the machine expects has to look in two opposite directions under the circumstances of business: labor becomes more expensive with each passing year, raw materials go up in price from time to time, and sales expenses too keep on rising steadily. Raising the selling price is one direction, and lowering the production cost is the other, for realizing a fair profit from the production and sales of the machine.

To be weighted against this dilemma is the fact that construction machines in general tend toward higher performance, larger flywheel horsepower and greater machine weight, inevitably raising the production cost. The manufacturer is required to improve the quality of the machine and to do his best in his attempt to lower or keep down the production cost. With all these adverse tendencies and conflicting requirements in view, we push ahead our efforts to work out methods of manufacture requiring less labor and less time in the manner illustrated in Fig. 4. We ramify the objective of our QC, with a control-data graph assigned to each ramification. Improvement toward the target is continued in each area.

In our pricing practice, in which the sales department has a final say, many existing conditions and their trends are taken into consideration. For reduction of user's cost, the following technique of pricing is often resorted to:

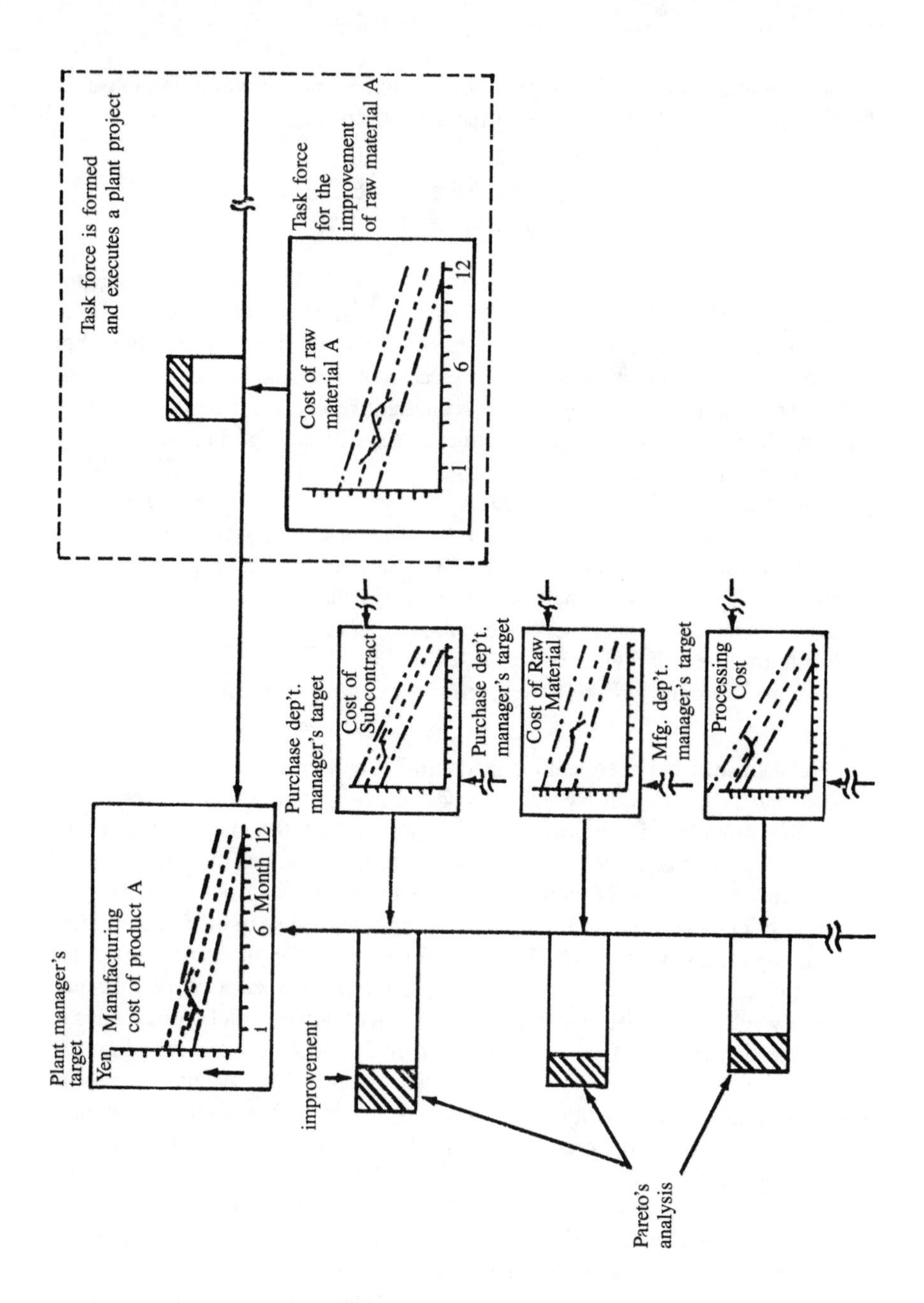

FIGURE 4 — Flag Shape Control System for Cost Reduction

Let the selling price be parameter "s", and the user's cost on the machine be parameter "U". Then

$$U = A s + B,$$

where "A" and "B" are the constants for a given model. Let "U_m" be the user's cost on an existing model, with "p" standing for its selling price. For the user's cost on a new product machine to be equal to the said "U_m", the selling price "q" of the new machine must be such as to fit to this equation:

$$U_m = A q + B$$

If the selling price of the new machine is equal to "p" (the selling price of the existing model), then the user's cost "U_n" on the new machine is determined by:

$$U_n = A p + B$$

Generally the change from an old model to a new model is so calculated as to make "U_n" lower than "U_m". The difference "$U_m - U_n$" may be regarded as a gain accrued to both the maker and the user. For the purpose of discussion, assume that this gain is shared on a 50-to-50 basis by maker and user, and also assume that the selling price resulting from this equal split of the gain is represented by "x". Then the following equation becomes valid:

$$\frac{U_m + U_n}{2} = A\frac{p + q}{2} + B = A x + B$$

Therefore, $x = \dfrac{p + q}{2}$. This is a price that takes user's cost into account.

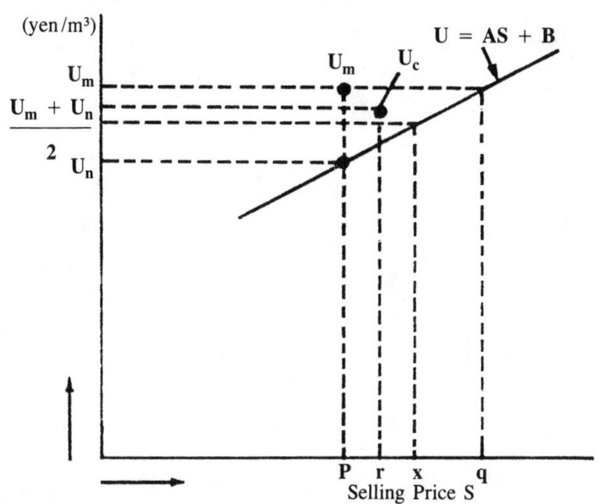

FIGURE 5 — Relation between Selling Price and User's Cost

The price may be determined by taking the user's cost "U_c" and selling price "r" of a competitor's machine as the reference.

4. SUMMARY

The QC activities being carried out in our Company, as outlined thus far, have lowered user's cost by 20% to 30% in each change from an old to its new model in recent bulldozers. Since we intensified our QC in 1961, our annual sales volume expanded 4.5 times in 8 years, tripling the amount of per-employee sales. At present the Company's share in the domestic bulldozer market exceeds 60%.

This high rate of growth may be measurably due to the rapid expansion of the national economy in those years but there is no denying that the intensified QC takes credit for a substantial part of the growth. Production and sales of KOMATSU machines meeting the full satisfaction of users everywhere is now firmly keyed to our company-wide QC activities.

PREDICTION OF HUMAN PERFORMANCE IN A QUALITY CONTROL SYSTEM: A STEP TOWARDS COST

A. Raouf, Associate Professor
University of Windsor, Windsor, Ont.

During the last two decades or so manufacturing processes have been considerably automated. The ever increasing cost of labour is perhaps one of the primary reasons for the acceptance of "automation" by the management.

Even in a highly automated manufacturing system labour is needed, at least in a surveillance capacity. The quality control system is not an exception from the above. Inspection of machined components can be automated thus resulting in reduced inspection costs. However, there are certain types of defects which can only be detected by inspectors.

The manufacturing system and the quality control system are essentially man-machine systems. For planning, controlling, and analyzing the production costs of a manufacturing system, industrial engineers amongst other techniques, use work standards, i.e. standard time for a given operation. However, industrial engineers did not use the same technique in the case of quality control systems. They might have attached priority to production costs as opposed to quality costs because production costs are invariably higher than quality costs. The other reason for not using work standards in studying the quality control costs is, perhaps, the fact that the technique of setting up standards for inspection tasks are not very well established.

The human performance in a manufacturing system differs from human performance in a quality control system. The motions made by an operator while performing in a production system are usually repetitive in nature and the sequence in which these motions are made remains invariable. The motions performed in an inspection task vary from cycle to cycle as the motions following the inspection depend upon the result of the inspection. For developing work standards for production operations conventional work measurement techniques can be used. However, such is not the case with inspection tasks. The time spent by an inspector in observing the defect and making a decision regarding the disposal of the component cannot be measured.

If an inspection task is examined at a "micro" level then in addition to various motions like reach, grasp, transport loaded, etc. the following activities take place:

1) stimulation of the retina
2) neural conduction of impulses from the retina to the brain
3) decisions or central processing of the signal
4) neural conduction of impulses from the brain to the muscles
5) muscle activation

This paper suggests a method for estimating the time required for above mentioned activities under various conditions. For estimating task cycle time when preview of the detail to be inspected is not possible, the estimate of the activities times is to be added to the manual motion time. The manual motion time is the sum of predetermined motion-times which form the task.

Similar activities take place when a subject performs in a choice-reaction task experiment. In a choice-reaction task experiment the subject is shown a signal from a set of signals. Each signal has an explicit response. The subject knows that a signal will come but he does not know which one it will be. On seeing the signal the subject makes the corresponding responses as fast as he can. The elapsed time between the signal coming on and the corresponding response made is measured. This time is call the choice-reaction time.

The choice-reaction time has been studied very extensively by psychologists as well as others interested in human performance. It has been shown that average choice-reaction time is given by (Hyman, 1, Grossman, 2, Edwards, 3)

$$CRT = a + b\,h \quad \ldots(5)$$

where CRT is the choice-reaction time, a and h are empirically determined constants and H is the information transmitted or processed in bits. This relationship has been shown to hold relatively well whether the value of H is varied by alternating the possible number of alternatives, the relative probability of each alternative, or the constraints on the sequence in which signals are presented.

It has also been shown that CRT is affected by the experimental conditions and these are

i) Type of input (signal)
ii) Type of output (response)
iii) Signal-response compatibility
iv) Signal discriminability
v) Amount of information transmitted

In most of the inspection tasks input is visual and output is in the form of manual motions. Signal-response compatibility depends upon the amount of decoding required to be done by the subject before making the response. In visual inspection tasks of simple nature, i.e. detection of a scratch, it can be assumed that practically no decoding is required by a skilled inspector. The signal discriminability is affected when the size of the detail to be inspected is varied. (Raouf, 4, Williams, 5). Weston (6) has shown that if the detail size is increased beyond 6 minutes of visual angle the performance is not improved significantly and inspection of details subtending angles of 1 minute or less is very difficult to be accomplished by unaided eyes. The amount of information transmitted in an inspection task depends upon the proportions of defective and acceptable components.

Supposing in an inspection task there are only 3 categories that the components can be sorted into and these are

1) accept
2) reject
3) rework

During the inspection 45% are found acceptable, 30% are rejected and the remaining 25% are declared reworkable. Then the average information processed by the inspector is

$$H = .45 \log_2 \frac{1}{.45} + .30 \log_2 \frac{1}{.30}$$

$$+ .25 \log_2 \frac{1}{.25}$$

$$= .5184 + .5210 + .5000$$

$$= 1.5 \text{ bits}$$

CRT equation by varying H and keeping the signal size invariant (20 minutes) has been developed (Raouf 7). This equation is labelled (4) in Figure 1. The task used for this had a very high signal-response compatibility. The response was made by pushing the buttons and the fingers travelled nearly 1/16 inch for making the response. For this experiment 17 subjects were tested and each subject performed in 2, 3 and 4-alternative task conditions, i.e. (H = 1, 1.58 and 2). The performance data used for developing equation (4) as shown in Figure 1 are provided in Table I.

Sternberg (8) has shown that if the signal is degraded then the increase in performance time is reflected in constant a of equation (5). Assuming that varying the size of the signal varies "a" and not "b" and using the data as developed by Raouf (7) for sizes of 1 minute, 3 minutes and 5 minutes, equations for estimating CRT for each condition can be developed. These equations are

$$T_{(1 \text{ minute})} = 12.28 + .81 \text{ bits} \quad (1)$$

$$T_{(3 \text{ minutes})} = 11.88 + .81 \text{ bits} \quad (2)$$

$$T_{(5 \text{ minutes})} = 11.68 + .81 \text{ bits} \quad (3)$$

Graphs of these equations are shown in Figure 1.

For estimating activities time for various size signals the following is suggested

1) use equation (1) when signals are 1   3 minutes
2) use equation (2) when signals are 3   5 minutes
3) use equation (3) when signals are 5   10 minutes
4) use equation (4) when signals are 10 minutes

These estimates are likely to include some elements of error. For example the data used in the development of these equations pertain to one condition, i.e. stimulation of one particular group of muscles, etc. Further research to determine the effect of motions following the decision made in an inspection task upon the activities time is currently under progress. Until the findings of the research are known it is suggested that the technique as mentioned may be used. Industrial validation of this methodology has yet to be undertaken.

The time standards can thus be established for inspection tasks which involve simple visual detections. The industrial engineer can then use the technique of controlling and analyzing production costs for quality costs as well.

REFERENCES

1. Hyman, R. "The Information as a Determinant of Reaction Time," *Journal of Experimental Psychology,* Vol.4.
2. Crossman, E.R.F.W. "The Measurement of Perceptual Load in Manual Operations," Ph.D. Thesis, Univ. of Birmingham, 1956.
3. Edwards, E. "Information Transmission," Chapman and Hall, 1964.
4. Raouf, A., and Hancock, W.M. "The Effect of Visual Angle and Contrast on Inspection Type Operations," Paper presented at the International Conference on Production Research, Univ. of Birmingham, 1970.
5. Williams, C.M. "Legibility of Numbers as a Function of Contrast and Illumination," *Human Factors,* Vol.9.
6. Weston, H.C. *Sight, Light and Work,* H.K. Lewis, 1962.
7. Raouf, A. "Study of Visual Performance Times for Inspection Tasks," Ph.D. Thesis, Univ. of Windsor, 1970.
8. Sternberg, S. "Two Operations in Character Recognition: some evidence from reaction time measurements," Paper read at Symposium on Models for the Perception of Speech and Visual Form, Boston, 1964. (Reported by Smith).

CRT FUNCTIONS FOR DIFFERENT SIZE SIGNALS

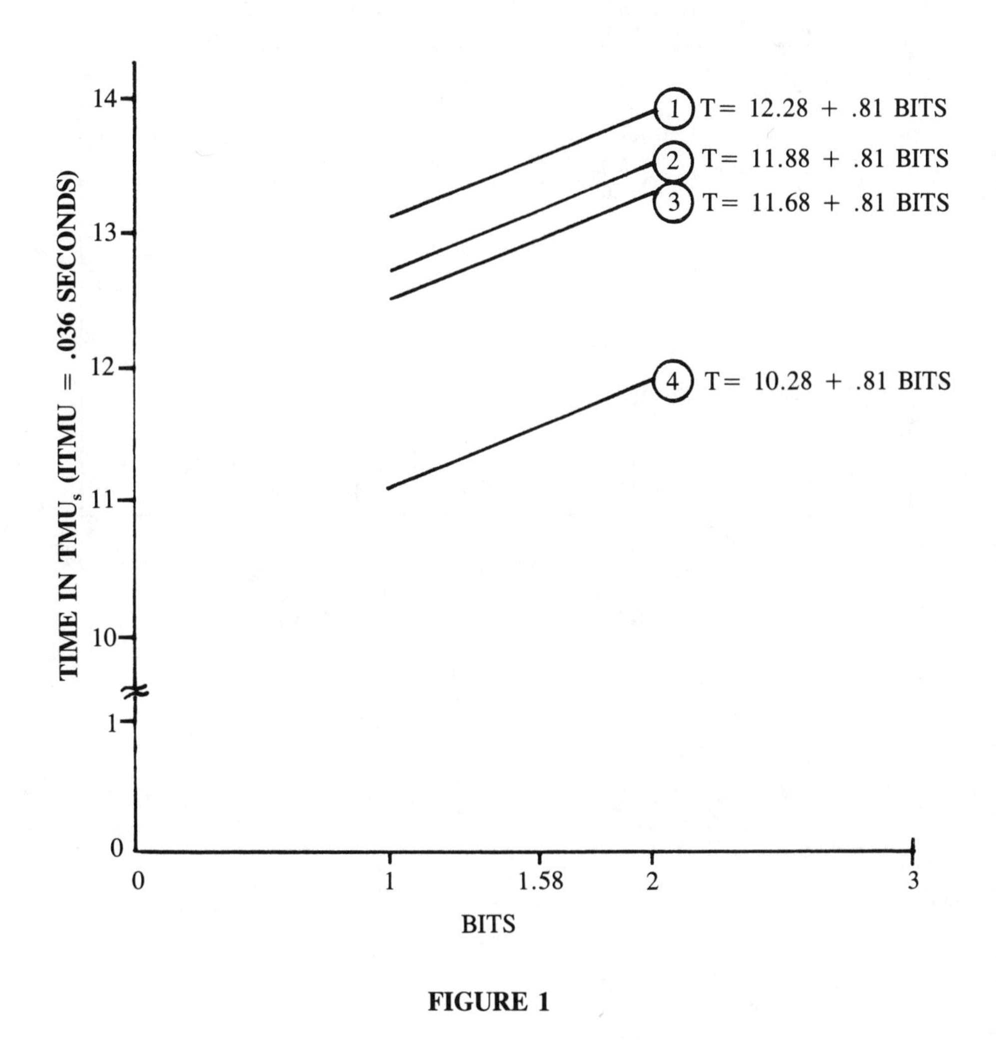

FIGURE 1

SUBJECT NUMBER	PERFORMANCE TIME EXPERIMENTAL CONDITIONS		
	2 Alt.	3 Alt.	4 Alt.
1	8.4	10.2	10.7
2	8.4	11.6	13.4
3	10.0	12	13.2
4	9.5	10.1	11
5	8.65	9.90	11.12
6	10.02	12.03	13.21
7	9.56	12.24	12.67
8	9.20	10.69	12.02
9	9.45	10.12	11.08
10	8.60	10.33	11.67
11	9.85	10.48	10.87
12	12.35	12.72	13.40
13	8.12	10.06	12.54
14	9.24	10.33	11.52
15	10.48	11.20	11.38
16	12.06	12.34	12.96
17	11.31	11.36	12.29

TABLE I
PERFORMANCE TIMES (IN TMUS) UNDER DIFFERENT EXPERIMENTAL CONDITIONS

NOTE ITMU = .036 Seconds

UNCOVERING THE HIDDEN COSTS OF DEFECTIVE MATERIAL

C.B. Rogers, Engineer — Quality Assurance, Camera Division, Polaroid Corp., Cambridge, Mass.

SCOPE

This paper analyzes the ways in which company costs are increased when defective materials are received. The concepts discussed in this paper provide a basis for developing a method which shows the true cost of purchased material for comparison with prices quoted by vendors.

Flow charts for both good and defective materials are shown, and specific techniques for determining costs at each stage of material handling are discussed.

DISCUSSION

Any attempt to gather cost data on items which pass through several departments is best approached via the flow-chart method. Figures 1 and 2 contrast in general terms the possible routes which defective material may follow with the route normally followed by non-defective material. As shown in the charts, defective materials can add considerable complexity to the flow system.

Table I analyzes the flow elements of Figure 2 into categories of excess costs. It is with the collection of these costs that the remainder of this discussion will be concerned.

Incoming Materials Inspection Procedure

When a lot of incoming material is rejected by the Materials Inspection department, special forms (Discrepant Material Reports) are completed and mailed to the concerned parties. Because a completed form by its very nature standardizes data, it becomes feasible to compute an average cost and, by merely multiplying this figure by the number of rejections per year for each item, an average annual cost.

Four Ways Costs Are Increased

Handling and storage costs are also increased when lots are rejected, although, in this case, items must be categorized into high, medium and low bulk items for obvious reasons.

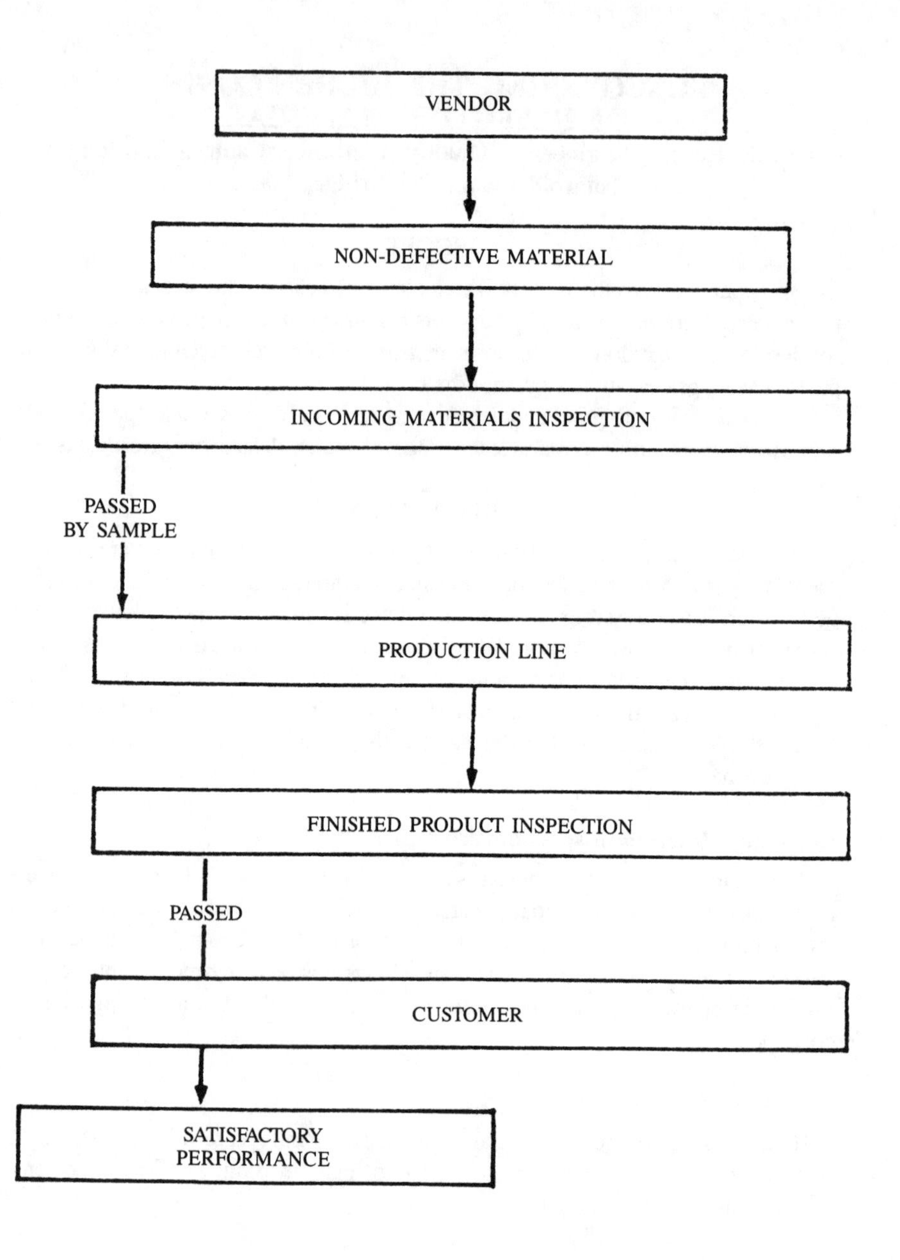

FIGURE 1 — Flow Diagram for Non-Defective Material

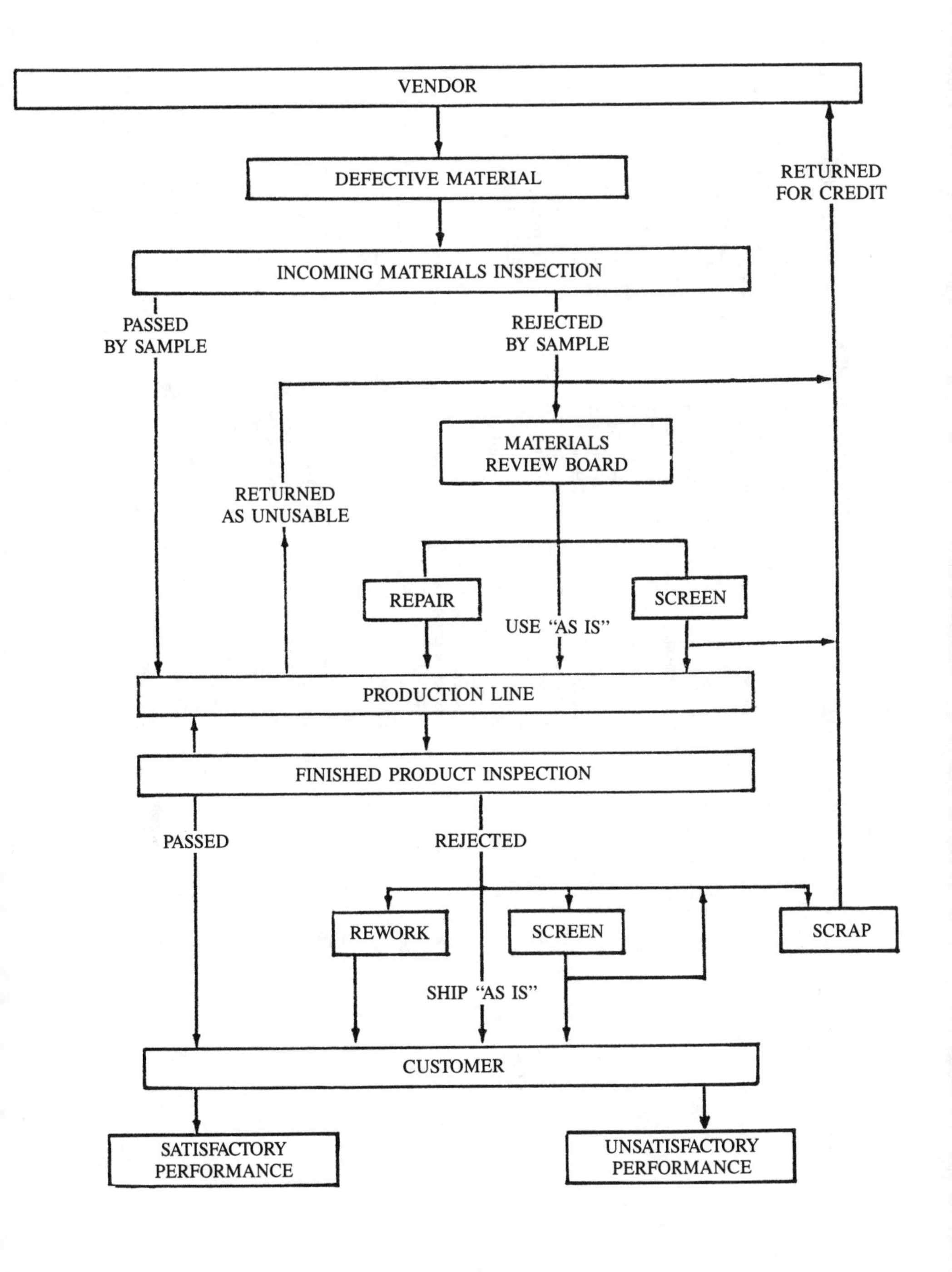

FIGURE 2 — Flow Diagram for Defective Material

TABLE I — Excess Costs for Flow Elements

Point of Detection	Action	Excess Costs	Method of Accumulating Item Costs/Year	Reporting Department
Incoming Inspection	Passed by Sample	None		
	Rejected by Sample	Paperwork	No. Rejected Lots	Incoming Inspection
		Handling	No. Rejected Lots	Materials Control
		Storage	No. Rejected Lots	Materials Control
		Inventory	% Safety Stock (Quality)	Purchasing
		Interest	$ x Days x 6%/365	Purchasing
	Materials Review Board	Meeting	No. Meeting Man-Hours	Quality
		Investigation	Engineering Man-Hours	Engineering
			Shop Work Order $	Accounting
	Repair	Labor & Material	Shop Work Order $	Accounting
	Screen	Labor	Shop Work Order $	Accounting
	Return to Vendor	Paperwork	No. Rejected Lots	Purchasing
		Handling	No. Rejected Lots	Stores
		Special Meetings	Hours	Purchasing
	Use As Is	Production Losses	$ (Estimated)	Production
Production Line	Investigate	Engineering Man-H		Engineering
	Return of Material	Paperwork	No. Rejected Lots	Materials Control
		Handling	No. Rejected Lots	Materials Control
		Production Losses	$ (Estimated)	Production
Finished Product Insp.	Use As Is	Production Losses	$ (Estimated)	Production
	Passed	None		
	Rejected	Paperwork		Quality
		Investigation	Engineering Man-Hours	Engineering
			Shop Work Order $	Accounting
	Ship As Is	None		
	Rework	Paperwork	No. Rejected Lots	Production
		Handling	No. Rejected Pieces	Production
		Labor & Material	Shop Work Order $	Accounting
	Screen	Paperwork	No. Rejected Lots	Production
		Handling	No. Rejected Pieces	Production
		Labor	Shop Work Order $	Accounting
	Scrap	Paperwork	No. Rejected Lots	Production
		Handling	No. Rejected Lots	Production
		Material	$	Production
Customer	Satisfactory Performance	None		
	Unsatisfactory Performance	Paperwork	No. Customer Complaints	Check Inspection
		Replacement	No. Customer Complaints	Customer Relations
		Lost Sales	No. Customer Complaints	Customer Relations

(A rejected lot of castings would require more handling than a lot of rivets if both lots contained the same number of pieces.)

In establishing an inventory policy, a certain portion of the "safety" or contingency stock is carried to protect against stock outs due to receipt of defective material. The resultant carrying cost can be obtained from estimates by the purchasing agent responsible for the specific items.

Along a similar vein, reject material ties up company funds which might otherwise be earning interest. This loss can amount to a considerable sum and can be readily computed by multiplying the dollar value of the reject material by the number of days the reject is not replaced times the 6% interest rate. (The percent chosen is the current interest rate of commercial money.)

Materials Review Board

The cost of holding Materials Review Board meetings is most often absorbed directly into the overhead accounts of the attending departments. A better approach would be to collect these costs in a separate account where they could be added to the cost of the items requiring the meetings. An average salary figure ($10.00 per hour) could be multiplied by the man-hours for each meeting. These costs could then be itemized at the end of each month.

Frequently, engineering investigation is necessary before a final decision can be reached on a rejected lot. These costs also are usually diverted to general overhead accounts. A more desirable method would be for each engineering department to separate the time spent on defective material investigation and report the man-hours spent in a manner which would permit allocating these costs to the total cost of the materials involved. If production labor and materials are required, they are normally obtained through the use of a Shop Work Order (SWO), and as such they are in a form which allows the charging of these costs to the material cost.

In cases where the defective materials require special screening or repair, the costs are usually covered by SWO's and can be charged to the material cost.

If it is decided to return the material to the vendor, the cost of the paperwork and handling should be determined and applied as a constant (as outlined above) to each lot which is returned. In cases where the value of the material is low (i.e., less than $20.00), the material is usually scrapped at company expense; this cost should also be added to the overall material cost.

When the value of the rejected material is high, the purchasing agent often finds much of his time occupied in special meetings with the vendors. Such man-hours spent should be added to the material cost. (One purchasing manager estimates that over 10 percent of his staff's workload could be eliminated if the defective materials flowing into the plant could be substantially reduced.)

When it becomes necessary to use defective material "as is," production efficiency, needless to say, suffers. Under current accounting practices, this cost usually is charged against the production department budget. Such a procedure causes a double distortion affecting both material costs and operating costs. Because the dollar amounts involved are often considerable, it is most important to correct this distortion by correctly identifying and charging those costs due to the faulty material.

Production Line (Items Covered Above Are Not Repeated)

When the defective material is found at the production line, it is usually returned to Materials Control as unusable. This procedure involves standard paperwork and handling which can readily be determined by using a fixed charge times the number of returned lots. As the sole receiving agency, the Materials Control Department is in the best position to report these costs, although a portion of the actual costs are incurred by other departments throughout the plant.

Finished Product Inspection

When the cost of the paperwork in rejecting finished products due to defective material is high, it should be separated from other costs and charged to the proper material account.

Rework and screening operations required as a result of defective material reaching the production line often entail additional paperwork and handling costs. These costs can easily be accumulated if sufficient forethought is given to allocating the cost of these added operations to the appropriate materials account.

When the defective material causes scrapping of finished product, this product cost should be added to the materials cost. Although it is proper to charge the selling price when an unlimited demand exists, it is generally best to use the manufacturing cost.

Customer

Whether or not costs resulting from unsatisfactory performance in the hands of the customer should be charged to the material account is debatable. When the percent of failures is low and reflects an expected return based on AQL's set for the company's suppliers, the costs should be charged to the material. When, however, gross failures occur, the costs should not be charged to the material because it is truly the responsibility of the company to prevent this type of failure from reaching the customer. In the latter case, the costs are more appropriately charged to the department responsible for defect prevention.

SIX STEPS TO HIGHER PROFITS

1. Initiate a pilot survey to determine the relative magnitude of the various cost categories listed above.
2. Applying Pareto's principle, isolate the major cost contributors.
3. Determine the true cost of several purchased items and contrast this with the company's standard cost.
4. Develop a system for collecting these costs for a given period of time.
5. Determine the dollar savings to be gained in setting up the proposed system.
6. Present the plan to top management.

SUMMARY

Many costs experienced by a company receiving defective material are usually not included in the cost of material account. Such costs, however, can be determined and, if properly reported, can be accumulated and charged to the various materials accounts. When this is done, the company knows the true cost of purchased materials and is better informed to make economically sound decisions.

METHODS FOR SELLING TOTAL QUALITY COST SYSTEMS

A. W. Whitton, Jr., Senior Corporate Quality Consultant
Abbott Laboratories, North Chicago, Ill.

The basic concept of Total Quality Cost as usually presented is deceivingly simple. The pathway to fellowship with cost responsible managers and company officers which it reveals is enticing. Prospects of swashbuckling about in the world of cost accounting with flashing sword slaying the dragons of waste is very compelling. The picture is, in fact, painted in such bright colors that the prudent person must become suspicious. Nothing can be that good! Undoubtedly many quality managers, being suspicious but unable to predict the pitfalls awaiting them, have declined to step upon the battlefield. Conversely, the unwary convert may charge forth armed with unfounded predictions of dollars which will be turned into profits and soon find that he has constructed his own prison. He will find that what he meant as suggestions of opportunities have been interpreted as promises to conquer. His imagined army of supporters will desert in favor of reinforcing the ramparts of their own castles. It will come as no comfort to discover that the defensive preparations were caused by the mistaken belief that the new weapon called Total Quality Costs would be turned against them once an enlarged kingdom were carved out for the Quality Control Department.

Each of the extreme situations just mentioned can be avoided by providing realistic answers to questions which have been prompted by presentations aimed at selling Total Quality Cost Systems. The questions are those of *all* executives whose operations will be affected, not only the top executive whose firm directive to the cost accounting department is solicited. Since we know that Quality Costs are everywhere, this directs us to every executive on the roster. We must remember that each has his own problems, suspicious, pet programs and defenses. Quality professionals must look at Total Quality Costs through the eyes of all supervision to predict and set to rest their all too human fears.

Figure 1 is the classical theoretical Quality Cost graph employed to illustrate how expenditure of prevention dollars (often construed as additional quality staff people) can reduce Total Quality Costs. It immediately prompts three questions:

1) I know how to measure cost, but how do you propose to measure quality?
2) How do we know the shape of our curves?, and
3) Where are we on the curve?

When these questions are asked, neither truthful answers nor skillful, evasive oratory are satisfactory. Beginning with a complete and relevant theory which realistically includes all possibilities is to be preferred.

Figure 2 depicts an expanded set of theoretical curves for Total Quality Costs. These new curves address themselves to the fact that product quality is not an absolute quantity. It is, rather, a complicated relationship of physical facts which can only be estimated in that mystical, ever-changing arena called the marketplace.

The abscissa of the graph has been redimensioned from just "Quality" to "Control of Quality" which is, after all, what the boss is begrudgingly paying his quality staff to achieve. Quality people have been chorusing, "we can't inspect quality into a product," for so long that few should have missed that message! Another feature of this diagram is its emphasis on matching control of quality to market need through specifications in an environment of cost awareness. While realistically pointing up limitations to achievement of a utopian state which face the quality organization, it does at once focus attention upon the responsibility of other organizational elements each as R & D, Market Research, etc.

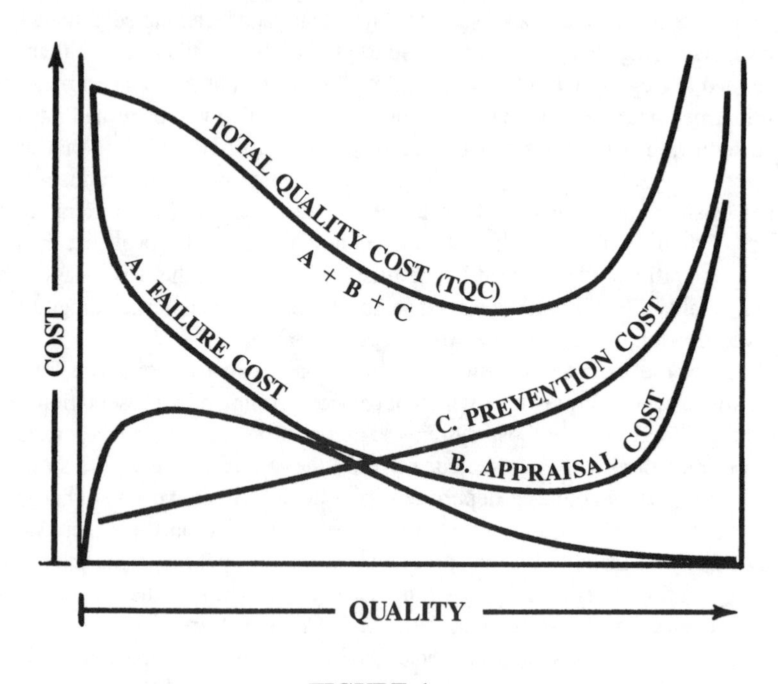

FIGURE 1

A note of caution is in order at this point. Don't yield to the temptation to advocate a product quality index system as a measurement of outgoing product quality. First of all, it is going to be tough enough to get cost accounting help for costs without adding an additional clerical and data processing burden. Secondly, emphasis upon the profit potential in making quality relevant to its market will be lost or greatly confused inasmuch as index systems are conformance oriented, not total quality oriented.

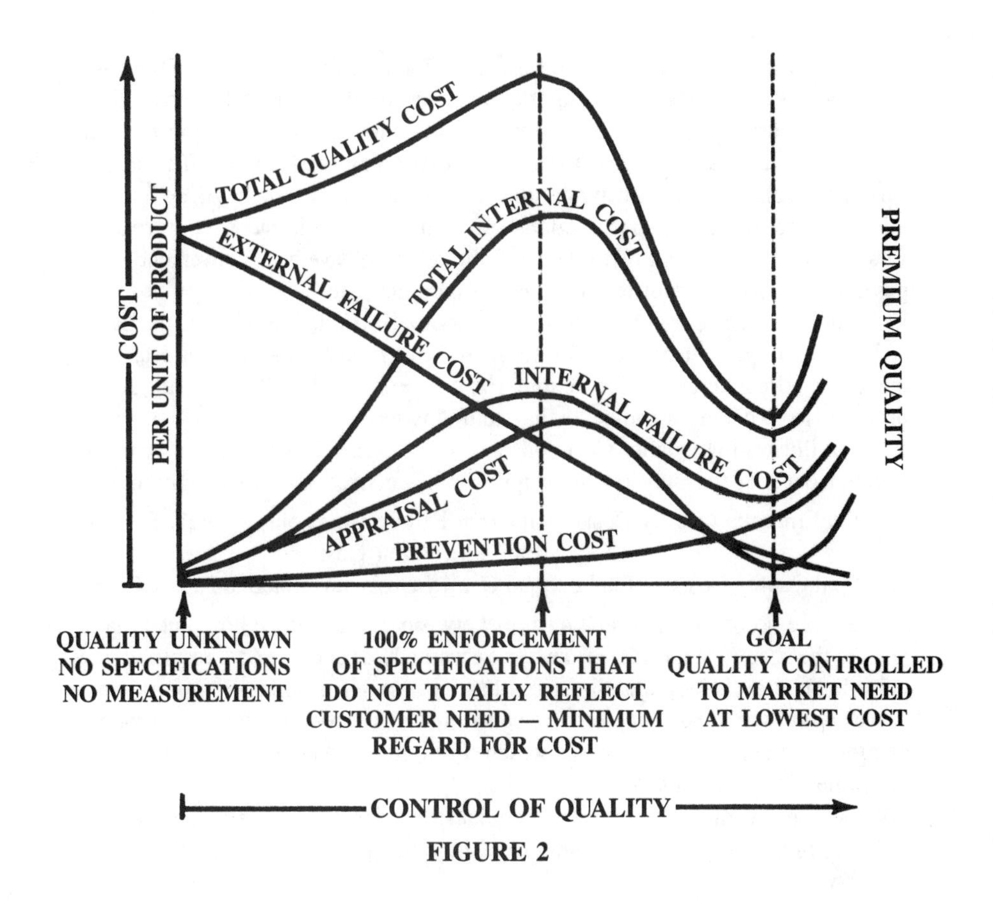

FIGURE 2

Woody Hayes, football coach of Ohio State University, has been given credit for stating, "Three things can result from an attempt at a forward pass and two of them are bad." The same sort of observation can be made about expenditures for control of quality. These expenditures can control the wrong quality, control the correct quality inefficiently, or provide economical control of the right quality. The odds (2:1) against spending money wisely is a telling argument with top management. Figure 2 illustrates this danger alongside of the usual glowing graphs of Total Quality Cost opportunity. A minus has been introduced to balance the plus, thereby comforting the cynic. With correct and forceful explanation of these curves, quality management can avoid being held solely responsible for producing results from quality cost accounting analysis. It is re-emphasized that the Quality Control organization should be charged with providing control leading to optimum quality at minimized cost, not with providing quality. Optimization of quality is measured indirectly by reduced customer complaints, increased salesman and customer enthusiasm, growing sales, or in short, marketing success. Minimization of cost can only be measured safely through a total quality cost program. The implication can be drawn that any top manager who does not insist that his people receive quality cost data and respond to it is likely to be guilty of negligence.

Some elements of the preceding discussion will please marketing and engineering executives. Recognition that the marketplace is everchanging and impossible to measure accurately in terms of quality requirements falls sweetly upon the ears of those whose careers directly depend upon expanding market share. The unseen side of the coin will be disturbing, however. Quantifying and comparing internal failure, external failure, specifications versus market requirement, conformance versus design, etc., might reveal vulnerabilities which have been conveniently impossible to separate from the information muddle. The balm to apply is found in the previous comparison to the game of football. As in that pursuit, the game of business requires teamwork for success. Constructive analysis of all facts is the key to effective corrective action. Expose problems? Certainly! But have a program for working together to find solutions too!

When illustrating Quality Cost concepts using Figure 2, care should be taken to point out that the only relationships which are fixed mathematically are:

Total Internal Quality Costs = Internal Failure Costs plus Appraisal Costs plus Prevention Costs

Total Quality Costs = Total Internal Quality Costs plus External Failure Costs.

There are no other fixed relationships between curves on the horizontal plane nor are the shapes of the various curves necessarily as shown. As a result, at any point in time each individual business has its own set of unknown control/cost curves with individual shapes and relationships. It is probable that each product or product line has its own curves which differ from other product produced in the same plant, division or corporation.

We set out originally to answer three questions for our executive benefactors and have arrived, by means of some handsome theoretical curves, at the following answers:

1) Quality is not economically measurable completely or absolutely, so we won't measure it as part of the Quality Cost System.
2) We don't know the shape of our curves, but
3) as a ray of hope, we can establish the magnitude of the quality cost elements at some point in time.

If this revelation doesn't result in an invitation to vacate the premises, it is possible that sufficient interest has been generated in the theory and its implications that further exploration will be fruitful.

The key word in Quality Cost reporting is "trend." This is also the key word in all business reporting. Corporate annual reports compare current figures to past performance using those classical management parameters, dollars and time. The *trend* of costs, sales and profits with time is the yardstick of importance. This is the ultimate measurement of performance used by every business executive. He is comfortable with it.

Given the possibility of measuring quality costs at any point in time and plotting trends, it is useful to apply the theoretical principles embodied in Figure 2 to prediction results which *might* be anticipated from effective quality management. The process is not unlike the popular ordeal known as profit planning. This is the sort of intellectual exercise many executives enjoy despite their cries of anguish. There are two major benefits of such activity. First of all, the participants get an understanding of the effects which relevancy of specifications to customer need and awareness of quality costs have upon Total Quality Costs. The second benefit lies in avoiding unrealistic estimates of potential savings.

To illustrate how anticipated cost/time curves can be developed, contemplate an old, established product in a rigid market — rigid in the sense of specification inflexibility. Figure 3 predicts the Quality Cost future of the product in response to professional quality attention. In an old, established market, assuming competitive quality performance and pricing, sales can be expected to rise with market growth. External failure costs are already low and do not offer appreciable opportunity for savings. The major remaining area for cost reduction lies with internal failure and appraisal costs. Since the situation is internally controlled, we can expect prevention investment to yield modest results in a relatively short period of time. The results can be considered modest as a percentage of Total Quality Costs; but, since we have assumed an established and competitive market with probable tight profit margins, the increase in profits could be very exciting.

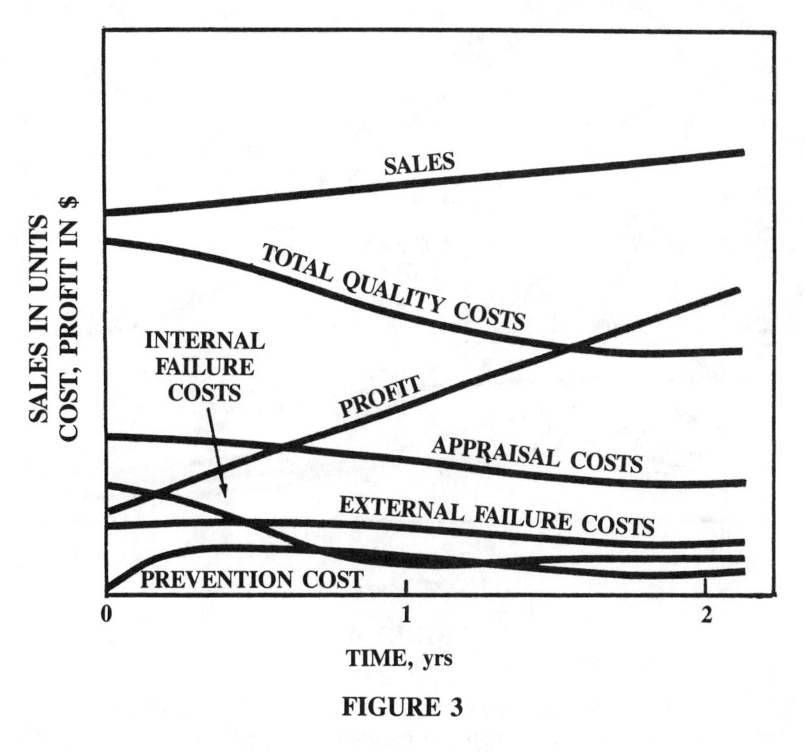

FIGURE 3

It should be noted that sales volume is plotted in units to eliminate effects of forecasted selling price changes. It is obvious that Total Quality Costs can be misleading unless compared to manufacturing/sales volume.

Figure 4 addresses itself to quite a different product situation. Sales have been dropping off and external failures in the form of returned goods have been rising. Profits have been dropping at a greater rate than sales. This is an all too frequently indulged habit of profits. Until something is done about whatever is creating the situation, it is reasonable to expect these trends to continue and our short — range — prediction as shown on the curves of sales, profits, and Total Quality Costs is drawn accordingly.

At this point many managements have made a mistake which is unfortunately too commonplace. With Marketing asserting that pricing and promotion is still competitive and Engineering standing behind a tried and true design, the assumption is made that quality (construed here as conformance) must have deteriorated. The quality control department is ordered to "tighten up." Whatever that means, appraisal is increased, specifications tightened, allowable defectiveness is reduced and a bad situation is made worse. The objective fact that the competitor's product is more defective in almost all measured characteristics often gets lost in this environment — put down as erroneous testing, perhaps.

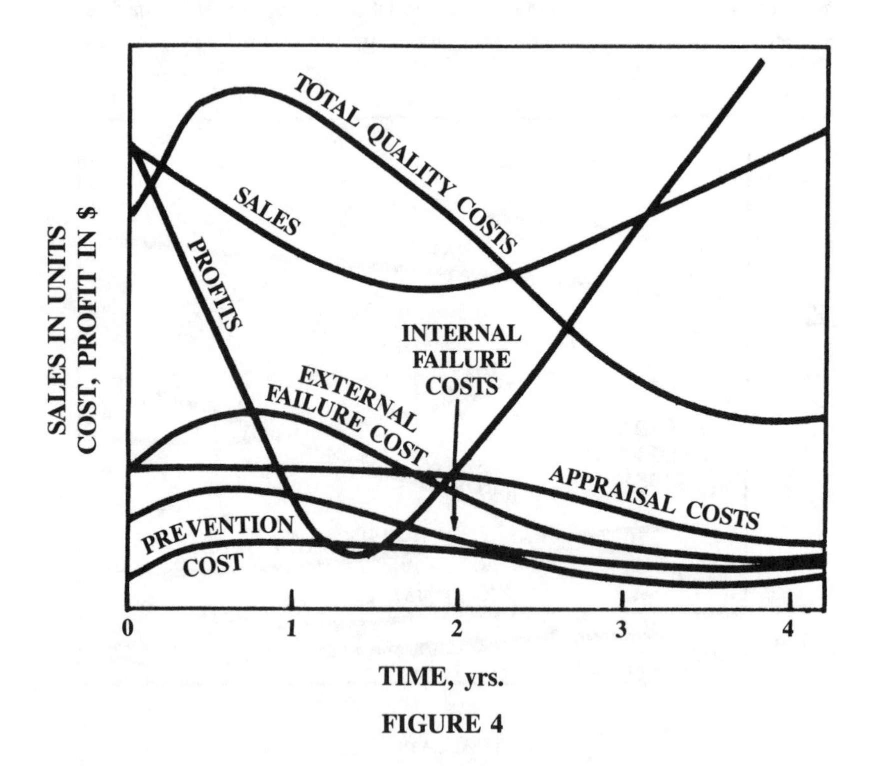

FIGURE 4

The Quality Cost concepts embodied in Figure 2 offer the real answer. The odds are that the market has changed so that product re-design and new specifications are required so as to once again reflect customer need. Spending money for more control of the wrong quality is not only wasteful of immediate profit but invariably delays efforts to uncover and solve the real problem.

Taking these factors into account, we can apply our prevention efforts correctly, hold the line on appraisal costs, anticipate some increased internal failure costs as new characteristics are brought under scrutiny and eventually rescue profitability. The word "eventually" refers to the extended period of time forecasted here in comparison to the situation previously posed. It does take time to re-design a product and rebuild a market.

Predicting results to be expected from prevention activities in a manner similar to the two hypothetical cases just discussed is obviously a valuable management exercise. It is especially so when actual, complete quality cost data is available as a starting point. An ongoing Quality Cost System will provide the data. An invaluable input to the sales and marketing organizations can occur if they have been helped to understand the TQC system thereby stepping out from behind emotional reactions to sales problems and arriving at the facts of customer need. The manufacturing people can be expected to breathe a little easier and say, "That makes sense!"

Unless a salesman is fortunately situated with a public utility enjoying a legal monopoly, he is faced with competition. He must know competitive products almost as well as his own. A trial lawyer spends much of his pre-trial effort preparing and predicting his opponent's case and methods. Competition also exists for the drummer for Total Quality Cost Systems. Since Quality Cost Systems are good management and good business, they should be compatible with other information systems and they are. The competition, therefore, is not system against system; but, rather for priority emphasis by top management. Before attempting to gain support for your program, look about to see what, where and who your competition is. Some of the labels they wear are standard cost system, Engineered Standards, value analysis program, value engineering, project or program management, product management, cost reduction program, management by objectives, participative management, or whatever else is in fashion at a particular point in time. Once they are identified, it should be possible to either find means to gain superior priority or join forces in a constructive manner.

For example, the proponent of a Total Quality System will often be challenged by the proprietors of a standard cost system. This system "belongs" to Industrial Engineering and Cost Accounting. Other functions such as Production and Equipment Engineering alternately use and are plagued by the data generated but, basically, the partners of the shop are the I.E.'s and accountants. They will react adversely to any new system with the word "cost" in its title. A cautious approach is counseled.

Figure 5 illustrates one approach to a standard cost system with an important addition. A quality cost standard has been added to the typical labor and material standards. The paramount fact here is that there is almost no change to the existing standards. This means, happily, no extra work for the I.E.'s. As an added extra bonus, however, this scheme removes the need to plan for inspection samples within the material standard. Placing the role of a non-conforming materials disposition activity inside the standards structure will please your cost accountants as they will appreciate the prospect of receiving good cause and result input data to back up "booked" scrap and rework costs.

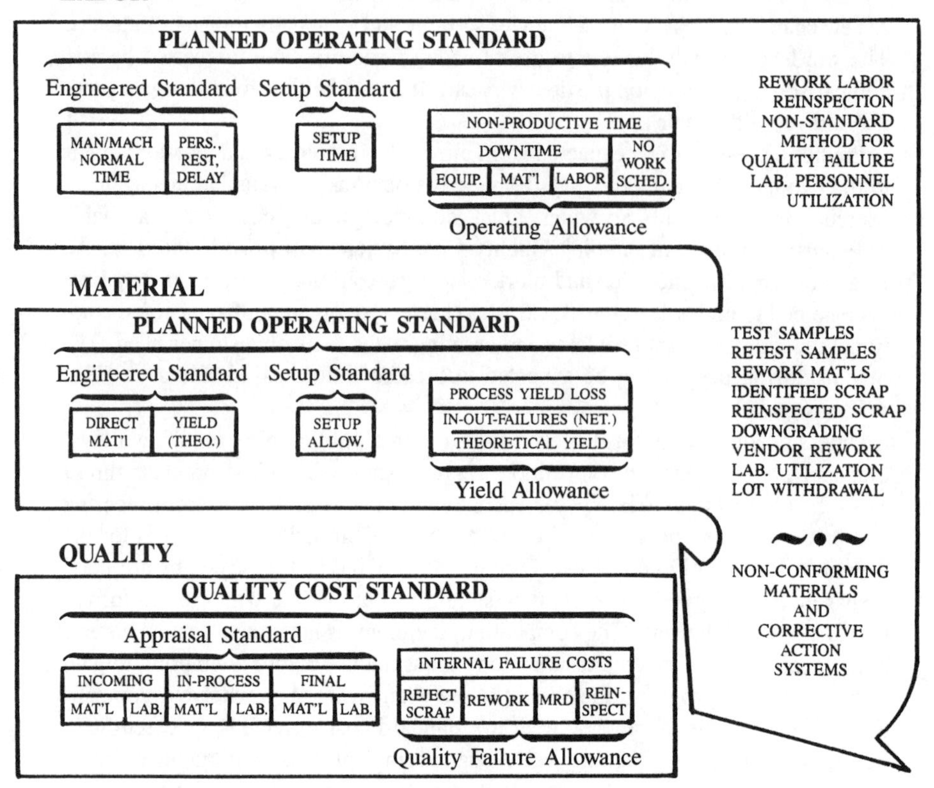

FIGURE 5

A most exciting aspect of using a quality cost standard is that cost items such as scrap, rework, QC samples, reinspection, etc., can be most easily differentiated from yield, schedule losses and production performance. In this manner responsibility for corrective action is accurately placed and is accompanied by substantiating data. The Quality Function becomes responsible for problem solutions and contributions to profits. Utilizing a Non-conforming Materials Disposition System, cost of "bad" quality can be efficiently managed and reduced. Production departments get full credit for performance reductions caused by working with out-of-specification materials without resorting to the popular, but unfortunate, practice of overall excessive "scrap" allowances. New, more economical inspection procedures and sampling plans can be instituted with credit placed where it will return the greatest motivation — with the quality staff. The entire engineered standards and allowances program is simplified. The question of "method" or "quality" is answered.

Beyond cost accounting systems which are basic, ongoing business essentials, the other programs and systems competing for attention have a basic purpose — the organization, motivation, and direction of the employees of an enterprise to achieve some special purpose such as lower cost, higher morale, or greater efficiency. They all have one initial fact in common and that is the need for good input data about operating status. This is in the finest tradition of the scientific method. In this phase a Total Quality Cost System is complimentary as we have just seen in cost accounting. The corrective action portion of Quality Cost activity is either non-conflicting or can be designed to be complimentary or integral with other schemes. At this point the question is one of allocation of resources in a manner which maximizes return on investment. On that battleground there should be no reason for yielding.

It is interesting to note, incidentially, that the literature advocating these other schemes generally mentions quality. This mention, however, is in the form of side benefits. The formalized procedures do not include quality considerations. The fact that Total Quality Cost theory and systems bring together quality, dollars, and people; vendors, customers, and employees from every department make this tool the most powerful of them all. Total Quality Cost is an overall operating philosophy — a way of life capable of encompassing any of the secondary systems.

A cardinal rule of salesmanship is to learn the customers' needs and demonstrate how your product fills them. Whether or not a product quality cost standard is ever to be instituted, the fact that you have taken the time to demonstrate compatibility with the existing Cost Accounting System must be appreciated. If your company uses direct costing, standard costs or what-have-you, you must learn it yourself sufficiently to demonstrate advantages to others. A potential for reducing the work of others will be a powerful incentive for them to learn more about your Quality Cost Theory. Equitable alignment of data with responsibility should always find acceptance. The author's experience has been one of enthusiastic interest from Industrial Engineering and Production when presented with the premise contained in Figure 5, for example.

Most current Total Quality Cost literature references are written by Quality Control people for Quality Control people. It is, therefore, structured to motivate that discipline. Implying that increased respect, income and advancement will be the prize captured by the Quality Manager clever enough to sell management on rearranging cost data is a disservice if such implication sends one of our number into the foray unprepared. The customer, when selling the benefits of a Total Quality Cost System, is not the Quality organization nor even top management. The customer is everybody in the company who has an interest in success but too much to do already. Prepare to sell the entire market.

REDUCING QUALITY COSTS

W. N. Moore, Staff Assistant, Headquarters Quality Control
Westinghouse Electric Corporation, Pittsburgh, Pa.

Despite all the talk in the last few years about quality costs, the many conference sessions devoted to it, and all the papers in the quality publications, many quality managers have not progressed beyond the stage of collecting a few numbers and slotting them into the four categories. I think we are missing the most promising boat to ever come along in the quality profession. In quality cost analysis techniques, quality control managers have the ability to speak a language that is common to all business managers — money. Let's face it, other managers don't understand AQL's control charts and statistical significance tests. They do understand money, but many quality control managers have not approached the problem of getting other people to do something about noncomformance problems in this way. Most quality improvement efforts require actions by others. Not many are of the type which are solvable by quality control alone. Quality costs programs must be cost improvement programs and involve the whole management team. No rational manager would expect a product cost improvement program to be carried out by one department and the same must be true of quality cost improvement.

What is needed is a managed quality cost improvement program which must consist of:

- Finding the high cost areas
- Setting targets for improvement
- Planning action programs
- Organizing to take action
- Measuring results

It doesn't matter whether we're going after appraisal or failure costs; only by managing quality costs reduction in the same way as other programs are managed can we expect to get the needed support from other parts of the organization and the results we hope for. Let's examine each of the steps in a managed quality costs improvement program in detail.

FINDING THE HIGH COST AREAS

It is really in this first step in the sequence that quality costs are used. To find high potential cost improvement possibilities, analyze the appraisal and failure costs. In what departments are the scrap and rework cost highest? What product lines have the highest scrap rework or warranty costs? Where are the highest inspection and test costs? As in product cost improvement, start with the high cost or high activity areas — there is where the highest return on the effort invested can be found. Use Pareto analysis — it's an old technique but still the most effective way to find potential cost improvements. Break down the costs to the point necessary to tell responsible management what the highest contributors in his area are and, if possible, what is causing the problem. Don't forget inspection and test costs. Better ways to test and inspect have come along in the past few years and if quality people are still doing things the same way they did five years ago, there's probably room for improvement. We must find ways to become more efficient just to stay even with some of the wage, benefit and raw material price increases.

SETTING TARGETS FOR IMPROVEMENT

What should quality costs be? Many people get hung up on this. It is only natural to wonder how your costs compare to others in the same business, but it is not really important to the business of structuring a cost improvement effort. Besides the fact that there is always room for improvement, quality costs are almost untapped sources of cost improvement. Many of them have not been thought of as preventable expenses. Most are not considered as direct product costs but as overhead and no one has led the way in establishing improvement programs. Because of this, reduction should be possible in most areas found to be high cost by the analysis. Targets should be established by responsible management considering the factors present in each situation. The important thing is that objectives be established and that responsibility for meeting the objectives be clearly assigned.

Although it seems that more people are talking about their costs now than did a few years ago, it is still difficult to find out what "average" costs in an industry are. Even if numbers are available, there is seldom enough information on the specific elements included to make meaningful comparisons — we're not always comparing apples with apples. Reasonable target setting is therefore hampered by insufficient basic data on what reported costs consist of and by a general reluctance to report them. Because of a wide variation in product lines, everything from consumer goods to aerospace equipment, from transistors to electronic equipment to electric power generation, Westinghouse Divisions present a pretty fair cross section of hard goods manufacturing. Some of these divisions have active quality cost reduction programs and others who wanted to get started needed some guidelines on what their costs ought to be. To answer this need, Westinghouse Headquarters Quality Control has prepared some cost target setting guidelines developed from the costs incurred in Westinghouse Divisions. These guidelines are presented here in the belief that they can profitably be used by quality control people in similar kinds of business.

Although it is not always possible to directly compare costs in one operation to those in another because of many differences in the way costs are reported, some patterns have become evident as costs in Westinghouse have been studied. Factors which appear to influence the economical minimum levels of quality costs are:

1. Production Rate
2. Customer Requirements
3. Product Complexity
4. Product Sophistication

The actual costs incurred will be influenced by many other factors such as the type of quality program (prevention oriented or inspection oriented), the attitudes of all personnel toward quality and quality improvement, and the capability of the manufacturing facilities. The latter set of factors also heavily influence the levels of quality costs but are ones which can be most affected by day-to-day management decisions on the operation of the business and by the business planning. They are most often the factors which are causing quality costs to be higher than they should be and, as a result, are the ones on which cost reduction effort should be concentrated.

Considering the first set of factors influencing costs, target or objective levels of costs can be estimated. The target costs obtained are the ones which are thought to be close to those obtainable with a good defect prevention oriented quality program, good attitudes toward quality and quality improvement, and facililties capable of manufacturing the product. The cost targets given here are for a quality cost reporting system which includes all the appropriate elements including in-process and final testing, rework, and remedial engineering. The target estimating technique consists of taking a base figure of 4% of gross sales billed and adding or subracting factors related to the basic influencers of cost. The modifying factors are as follows:

Production Rate

Units/Year:		
	Greater than 1,000,000	−1
	100,000 – 999,999	−.5
	1,000 – 99,999	0
	Less than 1,000	+.5

Primary Customer Requirements

Military/Government/Aerospace	+2
Utility	+.5
Industrial/Commercial	0
Consumer	−.5

Product Complexity

Parts/Final Products —		
	Greater than 1,000	+1
	300 — 1,000	+.5
	20 — 299	0
	Less than 20	−.5

Product Sophistication

State-of-the-Art	+1
New Design using conventional materials and concepts	+.5
Design in production but subject to periodic change	0
Stable product in production for long period	−.5

For example, for a division manufacturing consumer products with an average of 200 parts per assembly at a rate of 700,000/year, the target quality cost as a percentage of GSB would be:

$4\% - .5 - .5 + 0 + 0 = 3\%$

For a division manufacturing a product containing 100 parts for a utility customer at the rate of 1,5000,000 per year, the target would be:

$4\% - 1 + .5 + 0 - 0 = 3.5\%$

While it is thought that the factors considered in this technique are the major ones influencing costs, the estimates obtained by this method obviously do not consider all the factors which may influence cost levels in a particular situation. If other influences are present, target costs should be modified to consider those influences.

PLANNING ACTION PROGRAMS

Planning programs to reduce appraisal and failure costs should be basically the same as planning other management programs for improvement: identify high cost areas, target reductions, explore the alternatives, select a strategy from those alternatives and plan the steps necessary to implement the improvements. Following are some of the possible alternatives for the reduction of appraisal costs.

1. Could additional effort in planning profitably reduce appraisal or failure costs?
2. Could audits be used instead of 100% or lot-by-lot inspection?
3. Could improved vendor selection and control programs reduce total costs?
4. Are inspection points located to maximize the return on dollars spent for inspection?
5. Are inspection stations and methods engineered for the most efficient work accomplishment?
6. Could inspection and test operations be economically automated by using special purpose instrumentation, or tape or computer controlled equipment?
7. Could inspection and test record and data reporting functions be more efficiently performed using the computer or other modern data handling devices?
8. Is it possible to control processes sufficiently to prevent production of defectives and eliminate product inspections?
9. Could tests now being performed by outside laboratories be performed at less cost in-house or vice versa?
10. Are some tasks now being performed by highly paid inspectors or testers which could be performed by lower classification employees?

As can be seen, many of these things are ones that quality control department people can do something about. Failure costs are normally ones that other departments must do something to reduce. Following are some of the alternatives in reducing failure costs.

Internal Failure

1. Are the causes of high cost elements known and reported to those responsible for action?

1a. Are scrap and rework charges reported to the manufacturing foreman in such a way that he can identify the vital few high cost contributors and take corrective action?
1b. Are remedial engineering charges reported to the responsible engineering supervisor in a form which allows him to quickly find the high cost contributors and take action?
1c. Are vendor caused losses reported to purchasing and the receiving inspection departments so action can be taken?
2. Are defects found in assembly areas reported to sections or vendors responsible and is corrective action taken?
3. Is non-conforming material salvaged where economical?
4. Are there areas where a high rejection rate has become a way of life and little effort is being made to correct the situation?
5. Is the corrective action program broad enough to demand design or tooling changes if that is what is needed to prevent the production of defectives?
6. Is there a coordinated quality improvement effort involving all the necessary departments?
7. Does the quality improvement effort include regular meetings, docketed projects with completion dates, and written progress reports?
8. Does the program have the continuing involvement, interest, and support of the functional department managers and the general manager? If not, what must be done to gain that interest and support of the functional department managers and the general manager? If not, what must be done to gain that interest and support?

External Failure

1. Are causes of high cost field problems identified and is action being taken on them to prevent recurrence?
2. Are warranty charges audited for validity and action taken as necessary?
3. Are returned products economically handled, repaired and returned to the customer?
4. Are clear field performance objectives set to assure continuing controlled or reduced cost?

The most promising areas for improvement should be pursued with a step-by-step plan developed with the help and full knowledge of the people who must carry out the plan. We should organize for attacking high failure cost problems.

ORGANIZING TO TAKE ACTION

Most non-conformance problems are not ones that quality control can do much about. It takes a group of people, typically including line management, manufacturing engineering, design engineering and quality control, to attack a problem from all sides. Each problem requiring co-ordinated effort for solution should be individually docketed. The docket should fully describe the problem and its probable cause. Actual and projected dollar losses should be given and a corrective action plan documented. Responsibilities and due dates for the steps in the corrective action plan must be assigned and agreed to by the involved parties. Projected savings should be estimated and actual savings recorded for each docketed problem. Figures 1 and 2 are illustrations of a form used for docketing problems.

MEASURING PERFORMANCE

We should not attempt to measure quality cost improvement solely by looking at total costs or even at ratios. While eventually total costs should come down with the use of an effective program, short term changes in these costs are not good measures of progress. How many product cost improvement programs measure their impact only by comparing overall product costs from one period to another? Most measure success and each individual's contribution by measuring and totalling the effect of each individual improvement. In some situations total costs go up because of increases in wage and material costs, but the cost improvement program is a success because it prevents a larger rise.

If we don't actively pursue quality costs improvement, these costs will rise. On the quality problem docket shown in Figure 2, there is a spot for calculating savings on each docketed problem. Net savings should be accumulated as a part of the division or company cost improvement program.

In review, there are five steps in effective quality cost reduction:

Finding the problems
Setting Targets
Planning Action
Organizing to take action
Measuring results

Taken together these steps are an almost sure — fire managed quality cost reduction program. By effectively applying the principles, I believe any organization can reduce their costs of quality.

QUALITY PROBLEM DOCKET

Docket Number	Docket Title	
Product Affected	Departments Affected	Indv. Resp. for Coordination
Problem Description		
Estimated Date Problem Began:		
Probable Cause		

Loss Due to Problem	In-Plant	In-Field	Total
Loss to Date	______	______	______
Projected Loss if not Corrected	______	______	______

Corrective Action			
Required on:	Current Production	Inventory	Product In Field
Date Scheduled	______	______	______
Date Corrected	______	______	______

Initiated By______________________ Date ______________

FIGURE 1

QUALITY PROBLEM CORRECTIVE ACTION PLAN

Event No.	Corrective Action Event	Event Responsibility	Date Completed		Status Review	
			Schedule	Actual	Date	Comments

Quality Cost Savings		
Cost Category	Projected	Actual
Appraisal Cost Reduction		
Internal Failure Cost Reduction		
External Failure Cost Reduction		
Other (Explain)		
Total Savings		
Cost of Problem Solution		
Net Savings		

Signature of Resp. Individuals

____

____

____

____

____

FIGURE 2

HONEYWELL'S COST EFFECTIVE DEFECT CONTROL THROUGH QUALITY INFORMATION SYSTEMS

Ronald R. Cerzosimo, Supervisor, Quality Control
Honeywell, Inc., Ft. Washington, Pa.

Quality Information Systems permit cost effective defect control, and provide management with visibility of quality performance in manufacturing.

Does your management require you to provide visibility of quality performance? Has your management told you Quality Control must contribute to profits? If your company has a profit motive, the answer to these questions is YES. But, do you in fact provide the visibility necessary to contribute profits? Do you know the cost of your quality problems?

In recent years, numerous companies have implemented Management Information Systems to provide visibility of quality performance. It is important to note that Q.I.S. can contribute meaningful inputs to M.I.S. A prime mechanism to Q.I.S. is defect reporting and feedback aimed at cost effective defect control. Unfortunately, in practice, too many defect reporting and feedback systems aim for completeness, thereby taking attention from the vital few defects and transferring it to the trivial many. Also, too many systems are designed to exercise the statistical skills of the engineer rather than to provide swift, simple, sure understanding to Management — not to forget the operator. When these conditions exist, the real needs of Q.I.S. are violated, and you can be sure the quality organization is in fact perpetuating its own existence and thereby contributing to loss — not profit.

The best means to offset accusations which challenge controllable performance and product integrity is to assure they are both satisfactory. This is accomplished, in part, with objective evidence. Objective evidence can be obtained through a comprehensive defect reporting and feedback system.

The methodology presented here is a result of personal experience, and extensive discussions with other practitioners. The disciplines are idealistic and realistic. They do not advance the "state of the art" in reporting systems because they simply exercise the often overlooked fundamentals. These disciplines have proven successful results; and, when conditioned to satisfy your individual applications, they will be conducive to profit, thus enabling you to reap the full economic benefits of our rapidly advancing technology.

The general objectives of the method presented are to:

1. Provide management visibility of quality performance.
2. Improve defect prevention.
3. Optimize economic balance of quality.
4. Assure design and manufacturing problems are identified and corrected in a minimum time period.

It is really quite simple to satisfy the objectives of a comprehensive defect reporting and feedback system when there is recognition and understanding of the system's make up. The system's make up is as follows:

1. Manufacturing inspection and test reports.

2. Display of avoidable costs related to quality.
3. Data handling, analysis and review disciplines.
4. Problem identification summary and responsibility assignment.
5. Corrective action.

MANUFACTURING INSPECTION AND TEST REPORTS

Manufacturing must provide specific defect reporting and indicate expended (avoidable) time related to those defects reported. The expended (avoidable) time is related to one of three separate categories:

1. Production (Shop) Error— This indicates the hours expended to correct workmanship errors and rework of non-standard practices, each of which are the responsibility of Production.
2. Engineering Error — This indicates the hours expended to invoke engineering changes, and rework due to an error initiated by Engineering, each of which are the responsibility of Engineering.
3. Vendor Material — This indicates the hours expended to repair, rework or replace defective vendor material.

Accordingly, defect identification should be coded, where possible, and be comprehensive.

Figures 1 and 2 are examples of a typical format for a printed circuit board test report.

Figure 3 is an example for a typical subassembly or final product test report.

Figures 4 and 5 are examples for a typical visual inspection record.

In each case the report was designed for ease of reporting comprehensive information, as well as providing for the display of trends.

DISPLAY OF AVOIDABLE COSTS RELATED TO QUALITY

Each manufacturing employee who reports his daily activities on a job card or labor distribution card is required to identify that expended time which was previously described as avoidable. The information required from such a card is:

1. Expended time (normal)/job.
2. Expended time (avoidable)/job.
3. Employees' clock or identification numbers.
4. Job, sales, or work order number identifying a product, assembly, or part.
5. Employees' control unit, cost center or section number.

These labor distribution cards are processed by E.D.P. to provide date segregation. Each day or week, whichever is satisfactory to your application, Data Processing will report, to Quality Assurance, avoidable cost apportionment by job number within each control unit. In addition, the cumulative avoidable cost total will be displayed, as well as the total assembly or test cost for each job number. Also, the total avoidable cost will be shown as a percentage of the total assembly or test cost for each job number.

Figure 6 is an example of a typical TAB format.

DATA HANDLING, ANALYSIS AND REVIEW DISCIPLINES

Quality Assurance is responsible to analyze the Data Processing report and compare it to the inspection and test reports for a given period. The analyst should be knowledgeable of the products and manufactruing processes, and be able to differentiate sporadic and chronic type problems. Empirically, good common sense judgment is required for selecting the vital few defects from the trivial many that are reported.

The Data Processing report eliminates subjectiveness in choosing the products or jobs out of control or close to it. Once he determines the products with the highest avoidable costs, he can begin to extract his pertinent data from the inspection and test reports.

At this point of the analysis, it is extremely important not to hide the blackboard. If conditions exist (problems reported) that may have adverse consequences to any operation, don't wait for the formal summary to be prepared and distributed. Make sure the people involved are apprised of the condition so that contingent action can be triggered.

PROBLEM IDENTIFICATION SUMMARY AND RESPONSIBILITY ASSIGNMENT

This is the document that is distributed to Management and other responsible individuals. It summarizes the cost apportionment reflecting highlights of the avoidable costs realized in a manufacturing operation for a given time period. Emphasis is placed on identifying the significant high cost defects rather than the high quantity defects with low cost.

The summary report should display the pertinent facts about each control unit in the manufacturing operation. Brevity in reporting, is most desirable. You should begin your report with an introductory statement like, "The cost apportionment and related defects shown reflect avoidable costs realized in manufacturing the week ending ________." Then identify the first control unit. Show the total avoidable costs as displayed for the control unit in the Data Processing report. Next, identify the three most significant products or jobs which experienced the highest avoidable cost, and show the costs as displayed in the Data Processing report. Now identify the significant defects contributing to each of the three individual products or jobs you have identified as having the highest avoidable costs. If the control unit you are reporting is responsible for the defects indicated, enough said. But, if the control unit expended the avoidable time because the defects were the responsibility of a previous operation performed by another control unit, then you must identify which control unit was responsible.

Take the next control unit to report on, and display your data. Each control unit supervisor is responsible for obtaining corrective action in his area. Other functional groups, such as manufacturing, engineering, etc., must provide corrective action when they are responsible.

CORRECTIVE ACTION

Corrective action must be positive to close the loop in feedback. Perhaps manufacturing will exercise a problem analysis to determine cause and effect relationship. They may need to change their process, or acquire new tools. Since the avoidable costs have been displayed, justification of special tools and/or capital equipment is not nebulous. Operator training is another form of action that is triggered because of the data that can be provided.

Defect prevention can be improved by tightening inspection of specific procurred material. Possibly, the characteristics of a device that is causing problems are not tested adequately. If this is so, you should revise your inspection procedure or even modify your test instrumentation.

Design problems can be identified, and, when they are too frequent, the avoidable costs display can justify the need for more mature designs and improvements in new design control programs.

The cost effectiveness of a quality organization is a most serious consideration. This reporting scheme provides a cost barometer for Quality Assurance so that economic balance can be realized. When was the last time you heard someone in your company say, "Too much control, too little quality"? Each of you has a responsibility — don't lose sight of it.

In summary, a comprehensive cost oriented defect reporting and feedback system is, in part, the intelligence of the quality system. It provides the factual basis on which quality decisions can be made and action taken. There is no single defect reporting and feedback system that is correct for all applications. Each system must be developed and fitted for the particular application.

The design of a defect reporting and feedback system begins with deciding what is required and how the system should be structured to help you determine how your quality performance rates. How does your quality performance rate? Are you contributing to profit — or to loss?

REFERENCES

1. Cerzosimo, Ronald R., "Defect Reporting and Feedback," presented at the 14th Annual Symposium, Philadelphia Section A.S.Q.C., November 12, 1970, Philadelphia, Pennsylvania.

PANEL PRINTED CIRCUIT BOARD TEST REPORT

Board Type	________	Tested By	________
Part No.	________	Date	________
Lot No.	________	Total Tested	________
Lot Qty.	________	Total Rejected	________

MANUFACTURING WORKMANSHIP DEVIATIONS

Type	Tally	Total
Solder Shorts		
Lead Shorts		
Dirty Contacts		
Wrong Value/Comp.		
Missing Comp.		
Reversed Comp.		
Lead Not Soldered		
Solder Not Flowed		
Lead Out		
Transistor In Wrong		
Solder Under Trans.		
Wire Error		
Broken Components		
I. C. In Wrong		
VENDOR MECHANICAL DEVIATIONS		
Joined Track		
Open Track		

F-4340f (11/69)

REMARKS

Trouble Shoot & Repair Time (Tally)

Hrs.	91	92	93
.1			
.2			
.3			
.4			
.5			
.6			
.7			
.8			
.9			
1.0			
Tot.			

Code - 91 = Mfg. Error
92 = Eng. Error
93 = Vendor Responsiblity
Note: Forward to Q. C. Engineer when completed

FIGURE 1

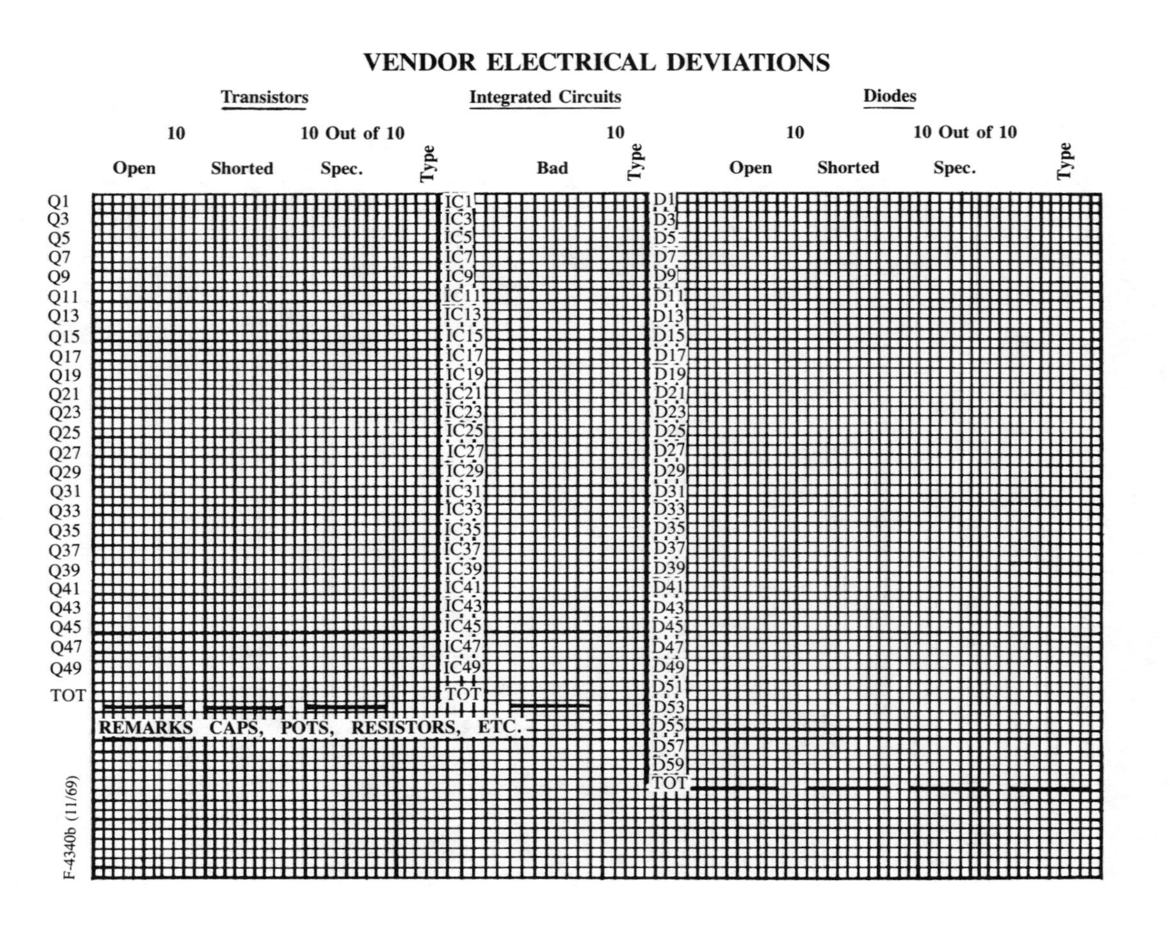
VENDOR ELECTRICAL DEVIATIONS
Transistors
10
10 Out of 10
Open
Shorted
Spec.
Type
Integrated Circuits
10
Bad
Type
Diodes
10
10 Out of 10
Open
Shorted
Spec.
Type
Q1 Q3 Q5 Q7 Q9 Q11 Q13 Q15 Q17 Q19 Q21 Q23 Q25 Q27 Q29 Q31 Q33 Q35 Q37 Q39 Q41 Q43 Q45 Q47 Q49 TOT
IC1 IC3 IC5 IC7 IC9 IC11 IC13 IC15 IC17 IC19 IC21 IC23 IC25 IC27 IC29 IC31 IC33 IC35 IC37 IC39 IC41 IC43 IC45 IC47 IC49 TOT
D1 D3 D5 D7 D9 D11 D13 D15 D17 D19 D21 D23 D25 D27 D29 D31 D33 D35 D37 D39 D41 D43 D45 D47 D49 D51 D53 D55 D57 D59 TOT
REMARKS CAPS, POTS, RESISTORS, ETC.
F-4340b (11/69)

FIGURE 2

PANEL PRODUCT/SYSTEMS TEST REPORT

Product P/N or Sales Order ________ X = Hours Sheet ______

Stock or Customer ____________ Y = Code X/Y of ______

Lot No.		Qty.		Code 91 = Shop Error 92 = Engineering Error 93 = Vendor Responsibility						
Date	Unit No.	No. Problem	Component Failure	Wiring Errors		Miscellaneous Defects	TS & Rep. Time (Hrs.)			Tested By
				Wiring Is	Should Be		Start	Stop	Elapse	
									/	
									/	
									/	
									/	
									/	
									/	
									/	
									/	
									/	
									/	
									/	
									/	

FIGURE 3

PANEL AUDIT AND/OR FINAL INSPECTION RECORD X/Y X = Item Y = Qty.

VISUAL

Date	Item Description	Lot No.	Serial No.	Characteristics Inspected									Total Units Insp.	Total Units Accp.	Total Units Rej.	Inspector
				A	B	C	D	E	F	G	H	J				
				/	/	/	/	/	/	/	/	/				
				/	/	/	/	/	/	/	/	/				
				/	/	/	/	/	/	/	/	/				
				/	/	/	/	/	/	/	/	/				
				/	/	/	/	/	/	/	/	/				
				/	/	/	/	/	/	/	/	/				
				/	/	/	/	/	/	/	/	/				
				/	/	/	/	/	/	/	/	/				
				/	/	/	/	/	/	/	/	/				
				/	/	/	/	/	/	/	/	/				
				/	/	/	/	/	/	/	/	/				
				/	/	/	/	/	/	/	/	/				
				/	/	/	/	/	/	/	/	/				
				/	/	/	/	/	/	/	/	/				
				/	/	/	/	/	/	/	/	/				
				/	/	/	/	/	/	/	/	/				
				/	/	/	/	/	/	/	/	/				

FIGURE 4

CHARACTERISTIC DEFECTS CODING

A SOLDERING

1. No Solder
2. Cold Solder
3. Rosin Connection
4. Solder Not Flowed
5. Insufficient Solder
6. Excess Solder
7. Solder Snort
8. Solder In Prong
9. Solder Point Draw
10. Solder Splashes

B INSULATION

1. Cut Insulation
2. Frayed Insulation
3. Insulation Back
4. Insulation in Soldered Lug
5. Burnt Insulation
6. Insulation Missing
7. Pinched Wire
8. Tight Wire
9. Wire Too Long
10. Frayed Wire
11. Broken Wire

C WIRING

1. Poor Wire Dress
2. Loose Cable Lace
3. Wrong Color Code
4. Incorrect Lacing Tie
5. Poor Crimp
6. Pin Not Seated
7. Wiring Error
8. Missing Wire

D P/C BOARD

1. Reversed Component
2. Damaged Component
3. Wrong Value Component
4. Improper Clearance
5. Improper Lead Config-uration
6. Cracked Board
7. Pads and/or Traces Lifting
8. Dirty Finger Contacts
9. Excess Warp or Twist
10. Flux Not Removed
11. Clipped Leads
12. Pencil Marks or Writing on Board
13. Solder on Contacts
14. Chipped Plating
15. Blistered Board
16. Open or Broken Trace
17. Shorted Trace
18. Missing Component
19. Defective Component
20. Hole Missing
21. Hole Undersize
22. Hole Oversize

E PNEUMATIC

1. Damaged Pipe (tubing)
2. Loose Fittings
3. Missing Fittings
4. Insufficient Support
5. Improper Flaring
6. No Flaring
7. Poor Weld, Braze, and/or Solder

F WIRE WRAP

1. Number of Turns
2. Overlap
3. Spacing of Turns
4. End Tails
5. Terminal Twist
6. Wire Damage

A	Soldering
B	Insulation
C	Wiring
D	P/C Board
E	Pneumatic (Piping)
F	Wire Wrap
G	Packaging
H	Metal Fabrication
J	Miscella-neous

G PACKAGING

1. Improper Container
2. Insufficient Packing Material
3. Damaged Container
4. No Labels
5. Improper Identification
6. Missing Items
7. Improper Papers of Authorization
8. Poor Handling

H METAL FABRICATION

1. Bad Weld
2. No Weld
3. Rough Finish
4. Poor Paint Finish
5. Cutout Over Size
6. Cutout Under Size
7. Cutout Wrong Location
8. Missing Cutout
9. Excessive Warp
10. Insufficient Putty or Fill-In
11. Wrong Finish
12. Stud Location Wrong
13. Hole Location Wrong
14. Tapped Hole Wrong/Missing
15. Hole Missing
16. Operation Missing

J MISCELLANEOUS

1. Loose Part
2. Wrong Part
3. Damaged Part
4. Non-Conforming Part
5. Missing Part
6. Foreign Material
7. Wrong Value Part
8. Potential Short
9. Not to Latest E.O.
10. Short
11. Wrong Assembly
12. Component Open

FIGURE 5

PRODUCT QUALITY AVOIDABLE COST REPORT

Date (week ending) ____________

CONTROL UNIT NO.	SALES ORDER/ POOL NO.	#91 SHOP ERROR HOURS	#91 SHOP ERROR DOLLARS	#92 ENG. ERROR HOURS	#92 ENG. ERROR DOLLARS	#93 VENDOR ERROR HOURS	#93 VENDOR ERROR DOLLARS	AVOIDABLE TOTAL HOURS	AVOIDABLE TOTAL DOLLARS	TOTAL HOURS	TOTAL DOLLARS	PERCENT AVOIDABLE
705	00000-00-000	47.2	175.98		.00	7.2	24.48	54.4	200.46	187.0	687.66	29.15
705	00000-00-001	32.7	121.65		.00	16.3	59.44	19.0	181.09	113.0	417.84	43.33
CONTROL UNIT TOTAL		79.9	297.63		.00	23.5	83.92	103.4	381.55	300.0	1105.50	34.51
706	11111-11-000	31.0	74.63		.00		.00	31.0	74.63	311.5	822.50	09.07
706	11111-11-001		.00	4.0	10.64		.00	4.0	10.64	44.8	110.72	09.60
CONTROL UNIT TOTAL		31.0	74.63	4.0	10.64		.00	35.0	85.27	356.3	933.22	09.13
GRAND TOTAL (Department)		110.9	372.26	4.0	10.64	23.5	83.92	138.4	466.82	656.3	2038.72	22.89

EXAMPLE: TAB RUN-OFF

FIGURE 6

OPTIMIZING ATTRIBUTE SAMPLING COSTS — A CASE STUDY

Edgar W. Dawes, Director of Quality
Dictaphone Corporation, Bridgeport, Connecticut

ABSTRACT

Together with the Acceptable Quality Level (AQL), Average Outgoing Quality Limit (AOQL), and the Lot Tolerance Percent Defective (LTPD) concepts, Bayesian statistics provides an additional tool to further improve sampling costs. This paper develops a Bayesian probability model which was applied to improve sampling cost on a vendor supplied item.

INTRODUCTION

The cost-conscious Quality Engineer will carefully derive his sampling plan, for he knows that fraction defectives in excess of a break-even point should result in a reject decision (the break-even point is defined as that fraction defective value on the operating characteristic curve where the penalty cost of using defectives equals the cost of rejection). He also knows that it is less costly if an accept decision accompanies fraction defective values below this break-even point.

But what of the various fraction defective values to be encountered in submitted lots; are they all equally likely to occur? The probablistic answer is "*No*"! Also, inasmuch as the penalty cost for using defectives is a function of the frequency of the various fraction defective values, is the *total cost* of the sampling scheme being affected by these unknown proportions? The answer is "*Yes*"! If these questions are familiar, the reader has probably had experience with the classic problem faced by many inspection functions, namely, "What is the (unknown) fraction defective in the lot being inspected?

It is to this point that Bayesian statistics can be of value.

BAYESIAN PRINCIPLES AS APPLIED TO A VENDOR SUPPLIED ITEM

The basic concept of Bayesian statistics involves the decision maker's belief concerning the likelihood of an event occurring. This is supplemented by (but more preferably derived from) some prior information. Out of these beliefs and information, a "prior probability distribution" is constructed and can be defined as "the estimated probability of occurrence assigned to various values of an unknown parameter based upon prior information or belief". Such estimates are helpful, (indeed essential,) in improving control over future events.

In our case, the future events were defined as the need to optimize sampling costs on a vendor supplied item. To facilitate this, a prior probability distribution was constructed which emperically derived the various fraction defectives present in previously supplied vendor lots Figure (1).

As a further extension of the prior probability distribution, Locke[1] has pointed out that *conditional* errors and losses related to faulty accept-reject decisions can be assigned to each of the various values of a lot fraction defective. Such errors and losses are *conditional* in that they depend on the value of the unknown lot fraction defective. Further, Locke cites that the discipline of determining these errors and losses is fundamental in deriving optimal minimum losses.

Finally Schlaifer[(2)] indicates that there is an "Opportunity Loss" associated with faulty decisions. He defines this as "The difference between the cost actually incurred because of the decision, and the cost which would have been incurred if the best possible decision had been made with full knowledge of the value of the parameter which actually exists".

The derivation of these opportunity losses for our vendor item first required an analysis of the break-even point for the use versus non-use of defectives.

NO SAMPLING ACCEPT-REJECT ECONOMICS FOR THE VENDOR ITEM

In our operation, the penalty cost for using a defective piece was easily derived. This equalled the labor cost of assembly and disassembly and was a function of the failure to find the defect until final test. This cost was \$.40 per occurrence. Conversely, the cost of inspecting each item was \$.025. This resulted in the following break-even calculation.

$$A = \frac{I}{p}$$

Where A = the penalty cost of using a non-detected defect
I = the cost of inspecting each piece
p = the sought for break-even point fraction defective

And in our case:

$$\$.40 = \frac{\$.025}{P}$$

$$.40\ p = .025$$

$$p_{bep} = .0625$$

The .0625 fraction defective therefore becomes our break-even point.

The reader will note that the costs for replacing failing parts and rectifying inspection are not included in the calculation. These costs were borne by the vendor. However, rectifying inspection costs were important in the cost analysis inasmuch as our labor (cross-charged to the vendor) was sometimes involved in rejected lot sorting. Also, total costs were being presented to the vendor to facilitate analysis of cost alternatives.

These costs alternatives were classic in nature:

1) *Do no sampling and reject all lots,* thus incurring added expense for those lots below a . 0625 break-even point fraction defective. (A Type One error).

2) *Do no sampling and accept all lots,* thus incurring added expense for those lots above a .0625 break-even point fraction defective. (A Type Two error).
3) *Utilize one of several sampling plans* where the .50 probability of acceptance was at the calculated break-even point fraction defective of .0625.

The costs for the no sampling-sort or accept alternatives were defined as:

No Sample-Sort all lots (Type One error)

$$L_1\ (p) = A \cdot 1000 \cdot (p_{bep} - p') \ ; \ p' \leq p_{bep}$$

No Sample-Accept all lots (Type Two error)

$$L_2\ (p) = A \cdot 1000 \cdot (p' - p_{bep}) \ ; \ p' \leq p_{bep}$$

where:

A = the penalty cost of using a non-detected defective piece
1000 = the lot size
p' = the fraction defective value under consideration
p_{bep} = the break-even point fraction defective

Figure (2) depicts Type One and Two error costs based upon the above formulas. The reader will note that, as expected, a Type One or Two error cannot be committed at the break-even point of .0625.

It should also be noted that no attempt has been made to weight the conditional loss for each fraction defective. It is at this point that we depart from the usual method of assessing losses and initiate the use of the prior probability distribution. Figure (3) shows the *expected* losses as a function of the probability of the various fraction defective values occurring. The expected loss without sampling is defined as

$L_1\ (p) \cdot P_p$ for a faulty reject decision

$L_2\ (p) \cdot P_p$ for a faulty accept decision

where:

P_p = the probability of occurrence as derived from the prior probability distribution.

As an example, the conditional loss for the zero fraction defective ($25 per lot) thus becomes $5.25 per lot based upon a .21 probability of occurrence for the zero fraction defective.

Summing the costs of the individual faulty rejection decisions below .0625 fraction defective gives a total of $12.70 for the various anticipated fraction defective values while the total of all faulty acceptance values above .0625 fraction defective yields $2.98. Our prior probability distribution model consequently appears to favor a no sampling-accept decision as less costly than the no sampling-reject alternative.

But what of the third alternative, namely, the use of sampling to further minimize the expected opportunity loss?

SAMPLING ECONOMICS FOR THE VENDOR ITEM

The optimum sampling plan for our vendor item must conform to the following requirements:

1. The .50 probability of acceptance on the operating characteristic curve must be at .0625 fraction defective. Experience teaches us that there are many such plans derivable from the Poisson distribution. Several had been considered and the three shown in figure (4) were chosen for cost analysis.
2. The total cost of sampling plus the expected opportunity loss must not exceed $2.98, the total no-sampling-accept decision.

The cost of inspecting a single unit was quoted earlier as $.025. To determine the total cost for sampling, we merely multiply the sample size (n) times the $.025 unit cost.

The expected opportunity loss associated with sampling is in the error of rejecting lots containing fraction defectives below .0625 and accepting lots with fraction defectives above .0625. This is directly determinable from the operating characteristic curves shown in Figure (4) and is defined by :

$$L_1 (p) \cdot P_p \cdot P_e$$

$$L_2 (p) \cdot P_p \cdot P_a$$

where:

P_e = the probability of sample rejection for true fraction defectives below .0625.

P_a = the probability of sample acceptance for true fraction defectives above .0625.

Figure (5) shows the results of applying the formulas immediately preceding. Once again, the reader will note that at the break-even point, there is no expected opportunity loss inasmuch as the cost of rejection equals the cost of acceptance. Summing the expected losses for the various fraction defective values and adding these costs to the previously cited sampling costs results in a total cost for each sampling scheme. These costs are summarized in Table 1 below and reveal that a sample plan with n = 42, c = 2 has the lowest total cost of the alternatives examined. This subsequently was selected as the optimum sampling plan for use on the vendor item.

TABLE I — COST ANALYSIS
SAMPLING VS NON SAMPLING —
EXPECTED LOSSES AND TOTAL COSTS

Cost Category	No Sampling		Sampling		
	Reject Decision	Accept Decision	n = 42 c = 2	n = 74 c = 4	n = 122 c = 7
Sampling	–	–	$1.05	$1.85	$3.05
Expected Loss	$12.70	$2.98	$1.17	$0.77	$0.52
Total Cost	$12.70	$2.98	$2.22	$2.62	$3.57

SUMMARY

The true benefit to the Bayesian approach is in formalizing and improving the decision maker's opinions concerning the probability that certain events will occur.

In our example, there was another benefit, namely, the economic analysis of the inspection imposed on the vendor item precipitated more thorough consultation with the vendor. Although not detailed in this article, the additional costs associated with rectifying inspection were prohibitive for both the vendor and the consumer. The total cost picture resulted in an extensive dialogue with the vendor and permanent corrective action on his part. We sincerely believe that this would not have been the case had we not applied sampling cost analysis, including the Bayesian probability approach to this particular item.

REFERENCES

(1) Locke, Lawrence G., "Bayesian Statistics", *Industrial Quality Control,* April 1964

(2) Schlaifer, R., "Introduction to Statistics for Business Decisions", New York: McGraw Hill, 1959

(3) Schlaifer, R., "Introduction to Statistics for Business Decisions", New York: McGraw Hill, 1961

BIBLIOGRAPHY

"Quality Control Handbook", Juran, Joseph M., McGraw Hill, 1962

FIGURE 1

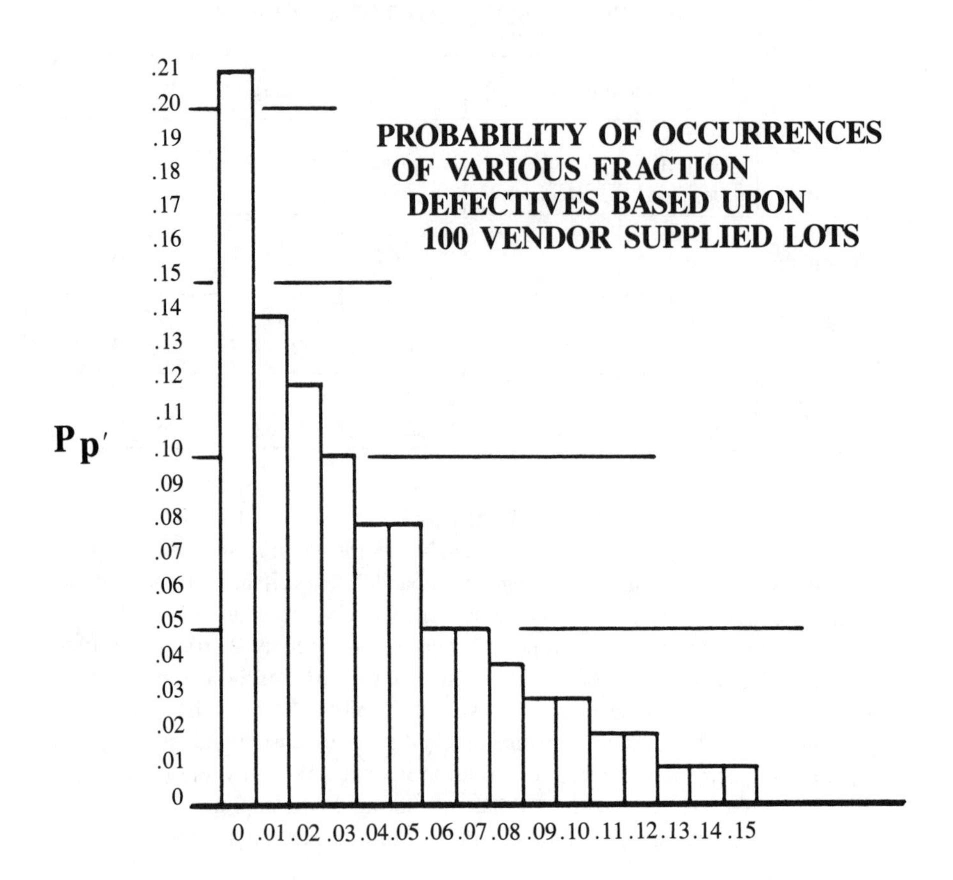

FRACTION DEFECTIVE p′

FIGURE 2
CONDITIONAL LOSSES WITHOUT SAMPLING — LOT SIZE: 1000 UNITS

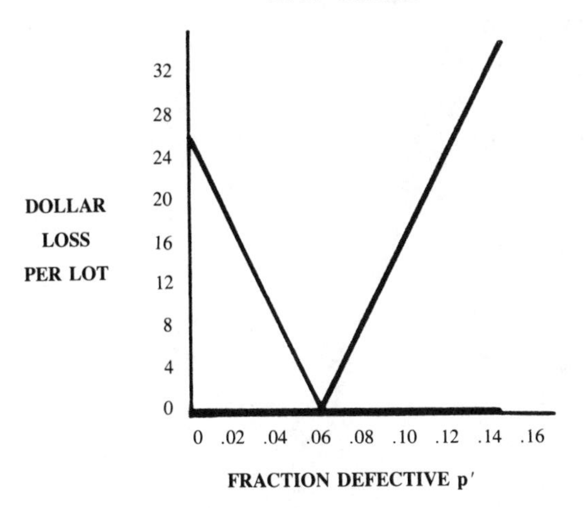

FIGURE 3
EXPECTED LOSSES WITHOUT SAMPLING — LOT SIZE: 1000 UNITS

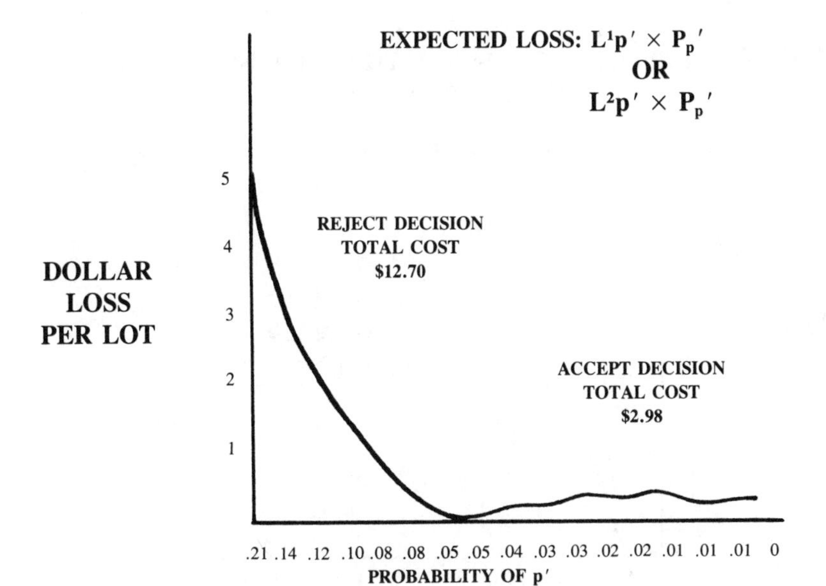

FIGURE 4
OPERATING CHARACTERISTIC CURVES WITH BREAKEVEN POINTS AT OR NEAR 0625 p

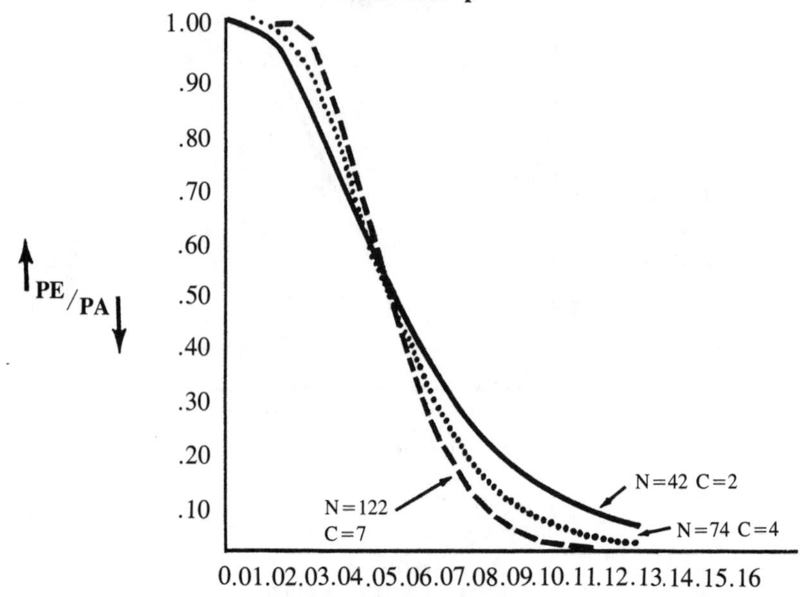

FIGURE 5
EXPECTED LOSSES WITH SAMPLING LOT SIZE: 1000 UNITS

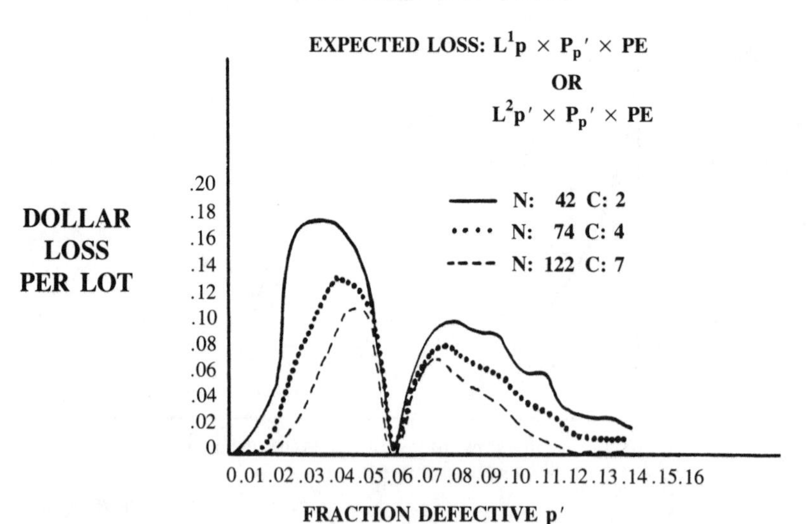

QUALITY COST MEASUREMENT AND CONTROL

Anthony M. Agnone, Manager, Reliability and Quality Assurance Administration
And
Clayton C. Brewer and Robert V. Caine
Quality Control Engineers
Aerospace Electronics Systems Department
General Electric Company, Utica, New York

Any reader of technical journals dealing with quality assurance or industrial management has seen an increase over the past few years in the attention being given to operating quality costs. Hardly a month goes by without at least one of the major publications presenting an article on this subject. Many of these articles contain what appears to be a consistent weakness — they typically present systems that give top management a report on the effectiveness of their quality program but provide little insight into specific problems that can be addressed effectively by first level management.

This paper will trace the evolution of one of these high level quality cost measurement systems into a working measurement and control system with visibility and corrective action at all levels. This system was developed at the Aerospace Electronic Systems Department of the General Electric Company.

Since the term "department" will be used frequently, we had better understand what it means. Within General Electric, a department is the lowest level in an organization having its own:

1. Product charter
2. Income statement (profit and loss)
3. Operating functions (e.g., Marketing, Manufacturing, Engineering, Reliability & Quality Assurance).

Within our department we design, manufacture, and market military and space electronics hardware and technology ranging in complexity from components to major systems with output varying from single technology reports through high volume production. Product scope covers several product lines organized within higher level business segments. The department has hundreds of active requisitions, usually representing over thirty programs, and experiences significant year to year fluctuations in sales volume and inventory levels. Most of the factors which make quality cost measurement and control difficult are evident within our operation.

General Electric pioneered quality cost measurement in the mid-fifties and has continually provided for formal reporting from department to corporate level. Quality costs include many of the variances which affect profit. Their analysis, like the income statement and the balance sheet, helps to determine the current and continued health of the business. Their periodic review provides insight into management's responsiveness to the changing business environment.

Examining corporate or department level reporting is like evaluating the success of a football team by examining its won/loss record. We would rightly assume, from their 1972 record of seventeen wins and no losses, that the Miami Dolphins had a highly successful season. We would also agree that the Philadelphia Eagles, with a record of two wins and twelve losses, had a dismal year. But examination of the record alone does not give insight into the nature of a team's strengths or weaknesses. When conducting evaluations at this level, what does management do if the results are unfavorable? They can worry; they can shout; they can inspire; and they may even perspire — but that's about all. Our managers objected to this type of quality cost report card since it gave them no assistance in problem isolation. We were, in effect, showing only a record of wins and losses; and they needed more.

If our football coach had known the scores of all the games in his 2-12 season, and found that in each game his team had scored over thirty points, he would know that his problem was primarily with his defense. The score would tell him something that the record of wons and losses could not. In the same sense we wanted to supply our managers with the kind of data they needed to pinpoint weaknesses and thereby change performance. To meet this need, we developed a quality cost measurement, control, and corrective action system.

To establish such a system it was necessary to separate operating quality costs from other related activities (e.g., reliability, test equipment design). It was also concluded that since quality costs are incurred at the lowest operating level, they should be accumulated at that level.

Figure 1 contains a portion of our Reliability and Quality Assurance organization. The left side of the chart shows a typical product line organization, each line having its own quality assurance, reliability engineering, and test equipment engineering units. The right side contains functions that support all product lines. It should be noted that test and inspection are not a part of this organization, but report to Manufacturing.

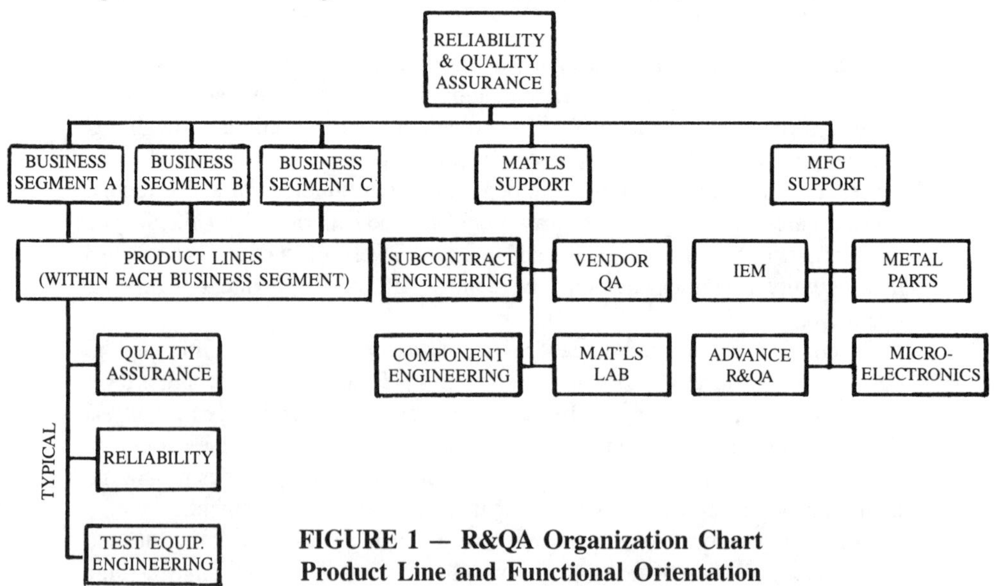

FIGURE 1 — R&QA Organization Chart Product Line and Functional Orientation

For quality cost reporting to be truly effective, it became apparent that from the same data base reports would have to be developed for each product line and each functional area that contributes to the manufacturing cycle. Contributing areas include the machine shop, the integrated electronics area, microelectronics, vendor material control, and the assembly areas. The optimum situation would be a matrix of each cost element by product line divided among the contributing areas. In other words the report would depict how much of a product line's appraisal effort had occurred in the machine shop or how much a business segment's prevention effort was being expended on multilayer board manufacture in the integrated electronics area. This segregation would give each manager the data necessary to evaluate quality cost performance by category for his own product line or functional area. Although a significant improvement over previous reporting, a major deficiency would still exist.

Such reporting would be much like looking at game scores and finding that the problem is with the defense. But where on the defense? With no additional data we might replace our linebackers when, in reality, nearly every score resulted from a long pass. Our response to high failure costs might be to work on our manufacturing processes, when, in reality, our failures are defective purchased components not found until they have been buried in higher level assemblies. In either case no improvement would be seen as a result of our corrective action efforts. Added visibility was a must if our managers were to be held accountable for quality cost trade-offs within their areas.

Bringing reporting down to this control level, however, posed some real problems because our quality cost system was not really an integrated system. It required that bits and pieces be extracted from a number of systems not designed for quality cost reporting. Figure 2 depicts the data sources for the various elements comprising quality costs. A number of different cost systems were involved in supplying data to the elements. Each system had to make its contribution to the reporting; but to extract meaningful data required the design of an extensive coding system.

Two types of coding schemes were developed: one would provide elemental break-downs within the classic quality cost categories by input data coding; the second would manipulate input data from the various cost systems into appropriate product lines and control areas by a coding operation, internal to the computer.

Table I depicts the R&QA and Test & Inspection labor codes developed to effect the detail level of reporting needed. Some forty codes were developed which allowed labor vouchering by R&QA and T&I personnel performing quality tasks. About sixty percent of the codes were for operating quality cost labor and the remainder were for test equipment design and reliability-related tasks. As can be seen, of those in the operating quality cost category, about fifty percent are for prevention, five are appraisal and seven are for failure.

To assure effective corrective action, scrap and rework reason codes were broadened so that each failure could be identified by cause and responsible organization (Table II). About one-third of the codes are operating quality losses, and the remainder are design losses, reliability losses, and production management losses.

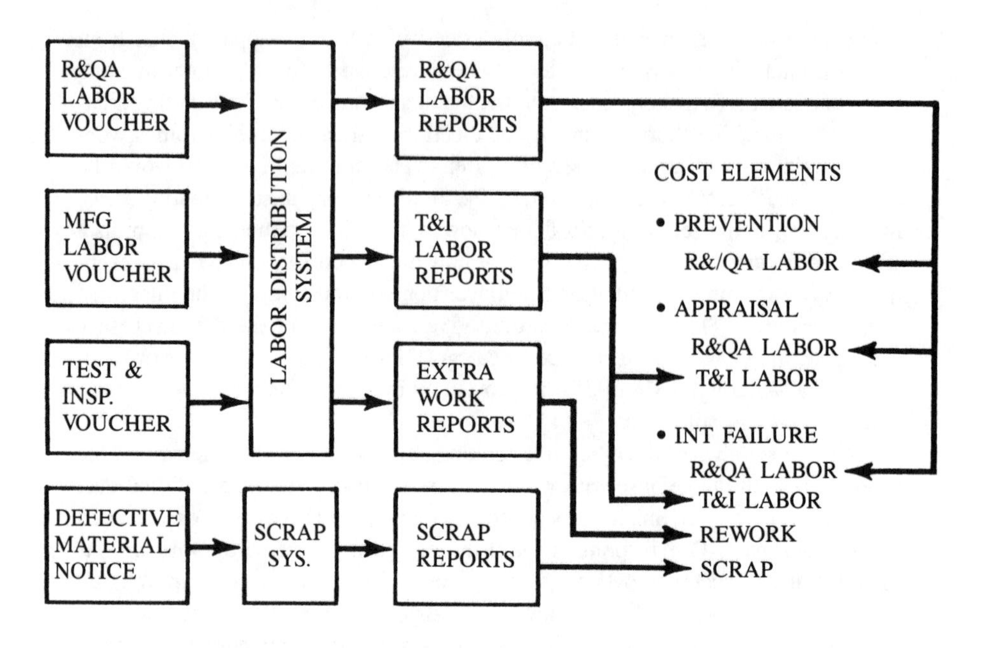

FIGURE 2 — Internal Quality Cost Reporting Sources
Quality Costs Derived From Many Sources

The coding scheme permits combining of R&QA labor, Test & Inspection labor, and rework/scrap cost in numerous ways, enhancing visibility into the underlying causes of apparent cost problems. For example, cost of test can be examined by combining all activities associated with the preparation for and the testing of the product. Costs associated with manufacturing proceses (e.g., process control losses, corrective actions) can be combined and analyzed. Quality costs related to vendor and material quality control activities can be separately assessed. Many additional combinations and segregations of cost activity necessary and valuable in cost control were made possible through these coding schemes.

Control area coding was designed around the existing manufacturing labor voucher system which already provided Test & Inspection labor by functional area. By adding one character to this code, we were able to segregate costs by control area without adversely affecting existing reports. This code was adopted for all quality cost source data. (Table III depicts control area sorting resulting from this coding.) Similarly, the addition of one character to the existing program code permitted grouping sub-program requisitions into product lines. The essential ingredients for a good game plan were now in place. Costs could be segregated to show the team's strengths and weaknessess. Effort could be intelligently applied where it was needed most.

Having elemental costs segregated by control area solved most of the problems; however, absolute numbers could not provide the proper trend information due to changes in business volume. An adequate base was required. An existing base,

TABLE I — Labor Codes R&QA and Test & Inspection

TYPE	CATEGORY	ELEMENTAL TASK	GENERIC COST
Oper Quality Costs	PREVENTIVE	Training	Systems Procedures
		Preventive Support	
		Program Quality Plan	
		Quality Audit	
		Inspection Plan	
		Test Plan	In-Process Planning
		Test Equipment Plan	
		Data Management	Process Control
		In-Process Assessment	
		Process Control	
		Incoming T&I Plan	Materials & Vendor
		Vendor Eval & Surveillance	
		Procurement Qual Assessment	
	APPRAISAL	Appraisal Support	R&QA Appraial
		Test Equipment Following	
		Lab Analysis	
		Good Test	T&I Appraisal
		Set-up	
	FAILURE	Failure Support	R&QA Failure
		Material Review & Disp-AES	
		Failure Analysis	
		Troubleshoot & Retest-AES	T&I Failure
		Troubleshoot & Retest-Vend	
		Field Failure	External
Test Equip Related		4 Codes R&QA	
Reliability (Rel./Dev.)		9 Codes R&QA 2 Codes T&I	

Net Sales billed, was not ideal due to billing peaks and valleys. Its use at product line level resulted in such large fluctuations that the measurement was clouded. It was apparent the base had to be free of extraneous variances.

An intensive evaluation of possible sales bases revealed that deviations caused by changes in inventory levels affected the quality cost to sales ratio approximately six times as much as the change in quality costs (Figure3). After evaluating the effect of inventory changes over a two year period, a cost of sales-input base was selected because it minimized uncontrolled variances in the measurement. By reversing changes in inventory balance and gross margin, this base reports sales as the costs are incurred, rather than when they are billed; thus leveling the enormous variations associated with the use of Net Sales Billed. It also correlated very closely with another input base — direct labor (Figure 4). With selection of the proper base, the system contained all of the elements essential for control of quality costs.

TABLE II — Losses Scrap and Rework

TYPE	ELEMENTAL LOSSES	GENERIC COST
Operating Failure Quality Costs	Not To DWG — Planning Error Not To DWG — Tooling Error Not To DWG — Tape Error Not To DWG — Workmanship Performance Failure — AES Damaged In Process	Process Control
	Test Prod/Insp Instr/Eq. Err Test Failure — To Be Isolated	Testing
	Vendor Resp — Rej At Incoming Vendor Resp — Rej In Process Damaged — Materials Custody Incorrect — Stock Release	Mat'ls & Vendor
Reliability	10 Codes	
Design	10 Codes	
Production Management	4 Codes	

TABLE III — Control Area Codes

Control Area
Function
Discipline

Code	Area	Code	Area
1	Assembly	3	Integrated Elec
110	Mfg Assy	330	IEM Mfg
150	Assy T&I	357	IEM T&I
180	Product Line Engr	391	IEM QA
191	Product Line QA	4	Micro-E Area
192	Reliability	440	Micro-E Assy
193	Test EQ Engr	455	Micro-E T&I
2	Machine Shop Area	480	Micro-E Engr
220	MS Fab	491	Micro-E QA
256	MS T&I	492	Reliability
291	MS QA	5	Materials Area
		552	Inc T&I
		570	Materials
		575	Vendor
		594	Vendor QA
		595	Materials Lab

Additional cost areas that should be mentioned are those related to overhead, vendor, and test equipment calibration and maintenance expenses. These are included in our department quality cost report but were not readily available by product line and functional area. After extensive study they were excluded from the lower level reporting for two reasons. First, their impact on the quality cost

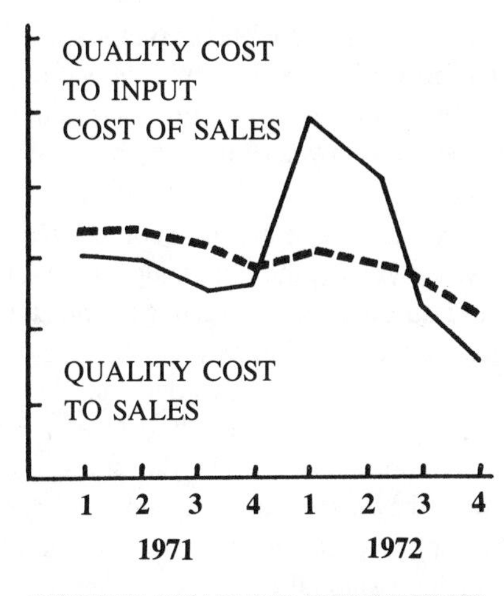

(SINGLE QUARTER REPORTING)

FIGURE 3 — Effect of Inventory in Base NSB vs Cost of Sales Input

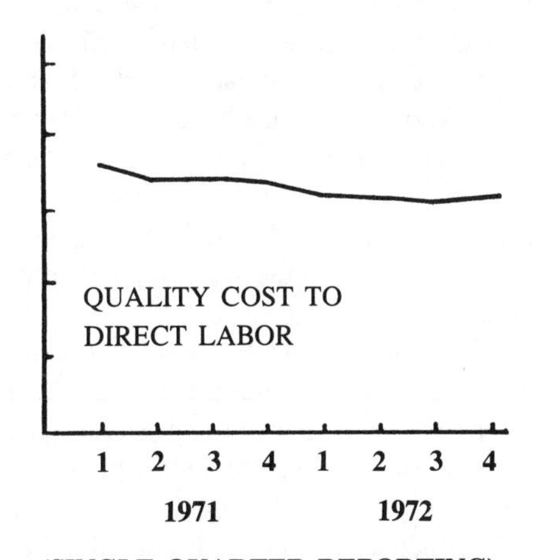

(SINGLE QUARTER REPORTING)

FIGURE 4 — Quality Cost to Direct Labor

trend lines was minimal and, in fact, their elimination made the reports very sensitive without detracting from the significance of the data. Second, to do so would require extensive prorating, which is an almost certain way to reduce the effectiveness of data for specific analysis.

To illustrate, suppose that our head coach realized that missed tackles were a major reason for his most recent loss? In an effort to isolate the problem, he instructed his defensive coach to provide information on who was responsible for the missed tackles. Having data which indicated a total of twenty-seven missed tackles, the defensive coach reported that each of the eleven men had missed 2.46 tackles. This information would be of little value in establishing a corrective action plan. If, on the other hand, instead of prorating he had reported that the left end had missed nine tackles, the middle line backer six, the left corner back five, and that no one else had missed more than two, he would have been providing highly meaningful data from which positive corrective action could be taken. Prorating has the same deficiency as any sort of average — it tends to mask individual performance.

With all coding complete and base selection accomplished a system was available that would indicate how many yards the opposing team gained by passing and by rushing, who made the important defensive plays, and who had lapses in their responsibility. The cause and effect picture was available — where more prevention was required and where it was already adequate. The game plan was complete. It could be determined what was required to win, and what went wrong if the team lost. An elaboration on the mechanics of the system will demonstrate some of the reports, how they were implemented, and some potential uses.

Raw data fed into the computer for each product line and functional area is summarized in output reporting. Manual effort is needed to extract the data in the format required. Cost ratios to selected bases are calculated and reports are issued to control level management.

An important step in implementing these reports, and an aid to selling the system was the collection of six quarters of data depicting actual trends. A written analysis was made by the system developers, who had a thorough knowledge of operating quality costs but a limited knowledge of the history of each program and the events which might affect costs. This strategy amplified to the contributors that the real analysis and associated corrective actions had to eminate from quality assurance personnel in each area. The extra effort associated with getting the end user involved paid handsome dividends as reflected in their willing, if not enthusiastic, acceptance of the system.

The reporting is further illustrated in Figures 5 through 7. The right side of Figure 5, identified as "department," shows the total of the combined costs of all product lines. The trends are similar to the previous department level reports which were made available for management control; however, compare the other three trend charts shown in Figure 5, which are the summary costs for the business segments comprising total department costs. It is hard to believe that these costs are in the same system, with two of the three moving in a direction opposite that of the department.

The surprises are even greater at the lower control level. The example in Figure 6 is a further segregation of Business Segment A shown in Figure 5. It shows that Product Line 1 is a major cause of the business segment's upward trend; and in Figure 7 we can see that the product line assembly control area is the major cause of the adverse cost performance. This series of examples makes clear why control area reporting is essential not only to provide the control manager with the visibility needed to take action; but it is also the only way that quality cost reporting will ever become an effective management tool.

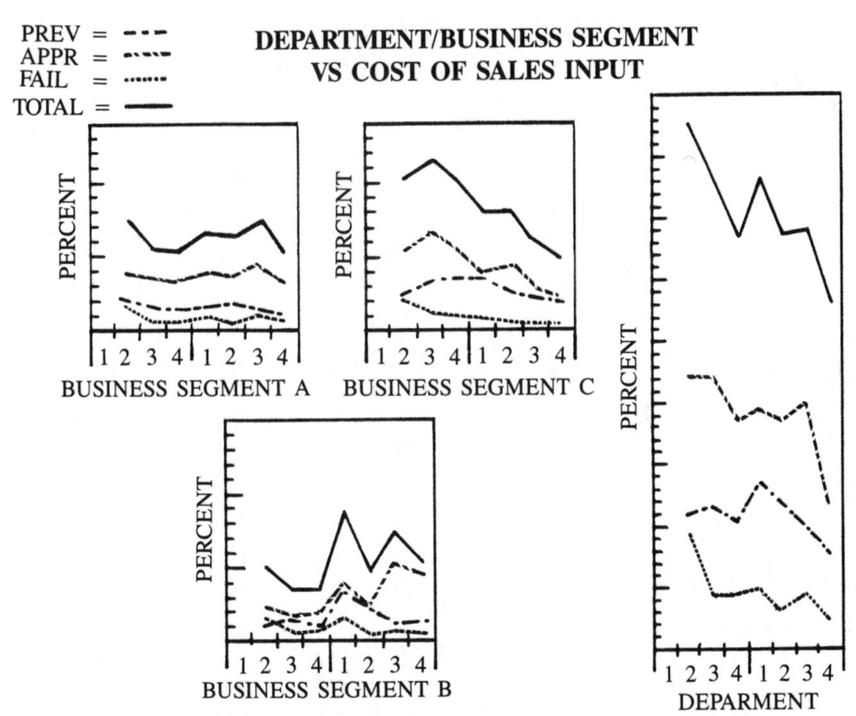

FIGURE 5 — Department/Business Segment Quality Costs

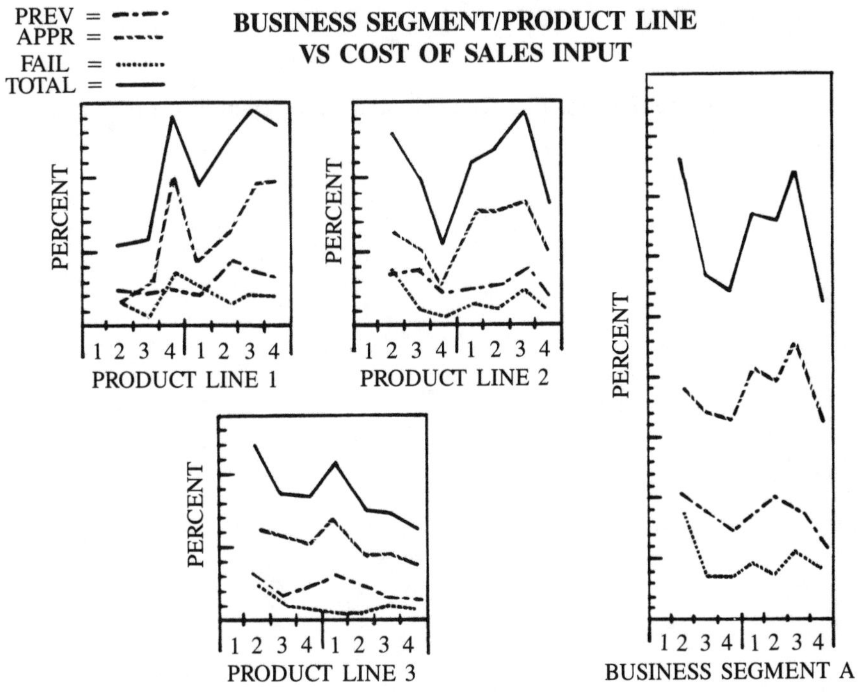

FIGURE 6 — Business Segment/Product Line Quality Costs

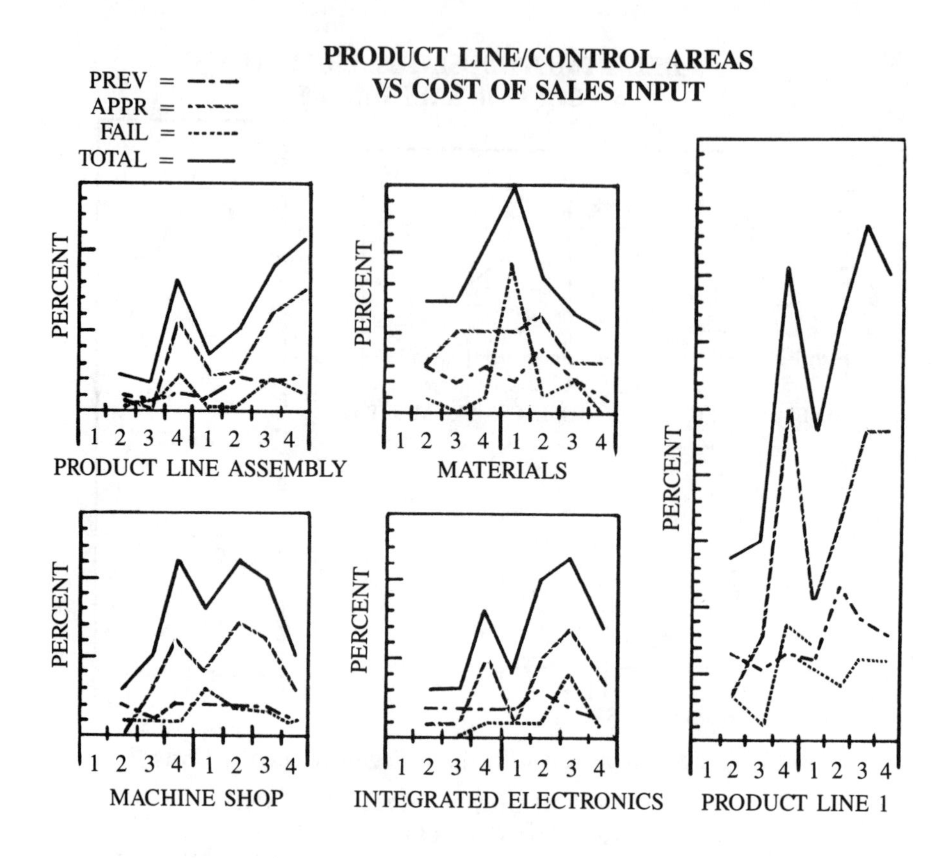

FIGURE 7 — Product Line/Control Area Quality Costs

QUALITY COSTS AT WORK

John T. Hagan, Director-Quality Measurement
ITT, New York, New York

In the short course of five years, ITT has saved hundreds of millions of dollars that may otherwise have been lost forever in the accepted operating cost of doing business. These particular savings all came from the elimination of scrap and rework and the reduction of excessive inspection and test. Once management was convinced that the potential existed, accomplishment was easy because the ITT Quality program is so practical. This program takes the abstract business of quality and converts it into hard cash. By so doing, it brings into focus the dollar impact of unproductive work or waste and it allows overall Qualilty improvement to begin. If, as many professionals seem to think, the seventies become the "decade of quality", there won't be a better time to learn about a Quality Cost system that really works.

The ITT system is based on defining the Cost of Quality and relating it to measurable performance in all areas of the operation that can affect end-product quality and cost. The concept here is that it costs a lot of money to make things wrong. Next in sequence is the analysis of performance in sufficient detail to identify the real obstacles to improved performance. With improved cost as justification, correction of these obstacles becomes the start of progress. Learning from this experience then allows for the prevention of the same or similar problems in the future, thus achieving new levels of operating performance and cost. ITT Management understands this. They have supported it, paid for it and continue to encourage it.

The Cost of Quality is a measure of appraisal, failure and prevention costs associated with the achievement of product quality. Quality, in this case, simply means "conformance to requirements". More specifically, Quality Costs are the costs of appraising product for conformance to design requirements and to market specification (e.g. product inspection and design qualification); the cost required by failure to meet requirements (e.g. redesign, rework, scrap and warranty costs); and the cost of preventing failures (e.g. design reviews, vendor qualifications and process capability studies). Detailed definitions and collection guidelines should be developed by individual companies as applicable to their special situations and needs.

At ITT, both the detailed definitions and the responsibility for Quality Cost compilation and reporting has been established as a function of the Comptrollers' Office. The Comptroller's procedure for Quality Costs provides ITT definitions and location of costs within the Manual of Accounts. It also defines which cost elements require the addition of associated labor benefits, full burdens or allocated costs in order to more closely represent "total cost to the company". Holding the Comptroller responsible for Quality Cost measurement has established three important standards for the overall program. It has kept collection costs within practical limits. It has given the stamp of financial validity to the results. And it has provided an unusual opportunity for effective teamwork between the Comptroller's and Quality organizations.

The philosophy of the Cost of Quality has been built on three premises. Failures are caused; prevention is cheaper; and performance can be assessed. It works on the principle of "cause and effect". As failures are revealed through appraisal actions, they can be examined for cause and eliminated through corrective action. The further along in the process that the failure is discovered (i.e. the nearer to end product use by the customer), the more expensive it will be to correct. As failure costs are incrementally reduced to new levels (i.e. new levels of performance have been achieved), appraisal efforts can then also be reduced in statistically sound increments. The experience of this improvement can then be applied, through prevention activities, to all new work. When the level of Prevention is such as to maintain a minimum of Appraisal and Failure costs, a state of "quality control" will exist.

As straightforward as this philosophy may seem, it cannot work without a basic quality measurement system to determine and clearly identify the correctable elements of failure that comprise the initial essence of cost improvement. A basic quality measurement system is a system designed to utilize the results of all inspections, tests and process control measurements as a measure of company performance and a reservoir of potential cost reduction items. This measurement is the most basic and the most important role of the quality function. The resevoir for cost reductions can be built around the failure costs of quality and some portion of appraisal costs. The exact potential for improvement cannot be known until these measurements are established and analyzed. The managements that don't recognize and seek this use of Quality data are still living in the Dark Ages of Quality Improvement.

As a minimum effort, raw results of all quality measurements in key operating areas (i.e., receiving inspection, fabrication, processing, assembly, test, etc.) should be collected and reported as a trend. The frequency of plotting can be daily, weekly or monthly — depending entirely on the sensitivity of the measurement. The trend charts will tell whether overall performance is getting better or worse. They will also initially tell what standards of performance have been tolerated up to the point of measurement. Finally, they will provide a direct path to determining overall improvement potential. Whether these measures are reported as "percent defective" or yield, which is the reciprocal of percent scrapped, the costs associated with these failures are the Failure Costs of Quality. Therefore, each improvement in these measures of performance will result in a direct reduction of Failure Costs. And Failure Cost reduction is the initial target for cost reduction in a Quality Cost system.

At ITT, monthly measures of performance in key operating areas are an essential ingredient of Qualilty Status Reporting. Since Cost of Quality is an "after the fact" accounting report, which is usually not available for several weeks after the close of books, performance measures are the only true indicators of current progress. In fact, there can be no real progress in the Cost of Quality without performance improvements in the key areas of company operations. That is why basic quality measurements are a requirement in the management reporting system of ITT's Quality Program. And why demands for performance improvement are the focal points for subsequent cost improvement and profit contributions.

Performance improvement is easier said than done. Except for obvious deficiencies that may have been tolerated and are now to be stopped, determination of specific improvement actions will depend on a detailed examination of performance results. That is, on the establishment of distributions of defect types, sources and causes for each area of performance measured. These distributions are but the first step in the overall analysis of performance that is necessary to determine profitable corrective actions.

The specific object of each analysis is to determine the most significant or the most frequent contributors to the undesirable situation being reported and to identify their elimination as the most profitable avenue to improvement. That is, as the areas which will provide the biggest payback for the expenditure of investigative and corrective action budgets. The step-by-step investigation and elimination of identified contributors to failure costs is the overall object of performance analysis. These steps then become a series of incremental cost reductions that add up to significant overall improvement. One shot, miracle solutions are few and far between these days.

The analysis of quality data to determine significant failure causes, or to identify where further investigative efforts are necessary, is not a job for the amateur. To learn the secrets of vast amounts of any information requires expert knowledge in the technologies of the business or system being analyzed and some idea of the expected distribution of key characteristics. The calibre of the technical talent (quality engineers) found in the Quality organization will determine, to a large degree, the extent of the technical contributions that will be made. This factor alone may well spell the difference between success or failure for the total program.

There is no tried and true method for the analysis of quality data. It takes knowledge of the products and processes involved, at least some knowledge of basic statistics and an inquisitive desire to find and eliminate the causes of defects. Once a cause in need of correction is identified, corrective action itself is not automatic. It must be individually justified on the basis of an equitable cost tradeoff (e.g. a $500.00 per week rework problem versus a $5000.00 solution). At this point, knowledge gained in measuring Quality Costs will be invaluable for estimating the payback on specific corrective action investments. Justifying corrective actions is a continuing part of the program. As experience is gained in correlating defects with costs, the justification for actions required will get easier and Corrective Action will become a legitimate, profitable and integral part of overall company operations.

To be truly effective, a corrective action system must be formal and all encompassing. That is, it must include a method for documenting, tracking and statusing the progress of all investigations and actions leading to the solution and elimination of identified problems. And it must not condone unauthorized work or allow too many changes to be made simultaneously and indiscriminately. Otherwise, important problems will fall in the crack or too much activity will camouflage actual results.

An effective corrective action system will point out very clearly that human error is not limited to the operator. Errors that result in scrap and rework are committed by product and process design engineers, by both the designers and fabricators of tools, by those who determine machine capabilities and by those who provide the written instructions for the operator. Obvious errors that affect product can also be committed by the calibration technician, the maintenance man and even the material handlers, who can put the parts in the wrong bin or deliver product for processing out of sequence to the manufacturing plan. Almost anyone within the total operation can contribute to failure costs. Effective corrective action, therefore, can and will take many avenues throughout all operating organizations.

Some errors will become fairly obvious when defect causes are investigated as a routine part of formal corrective action. They will just as easily be fixed (e.g. the re-instatement of discontinued preventive maintenance practices). Others are more insidious (such as a marginal condition in design, tooling or processing) and are almost never discovered without the benefit of a well organized and formal approach. Marginal conditions clearly represent the kind of problem that can easily become lost forever in the accepted cost of doing business. This is often the case when the unit cost of product does not follow its expected cost curve. Having an organized corrective action system can prevent such conditions from remaining lost forever to management's visiblity. It is the one way to assure that step-by-step progress will happen and will achieve the full potential that is available. It is the kind of experience that contributes to and proves the value of prevention activities.

Learning from corrective action experience is something that just makes good sense. The whole idea of prevention is to push the discovery of defects back from their most expensive areas, such as in the hands of the customer or at final test, to less costly areas, such as final inspection or in-process inspection, all the way back to the minds of the people who can prevent the defect from being born in the first place, such as the operator, the inductrial engineer, the designer, etc.

For reasons of prevention, corrective action often seeks a temporary solution, which is immediately less expensive to the company, while it searches for the final solution. It is certainly not uncommon to increase final testing of products to preclude further shipments of newly discovered field problems. In a like manner, increased inspection or test anywhere within the operation can be used while final solutions are being sought in such areas as tooling, design, or at the vendor's plant. While most of these additions are temporary, certain portions may have to remain as permanent assurance of the final fixes. In extreme cases, such as the introduction of qualification testing, necessary appraisals of design will be both introduced and maintained. For it is cheaper to find design problems through a qualification test program than to discover them in production, or worse still, in the use of the customer. Qualification test, therefore, is a prevention activity when viewed from Manufacturing, or customer use.

Prevention activities, in general, are those elements of the management system that are built-in and scheduled as a result of learning from our mistakes, those particular mistakes that allow the birth and propogation of defects. Building these new standards into the management plan and enforcing their execution is the right way to close the loop of quality measurement and corrective action experience and to cement your victory over defects. That is, to maintain the lowest level of operating Quality Costs. The defects that are prevented don't cost. And the Quality Program that is built on prevention is a program that is alive to the real needs of business.

Prevention has become a way of life at ITT. The overall program of Quality Improvement based on the Cost of Quality got off to a good start primarily because of management involvement and the fact that it really works. Progress has then been maintained primarily through training programs (the ITT Quality College) which emphasize the important roles of all key functions. There have also been some rather unique management seminars. These seminars were organized to present Quality Cost success stories to large groups of ITT executives. The presentations, however, were not prepared or conducted by Quality personnel. They were, in fact, the personal success stories of ITT executives who have profited from active participation in the ITT Quality program.

Future plans for ITT call for the continued refinement of Quality Cost data for more detailed comparison purposes. A computer program has been developed to support this task and it is expected that there is more to be gained from the data that already exists. One of the principle objectives of this continued study will be to develop a clear identification of the total cost reduction potential that is available in the Cost of Quality throughout the entire ITT system. This potential is like money in the bank but it will take a lot more Quality Cost experience before valid and useful Bank Books can be prepared. If significant Quality Cost reserves exist in your company's accounts, this program can help you. Whether or not it is sold to management, in preference to other investment choices that they always have, is a measure of your personal commitment.

REDUCING CLERICAL QUALITY COSTS

William J. Latzko, Manager, Quality Control
Irving Trust Company, New York, New York

I INTRODUCTION

Quality is an economic consideration; it costs money. And few managers are aware of just how much money is involved. This is particularly true in the service industry which is so heavily dependent upon clerical operations.

A good deal of work has been done with the determination of quality, the controlling of quality costs in manufacturing operations. Indeed, Dr. Walter Shewhart's landmark book in quality control (6), was entitled, "The Economic Control of Quality of Manufactured Product".

It is easy to see why early efforts at controlling quality cost were directed towards manufactured products. The output of the manufacturing process can be easily measured and lends itself to statistical treatment. Generally, the product is made entirely, or to a large extent, with mechanical devices.

II CLERICAL QUALITY

The output of clerical efforts usually brings a less tangible result than is found in the manufacturing process. The quality characteristics of correspondence, telephone calls, customer contacts, etc., are less easily defined and, therefore, less easily measured.

Only recently has there been an awakening to the fact that clerical output can be measured and controlled. In spite of the subjective nature of the work, applicable measures have been derived and statistical techniques applied to clerical efforts. The major distinction between manufacturing processes and clerical ones is the degree of variability which each exhibits. Manufacturing processes tend to remain stable once they are in control. Out of control conditions normally occur after a gradual drift of the system has taken place. Once put of control, the manufacturing process tends to remain in this state until correction has been achieved. Not so with clerical operations. Out of control conditions occur suddenly, without prior warning and frequently disappear as rapidly as they started.

The nature of clerical operations as such that efficient measures of variables are rarely possible. As a result, clerical controls have to be designed using the less efficient measures of attributes. It is usual to measure the percent defective of a sample of work. The sample consists either of a predetermined number of clerical outputs or the number observed in a set period of time.

The problem of determining what constitutes a unit of sample and what constitutes a defective seems disarmignly simple. Yet this is often one of the more difficult tasks in developing a good clerical quality control system. For example, in keypunching and verifying name and address masters, the determination of a unit is quite simple: a card. But what constitutes a defective? Each error found on a card is a defect. The card which has an error must be re-keypunched. It must be re-keypunched regardless whether there exists one or more defects on a single card. A defective is considered a card that must be done over. As a rule of thumb, a unit is that amount of work that is to be repaired if one or more defects are found. A defective is a unit requiring repair.

The classical approach to controlling clerical operations is to check the output. Sometimes layers of cherckers are used. The slogan appears to be "Quality at Any Cost". Well, the cost is there, but improvement in quality is not there. Fortunately, there are economic alternatives available.

III THE COST CONCEPT

The cost of checking or appraising is not the only cost of quality. The ASQC Committee on Quality Cost has identified four distinct elements. These are:

1. Appraisal Cost
2. Internal Failure Cost
3. External Failure Cost
4. Prevention Cost

The total cost of quality is the sum of these four costs. The profit oriented manager is interested in minimizing his total cost.

Appraisal Cost is the cost of checking, inspecting, verifying, signing, etc., the clerical output to determine that it is executed correctly. Any effort along this line is an appraisal cost. The distributer in a typing pool who checks the work for typographical errors is included in the appraisal cost. The supervisor's time spent in routine checking of typed work should also be included in appraisal costs.

The signer, who attests by his signature that the work or document is correct, is in fact expected to perform the act of appraisal before signing. Costs of this nature are all too often classed with other costs of doing business, yet they exist only to provide a guarantee of quality.

While it is not proposed to eliminate the appraisal function, possibilities are discussed later for doing this work more efficiently and so minimize this cost.

Internal Failure Costs are the result of defective items found in appraising the work. This cost includes all spoiled forms, lost computer time, re-work of the item and other related costs. This constitutes the hidden cost of bad quality. Seldom are these costs broken out or controlled separately. Most cost studies tend to average or factor this cost into the standards. Such a technique is as bad as ignorance of the cost and gives no incentive for attacking this rather substantial drain on the corporate resources.

External Failure Costs are incurred when appraisers fail to identify defective work. This cost includes expenses similar to internal failure costs as well as the additional outlay of investigation or research of the problem and penalty payments. External failure is the one cost of quality that is most evident and obvious to managment. It is the area on which much effort and concentration is placed.

Frequently, management will overreact to external failure. Rather than systematically and scientifically attacking the cause of the failure, management will add further appraisers increasing both appraisal and internal failure costs without coming to grip with the real problem. Often the appraisal cost and internal failure cost are raised beyond any possible savings in external failure costs.

Prevention Cost is an investment that management makes to minimize the total cost due to bad quality. In clerical operations, the prevention cost is the cost of the quality assurance function within the organization and its related expenses. It is the function of the quality assurance group to employ such modern techniques of quality control (mostly statistical) as will minimize the total cost of quality. The efficiency of this group can be measured by the extent to which they achieve these savings.

Relatively few organizations in the service industry have invested in prevention. In those cases where such an investment was made, and where technically qualified personnel was used, the returns have been a manifold savings.

IV DISTRIBUTION OF CLERICAL QUALITY COSTS

Just how much is the cost of clerical quality? The results of a recent study of a department in our bank are probably representative; they are in line with other studies of this nature. The cost of quality in this study was found to be nearly 40% of the total cost of operating the department. A substantial sum!

These costs are distributed among themselves as follows:

1. Appraisal Cost	28%
2. Internal Failure Cost	41%
3. External Failure Cost	29%
4. Prevention Cost	2%
Total	100%

An investment in the study of quality costs and in the methods of reducing these costs is very much in order. Relatively small percentage savings can yield substantial profits.

V REDUCING THE TOTAL COST OF QUALITY

With the exception of external failure costs, managements are generally unaware of the immense impact of quality costs on profits. Effective reduction of internal failure costs through improving the quality of the output will reduce both appraisal and external failure cost. It is recommended that management invest resources in measuring the present state of affairs and pursue a policy of reducing all quality costs to their minimum.

How can the total cost of quality be minimized? By the application of modern quality control techniques. This implies statistical quality control since these are the most efficient quality control techniques available today.

There are two dimensions to quality. *Design quality* is that characteristic of the product or service that makes it salable. If a customer requires that transactions concerning him be delivered by messenger rather than mail, and on that basis he will deal with the organization, then this requirement is the design quality.

Production Quality is the quality of the work as executed. If transactions are indeed delivered by messenger in the above case, the production quality is good. However, if the transaction should be mailed through an error, the production quality is bad.

Quality Control is the determination that production quality conforms to the design quality. There are a number of ways to exercise quality control. The goal is to obtain maximum quality effectiveness at the minimum cost. Statistical quality control is the best tool to use to achieve this objective.

The approach recommended here is to study the system to determine the level of quality that is capable of producing. Clerical operations are a system, a series of actions which are interrelated to produce a result. As long as the relationships remain unchanged, a given level of quality can be achieved.

Until the level of quality, or the process capability, has been determined, little else can be done. Once the process capability has been established, several actions are possible. If the system is not functioning at the process capability, immediate improvement in quality should be possible.

VI QUIP — A METHOD TO REDUCE CAUSES OF FAILURE

In a previous paper (4) a method was described to achieve the process capability. This method was given the acronym, QUIP which stands for Quality Improvement Program. The concept is simple; the supervisor of the department looks at the output of the clerks who work for him. (He is supposed to do this anyway). In the QUIP system, he records all good work as well as any defective work. A separate record is maintained for each clerk. If defective work is found in the sample, the supervisor attempts to determine why the defect occurred. He does this with the attitude of, "fix the problem not the blame." The reason for the error in clerical operations generally falls into one of three catgories: 1. lack of training, in which case on the spot effective training is to be done; 2. intermittent systems problems, in which case the supervisor corrects these or gets help; and 3. operator failure. In the latter case, action is taken based upon the record of the clerk and whether the clerk is operating in or out of control.

The supervisor's results are accumulated and used in two ways. First of all, the data is statistically analyzed and used to determine the process capability (5). Secondly, the data for each clerk is compared to the process capability and out of control conditions are noted. The supervisor now concentrates on these clerks and uses his quality records and his skill in improving the quality of those clerks who are out of control.

The results of this technique have been most satisfactory. Generally, internal failure is reduced by 50%-75%. The external failure rate also is proportionally reduced. The savings have been substantial in both out of pocket costs as well as improved customer relations. The latter while intangible can, nevertheless, be quite real. When quality improves to such an extent that senior management gets feedback from customers, then it can be assumed that sales efforts have been helped.

The system under which clerks operate, their environment, determines their output in both numbers and quality. Clerks have no control over the system. If they operate at the level of quality determined by the system they are doing the best that can be expected of them.

If the system is operating at the process capability, improvements can be achieved only by changing the system, a managerial responsibility. Systems changes such as better training, new and/or better equipment, new methods, improved working conditions, etc., can be instituted only by management. Therefore, only management can improve the process capability; the clerical staff is not in a position to make such changes.

To evaluate a cost reduction plan it is necessary to determine whether operations are carried out at the process capability. If not, then steps must be taken to achieve this quality level. Once the quality of the process is at the level it can achieve within the system and it is under control then it is possible to consider systems alternatives. Each alternative represents some investment which is balanced against the change in the total quality cost under the alternate system. That system is chosen which gives the maximum incremental benefit after accounting for the incremental system cost.

VII REDUCING APPRAISAL COSTS

Sampling procedures are the simplest form of achieving reduction in appraisal costs. There are some basic principles which should be noted. First of all, the process must be controlled. By that is meant that the process capability must be known and achieved. A controlled process is one that will operate in a predictable manner. All acceptance sampling procedures are based on this principle.

A second point is that 100% inspection is rarely 100% effective. On this basis, sampling is often no more risky than a complete check of the work.

The third consideration is that sampling can be stratified. In paperwork situations it often happens that the transactions in a given department are of varying values ranging from those of low importance to those of high priority. Often checks are performed on every item regardless of its value. This is inefficient inspection. Work of low importance can be sampled on a batch basis while high value work can be more rigorously screened. For instance, in a department handling money transactions, the value of the transactions ranged from a few dollars to many millions of dollars each. Items of $50,000 or less represent a relative low risk of external failure yet represent 70% of all transactions. By using a batch acceptance sampling plan for items of $50,000 or less and completely screening all other transactions, substantial savings in appraisal cost are made without any real risk in increase in external failure costs. Indeed, because high dollar value items contribute heavily to the external failure costs and because the sample plan allowed more time to be spent on these items, the external failure costs are reduced.

The problem of economic sampling plans considering cost alternatives has been explored by a number of authors. An extensive bibliography has been prepared by Quality Cost Committee of ASQC (1). Dodge and Romig (2), Hald (3) and others published articles relating appraisal to its various costs and developing methods for minimizing these costs.

REFERENCE

1. E. W. Dawes, "Literature Search", *ASQC Quality Cost Committee,* May 1, 1973.
2. H. F. Dodge and H. G. Romig, *Sampling Inspection Tables — Single and Double Sampling,* 2nd. ed. New York: John Wiley & Sons, Inc., 1959.
3. A. Hald, "The Compound Hypergeometric Distribution and a System of Single Sampling Inspection Plans Based on Prior Distributions and Costs", *Technometrics Z,* Aug. 1960, p. 275-340.
4. W. J. Latzko, "Quality Control in Banking", *Transactions of the Rutgers Conference,* 1972.
5. W. J. Latzko, "Process Capability in Clerical Operations", *Rutgers Conference,* 1973.
6. W. A. Shewhart, *Economic Control of Quality Manufactured Product,* New York: D. Van Nostrand Co., Inc., 1931.

QUALITY COSTS: WHERE ARE THEY IN THE ACCOUNTING PROCESS?

A. F. Grimm, Product Assurance Manager
Eaton Corporation Brake Division, Southfield, Michigan

Quality Control practitioners consider analysis of quality cost data to be an indespensable tool in managing the Quality Control function. Detailed instructions for preparing and presenting quality cost data are found throughout the quality control literature i.e., "Total Quality Control Management and Engineering" by A. V. Feigenbaum through, "Quality Costs, What and How", prepared by the Quality Cost Committee of the ASQC. There is no lack of information on what quality costs are and how to categorize them. The structuring of these data for practical quality cost management is "cook booked" in such detail, that most people would have no trouble setting up a program.

The QC practitioner soon finds out that he must eventually meet with the accountant to initiate a quality cost reporting program. Many times, he finds that the accountant is not readily sympathetic to a program that he visualized as the great saving device for the firm. One reason for an accountant's lack of enthusiasm may be the QC practitioner's ignorance of accounting practices. This paper proposes to provide an understanding of how the accountant views costs. Hopefully, this discussion will help the QC man understand those accounting areas containing the quality cost data, how the accountant views these costs, and how the accountant would treat quality cost data for analytical purposes. With mutual understanding, the QC practitioner, with the help of the accountant, may achieve his ultimate goal, the introduction of a quality costs reporting and control system.

A starting point for relating to the accountant's view of costs is an understanding of the accounting process. The system is divided into external and internal accounting functions. External accounting deals with reports to stockholders, government and other outside parties. Internal accounting is concerned with preparing information for managers for use in planning and controlling current operations and for strategic planning purposes. Quality cost reporting would be viewed as part of the internal accounting system. However, "Nader type" pressure could lead to outside reporting.

Within the internal accounting framework is the area associated with cost analysis. Figure 1. shows the relationship between costs, revenue and volume for a typical firm. The cost accountant views costs as fixed or variable. He relates these two cost categories to revenue. A successful business is one where the sum of the cost lines is maintained below the sales line. Accountants use these elements to analyze the cost behavior of a business. This method is known as cost-volume-profit analysis.

A fixed cost, on a per unit basis, varies inversely with activity or volume changes but is constant in total dollar amount. Fixed costs may be either committed or discretionary. Committed fixed costs arise from possession of plant and equipment and a basic organization. They are affected by long run decisions as to desired level of capacity. Some committed costs are rent, depreciation, insurance and property taxes. Discretionary fixed costs arise from periodic appropriation decisions that reflect top management policies. Discretionary costs include advertising, research, training, public relations, sales promotion, donations and the size of sale and engineering forces.

A variable cost is constant per unit and its total dollar amount changes proportionately with changes in activity or volume. Examples of variable costs are manufacturing costs which include direct material, direct labor and variable indirect manufacturing costs including power, supplies, idle time etc. Other variable costs would be nonmanufacturing costs such as selling and administrative costs, i.e. sales commissions, packaging supplies, warranty service, clerical costs, etc.

The cost accountant in his cost-volume-profit analysis chart would aggregate these costs, classify them into the fixed and variable cost categories and determine the contribution margin. The contribution margin is the excess of the sales price over the variable expenses. It may be suspected that quality costs are included in all cost categories and may influence overall cost-volume-profit analysis.

Concentrating on costs which concern the accountant will provide clues for the quality control practitioner to set up a quality cost report with the accountant. The accountant's purpose in preparing the cost-volume-profit analysis is to provide management with knowledge of the impact of expanding or contracting volume on the profit contribution of the business. Within the "relevant range", figure 2, a period of volume activity, one will find that fixed costs remain relatively unchanged. However, variable costs may change depending on the volume increase or decrease. The accountant is interested in the size of the contribution margin which is a function between sales volume and variable costs, figures 3 & 4. C-V-P analysis includes other limiting assumptions such as: the revenues and expenses are linear over the relevant range; efficiency and productivity will be unchanged; sales mix remains constant; and the differences in inventory level at the beginning and end of a period is insignificant. These notions may appear detailed, but it is important for the quality control practitioner to understand these elements so that his approach to the accountant will be rewarding for both. We may conclude that the accountant's interest is concerned with overall business activities, and rightly so. It is his job to record and analyze these activities and keep his management informed of the business' performance. He feels that his activity must measure the whole business. When one manager approaches him for a specific report, he would view the request in light of its significance to the total business structure. If in the case where business considers quality costs as part of the cost of manufacturing, the quality control practitioner may expect difficulty in convincing an accountant that a quality cost reporting program is needed. However, if the quality practitioner is aware of the accountant's view of costs, he will be able to relate the value of quality cost behavior in the overall cost structure. We will explore these relationships in this paper. These ideas may help create a better understanding for all parties involved in cost analysis.

The quality control practitioner is well acquainted with the quality cost categories of Prevention, Appraisal, Internal Failure and External Failure. He has been conditioned to these segments through literature, conferences and the like. His concept of quality costs is somewhat different than the accountant's since these costs are more closely structured for him. He is more concerned with these integral costs than the accountant since they are directly associated with those value elements of product quality. Unlike this view the accountant is interested in analysis of all business costs.

Specifically identifying quality costs elements in each of these categories will later lead to correlating identifiable costs that accountants know and costs that the quality control practitioner recognizes. One of the accountant's functions is to provide segment reports for departments within the business. The Quality Cost Report is a segment report needed for the management of the Quality Control Department. It is needed for designing department objectives that will coincide with the total business' objectives as well as controlling internal departmental operations. One premise adopted by this paper is that the accountant retains the responsibility for preparing the Quality Cost Report since it is a segment report. However, the paper attempts to help the quality control practitioner and the accountant to design the format that would be best for both. Also, the format becomes important so that other managers who have interest in product quality may understand the magnitude of quality cost impact. In the more traditional accounting reports, these costs are included with others and lose identification.

In initial attempts at segregating quality costs, most companies are surprised at the low proportion of funds allocated to prevention of nonconforming product. Low expenditure in this segment is usually associated with a poor quality image. Of course, under the traditional approach to cost reporting, these expenditures were blended to the point of escaping notice and subsequent control.

Of the four quality cost categories, it seems that prevention is the least understood by management. Initial analysis of quality costs inevitably identifies prevention as the lowest expenditure. Elements that comprise this category are salaries for Quality Control Engineers and expense budgets including such activities as vendor surveys prior to order placement, program costs for analysis and diagnosis of quality problems which lead to improvements and the costs of designing and developing quality measurement. Cost elements in prevention necessarily entail expenditures in a high technology field that many companies originally feel are unnecessary, until the costs are broken out of the traditional cost accounting records and examined for merit.

Appraisal costs are most familiar to accountants and management as being directly associated with the quality control function. The majority of expenditures to control quality are made by management in this category. These costs are compensation received by inspectors and testers working with raw material and purchased products, in process quality control work, and final product evaluation including the material consumed or destroyed in performing these inspections and tests. Product endorsement costs by outside agencies, calibration and maintenance of inspection and test equipment, field performance testing costs, EDP quality report costs and materials consumed as part of inspections and tests are also included.

The third category is Internal Failure Costs. Included are the familiar scrap and rework costs. However, other common costs include trouble-shooting and failure analysis costs, reinspection and retest costs associated with reviewing nonconforming product, cost of operating nonconforming material review, and diffenence in product selling prices (lost revenue) due to product downgrading (factory seconds) because of expected quality standards not being achieved in product manufacture.

The last quality cost category is External Failure Cost. Cost elements found in this category include complaint adjustment costs, service charges involved with correction imperfections, cost of handling rejected returned products, warranty cost for replacing returned or failed products and cost of replacing material caused by marketing error, engineering error or factory and installation error.

To summarize the meaning of those four cost categories, we may state the categorical definitions in "Quality Costs-What and How".

1) *Prevention Costs.* Costs incurred for planning, implementing and maintaining a quality system that will assure conformance to quality requirements at economic levels.
2) *Appraisal Costs.* Costs incurred to determine the degree of conformance to quality requirements.
3) *Internal Failure Costs.* Costs arising when products, components and materials fail to meet quality requirements prior to transfer of ownership to the customer.
4) *External Failure Costs.* Costs incurred when products fail to meet quality requirements after transfer of ownership to the customer.

To understand quality costs relative to overall business costs, a fictitious example is prepared to explore the notions discussed previously. The company we will study is FCA (Fictitious Corporation of America) which produces a highly indispensible household appliance, the widget. The selling price of the widget is $11.95 and FCA holds 35% of the market. This share indicates that FCA is the leading producer. Annual industry sales are 10 million units. FCA sells the widget to distributors for $6. apiece. Therefore, FCA's annual revenue for the widget is $21,000,000 (i.e. 10 million units × .35 × $6.). We will assume that there is a fairly steady volume of units sold that is not seasonally influenced. Our intent is to illustrate the proportion of quality costs in a static synthesis rather than to analyze periodic changes to the business. Also, we will not try to establish a precise classification system to itemize the cost dollar allocation. The attempt will be to establish an idea of the proportional relationship of quality costs to overall costs. Figure 5 illustrates the organization structure of FCA's widget operation. Even though such simplicity is usually not the case, we will arbitrarily state that there are seven salary and three hourly grades. It is assumed that each person in the same grade will receive the same annual compensation. Grade 1 is the highest area and is the salary received by the President. Figure 6 represents the salary and wage scale per grade. The number, on the lower right side of each box of the organization chart in figure 5, is the grade assigned to that position. The total salary and wage cost is $2,781,000 (see figure 7).

This paper will not consider more complex costs, such as mixed costs, or enter into the area of budgeting. Even though the example neatly categorizes costs into clearly defined categories, in actual practice this condition becomes harder to define as the business becomes more complex. In order to handle these complexities, the use of semifixed and semivariable costs becomes more helpful. To further complicate the matter of cost categorization for the accountant, a manager's actions may attempt to make costs either variable or fixed. Our example contains an incident of this nature. Note that fringe benefits for other QC Department personnel in figure 11 is carried as a fixed cost since total fringe benefits are carried as a fixed cost in figure 10. However, the case is that these particular QC personnel are carried as indirect labor which is a variable cost carried in Variable Factory Overhead in figure 12. To neatly categorize this condition for reporting purposes has proved difficult when trying to extract the information for developing the Quality Cost Report. Aside from acknowledgement of this qualification concerning categorization of costs, let us continue. Construction of the cost-volume-profit charts will assume linearity over volume.

Annual fixed costs our business must arrange accounts for are; building rent which is $300,000/yr. (100,000 sq. ft. × $3/sq. ft.); Electricity — $50,000; Water — $5,000 and Gas — $30,000. Also included in fixed costs are payroll fringe benefits such as group health and life insurance, workmen's compensation insurance, employer unemployment benefit contributions, etc. which compositely amount to $100 per employee, annually (a debatable assumption). This cost is assumed equal for all employees. The annual payroll fringe benefit cost is $28,500 ($100 × 285). Since this company is in the commercial home appliance field, it accepts an annual advertising cost of $1,500,000. Management salaries, Grades 1, 2 and 3 amount to $415,000, and R & D costs $1,000,000.

Variable costs are treated next. They are separated into three categories, i.e.: Prime costs, Variable Factory Overhead, and Selling and Administrative costs. These costs are listed in figure 8.

We now have a summary of the three elements needed for C-V-P analysis. The cost summary, figure 8, is the outline used by the accountant for the chart. Figure 9 is the particular chart for these costs.

The quality control practitioner has little use for this analysis since it does not show the influence of quality costs. The approach used is to take the traditional quality cost categories and identify these for each of the accountant's cost categories. Figure 10 shows the identification of fixed costs in tabular form. Assuming that quality control activities participate at a 10% share of costs. QC costs account for 2.9% of total fixed costs ($96,500/$3,328,500). Further breakdown into quality cost categories is indicated by expense categories, figure 11. Two observations that can be made from this table are that no failure costs are fixed costs and that a relatively small proportion of quality costs are fixed, the bulk of quality costs being variable. If true, it may be concluded that quality costs play an important role in influencing the contribution margin.

The next step is to examine the proportion of quality costs in variables costs. Examination of prime costs reveals that quality costs are not included. There are quality costs involved indirectly such as the cost of training direct labor to measure work. In general, prime costs will not be considered as a source of quality cost information. The situation changes when analyzing quality costs in Variable Factory Overhead. Figure 12 indicates the cost categorization for Variable Factory Overhead. A recapitulation of Variable Factory Overhead quality costs by quality cost category is covered in figure 13. The next analysis identifies the quality costs that are contained in the Selling and Administrative Costs of the business, figure 13a.

A summary of quality costs interspersed in the accounting cost structure may be reviewed in figure 14. Some conclusions concerning these results at FCA are:

1) Quality Costs are predominantly variable costs.
2) Quality Costs are a significant determinant of the contribution margin. This conclusion is reached since quality costs in this example are 9% of total costs, ($1,045,400/$11,725,000).
3) Much of the quality cost is in the Appraisal category and External Failure. This indicates that the system is not providing assurance of product quality.
4) Fixed costs are proportionately low relative to variable costs. Since prevention costs are primarily fixed, management is not allowing for defect identification and control.

Results are not too different than what may be expected when most firms analyze quality costs for the first time. The quality cost participation in the overall effect of variable costs on the contribution margin is the final step in constructing the communication bridge between the quality control professional and the accountant.

In actual practice, volume would not start from zero unless the C-V-P chart represented a new business starting out with a new product. For a C-V-P analysis, the relevant range of certain volumes would be used in chart construction. Figure 15 is the representation of quality costs contained within the total cost lines of the business. The total costs contribution margin is $12,603,500 ($21,000,000 −$8,396,500) at the maximum volume level. The contribution margin for variable quality costs (VQC) to revenue is $20,051,100 ($21,000,000 − $948,900). The difference between the two contribution margins indicates a relative margin of quality costs to the total contribution margin structure. This margin figure is $7,447,600 ($20,051,100 − $12,603,500).

Presuming that the variable prevention cost is increased 10% and a resultant 75% reduction is experienced in Warranty Service Costs, we may ask, "What would be the change in relative value of quality costs to the total variable cost structure under these new conditions?" The new variable cost would be $8,086,850. This new figure is obtained by calculating the new variable prevention and the new warranty service costs and making the proper addition and subtraction. The new VPC is calculated from the old VPC of $53,500 i.e; $53,500 × 0.1=$5,350, or $53,500 = $5,350=$58,850. The new WSC is likewise calculated from the old WSC of $420,000, i.e; $420,000 × 0.75=$315,000 or $420,000 − $315,000=$105,000. These two new cost values may be substituted back into the previous tables, but the simplest method is to directly add and subtract the change values to and from the old total variable cost of $8,396,250, i.e.; $8,396,500 + $5,350−$315,000=$8,086,850. Concurrently, the new variable quality cost becomes

$639,250, ie; the old variable quality cost of $948,900 + $5,350 − $315,000=$639,250. The new respective contribution margins are now $12,913,150 ($21,000,000 − $8,086,850) and $20,360,750 ($21,000,000 − $639,250). The relative margin between total variable cost and variable quality cost remains the same.

A large influence is obviously exerted on the quality cost analysis of the business by this change. Quality practitioners have experienced such changes when additional prevention measures have been applied. However, from the total cost point of view, there has been a relatively small change in the CM. On the one hand where there was a significant change in quality costs (and in this case of warranty service, a probable increase in good will), there did not appear to be a significant change in total costs. For the business as a whole, this improvement was obtained without significantly increasing total costs. Figure 16 illustrates the new condition where prevention quality costs were increased.

The quality practitioner now has a tool with which he can communicate with the accountant and also develop analyses of current quality cost needs. By identifying the quality data, benefits in specific cost areas may be obtained.

Some quality practitioners may be disappointed that a very important quality cost, nonconforming material scrap was not included in the example. The writer admits that many quality cost categories were left out of the example. However, the object was to study the basic elements as understood by both the accountant and the quality control practitioner. The example was used to form a bridge between the two disciplines in order to provide the quality control practitioner with a means of communicating with the accountant. The expected result of a successful communication would be the development of the quality cost report and analysis.

Unless interest in quality costs is expressed by management, the accountant would not necessarily break them out. One reason an accountant might look at quality costs would be when the cost-volume-profit analysis indicates a breakeven status or worse, a loss situation. He may then analyze the variable costs to see if some were out of line. But, as we saw in the example, he might not be able to specifically identify quality cost as the extreme variable since the various quality costs may be spread over several categories by the accounting process.

Another point worth repeating is that significant opportunities for cost reduction in one type of cost may not mean a significant opportunity for reduction in total business costs. It was assumed in the example that for an additional small expenditure for prevention costs a significant reduction in Warranty Service Costs might be obtained. However, the effict on total costs would be considered relatively insignificant. When the quality control practitioner wonders why such a significant impact on his specialized area is not accorded the expected attention by management, he must recognize the sphere of his parochial view and then view the condition from the management viewpoint. This is not to say that management is not interested in significant cost reductions in specific cost areas, but that the specific area manager's perspective should include his efforts in the overall picture and recognize his contribution as important but not necessarily earthshaking. Management does recognize that good cost control in all parts of the business is vital to survival. This is the essence of the internal accountant's role, i.e. to keep management informed of the financial condition and progress of the business.

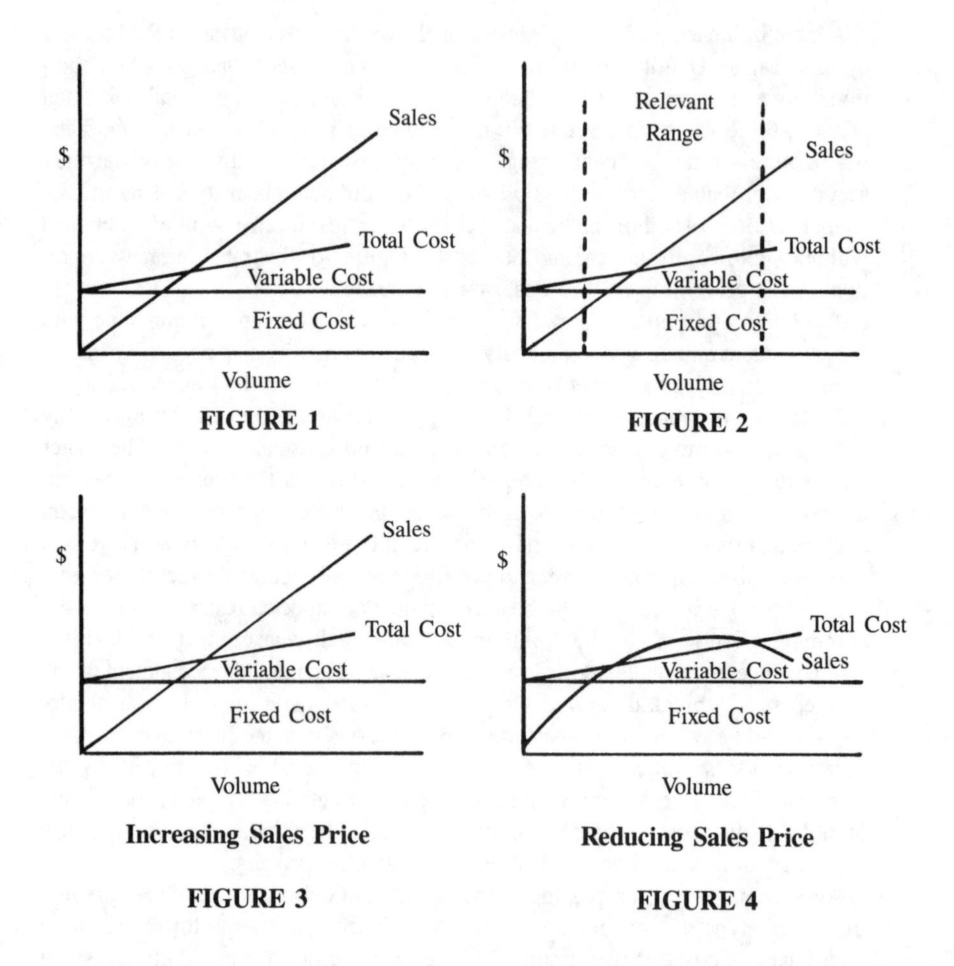

ORGANIZATIONAL CHART — FCA

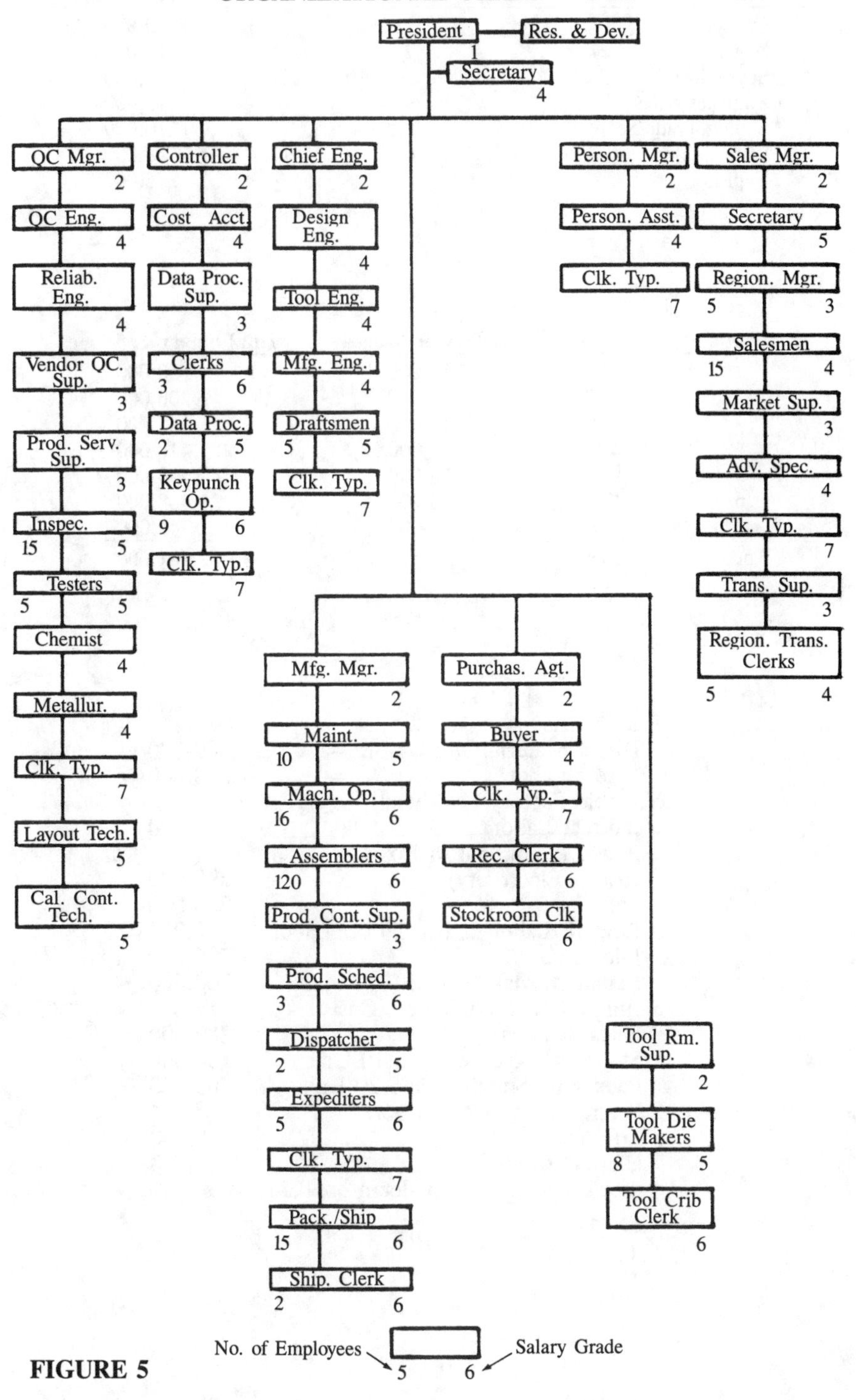

FIGURE 5

Title	Grade	Annual Salary
Officer	1	$50,000
Manager	2	25,000
Supervisor	3	15,000
Engineer/Sales	4	13,000
Skilled Trade/Secretary	5	10,000
Semi Skilled	6	8,000
Typists	7	6,000

FIGURE 6

Grade	No. People	Salary & Wages	Total Salary & Wages
1	1	$50,000	$ 50,000
2	8	25,000	200,000
3	11	15,000	165,000
4	32	13,000	416,000
5	50	10,000	500,000
6	176	8,000	1,408,000
7	7	6,000	42,000
Totals	285		$2,781,000

FIGURE 7

Prime costs:	
Direct Material @ $1/Unit =	$3,500,000/yr.
Direct Labor* =	1,258,000/yr.
Variable Factory Overhead:	
Indirect Labor*	789,000
Power (additional to Fixed cost of Electricity)	290,000
Supplies	250,000
Rework Labor @ 5% Direct Labor	62,900
Idle Time*	206,000
Repair & Maintenance	200,600
Selling and Administrative Costs	
Sales Salaries*	289,000
Shipping Expenses @ $.10/Unit	350,000
Packaging Supplies @ $.20/Unit	700,000
Warranty Service* @ $6/ Returned Unit	420,000
Clerical Costs	81,000

* (See Appendix A for detail breakdown)

FIGURE 8

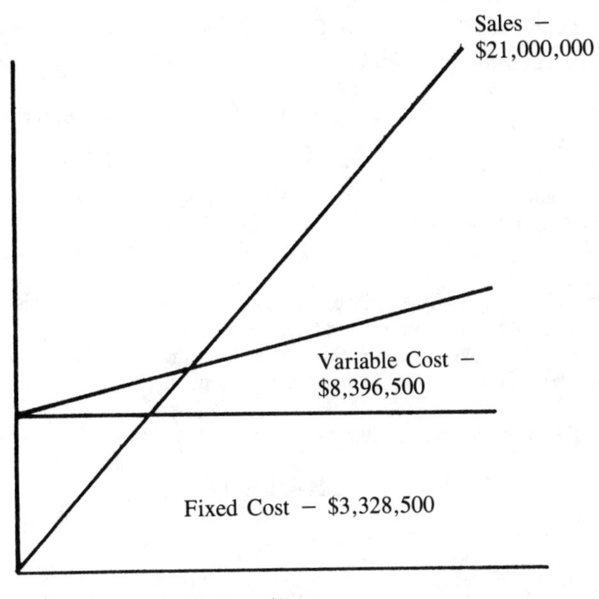

FIGURE 9

Fixed Costs	Accounting Total*	Quality Participation
Rent	$ 300,000	$30,000
Electricity	50,000	5,000
Water	5,000	500
Gas	30,000	3,000
Fringe Benefits	28,500	3,000**
Advertising	1,500,000	–
Management Salaries	415,000	55,000***
R & D Costs	1,000,000	–
	$3,328,500	$96,500

* See report, 3rd page
** Based on 30 employees assigned to the QC function
*** Salaries of the QC Manager and 2 supervisors

FIGURE 10

Category A: Prevention Costs — Fixed Costs			**Category B:** Appraisal Costs — Fixed Costs	
5. Other Prevention Expenses			**1.** Receiving or Incoming Test and Inspection	
Rent	$30,000		Fringe Benefits***	$2,400
Electricity	5,000			
Water	500			
Gas	3,000			
Fringe Benefits*	600			
Management Salaries**	55,000	94,100		

* For the QC Manager, QC Engineer, Reliability Engineer, Clerk Typist, Layout Technician and Calibration Control Technician.

** For the QC Manager, Vendor QC Supervisor and Product Service Supervisor.

*** For the balance of the QC Department personnel.

Note: Category Letters and Numbers — refer to "Quality Cost — What and How."

FIGURE 11

Category A: Prevention Costs — Variable Factory Overhead	
1a) Quality Control Engineering	
Indirect Labor (QC Eng., Reliability Eng.)	26,000
3) Quality Planning by Functions Other than QC	
Indirect Labor (¼ Design Eng., ½ Tool Eng., ½Mfg. Eng.)	18,250
4) Quality Training	
Indirect Labor (¼ Personnel Asst.	3,250
5) Other Prevention Costs	
Indirect Labor (Clerk Typist in QC)	6,000
	53,500
Category B: Appraisal Costs — Variable Factory Overhead	
1) Incoming Test and Inspection	
Indirect Labor (5 Inspectors, 1 Tester)	60,000
2) Laboratory Acceptance Testing	
Indirect Labor (Chemist, Metallurgist)	26,000
3) Inspection and Test	
Indirect Labor (10 Inspectors, 1 Tester)	110,000
6) Inspection and Test Material	
Supplies (25% of Total)	62,500
7) Product Quality Audits	
Indirect Labor (1 Tester, Layout Technician)	20,000
9) Maintenance and Calibration of Test and Inspection Equipment Used in Control of Quality	
Indirect Labor (Calibration Control Technician)	10,000
12) Internal Testing and Release	
Indirect Labor (2 Testers)	20,000
15) Data Processing Inspection and Test Reports	
Indirect Labor (1 Accounting Clerk, 2 Key Punch Operators)	24,000
	352,500
Category C: Internal Failure Costs — Variable Factory Overhead	
2) Rework Labor	62,900

Note: Category Letters and Numbers — refer to "Quality Costs — What and How."

FIGURE 12

Variable Factory Overhead Costs for Quality

Category:	
A: Prevention Costs	53,500
B: Appraisal Costs	352,500
C: Internal Failure Costs	62,900
	468,900

FIGURE 13

Category D: External Failure — Selling and Administrative Costs	
1) Complaints	
Salesmen's Time (⅕ of all Salesmen)*	39,000
2) Product Rejected or Returned	
Warranty Service**	420,000
Shipping Expense on Warranty Service	7,000
Packaging Supplies on Warranty Service	14,000
Total Category D Costs	480,000

* See Appendix A, IV
** See Appendix A, V

Note: Category Letters and Numbers — refer to "Quality Costs — What and How."

FIGURE 13a

	Fixed Costs*	**Variable Costs**		**Total**
		VFO* **	**S&A*** **	
Prevention	$94,100	$ 53,500		$147,600
Appraisal	$ 2,400	$352,500		$354,900
Int. Failure		$ 62,900		$ 62,900
Ext. Failure			$480,000	$480,000
Total	$96,500	$463,900	$480,000	$1,045,400

* Fig. 11
** Fig. 13
*** Fig. 13A

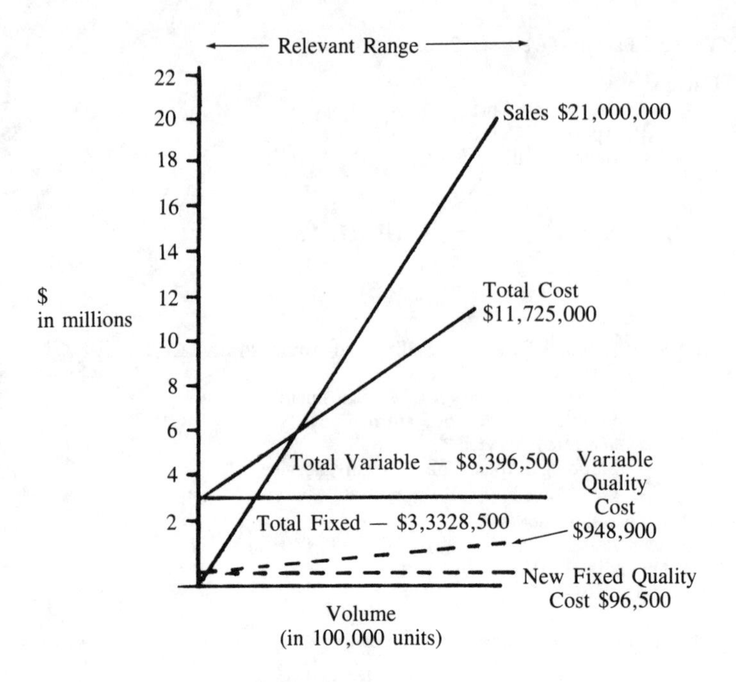

FIGURE 15

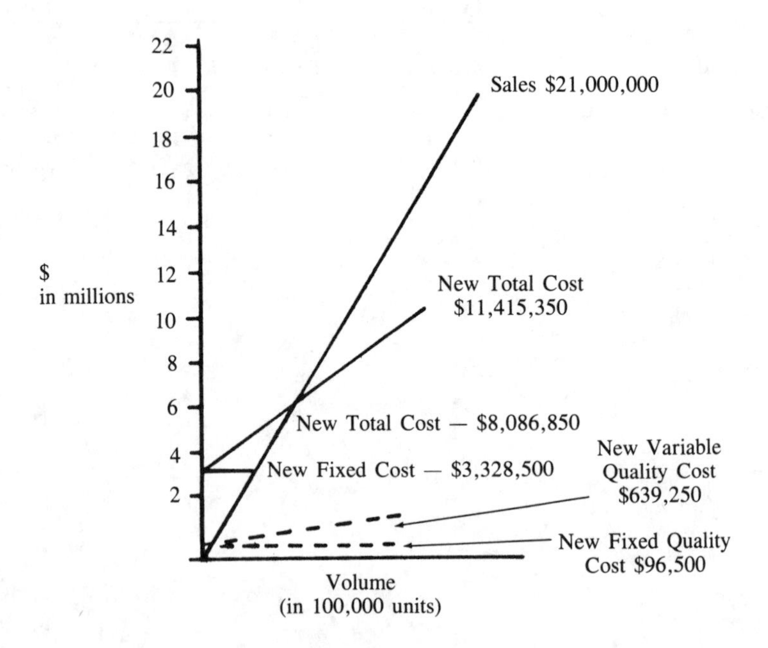

FIGURE 16

Appendix A
Variable Cost Category Accounts Detail

I Direct Labor

Job Class	Job Title	No. People	
6	Machine Operators	16	$ 128,000
6	Assemblers	120	960,000
6	Packer/Shippers	15	120,000
5	Tool & Die Makers	8	50,000
		159	$1,258,000

II Indirect Labor

Job Class	Job Title	No. People	
4	QC Engineer	1	$ 13,000
4	Reliability Engineer	1	13,000
5	Inspectors	15	150,000
5	Testers	5	50,000
4	Chemist	1	13,000
4	Metallurgist	1	13,000
7	Clerk Typist	1	6,000
5	Layout Technician	1	10,000
5	Calibration Control Tech.	1	10,000
4	Cost Accountants	1	13,000
6	Accounting Clerks	3	24,000
5	Data Processors	2	20,000
6	Key Punch Operators	9	72,000
7	Clerk Typist	1	6,000
4	Design Engineer	1	13,000
4	Tool Engineer	1	13,000
4	Manufacturing Engineer	1	13,000
5	Draftsmen	5	50,000
7	Clerk Typist	1	6,000
5	Maintenance Men	10	100,000
6	Production Schedulers	3	24,000
5	Dispatchers	2	20,000
6	Expediters	5	40,000
7	Clerk Typist	1	6,000
6	Shipping Clerk	2	16,000
4	Buyer	1	13,000
7	Clerk Typist	1	6,000
6	Receiving Clerk	1	8,000
6	Stock Room Clerk	1	8,000
6	Tool Crib Clerk	1	8,000
4	Personnel Assistant	1	13,000
7	Clerk Typist	1	6,000
4	Secretary	1	13,000
		83	$ 789,000

III Idle Time

Job Class	Job Title	No. People	
5	Inspectors	15	$ 150,000
5	Testers	5	50,000
5	Layout Technician	1	10,000
5	Calibration Control Tech.	1	10,000
5	Maintenance Men	10	100,000
6	Machine Operators	16	128,000
6	Assemblers	120	960,000
6	Expediters	5	40,000
6	Packer/Shippers	15	120,000
5	Tool & Die Makers	8	80,000
			$ 1,648,000

Idle time = 1 of 8 hours worked = .125
$1,648,000 × .125 = $206,000

IV Sales Salaries

Job Class	Job Title	No. People	
4	Salesmen	15	$ 195,000
4	Advertising Specialist	1	13,000
7	Clerk Typist	1	6,000
4	Regional Transport Clerks	5	65,000
5	Secretary	1	10,000
		23	$ 289,000

V Warranty Service

Warranty Returns are 2% of total units shipped annually

3,500,000 total units shipped annually
.02
70,000.00 total units returned

Each unit sold by FCA = $6 (also return cost-transport)

70,000
$ 6*
$420,000

* The assumption is made that each returned unit has a net $0 scrap value since it is completely replaced and no salvage or repair is attempted. The assumption is continued that there is further demand for the product. Therefore, there are no recovered costs from the returned goods.

REFERENCES

1. Feigenbaum A. V., *Total Quality Control Engineering and Management* McGraw-Hill Book Co. New York 1961.
2. Horngren C. T., *Accounting for Management Control — An Introduction* Second Edition Prentice-Hall, Inc. Englewood Cliffs, New Jersey 1970.
3. Quality Cost-Cost Effectiveness Committee, ASQC *Quality Costs — What & How* Second Edition American Society For Quality Control, 310 West Wisconsin Ave., Milwaukee, Wisconsin 1971.

ATTACKING QUALITY COSTS

Robert G. Stenecker, Director —
Corporate Quality Systems Engineering
Abbott Laboratories, North Chicago, Illnois

During recent years, there have been extensive tutorials and lecture sessions on Quality Cost Systems. For the most part, these have attempted to deal principally with the broad concepts of Quality Cost Systems, general management considerations, and very heavily with the nature of the data to be collected and the methodology for collecting it. There is little question that such sessions have been most effective in getting Quality Cost Systems off the ground in many many companies.

Perhaps this is a good point-in-time to emphasize the fact that the Quality Cost System is far more than just the measurement of Quality Costs, but that the ultimate objectives are rather in the "attack" on and the minimization of such costs without a reduction in product quality. All too frequently, during my involvement in lecturing and teaching on Quality Cost Systems, I have been asked the questions by the audience of "How do I get started?" and "How do I attack the more significant Quality Costs?". It is the intent of this discussion to review with you some of the more practical aspects of the Total Quality Cost System based on my own experience in initiating and managing such systems.

"KICKING OFF" A QUALITY COST SYSTEM

There are a number of ways bringing a company into a Quality Cost Program, but in general we might discuss the formal approach versus the informal approach.

The formal approach is the one best understood and one that has been much more comprehensively covered in previous lectures and discussions on Quality Cost Systems. This, in fact, is the approach that we have used at Abbott Laboratories. It assumes the ability of the Quality operations to get complete top management commitment. In the case of Abbott, for example, we faced the problem of convincing management that we were *not* trying to equate Quality with dollars where health care products are involved. However, we were able to provide justification that, with a properly run Quality Cost Program, we would be able to make a more efficient use of the Quality dollar and that the program would thus result in better utilization of our resources at no compromise in product quality

Once top management commitment is obtained, it is still necessary to follow through in "selling" the program to the operating personnel, as for example, division management, the controller staff, cost accounting personnel, production, quality control and engineering functions. I would like here to emphasize the importance of this second step. A Quality Cost Program is a comprehensive one which involves all the disciplines of the company. Unless all of those personnel involved are thoroughly familiar with the concepts and the objectives, a maximum program is not possible.

Concurrent with these steps, is the detailed job of designing the Quality Cost

System package — the reporting formats, and the identification and characterization of quality costs. Once these steps are accomplished, goals of performance must be established in close cooperation with the plant and/or divisions. At Abbott we have followed a philosophy of moving a step at a time. For example, initial reports required from the divisions were to reflect only those quality costs which were readily accessible in the existing accounting system. Accomplishing this objective took considerable time, but it was time which was well spent in turning up problems in the collection of information, simplification of reporting, and in bringing all of the operating personnel "on board." In addition a number of areas of high quality costs were highlighted and attacks on these costs initiated. Schedules also called for 90% completeness in Quality Cost Reports within the following six months, periodic analysis and reviews with top managements of all the divisions by the divisional Quality management, and expansion of corrective action activities. Cost reduction goals were established with the divisions.

So far, because this area has been reasonably well covered in the past, I have dealt very briefly with this formal approach in establishing a Quality Cost Program. Let us now consider an "informal" approach. For this, minimal top management commitment is required. A significant area where good quality cost data is available is selected, a task team (or teams) is set up with representation from other pertinent functions, and initial successes are used as a fulcrum in soliciting additional management support toward expanding the program into all the quality cost areas. A typical program might proceed as follows:

a. Set up the task team. Chairman: a member of Quality Control management; Members: Production Engineer, Industrial Engineer, R & D, and cost accounting (part time).
b. Establish approach, reporting and schedule of activities.
c. Ensure necessary data is provided to the team.
d. Analysis of scrap causes and institution of corrective action.
e. Develop and implement motivational programs (use team).
f. Monitor and provide support.
g. Use the success in the initial area of attack in expanding into other areas of Quality Costs.
h. Eventual formalization after the "developmental" efforts are complete.

As an example, I started such a program many years back prior to my entry into the Health Care Field and in the early days of the Quality Cost System using scrap costs as the fulcrum for initial action. In this instance, getting started involved my meeting with the president of the company, reviewing with him all those areas of expense which required his personal signature (new hires above certain salary levels, purchases above $15,000, etc.) and in conclusion, pointing out to him that division "A" had expended in the previous year somewhere between $400,000 and $500,000 in scrap against gross sales of approximately $30,000,000 and asking him, "Who signed off for that?". The Quality Cost Program for that division was instituted the following morning by a personal and emotional memorandum from the president! The discussion that follows as well as the illustrations which are part of this presentation reflect some of my experiences with that Quality Cost Program, some years prior to my joining Abbott Laboratories.

"ATTACKING" QUALITY COSTS

Scrap costs almost always constitute a major area of identifiable quality costs and one that is most susceptible to improvement. In general, an attack on two fronts will be found advisable. The first is in the setting up of teams to apply Pareto principles in attacking specific areas of costs and the second in instituting motivational programs to reach all individuals in the company who are involved in any way with creating or causing scrap. The first step taken in the informally initiated program discussed above was setting up a team, which was identified as the Quality Cost Improvement Team, and, which consisted of a Quality Control Manager as chairman, with representatives of production engineering, industrial engineering, R & D, and cost accounting. The directions given to the team were to use Pareto principles in identifying and scheduling projects for scrap reduction, maintaining a running log of these projects, identifying anticipated (and eventually, actual) cost savings, and establishing investigational responsibilities and deadlines which were continuously reviewed by myself and general management (FIGURE 1). In this instance we thought we had a fairly good reporting system for scrap; and scrap reports, costed out, were available broken down by cost center, by cause, and by product item in three monthly IBM summary reports (FIGURE 2). It is worthy to note that the first bit of information coming from the program was that our quality costs in the area of scrap were closer to $1,000,000 than to the $500,000 originally reported and that a considerable amount of scrap costs were being lost or buried prior to the initiation of the Quality Cost Program. This, of course, made it difficult to demonstrate the effectivity of the program on the basis of year-by-year comparisons and emphasize the advantages of the "separate project" approach in giving an immediate and continuous measurement of effectivity. For example, periodic reports as illustrated in FIGURE 3 were provided to management.

MACHINE SHOP SCRAP CORRECTIVE ACTION PROGRAM

Proj. No.	Date Entered on Log	Part Number and Problem	Ass'd to & Date	Estimated Completion Date	Actual Completion Date	Estimated Savings	Actual Savings	REMARKS
1 Compl.	4/9/68	51408600011 "L" Holes Break Out	A. Kurzweill R. Gliebe D.Bonner 4/11/68	4/20/68	4/16/68	(1) $10,000 (2) 2,000 (3) 3,000	(1) $11,168.25 (2) 2,758 (3) 3,185	(1) Yearly Scrap Reduction (2) Immediate Savings (3) Yearly Reduced Machining
16	7/26/68	512800400010 Parts Become Loose In Fixture Legs Not Perpendicular	A. Kurzweil 8/22/68	8/30/68	8/27/68	$ 250	$ 352.06	
17	7/26/68	52414400030 Part distortion and Various Dim. Discrepancies	R. Dietz 8/22/68	9/1/68	8/27/68	$ 1,500	$ 2,023	It Will Take Approx. 1 Year And A Half to Pay For New Tooling Before Realizing Savings
31	11/30/68	56413201030 Dim. Not Being Maintained	W. Bellion 11/30/68	12/2/68	12/3/68	$ 5,000	(1) $23,284.80 (2) 7,938	(1) Yearly Scrap Reduction (2) Yearly Reduced Machining
33	12/12/68	Instrument Cases Depth of Pads	D. Bonner 9/10/68	12/1/68	12/16/68	$10,000	(1) $1,531.80 (2) 12,000	(1) Yearly Scrap Reduction (2) Yearly Reduced Machining

FIGURE 1

SCRAP COST CENTER RESPONSIBILITY

19 ___

Code Center No.	Date	July/August		September		October		November		December	
	Dollars Scrapped	11382		3458		40979		7020		30066	
	Departments	Cost	%	Cost	%	Cost	%	Cost	%	Cost	%
372	Mech Lab.	283	2.4	–	–	–	–	–	–	11	–
563	Purchasing	961	8.4	224	6.4	2395	5.8	51	.7	1262	4.1
807	A.D.C. Ass'y	3821	33.5	135	3.9	49	.1	–	–	431	1.4

SCRAP CAUSE CODE RESPONSIBILITY

19 ___

Cause Code No.	Date	July/August		September		October		November		December	
	Dollars Scrapped	11382		3458		40979		7020		30066	
	Departments	Cost	%	Cost	%	Cost	%	Cost	%	Cost	%
11	Supplier (Lim.Value)	775	6.8	–	–	41	–	5	–	2	–
12	Supplier (Latent Def.)	856	7.5	222	6.4	2289	5.5	16	.2	492	1.6
13	Engineering Changes	179	1.5	–	–	4233	10.3	–	–	3709	12.3

AREA SCRAP COSTS

19 ___

Date	July/August		September		October		November		December	
Dollars Scrapped	11382		3458		40979		7020		30066	
Description	Cost	%	Cost	%	Cost	%	Cost	%	Cost	%
Manufacturing Scrap	9484	83.3	3033	87.7	29145	71.1	6949	98.9	24390	64.4
Tooling	–		–		265		3		50	
Nature of Process	8207		1271		10172		1803		5187	
Machine Capability	451		–		571		29		–	

YEARLY SAVINGS RESULTING FROM SCRAP CORRECTIVE ACTION COMMITTEE ANALYSIS
Period: 5 Months, April 9 thru August 1968

Invest. No.	Part No.	Descr.	Ancillary Savings	Yearly Scrap Red'n
GSD 1	51-40860-0011	Mech. Body	$3,185.	$11,668.
SRE 101	28-41860-0030	Window		4,456.
SRE 107	85-25362-0021	Diaphragm Assy. (VVI)		13,956.
SRE 103	36-25362-0020	Pinion Handstaff		11,757.
GSD 18	50-41430-0021	Case		2,810.
	Subtotals — Savings per Year		$3,581	$59,357.

FIGURE 3

In this example, we see the results of the first five months of the program. The report indicates that, in addition to savings created by reduction in scrap, the program was achieving other savings as a by-product of the investigations. In the three months following this reporting period, total savings were increased to approximately $300,000 per year of which close to $100,000 were in ancillary areas other than scrap. This was a very significant return on investment, particularly since the committee did most of its work in addition to the normal duties of the various members. A typical management presentation graph for one plant of this multi-plant operation is shown in FIGURE 4. This covers the first eight months of the program and illustrates the large variability in the reporting of scrap which was initially encountered and which can be seen to be "smoothing out" as the program took hold. The scope of the "least squares" linear trend line shows that in this plant savings in scrap reductions were being generated at a rate of approximately $1500/year for each month of effort.

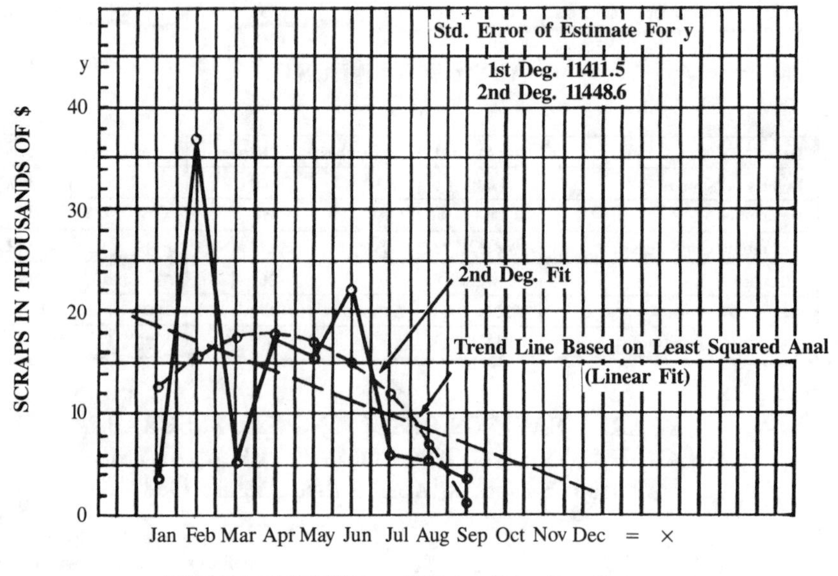

SCRAP LOSS TREND — 1968 (as of Jan thru Sep)
PLANT"A"

In addition to and concurrent with its attack on specific quality cost areas, the team participated in developing several types of motivational programs. To begin with a presentation was prepared which was given to all of the company's employees in separate groupings, and which involved a discussion of the magnitude of the scrap problem in relation to the company's profits and procedures. For example, the $500,000 scrap per year was shown to be equivalent to pure profit money that was thrown away and which represented a 10% profit on a $5,000,000 contract which might take several years to accomplish. Specific examples of scrap were hung on a large board and members of the audience were asked to guess at the cost of the scrap at the time the scrap occurred, which, of course, included not only the material costs, but the dollars in labor put into the piece before the scrap was created. Actual costs were then identified to the audience.

A second motivational program which was instituted was surprisingly effective in reducing scrap. In this case, a weekly committee was set up consisting of various shop operators, i.e., drill press operators, boring mill operators, deburring operators and the like. Also as part of the committee were the Manager, Manufacturing Engineering; a Quality Control Manager; and myself. In addition, at their request, a representative of the shop union also participated. In order to ensure complete freedom of expression from the shop personnel, no members of their management were present on the committee. The committee met for two hours each week for a period of two weeks and the shop personnel were replaced by four new shop members every two weeks. Reports were issued to each of the members which detailed the specific recommendations, the name of the man making the recommendation, and eventually the results in savings. Although originally intended as a motivational tool, the output of the committee was found to be extremely effective in getting at many specific causes of scrap, particularly in the areas of machine refurbishment, fixture inadequacies, documentation inadequacies, and the like.

Rework quality costs are usually found to be considerably higher than those occurring from scrap, but the data is correspondingly more difficult to retrieve and is rarely readily available. However, in this instance, during the intensive efforts on scrap costs reduction in the first year of the program, we had a breathing space of time in which to start collecting re-work costs and gradually expand these into the work of the committee. From the viewpoint of cause and effect relationship, of course, the rework costs were essentially of the same nature and required the same techniques as we had been applying for the scrap reductions, and similar success was encountered.

So far we have been discussing what is categorized as "uncontrolled" quality costs as compared to "controlled" costs, as for example, inspection and test, and this area should certainly not be neglected, particularly by those companies involved in high volume, continuous production. Furthermore, these costs can be and should be attacked almost completely within the Quality Control and Inspection organizations so that the task team is one that is internal to the Quality function. Their primary weapons, of course, are the tools of Quality Control Engineering; the principles of statistical quality control.

We sometimes need to be reminded that the original motivation for the quality control function was more toward the application of scientific principles in reducing the costs of inspection and test through sampling and the control of the process, rather than in attempting to improve product quality — although we now know that this was a very big by-product which followed the establishment of the quality control function. Any form of detailed discussion of the economic principles underlying statistical sampling and process controls would require far more time than I am permitted in this brief discussion. However, I would at this time like to highlight from my own direct experience a few of the more obvious areas for potential cost reduction. As I have already discussed, modern quality control can probably be traced to the development by Dr. Shewhart of the control chart, and the development of the early sampling plans by the Inspection Engineering Department at Western Electric in the early 1920's. This earliest sampling plan was essentially the Lot Tolerance Percent Defective Plan of Dodge and Romig, and was based on the Average Total Inspection concept (ATI). The average total inspection equation which was built into the mathematical model used in the LTPD system development is shown on FIGURE 5.

THE ATI CONCEPT
(Average Total Inspection)

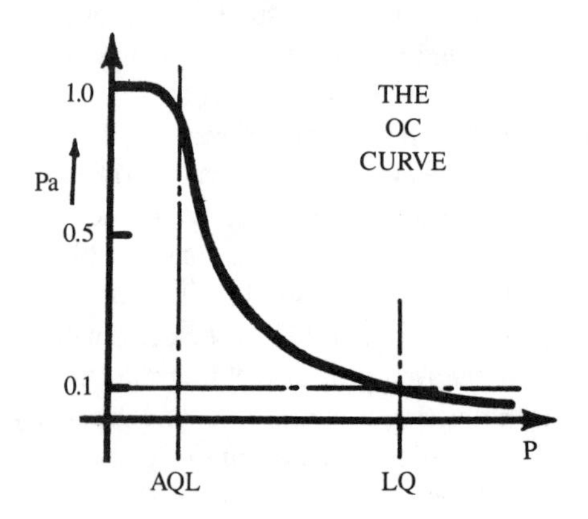

h = SAMPLE SIZE
Pa = PROBABILITY OF LOT ACCEPTANCE
N = LOT SIZE
P = PROCESS AVG. (% DEFECTIVE)
AQL = MAX. % DEFECTIVE, SATISFACTORY AS PROCESS AVG.
LQ = QUALITY ACCEPTABLE 10% OF TIME
C = (MAX. ACCEPTABLE NUMBER OF DEFECTIVES/SAMPLE)

$$ATI = n + (1 - Pa)(N - n)$$

(ASSUMES REJECTION RESULTS IN 100% SCREENING)

FIGURE 5

FIGURE 6 shows examples of three sampling plans which are essentially equal from a LTPD or limiting quality point of view but which result in significantly different total inspection costs if we assume that a lot rejection is followed by a 100% screening operation. It is obvious that the use of Plan I, which uses the smallest size sample, will in fact increase the cost of inspection by a factor of 20 to 1.

ATI EXAMPLES — CASE HISTORY

N = 20,000 PROCESS AVG. = 2% DEFECTIVE LQ = 10%

PLAN 1: n = 39 C = 1

ATI = 39 + (1− 0.81) (20,000 − 39) = 3832

PLAN 2: n = 120 C = 7

ATI = 120 + (1 − 0.997) (20,000 − 120) = 180

PLAN 3: n = 315 C = 21

ATI = 315 + (1 − 0.999) (20,000 − 315) = 334

FIGURE 6

However, in the example just discussed, we have not been looking at total quality costs, but only for those costs involved in the sampling and screening inspection. We have ignored "failure costs" — downtime, added labor to product, material review board activities, and the like. Furthermore, we haven't considered the value of the defective material which does not get rejected. When we fail to build visibility, flexibility and feedback into our inspection operations, we waste large sums of money, and more importantly, we fail to spend the inspection dollar wisely and effectively and therefore, lose quality. A review of quality records, if they exist, will show us that probably in excess of 80% of the characteristics we inspect result in virtually no rejections, and that some amount of less than 20% results in continuous rejections. Economic considerations certainly demand that our inspection systems be designed to recognize and take advantage of this fact. FIGURE 7 illustrates, in a perhaps over-simplified fashion, an economic model for making the decision of sampling versus no inspection by including those quality costs resulting from our failure to pick up defective material at a prior inspection point.

ECONOMICS OF SAMPLING VERSUS NO INSPECTION

C_{uf} = FAILURE COST PER DEFECTIVE UNIT ENTERING NEXT PRODUCTION LINE

N = LOT SIZE

P = PROCESS AVG. (% DEFECTIVE)

C_{ui} = COST TO INSPECT A SINGLE UNIT BREAK EVEN WHERE

$$C_{uf}\,N_p = C_{ui}[n + (1-Pa)\,(N - n)] = C_{ui}(ATI)$$

FIGURE 7

The next "Quality Case History" which I will discuss with you illustrates the economic advantages of applying Process Control techniques rather than Acceptance Sampling in any situation where we have access to the process (FIGURE 8). In this instance, a small plant was providing plastic sub-assemblies to a sister division which stipulated the requirement of 1.5% defectives or less. At the time this analysis was performed, an acceptance sampling operation was occurring at the end of the line every ten minutes and, when rejection of the sample occurred, the entire line was stopped and the operators performed 100% inspection in place. Then the line operation was resumed. Analysis of the data showed that the process was "in control" which is to say that the line was consistent in turning out product at a relatively constant level of approximately 1% defective. An analysis of the sampling plan indicated a probability of acceptance at the actual average defect level of 77%. Thus, they could expect to reject a lot roughly 23% of the time, even though the quality was consistently considerably better than required. A review of the quality history of the production line which we are discussing indicated that we were indeed rejecting lots 25% of the time. Obviously, the material rejected was of no different quality than the material accepted. Therefore, the sampling operation and its incident costs, including downtime on the line, was accomplishing absolutely nothing. The actual improvement in quality created by the 100% inspection actually improved the quality of the batches by no more than a few hundredths of a percent. By instituting process control methodology, it was possible not only to save approximately $16,000 per year per shift on this line, but also by directing attention at the process, there was eventually a considerable improvement in the quality of the product.

QUALITY COST CASE HISTORY

PLASTIC SUB-ASSEMBLIES **REQUIRED:** NO MORE THAN 1.5% DEFECTIVES

ACCEPTANCE SAMPLING PLAN: n = 24 C = 0	SAMPLE INSPECTED EVERY 10 MINUTES STOP LINE & 100% INSPECT IF SAMPLE FAILS
ANALYSIS:	PROCESS IS "IN CONTROL p = 1.06% Pa (for p = 1.06%) = 77%
QUALITY HISTORY:	REJECTIONS APPROX. 25% OF THE TIME
RESULTS:	CHANGE TO CONTROL CHARTING
SAVINGS:	$16,100/YR./SHIFT

FIGURE 8

In summing up, I have attempted to give you some practical suggestions for starting a Quality Cost Program; methods which I have tried and which have worked. We have discussed some actual programs for attacking the major areas of quality costs: scrap, rework, and inspection and test costs. However, the most important advice I can give you is to "start!" There is a very old and very well known story, which bears repeating, of the sales manager who called a meeting of his area sales personnel. Shortly after the meeting started, each of the salesmen was told to stand up and look under the cushion of his chair. When they did, they each found a dollar bill. The moral of the story which was given to them and which I pass on to you is: "If you want to make a buck, you have to get off your derriere!"

QUALITY COST MANAGEMENT FOR PROFIT

Richards K. Dobbins, Principal Quality Assurance Staff Engineer
Honeywell Inc. — Process Control Division/Ft. Washington, Pa.

BUSINESS STRATEGY FOR QUALITY ASSURANCE

Management Relations

Most quality professionals believe that optimum economics will follow implementation of an effective system for total quality control. The main problem seems to be convincing executive management that there is any positive connection between a total quality control program and improved company profits. Historically, quality professionals have tried to sell the concept of quality control to their superiors by:

(1) educating them on the mechanics of Q.C. principles;
(2) persuasion that "it is the way to do things right"; and
(3) recognition of external pressures (contractural obligations to customers and regulatory agencies, consumer demands, etc.)

Gentlemen, we have not approached this dilemma in the right manner, and our lackadaisical success in winning strong executive support proves it!

Instead of trying to teach others what we mean *in Q.A. jargon* (AQL's confidence limits, process averages, sampling plans, etc.), we should be communicating directly *in business terms* (costs, profits, inventory levels, delivery delays, etc.) which management can readily understand. As opposed to "doing things right" for academic principles, we should be more concerned with "doing the right things" from a good business sense. Rather than minimal conformance to external pressure requirements, we should be aggressively pursuing control programs to attain long term optimum operating conditions. In short, *we must actively demonstrate that a meaningful total quality control effort makes good business sense with a profit payoff.* A quality control analysis program can be the vehicle for Q.A. to prove quality needs, identify target problems, track progress towards the goals, and document the contributions made to company profitability.

Many companies regard their Quality Assurance/Inspection Departments solely as indirect or burden type activities, whose salaries and operating expenses are called quality costs. If their concept of quality costs stop at this point (or only include other obvious items such as scrap and rework), the quality effort within such companies will be in constant jeopardy. In lean business times, the "economy axe" will pare such overhead activities to the bone, ostensibly to reduce expenses and maximize profits. Although this may produce results exactly opposite to that intended, these companies will seldom realize that their actions actually hurt their profit stance. Quality departments in such companies will always be on the defensive trying vainly to justify their existence.

Business operating policies and decisions are, quite correctly, greatly influenced by economic considerations. This means that total quality costs must be *managed* in a fashion compatible with the executive policy decisions for the company concerned. Everyday good management practices require that:

To "manage", we must *control.* To " control", we must *measure.* To "measure", we must *define.* "In defining", we must *quantify.*[1]

Need for Quality Cost Analysis

There is a need, therefore, for a *total* quality cost analysis from at least three different viewpoints, as follows:

1. *From the Q.A. Manager's standpoint,* to remove arbitrary constrictions on effective quality efforts; and to provide optimum product quality at minimum total quality cost.
2. *From an executive management standpoint,* there is a need for objective assistance in: measuring, analyzing, budgeting, planning and programming; and determining marketing strategy.
3. *From a company finance standpoint,* there is a need: to reflect the direct tie-in between quality costs and profits; to deal in "big hunks", going beyond departmental boundaries; and to take advantage of existing accounting systems.

A well planned Quality Cost Analysis program should have several major provisions. It should provide overall measurement of the "Quality Health" of your business. It should provide measurement of the effectiveness of the Quality Assurance Manager in "pulling it all together". It should provide the means to establish long and short range quality efforts. It should provide the means to establish specific quality priorities. It should provide the means to establish effective checks and balances on operating conditions influencing product quality. And most importantly, it should translate the various quality theories and principles into the language of management-dollars. Putting this last provision another way, it should provide visibility to operating intangibles associated with the quality efforts, by using the common denominator of dollars.

TOTAL QUALITY COST ANALYSIS

Definition of Quality Costs

It is necessary to be quite explicit in the definition of quality costs. *What are the quality costs?*

> The costs that exist in the design, manufacture, distribution and servicing of a product because quality deficiencies *do* or *can* exist, are "Quality Costs". *They are costs that would disappear if the quality deficiencies disappeared,* or if perfect control of material, people, and processes were possible.

As used in this presentation, all operating quality costs can be defined in one of four broad categories: Prevention; Appraisal; Internal Failures; and External Failures. The first two categories may be thought of as *"quality expenses",* meaning dollars *invested* to avoid or control losses. "Prevention costs" are costs for activities to keep defects from occurring in the first place. They include such elements as quality control engineering, employee quality training, quality planning, design of measuring equipment, etc. "Appraisal costs" represent expenses for maintaining company quality levels by means of formal evaluations of product quality. This includes all inspections and tests (whether performed by production, inspection or other personnel), maintenance of test and inspection equipment, quality audits, and cost of product consumed during destructive tests.

The last two categories may be thought of as "*quality losses*", meaning dollars *wasted* because perfect control does *not* exist. Both internal and external failure costs are associated with defective items which fail to meet quality requirements. "Internal Failure" costs concern defective products, components, or material which result in *operating manufacturing losses.* "External Failure" costs are generated to the company as the result of the *defective product being shipped* to the customer. Included within Internal Failures are such cost elements as: scrap; rework; troubleshooting; material review board; reinspection and retesting of previously rejected material; and the price differential between normal and reduced selling price due to downgrading. External failures include costs associated with: rectifying valid customer complaints; field service and replacement material for warranty reasons; processing the return of defective material and its repair; field retrofits and/or recalls; and any liability costs associated with supplying defective material or services.

Magnitude and Distribution of Quality Costs

"Quality Expenses" should only represent *intentional management investment* in quality efforts to avoid or minimize restoration or failure costs, known as "Quality Losses". *Quality losses sometimes approximate 5 to 15% of total sales billed;* much of it centered around warranty adjustment costs. Quality losses from $500 to $1,000 per year per productive employee are not uncommon.(2) Elimination or reduction of these quality losses will be directly applicable to the bottom line of the profit and loss statement for the company. Consequently, properly presented and documented programs to control or minimize these losses are sure to receive management's attention.

The distribution of total quality costs typically show that approximately 70% of the total quality costs are associated with Internal and External Failures, while appraisal costs usually range in the neighborhood of 25%. This means that typical Prevention costs do not exceed 5%. Such distribution suggests that companies historically are spending their quality dollars the wrong way! The vast majority of their quality costs represent money "down the drain" because of product failures; while another sizeable sum is required to "sort-the-bad-from-the-good", to try to keep too many bad products from going to customers. Accordingly, there is little money available for true defective prevention technology, that can do something about reversing the vicious upward cycle of higher quality costs and less reliable product quality.(3)

Documentation of quality costs according to the four categories above is the starting point for any total quality cost management program. Operating quality cost elements must be correctly categorized to achieve this end. Charts showing typical operating quality cost items categorized as to prevention, appraisal, internal and external failures are published in "Quality Costs — What and How" prepared by Quality Cost Committee, ASQC.

Quality Cost Analysis Strategy

The strategy involved with planning and executing an effective quality cost analysis program, when reduced to skeleton form, would be as follows:

1. List applicable quality cost elements and categorize each with respect to the four quality cost segments.
2. Define each element as it applies to your operations and accounting set-up.
3. Collect, measure, and estimate past and present costs associated with each element of concern.
4. Analyze results and estimate recoverable amount within practical time limitations. Establish goals to achieve same.
5. Map out basic program by which such goals are to be attained. Indicate the responsibilities shared by the various operating departments in achieving such goals.
6. Sell basic program to management — both executive management and the individual managers of the operating departments affected and concerned.
7. Organize for the improvement effort attack, and start implementation program.
8. Provide for regular reporting to management to show results of the basic improvement program.

The skeleton program above represents the ideal situation involving quantification, definition, measurement and control in order to achieve management of quality costs. When starting from scratch, however, there are several pitfalls to be avoided.

Pitfalls in Establishing Program

In reviewing all possible quality cost elements, frequently little documentation exists to show what those elements have previously cost. Capturing some cost elements for a historical base is a tedious process, and, there is a tendency to become much too detailed in the initial effort. Test elements that affect your operations by seeing what the past costs are, or in "questimating" what they must have been. Go after the big ones first, and as a first-pass effort, ignore elements which only contribute a minute portion of the total dollars involved. Look at the entire company's operation, however, and do not limit yourself solely with the quality/inspection departments. Recognize that job functions and titles vary widely from industry to industry and company to company. Apply the test of the quality cost definition itself. If the cost would disappear under the utopian situation that perfect control of people, material and processes existed, then that is a true quality cost and should be considered. This means that certain portions of the engineering budget and of manufacturing will undoubtedly be included along with quality, inspection and field service costs.

Use care not to paint with too broad a brush, or you very likely will categorize all salaries associated with your quality control effort with that of prevention. Study will indicate that a significant majority of quality control personnel's time concerns day-to-day appraisal efforts, handling of floor problems caused by internal failures, or analyzing reports from the field concerning external failures. The reasons for care in this segregation should be obvious. Anything that is slotted in the ''prevention'' or ''appraisal'' categories should represent an intentional company investment or asset. Anything that is slotted into ''internal'' or ''external'' failures represents a quality loss or debit. Reduction of such losses is the primary target of the quality cost analysis program, and must be considered on a company-wide basis. Do *not* include costs associated with marketing policies where no-charge service may be extended to attain or keep customer good will, as opposed to instances where there is faulty design or workmanship. Do *not* include overshipments or inventory returns due to changes in product design evolution. Do *not* include the cost of test equipment, whether purchased or made in-house, but *do* include the cost of any design or specification establishment for such test equipment.

Side Benefits

Although free service extended as a result of marketing discretion is to be excluded from the quality cost analysis, it may be desirable to separately show the results of such policy decisions along with the tracked quality costs. As with any detailed study, the initial analysis of the quality cost elements and associated costs will very likely be surprising to company management. In defining what needs to be tracked, the measurement and recording process of the individual segments will probably be improved. Therefore, a bulge may appear *outside* the quality cost analysis program as a direct result of more detailed attention as to what is charged to quality cost accounts. When this occurs, it will permit executive management for the first time to measure the cost of certain policies and practices the company has been following, and determine whether those policies need to be redefined and practices altered. In such a situation, vast reductions are possible in what was previously considered ''field warranty costs'' merely through the visibility and separation of true external failure costs and marketing policy costs.

Reporting and Charting Techniques

After the various quality cost elements have been selected and defined, be specific in the manner in which those costs will be collected and reported. Use care that all costs are collected on the same basis, with overhead and burden applied equally to all accounts. In reporting, use charting techniques wherever possible dealing with only the four major cost segments. Then, as necessary, identify the few significant contributors to each of the four categories. Avoid the tendency to publish complete tabular data as opposed to brief summaries complete with analysis of levels, trends, and goals. Each of the four major cost elements should be analyzed in relationship to the other elements, and the total of all quality costs. It is necessary to include time-to-time comparison: that is, comparing one month's operation with the previous period of several month's operation, or one quarter with the previous several quarters.(4) Dealing in absolute dollars by itself requires comparison to some other measurement base so that variables of production levels, inflation, product mix and other factors are taken into consideration in making the analysis. It is suggested that operating quality costs be related to at least two (and preferably three) different volume bases. The bases selected will vary depending on the product and/or type of manufacturing for a particular business. Examples of volume bases that should be considered are:

A. *A labor base,* such as total labor, direct labor, or applied labor.
B. *A cost base,* such as shop cost, manufacturing costs or total material and labor.
C. *A sales base,* such as net sales billed or sales value of finished goods transferred to inventory.
D. *A unit base,* such as the number of units produced.

Caution: Measurement bases are only as good as the methods for keeping them consistent. If major changes affect such a base, consideration must be given and (adjustments made) to correct previous data to agree with future data.[5]

With businesses having more than one product line, it is usually advisable to maintain quality costs for each product line as a separate business. In comparing two product lines, consider the difference in product complexity, applicational requirements, process stability and volume. A good comparison can be made with relative improvements in total quality cost from one point in time to another, usually compared to the last year. A quality cost breakdown by product line is another method for furnishing good information on allocation of total investment, compared with cost of quality losses. Use of the various charting techniques may help gain a better understanding of what is happening to the business. *Obvious benefits that can be obtained by charting techniques are:* (a) Comparison of prevention and appraisal investment in relationship to total failure losses. (b) Appraisal as a percent of direct labor. (c) Purchased material scrap and rework in relationship to purchased material evaluation costs. (d) End-of-line test and inspection costs in relationship to in-line testing costs. (e) Application of Pareto method for singling out the "vital few" in comparison to the "trivial many".[6]

Program Maintenance

Maintenance will be required for any effective quality cost analysis program. The program will only be effective if it factually and actually continues to measure the true quality costs within the organization. Periodic audits of the program are required to determine: (1) if the system is still conceptually adequate; and (2) if the system is functioning as it was designed. Under the *conceptual review* of the system, the following questions should be answered during the periodic audit:

Are the costs reported complete?
Are the costs reported properly classified?
Are the costs reported truly quality costs?
Is the measurement base still appropriate?

Under the *functional review* of the system, the following questions should be asked during the periodic audit:

Are the actual costs sustained truly reflected by the quality cost report?
Is the measurement base accurately reported?

The ASQC Quality Costs Committee recommends an annual conceptual review of the total quality cost system, and a monthly functional audit of one of the main divisions of this system. (That is, one month audit a specific department, another month a specific cost elemet, etc.)[7]

Determining Optimum Operating Range

After a quality cost analysis program has been in effect for some period of time, a difficult matter is determining when a company is nearing (or has passed) the optimum operating point. That is, whether the quality losses can be effectively reduced further by additional investment in the quality expenses. As a rule of thumb, a company is usually *short of optimum* operating conditions if the failure costs exceed 70% of the total quality cost and the prevention share of quality cost is significantly less than 10% of the total. In this range, consideration to normal quality control principles should still yield sizable returns. The company is operating *in or near the optimum range* when failure costs approximate 50% of total quality costs, and the prevention share is approximately 10% of the total. In this range, most efforts should be directed towards searching for techniques that would lead to breakthroughs to more effective operating levels; otherwise be satisfied to maintain the status quo. *Your company has probably gone overboard* if you find that your failure costs are much less than 50% of total quality costs and prevention expenses amount to 10% or more of the total. In this region, economies on the quality operation side can usually be achieved without jeopardizing overall quality levels of the end product.

APPLYING TOTAL QUALITY COST ANALYSIS

Views of Mr. Crosby, ITT Corporation

According to Phillip B. Crosby, Vice-President Director of Quality, ITT Corporation, there are three different beliefs which must be eradicated in order to make significant progress in Quality Assurance. These are:

(1) Quality means goodness or elegance.
(2) Standards should be based on acceptable quality levels.
(3) Quality costs more.

Mr. Crosby's feelings about these three myths are that:

(1) Quality means conformance. The product or service is either like the published requirements or it isn't.
(2) There are *no* levels of quality. What has been an inspection technique (meaning AQL) is being used as a performance standard. Zero defects is the only proper standard.
(3) There are *no* economics of quality. It is always cheaper to make it right the first time.

Again, according to Mr. Crosby, the problem of consumer dissatisfaction is the result of company management not taking quality seriously, and not recognizing that quality is *not* the responsibility of the quality department which merely measures, reports and requires corrective action. All of this is after-the-fact. Quality is the responsibility of management and even of the customer. Non-conformance *always* costs more. Expense of scrap, rework, warranty, inspection, test and other items exceeds 15% of sales for a typical company. Normally no more than 3%

should be expended if management concentrates on defect prevention and appraisal, and most of that goes for measurement.[8]

Assistance in Budgeting

The wall of many managers in industries is that they fail to receive sufficient funds for operating their departments at maximum effectiveness. Consequently, unreal budget requests are submitted to management in hopes that, after "butchering", enough money will be available to allow them to at least operate adequately. On the other hand, top managements have experienced, at considerable costs, the initiation of quality systems and purchase of instrumentation that were supposed to upgrade product quality and increase profit, yet the intended monetary benefits never materialized. The reasons are many and varied. It may have been due to the usual lament "I didn't receive management's full support", or possibly it was just an ill planned venture. Then too, it could have been that the program had no visible and meaningful method for measuring quality cost reduction.[9] An effective quality cost analysis program can prove the need for quality investment, and document the progress being made toward cost reduction and profit enhancement.

Procurement Quality Cost Payoff — Joy Manufacturing Co.

The experience of the Michigan city plant of the Air Power Division of Joy Manufacturing Company is an example of the cost reduction potential that is there, and the practical results that can be obtained. In March 1969, Joy Manufacturing Company started an extensive program with their chief suppliers, having a goal of reducing costs to both Joy and its suppliers of at least $200,000 within the next year. It took a lot of frank and candid communication between Joy and its suppliers about practical standards and inspection methods, test results and defective classification. Throughout the next year, Joy Manufacturing and its suppliers maintained detailed records of their individual problems and the cost of handling them. At the end of the year, the visible savings that had accrued to Joy Manufacturing and its suppliers had reached the fabulous sum ot $529,000. Remember, this was only the *visible* documented cost reduction. It is believed that there is an approximate 2 to 1 ratio applicable to this type of operation, and if so, the total of the visible and hidden benefits to Joy Manufacturing and their suppliers amounted to well over one and one-half million dollars in one year. [10] In order to gain these savings, work was concentrated strictly according to Pareto principles of going where the gold is; meaning, tackling the problems that contributed most to quality costs.

Warranty Cost Prediction — Chrysler Corporation

Another company that has made considerable progress using quality cost analysis techniques is the Chrysler Corporation. In the automobile industry, warranty costs make up a considerable portion of their external failure costs and of their total quality cost experienced. The inherent delays after introduction of a new car model until sufficient data has been received on their actual warranty cost trends could be devastating to profit pictures in cases of serious problems. Not only are delays due to length of time before the troubles develop within the automobiles, but also due to inventory build-up before initial sales for a new model or special sales programs. Then too, products are not sold in the same sequence that they are manufactured, which can create a false sense of security with budget shattering results. Chrysler turned to warranty prediction, therefore, using a stringent outgoing quality audit to evaluate characteristics and specifications of finished units which would likely contribute to customer dissatisfaction. Keeping track of numerical values assigned to each audit to indicate a degree of acceptance (and separating all data by month of production), accurate lifetime expenses were isolated and used to develop correlation curves between audit results and acutal warranty expenses. Chrysler experienced an 87% degree of correlation, and uses these audit results to effectively predict actual warranty expenses for all models produced in their various plants for each month of production. Such a high degree of correlation provides an effective means of emphasizing subsequent field failure costs due to current practices to operating plant management *while they can still do something about it.*[11]

Motorola Inc. Cost of Quality Program

According to Mr. Adolph Hitzelberger, Manager of Quality Control, Motorola Inc., companies that do not know their true cost of quality are often surprised to find it runs about 10 to 11% of sales. On the other hand, the industry average for commercial operations with a cost of quality program is 4% of sales, which breaks down to scrap 0.4%, rework 0.6%, warranty costs 0.5%, and Q.C. costs 2.5%. Thus, any company that is running 10% of its sales dollar in quality costs and reduces it through a cost quality program to 8% can automatically increase its profits 2%. In addition, the company's customers are likely to benefit from a better product. *One simply does not reduce quality costs by reducing product quality.* Three truisms underline a successful program to reduce quality costs: (1) *Excessive quality costs are caused by defective products.* The reduction of these costs to the economic minimum is possible only if the product is designed right, manufactured correctly, and serviced properly. (2) *The most useful common denominator for quality costs is the dollar.* It is necessary to cost out all categories where direct dollar loss is incurred and estimate the indirect dollar losses because of faulty product. (3) *The detailed elements making up quality costs are not distributed evenly.* Ranking all elements from the largest to the smallest and separating them into two categories — the vital few and the trivial many. This assures maximum return by improving specific elements. It is *not* necessary to have the cost of quality in exact terms to get a quality cost analysis program started. Pulling out the known items and estimating the rest should be sufficient to identify specific areas to management and accounting, and thereby solicit their help in developing a meaningful report. The preparation stage can be done with or without a management commitment. However, if the Q.A. Manager has done his homework, it should not be difficult to get active management participation. Note the word "participation", not "support". Support says, "Go ahead and try it, but don't bother me too much." Participation is just as the word implies; "getting with it" to : (1) Set a goal. . .to make the company's product the quality standard of the industry. (2) State a policy. . .which should have no number and should stipulate product performance exactly as the requirement, or cause the requirement to be changed officially to what the company and its customers really need. The "no number in the policy statement" is significant. A number would indicate a tolerance of defects, and people would start fishing immediately for the bosses' tolerance level. They would soon find out how far beyond the stated level they can go without unpleasentries. (3) Set a standard. . . An activity centered around defect prevention, not fire fighting. In this case, the standard is the "quality charter".[12]

Defense Against Backlash

An effective quality cost analysis program will do one other significant thing for the Quality Assurance Department. It will prevent you from suffering a penalty as a reward *for improving quality* performance. Look at a situation this way. Assume you launch a vigorous quality improvement campaign. Your Q.C. analysts identify the operations producing the most defects and indicate the probable causes; quality and manufacturing engineers advance against the sources of poor quality; they are successful in substantially reducing the problem areas. Now it is a year later, and guess where you are? You are on the carpet trying to explain why the ratio of quality control to manufacturing cost has *increased.* Yes, that is right. As a result of your efforts, scrap has been reduced, saving production labor; scheduled emergencies have been minimized, reducing production overtime. Therefore, manufacturing costs have been reduced considerably, but your costs are where they were in the beginning of the drive. You look worse, costwise, than when you started! It is important to *remember that money not spent is not recorded,* so when you set out to improve quality, and incidentally to reduce manufacturing costs, make sure you keep a record. Know your starting point: for instance, the average costs of rework and scrap for a given product line, and the average rejections per month. Then keep track of the number of subsequent rejections which require rework or scrap, and put a dollar figure on each. As your rejections are decreased, you can document the savings to the company as they accrue. The start figures may be the average of the previous six months ahead of your quality improvement campaign. Only by keeping such records can you counteract the paradox that the better you operate quality-wise, the worse you may appear to be costwise.[13]

Quality Versus Profit

Now let's return to the problem posed at the start of this discussion, and consider the views of R. J. Hendrickson, Chief of Procurement, General Dynamics, Convair Division. How does one sell the concept of quality to those outside the domain of the quality professional? The traditional quality sales approach has met with about as much success as trying to bowl with a tennis ball. In an across the board analysis, how many managements, outside the realm of the quality organizations themselves, have grasped the significance of the quality control concept. How many religiously practice it, how many go through the motions of practicing quality control only for propaganda purposes or because the contract requires it. Let's face it, to the average industrial manager, *Quality Assurance* (or whatever it may be called) *is considered a necessary evil and parasitic by nature.* Consider semantics and how the name has changed over the years. Starting with Checkers, Inspectors, Inspection, Quality, Quality Control, Quality Engineering, Quality Assurance, Reliability Control, Reliability Assurance, Production Verification, Product Assurance, ad nauseum. The name changes go on, but the scent is the same to the inner-circle outsider. We tend to make things too expensive, to nitpick, find fault, and all the while defective products *still* get through.

We might be better off if we quality professionals discarded the word "quality" completely from our title, and select instead, the term "Profit Assurance". Isn't this really the objective of the technology of which we claim to be a key professional element? Then why not stake our claim and let the world know our real identity? It is about time we do, instead of wasting half of our time defending the ambiguous title we are now wearing. The key to survival in the new role as profit assurance professional will require identifying parasitic profit consumers, and then reorganizing those with the biggest appetites. We cannot spend most of our allocated budget chasing gnats, while buzzards feast or we will have to scurry for *another title to hide under.* A total Quality Cost Analysis program is an ideal start to attain the level of management encouragement and participation that we never have experienced before. The budget now can be effectively reduced to its proper perspective by obtaining the numbers that relate to the company's profit loss, either by program or time period, as a total consequence of non-conformances and failures. These numbers, when compared to those required to recapture a significant portion of the company's projected profit loss, make the potential value of the proposed quality effort understandable to almost any responsible business manager.

Insurance Policy

An indication of whether the quality manager has arrived will have occurred when enough confidence has been attained to stake the works for the next year or program on a simple sporting proposition like this:

> *Insurance policy.* For the sum of ______ dollars for 19____, the Profit Assurance Department will insure the recapture of ______ dollars of profit normally lost through non-conformance/failure consequences by the John Doe Company. Signed I. M. Jones, Manager, Profit Assurance.[14]

Summary: Profit Payoff

The experience of other companies has documented the profitability of the total quality cost analysis program. It has not been uncommon for companies to reduce their failure losses by about $9.00 for every $1.00 invested as a quality appraisal expense, and reductions of up to $15.00 in failure losses have been experienced for every $1.00 invested in quality prevention expenses. The actual figures, of course, are going to vary widely from company to company. To a large extent, they will reflect the starting position, as well as the effectiveness of the program and personnel involved. More chief executives are going to reflect the attitudes expressed by Mr. David C. Scott, President and Chairman of Allis-Chalmers, "We cannot afford the extra capital cost that quality problems can create. . . *Quality is one of the biggest untapped profit producers of most American companies.*"[15] Quality Cost Analysis permits changing losses directly into new profits. Profit improvement of $500,000 can be equivalent to increased sales from 3 to 10 million dollars. *Gold waiting to be mined!*

REFERENCES

(1) "Systems Analysis in Management", D. Burchfield, *Quality Progress,* January 1970, page 18.

(2) *Quality Costs — What & How,* Quality Costs Committee, ASQC, Milwaukee, 1967, page 55.

(3) *Total Quality Control,* A. V. Feigenbaum, McGraw-Hill, New York, page 84.

(4) Ibid, page 89.

(5) *Quality Costs — What & How,* Quality Costs Committee, ASQC, Milwaukee, 1967, pages 41-42.

(6) Ibid, page 48.

(7) Ibid, pages 73-75.

(8) "U.S. Quality. . . Major National Concern", *Quality Management & Engineering,* August 1971, page 13.

(9) "Semper Paratus. . . How About You?", L. M. Walsh, *Quality Management & Engineering,* March 1971, page 10.

(10) "Remember 200K? Would You Believe 529K?", R. A. Maass, *Quality Assurance,* October 1970, pages 24-26.

(11) "Warranty Prediction: Putting a $ on Poor Quality", V. P. Burns, *Quality Progress,* December 1970, pages 28-29.

(12) "Having $ Problems?. . .Try These Cost Controls", A. Hitzelberger, *Quality Assurance,* April 1969, pages 24-26.

(13) "Can You Prove You Did Any Good?", F. H. Squires, *Quality Assurance,* May 1969, pages 8-9.

(14) "Let's Forget About Quality. . . Let's Go For Profit", R. J. Hendrickson, *Quality Management & Engineering,* March 1971, pages 20-21.

(15) "The Quality World of Allis-Chalmers", D. C. Scott and W. F. Schleicher, *Quality Assurance,* December 1970, pages 10-17.

COST EFFECTIVENESS OF CORRECTIVE ACTION

Tutorial: QUALITY COSTS — A PLACE FOR DECISION MAKING AND CORRECTIVE ACTION

Richard K. Dobbins, Principal Quality Assurance Staff Engineer
Honeywell Inc. — Process Control Division/Ft. Washington, Pa.

PROBLEMS AND RESPONSES

Text Omissions

Not too long ago, Frank H. Squires dedicated one of his "On The Quality Scene" columns to the omission of the term "Quality Control" and allied subjects from texts on modern management systems. To be sure, most management books fail to include any significant references on Quality Control, Inspection, Product Quality or Statistical Techniques. This is an unrealistic approach to industrial management in ignoring the common, everyday occurrences which cry out for professional quality control techniques. This paper, therefore, is dedicated to the dollar effectiveness of corrective action. Shads of Robert S. McNamara, that sounds impressive! But what does it mean?

The idea of system effectiveness was promoted under McNamara when he was U.S. Secretary of Defense, and stood for:

> "availability plus reliability, plus the cost of maintenance and all relevant cost for acquisition and use".

Cost was an element of each one of the conditions included, and thus system effectiveness and cost effectiveness became interchangeable. Bear in mind that *optimum conditions* are being sought, and *not necessarily minimum costs.* In other words, it is "getting the biggest bang per buck."

Now, what about the other half of this paper's title? *Corrective action. . .everyone* knows what that is. Surely every quality professional knows what that is. (At least, let's *hope* that quality professionals know what "corrective action" means.) Unfortunately, like the omission that Frank Squires reported about management texts, the term *"corrective action" is not defined nor adequately described* in some of the major texts on Quality Assurance and Quality Control. This is really paradoxical, for corrective action should be a Quality Control Engineer's "bag". . .the "thing" which really "turns him on"!

In an attempt to clarify the issue, I have borrowed some definitions from the Kepner-Tregoe method of problem analysis (with slight alterations from me to emphasize Quality Control interests). In simplest terms, a desired cause/effect relationship might be diagrammed as a straight line, where the end result is directly attributable to the initiating effort. If an unplanned influence intrudes before the process is completed, a deviation from the desired end result will occur, which is called "a problem". (See Figure 1)

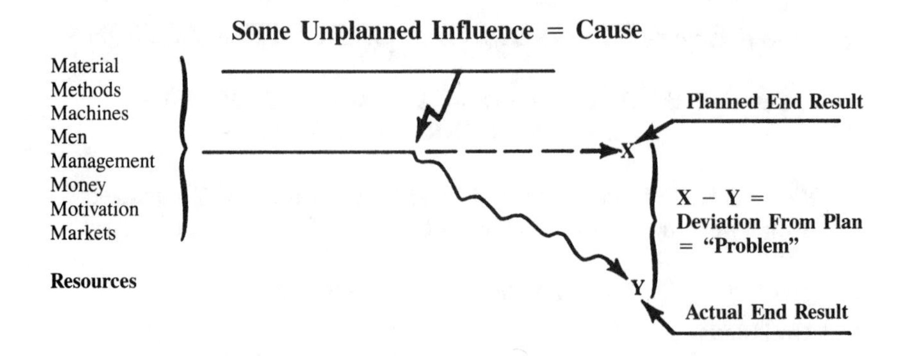

FIGURE 1 — ANATOMY OF A PROBLEM

Action Response

Action taken in response to (and treatment of) this problem is usually known as "corrective action", but I would like to make this general term a little bit more specific. The action response may actually be one of three different types of action, as follows:

1. *Interim Action — Action taken to "buy time" before permanent corrective action can be taken.* This is usually a stopgap measure and can be quite expensive. It permits a manager to keep the operation going under crisis conditions, and is usually the first response to any new major problem. Quite often, interim action is taken before the cause of the problem has been found, let alone a cure for the cause.
2. *Adaptive Action — Action taken to compensate for the effects of the problem,* in order to produce a tolerable result, even though different from the original, planned result. In other words, adaptive action enables one to live with the effects of a problem, and to minimize the undesirable aspects of same. This action may be the only practical alternative to an ineradicable cause, or to highly unfavorable economic situations. Adaptive action is usually taken after the cause of the problem is known and analysis of the various alternatives has been completed.
3. *Corrective Action — Action which gets rid of the known cause of the basic problem.* It eliminates the deviation by eliminating the cause that produced the problem, and as a general rule, is the most efficient action of the three types discussed here. *It is only possible, however, where the cause is known.*(1)

In particularly nasty situations, true corrective action is the epitome of "control" in the term "quality control" and can produce astounding economical results. It should be noted that interim, adaptive and corrective actions all refer to responses to existing (current or past) problems. There are two other action responses which we should include in our definitions for sake of completeness. These are:

4. *Preventive Action — Action which removes the possible causes of future problems,* reduces the probability of their occurrence or the magnitude of their influence.
5. *Contingency Action — Action which provides for standby arrangements* to offset or minimize the effects of a serious potential future problem.[2] Both preventive and contingency actions concern problems not yet experienced, but deemed to be a serious potential risk. Preventive and contingency actions associated with possible recurrence of existing problems are included under the definition of "corrective action", as used within this paper. Let us pursue this further.

A problem cannot be solved unless its cause is known, and failures in problem solving trace back to that basic fault. A problem is an unwanted effect, something to be corrected or removed. It was brought about by some specific event, or combination of events. To get rid of the effect efficiently and permanently, you must know how it came to be. Any decision based on a false cause is going to be ineffective, wastefully expensive, and sometimes downright dangerous. Every problem has only one real cause. It may be a single event that produces the unwanted effect, or it may be a combination of events and conditions operating as if they constituted a single event.[3]

Is It Operator-Controllable?

Normally we think of various defects according to common terms relating to their classical origin, such as: design; workmanship; material handling; etc. I suggest that it is more pertinent to effective corrective action to classify them as either "operator-controllable" or "management-controllable".

A defect is operator-controllable only if means have been provided so that the operator:

1. *Knows what he is supposed to be doing.*
2. *Knows what he is actually doing.*
3. *Can regulate the process.*

Lack of any one of these three prerequisites makes the defect *management-controllable.*[4] And should be so recognized in determining root causes of problems.

Now, back to basics. Lest we fall prey to the Madison Avenue influence, let's define "quality", again. In simplest terms used by ITT Corporation, *"quality" means conformance to requirements.* There are no economics of quality. There are no levels of quality. A product either conforms to requirements and is a quality product, or it fails to conform to requirements and fails to be a quality product. Do not confuse the term quality with degree of goodness, power, shininess, number of features, durability, luxuriousness, etc. All of these may be various aspects of performance levels and tolerances established in appropriate specifications. *Quality means conformance to those specifications. Quality is a noun, not an adjective.*[5]

Three additional definitions will be useful in considering our chosen topic for this paper. These definitions pertain to the principal functions of Inspection, Quality Control and Quality Assurance. As used herein:

1. *"Inspection" means appraisal or measurement* to specifications.
2. *"Quality Control" is quality engineering,* which is primarily concerned with the detection and correction of the causes for the defects, and, most importantly, the prevention of the production of defects.
3. *"Quality Assurance" is the management of Inspection and Quality Control,* plus the coordinative dissemination of quality information throughout the organization, with the intent of involving all levels of managment in a profitable quality program.[6]

GENERAL MANAGEMENT RESPONSIBILITIES

Quality with Profits

You noticed, of course, the word "profitable" in the last statement. Should profit be included in a definition of Quality Assurance? Most assuredly, it belongs there! Remember, *all managers,* regardless of their title, function within the company, departmental affiliation or professional discipline, *have three basic responsibilities.* These are *costs, schedules and quality.* A good manager cannot ignore nor slight any of the three. With the proper balance of good cost controls, together with dependable delivery schedules and reliable quality conformance to established standards, respectable profits should materialize. Although the Quality Manager is undoubtedly going to be biased in favor of quality (as he should be), he must *not* let his quality zeal result in drives for absolute perfection or better-than-specified performance standards. If the established standards are not good enough, get the standards changed officially and publicly — and then enforce them energetically!

Now, let's try to think specifically in terms of cost effectiveness of corrective action. A *quality cost analysis* categorizes all operating quality costs into one of four broad categories: *Prevention; Appraisal; Internal Failures;* and *External Failures.* The first two categories may be thought of as quality expenses, meaning dollars invested to avoid or control losses. The last two categories may be thought of as quality losses, meaning dollars wasted because perfect control does not exist. Internal failure costs pertain to operating manufacturing losses experienced before the product is shipped, while external failure costs pertain to losses experienced as a result of defective product being shipped to the customer.[7] *Both internal and external failures may be considered losses due to the cost of nonconformance to requirements.*

Now for the "hooker" in applying these definitions! It was stated that quality means conformance to requirements. Ideally, the requirements should be in the form of a written specification, with suitable definitions, means of measurement, and acceptable tolerances. Barring interpretation and measurement difficulites, conformance on items so specified should be able to be readily determined. Attributes not mentioned, being unspecified, are vulnerable to widely varying, subjective interpretations. *External failure costs* present an even greater problem, since the *requirements standard now consist of what the customer believes he has been promised, and by what is currently provided by competitive sources.*

A Chief Executive Speaks

Consider the views recently expressed by Mr. Harold S. Geneen, president, chairman of the board and chief executive officer for International Telephone and Telegraph Corporation for a candid, unusual admonishment for his executive management peers.

> "There isn't really much that the most sophisticated technical quality control department can do in terms of creating quality itself, because they only look at the product or service when it is a partial or complete entity. Thus, the people who are charged with upgrading the specific quality are those who have nothing to do with its creation. *The people who cause this situation are ourselves — top management. We are also the ones who can correct it".*[8]

The problem for the Quality Assurance department is to diplomatically, but firmly, get these facts before executive management, and to initiate the corrective action necessary. Here is the rest of the picture as described by Mr. Geneen:

1. Operating managers understand specific standards (expressed by that company's top management) from a practical standpoint. They know that the most important thing to top management is profit or loss. They don't *really* believe they are expected to make a profit and still effect "quality". They believe that "quality" is more expensive.
2. Top management refers to each non-conformance as a "quality problem". This phrase lets everyone else off the hook since it implies that the quality control department has been inadequate. Yet, it is not common to refer to a lack of profit as a "finance problem", or the inadequacies of a specific manager as a "personnel problem". . .*Defining all non-conformance as a "quality problem" provides the operating managers with a built-in excuse.* They then can plead an act of God, physics or an inadequate Quality Control manager.
3. *Operating managers* know they cannot get into deep trouble for creating non-conforming products or services. They will be frowned at for these difficulties, but they can be really put down only for profit loss. Therefore, they *concentrate on financial and schedule matters. Quality is third.* They may knowingly deliver defective material to the customer in order to meet billing commitments. They know that they will smooth this over somehow with the customer, and the material return item can be buried to be dealt with later.
4. Quality improvement efforts usually are accomplished on a limited scale and at a low level in the organization.[9]

Another point not mentioned by Mr. Geneen is that most *Quality Assurance departments* "spin their wheels" in routine matters and items of trivia, and *do not make the time to attack the root causes of the largest quality losses.* Now, What can be done about it?

QUALITY ASSURANCE OPERATING STRATEGY

Overcoming The Adverse Standard Operating Procedures

Face the issues squarely! Strip the subterfuge from all existing reporting to executive management, and tell it like it is! A credible quality cost analysis program is a most plausible way to convince your executive superiors about the present situation. Expose the fact that 100% inspection in certain places is required to provide a screening operation for inefficient manufacturing methods, or too stringent engineering requirements. *Tackle your major problem areas according to Pareto analysis.* This is sometimes known as the 10-90 effect, since it recognizes that approximately 10% of the things that may go wrong in a given situation are likely to account for 90% of what actually does go wrong.

Stop doing other people's work for them, and concentrate on your major responsibility of assuring quality conformance. Recognize and identify the interim, adaptive and corrective actions that are taken in response to identified internal and external failures. *Insist upon the formality of Material Review Board action* to dispose of non-conforming material which cannot be satisfactorily worked to original specifications. Work with your Accounting department to enable chargebacks of scrap, salvage and rework costs to the appropriate responsible operating departments. Investigate (again, if necessary) the principal loss leaders of quality costs. Recognize problems as deviations from the desired cause-effect relationship, and apply the appropriate action response to same. This may be done both with the identified internal failures (such as scrap, rework, reinspection, etc.), and with external failures (such as warranty costs, field retrofits, shipping damages, etc.). The same principles may also be applied to the mangement functions of Quality Assurance, as can be demonstrated by thoughtful consideration of the following questions.

Questions for the Quality Manager

How is your quality assurance budget being expended? How *should* it be spent to materially upgrade performance and profitability of your department and your company? Are internal failures charged back to the responsible operating department? Do they show on the functional budget of such department? *(Shouldn't they appear there?)* Are external failure costs charged back to the operating department that produced them? Probably not, if your company is like most. But, *shouldn't they be so charged back?* If they were charged back, would that cause a product line to appear unprofitable, or barely marginal? (If so, isn't that the *proper* financial assessment for that product line?)

Are operating goals established with respect to both internal and external failures chargeable to a given product line, or are such costs merely compared with the previous year's performance? If so, this builds up an ever increasing "threshold of pain", for *everyone* knows that the cost of living (and business operations) goes up each year, and this year isn't that much worse than last year. (*Or is it?)* Is your operating system good enough if such internal and external failure costs are routinely being experienced without being challenged? Aren't quality failure costs just as much a part of the product costs as the direct labor and material that went into the product intentionally? Should you be content with generating losses that are about the same as last year? What loss levels appear reasonable for your new products? Might it not be practical to add to existing in-plant testing (and thus direct production costs) in order to reduce internal or external failure costs? Is a field retrofit program the possible answer to an existing field failure problem?

The answers to these questions could be determined using the approach of failure analysis to ascertain the root causes of major problems, and a study of the various alternatives open for interim, adaptive and corrective action. As a manager, you must *consider the cost aspects according to the total cost involved.* As shown upon the accompanying graph (see Figure 2), this includes both the cost of *delivered quality* (which gets higher as you approach the higher in-plant confidence levels), *added to the cost of field correction of defects* (which becomes smaller as the in-plant confidence levels are raised). The total cost curve, being the composite of the other two, has a very significant dip and reaches a minimum somewhere in the neighborhood of the crossover point between the cost of delivered quality and the cost of field correction. In the wide range to the left of the crossover point (zone 1), it is more economical to screen and rework items within the plant in order to prevent grossly deficient products from reaching the field. In the narrow range to the right of the crossover point (zone 3), field failure costs are kept at

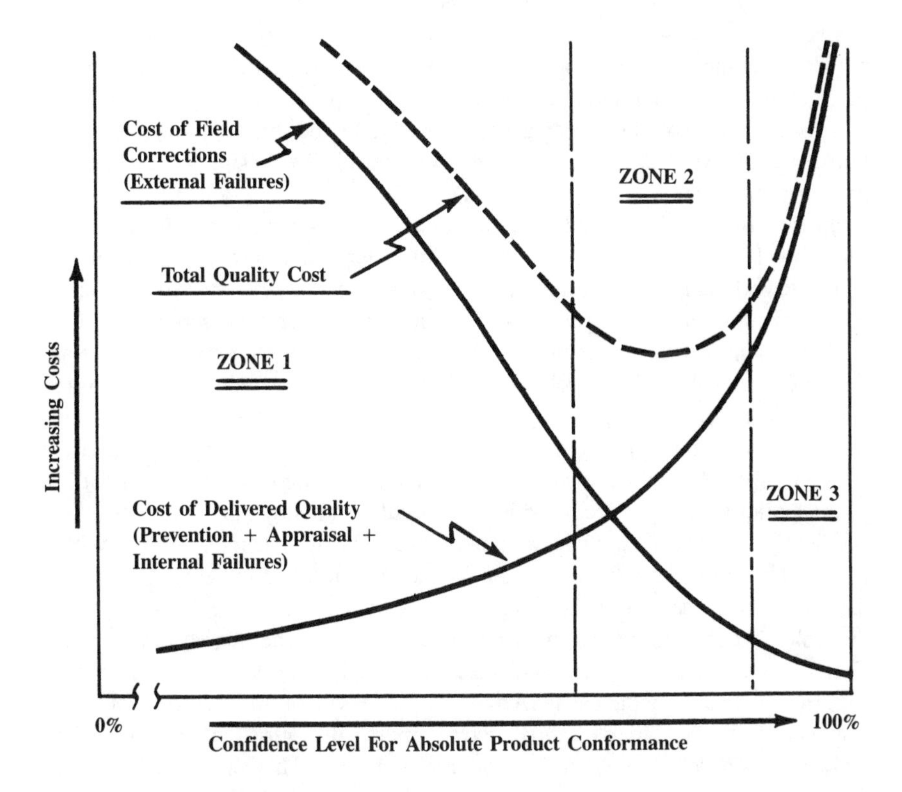

FIGURE 2 — QUALITY COST ZONES

Zone 1: Commercially Non-Competitive Due To Excessive Field Costs.
Zone 2: Optimum Quality Cost Zone; Most Aggressive Competitive Position.
Zone 3: Commercially Non-Competitive Due To Excessive Internal Costs.

a low level, but only due to the nearly exorbitant cost of delivered quality in-house. *There is a small band near the crossover point (zone 2) where optimum total quality costs are achieved.* When operating within this range, internal unit quality costs will still be reasonable, as will the cost of providing prompt and courteous service on those products which develop a field performance problem.

QUALITY ASSURANCE PROFIT IMPROVEMENT

Problems or Opportunities?

By continually analyzing the leading internal and external failure costs and performing the required failure analysis to determine root causes, a constant variety of opportunities will be presented to materially reduce quality failure costs. For instance, look at your Receiving Inspection records. Do you have some parts which have had a history of *repetitive rejection* for non-conformance to particular criteria, only to have these products *accepted on deviation?* Why not establish a policy that will only permit such acceptance two times on a temporary deviated basis? In order to accept such parts a third time with the same non-conformance, the engineering drawings or specifications *must be permanently changed* to permit a wider tolerance. This approach will help to reduce the credibility gap between Engineering specified tolerances and actual product requirements. This approach will also result in speedier acceptance and availability of products, fewer reinspections, reduced M.R.B. costs, and might even permit Purchasing to obtain better pricing consideration from the vendor.

On the other hand, are there some parts which have gone through repetitive rejections for non-conformance to particular criteria, and have required sorting, and/or reworking to specification by either yourself or the vendor? *Have you made a process capability study yet?* From your next rejected lot in such a case, draw a random sample of, say, 50 pieces, and measure and tag each piece with the actual measured value of the problem attribute. Then plot a frequency distribution curve for the random sample, and determine the plus and minus three sigma limits for the process and compare them to the tolerance limits for the attribute. If the six sigma spread is *greater* than the total tolerance allowed by your specification, your supplier is using a process which is *incapable* of holding to the proper limits, and both you and he should be aware of this fact. The displacement for the average of the distribution from the nominal attribute measurement indicates the amount that the setup for the process missed the nominal value. The degree that the distribution follows expected normal distribution will indicate the degree that the supplier's process is varying due to chance causes only, as opposed to distribution patterns due to assignable cause. Armed with that type of information, your Q.C. engineer should be able to establish good rapport with your supplier in determining the root cause of the repetitive non-conformances. This should enable effective corrective action to be taken by the supplier, thus *preventing avoidable future quality costs* and minimizing disruptions to delivery schedules. Now, why not do the same thing with respect to your other internal failure cost generators? Such an approach will take the mystery out of the "in-again, out-again" troublesome production jobs with considerable economic rewards.

Use Existing Records

By going after existing and documented internal or external failure costs, and using Pareto analysis to concentrate on those that are most significant, the cost effectiveness of any resulting corrective action can be most readily proven. You already know the cost of non-conformance in the past for at least as many incidences as are documented. Comparing this to production rates, it is possible to determine the approximate cost per hundred units for the non-conformances which you have been experiencing. If truly effective failure analysis and corrective action can be applied to reduce your losses to one tenth or one third that experienced previously, *you now have a documented case of quality cost avoidance.* Note that this money comes right off the top! It transfers losses directly into profits! There is also a magnifying effect of such reversals, because your present production problems and low yields require higher-than-normal production runs and inventory build-up in order to meet the demands of your production pipeline. Improving the yield on the parts initially, and avoiding in-plant failures at higher assembly levels frees men, machines and dollars for more productive purposes. In dealing with external failure costs the magnification factor is even greater, because the spare parts pipeline now has to stretch all the way out through the distribution chain of your worldwide markets and into your customers' storerooms.

Quality Management Opinion Polls

Results of recent opinion polls conducted by Quality Management and Engineering indicate that 70% of the respondents believe that design tolerances frequently exceed practical requirements.[10] Another poll reveals that 61% reply that their management has not explored (or makes no effective use of) Quality Assurance data in relation to cost effectiveness in the manufacturing cycle. These poll results suggest the presence of "*walls of indifference*" between Design, Manufacturing and Quality Assurance. Only 10% of those polled felt that quick and effective action resulted from Q.A. recommendations for improvements in production machines and processes. 81% felt that conditions varied from "no action" to "corrective action being initiated too late".[11] Similar walls of indifference exist between Marketing, Engineering and Quality Assurance. This is typified by field failures occurring on product lines which do not experience similar difficulties during their in-house testing and production cycle. *Differences between real world practical applications* of your product *and the original Engineering design specifications might become evident as a result of diligent failure analysis of field failures.* Adequate control of such performance attributes can only be achieved once they are identified and specified in measureable terms.

According to Mr. Harold S. Geneen,

> "The Quality Department should *not* be held responsible for quality. Its job is to measure quality, report it, and worry about it! But in reality, they do none of the functions that effect it! Don't let yourself be caught in a commitment gap. The chief executive, and he alone, establishes the personality of a company's management. If the product of service

is not what the customers think it should be remember, *the output looks exactly like the management.*"[12]

SUMMARY

Management Commitment

Now, Quality Managers, do your own thing! Obtain that commitment from your executive superiors, and use cost effective corrective action as a means to gain top management's attention and command their respect. If you demonstrate that Quality Assurance can be a dynamic, positive influence upon corporate profitability, you will find that your company's position will change to assure equal billing of quality considerations together with cost and schedules. Let's get back to the basics — *Quality means conformance to requirements.* A significant portion of your *quality costs are generated by a limited number of causes.* Application of appropriate corrective action will generate quality cost improvements.

No longer should you require an interpreter to talk to executive management, for *you have just learned to speak their language fluently — DOLLARS!!!*

REFERENCES

(1) *The Rational Manager,* C. H. Kepner & B. B. Tregoe, McGraw-Hill, New York, 1965, pages 132 and 175-176.

(2) Ibid., pages 132-133.

(3) Ibid., page 17

(4) *Quality Control Handbook,* J. M. Juran, Second Edition, McGraw-Hill, New York, 1962, page 11-7.

(5) "Fourteen Steps to Quality", H. S. Geneen, *Quality Progress,* March/April 1972, page 33.

(6) "What's the Difference Between Quality Assurance and Quality Control?", F. H. Squires, *Quality Management & Engineering,* August 1971, pages 20-21.

(7) "Quality Cost Management for Profit", R. K. Dobbins, *29th Annual Technical Conference Transactions,* A.S.Q.C., May 12-14, 1975, page 359.

(8) "Fourteen Steps to Quality", H. S. Geneen, *Quality Progress,* March/April 1972, page 33.

(9) Ibid.

(10) "Jungle of Gibberish", L. M. Walsh, *Quality Management & Engineering,* June 1972, page 6.

(11) "When the Walls Come Tumbling Down", L. M. Walsh, *Quality Management & Engineering,* March 1972, page 8.

(12) "Fourteen Steps to Quality", H. S. Geneen, *Quality Progress,* March/April 1972, page 35.

QUALITY COSTS — A PLACE ON THE SHOP FLOOR

Edgar W. Dawes — Quality Control Manager
Digital Equipment Corporation — Westfield, Mass.

INTRODUCTION

Beginning in the 1960's, product fitness for use has commanded more and more attention of business enterprises. Managers in these enterprises realize that, for their companies to improve profits, they must be increasingly effective in their response to customer quality opinions.

But with the ever mounting complexity of goods and services, the customer's quality expectations have also risen. Not only do they consider value received for acquisition cost (original purchase price), but they carefully evaluate the added logistic cost of their purchase. J. M. Juran and F. M. Gryna define logistics cost as those costs "related to putting and keeping the product in service. Both acquisition and logistics costs form life cycle costs which vary as a function of the product's fitness for use".(1)

This concept has prompted some constructive discussions in industry. For example should the factory and quality control managers respond to customer fitness for use as they strive for conformance quality (the maintenance of design quality by manufacturing)? To answer that question, we must present some additional cost principles.

FITNESS & USER OPTIMUM COSTS

Figure one shows the relationship between the optimum user cost and fitness for use and the elements contained in acquisition and logistic costs.

Grossly simplified, we can define acquisition cost (the cost of purchase) as containing material, direct labor, and overhead costs plus and profit for the manufacturer.

The reader will immediately (and correctly) sense the presence of a great many factors in the overhead category (i.e. factory repairs, product distribution costs, supervision costs, tests and inspections not in direct labor. . .etc.).

Again grossly simplified, we can define logistics costs (the cost of putting and keeping a product in service as containing maintainability costs, failure costs, user inconvenience costs, loss of ROI. . .etc.

Some of the reasons for customer detected failure, either upon initial product installation or in use, can be the result of:

1. Insufficient product protection in shipping
2. A life related failure of the infant mortality, steady state, or wearout variety (it should be noted that, in terms of user optimum costs, the consumer looks at product *total* life, not just warranty).
3. A customer quality expectation level higher than factory standards.

By aggressively solving these and similar customer complaints, the manufacturer can transfer substantial logistic costs to less costly acquisition costs et al to manufacturing conformance costs and markedly improve market position and company profits. The assumption is made that the manufacturer must pass the added conformance quality cost to the consumer in the form of higher acquisition cost. This may not be the case. It is the writer's opinion that "there is an efficiency level which varies from organization to organization in its attack on failure costs. Some organizations attack such costs aggressively, make progress and dramatically improve conformance quality and the cost of obtaining that feature".[2]

Juran and Gryna further state "the opportunities for manufacturers to use the concept of optimizing the users' economics are most intriguing because the effect is normally to achieve a large increase in income by a small increase in cost".[3]

At this juncture, the vital need for the factory to know the users' acquistion and logistics quality costs should be evident. However, there are other implications which are more profound. We should be sensing the interdependency of many of the major business functions (i.e. design engineering, manufacturing, and marketing) in developing a synergistic quality system, one where the total system is stronger than the sum of its individually structured parts. This becomes a total organization assignment. More especially, this is the coordination task of a skilled quality manager.

But what of factory quality costs under this broadened definition? Is there a place for quality costs on the shop floor? The answer is yes.

THE FOUR SHOP FLOOR QUALITY DECISIONS

There are four types of regularly made shop floor decisions, all of which affect quality costs. These are:

Type	The Decision Is:
Accept defectives	Incorrect
Accept conforming product	Correct
Reject conforming product	Incorrect
Reject defectives	Correct

To eliminate the incorrect decisions, there are an increasing number of techniques which can be applied. These will be developed in the latter portion of this paper.

Developing Fitness for Use

Figure two indicates that an appropriate question in each of the four decision areas might be; is the product fit for use?

Depending upon the "State of Health" of the quality system (i.e. the compatability and completeness of conformance and design qualities to customer needs), substantial changes in the shop floor quality decisions may have to be made. In a poor quality system, as much as 50% of the failure costs may be related to incorrect decisions on quality. The ITT definition of this situation says it quite well, "the product is to perform exactly like the requirement. . .or cause the requirement to be officially changed to what we and our customers really need".[4]

The Cost Traceability Concept

Let's assume that we are dealing with correctly made but extensive rejection decisions in the shop; we need to supply the General Manager with both historical records of the losses (trend charts) and a ranking of losses (Pareto-Lorenz Maldistributions). The concept of *prior* cost traceability to the major loss contributors becomes critical in properly allocating corrective action resources. This is shown in figure three.

THE CORRECTION OF INCORRECT DECISIONS

Figure two (previously mentioned) shows the decision sequence and tools which can be brought to bear on incorrect quality decisions. There are two groups of these, the first of which addresses the fundamental control expected on each shop operation. The following items should be reviewed before there is detailed economic analysis:

1. Is there sufficient precision and accuracy of measurement?
2. Do the personnel understand and properly use the measuring system?
3. Are there adequate specifications and measuring criteria?
4. Has there been proper analysis and interpretation of the data?
5. Is there clear identification and segregation of materials?
6. Is there sufficient control over the fatigue and error factors associated with high volume monotonous inspection?

Having assured satisfactory levels on those factors, we should then attempt to improve the economics of the statistical controls used with incorrect quality decisions. This part of the quality technology has both new and old techniques.

Optimum Inspection Intervals

Careful reviews of the measuring intervals on a process and a comparison to the process fraction defective and time frequency of occurrence sometimes reveals grossly insufficient measuring effort.

A more equitable cost balance between increased inspection and earlier detection and prevention of defects often solves the uncontrolled-undetected process producing too high a rate of defectives.

Economics of Control Charts

A basic hypothesis of a control chart is that the process is in statistical control. Again we encounter the classic errors of rejecting this hypothesis when it is true and accepting it when it is false. We must balance these two costs by evaluating control limits, sample size, and sampling interval.

Inspection Decision Analysis

In high volume operations containing many identical inspection points and large numbers of inspectors, analysis of inspection decision times (a decision is defined as a single accept-reject determination on the basis of sampling) can be statistically analyzed and reveals patterns related to productivity, motivation, and procedural difficulties. Cost improvement results from a resolution of these problems.

Although it is beyond the scope of this paper to present case histories related to each of the aforementioned techniques, the reader is referred to the reference and bibliography for the sources of such case histories.

SUMMARY

In summary, the concept of optimizing a company's quality costs must include *known* user optimum costs. It becomes the responsibility of the manufacturing organization to factor this cost into their conformance quality and to address and solve the problems of incorrect quality decisions, no matter what the source.

REFERENCES

1. Juran, J. M. and Gryna, F. M. Jr. *Quality Planning & Analysis* McGraw Hill Publishing Co., N.Y., N.Y., 1970 pp 47-52.
2. Dawes, Edgar W. "Quality Costs — A Tool for Improving Profits", *Quality Progress,* Volume VIII — No. 9 pp 12-13, Sept. 1975.
3. Opus Cit (1) pp 50.
4. Groocock, J. M., *The Cost of Quality,* Pitman Publishing Co., N.Y., N.Y., pp 48.

BIBLIOGRAPHY

Kirkpatrick, Elwood G. *Quality Control for Managers and Engineers,* John Wiley & Sons, Inc., N.Y., N.Y., 1970

Meske, A., "A Management Standard for Economic Inspection, *Quality,* January 1976, pp 28-30.

FIGURE 1 — Life Cycle Economics

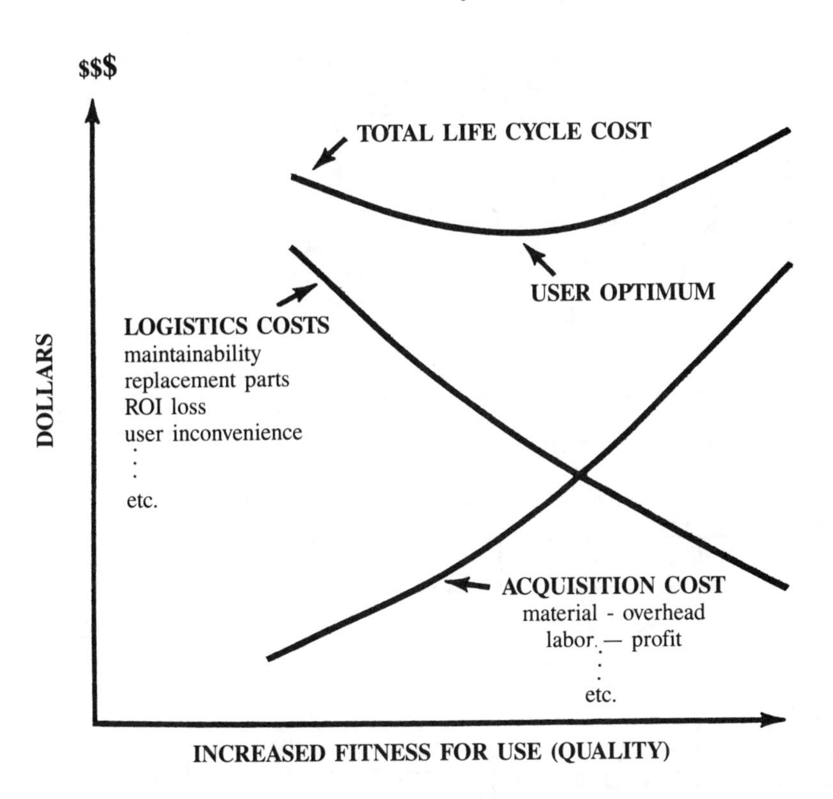

FIGURE 2 — Developing Fitness for Use and Improved Quality Decisions

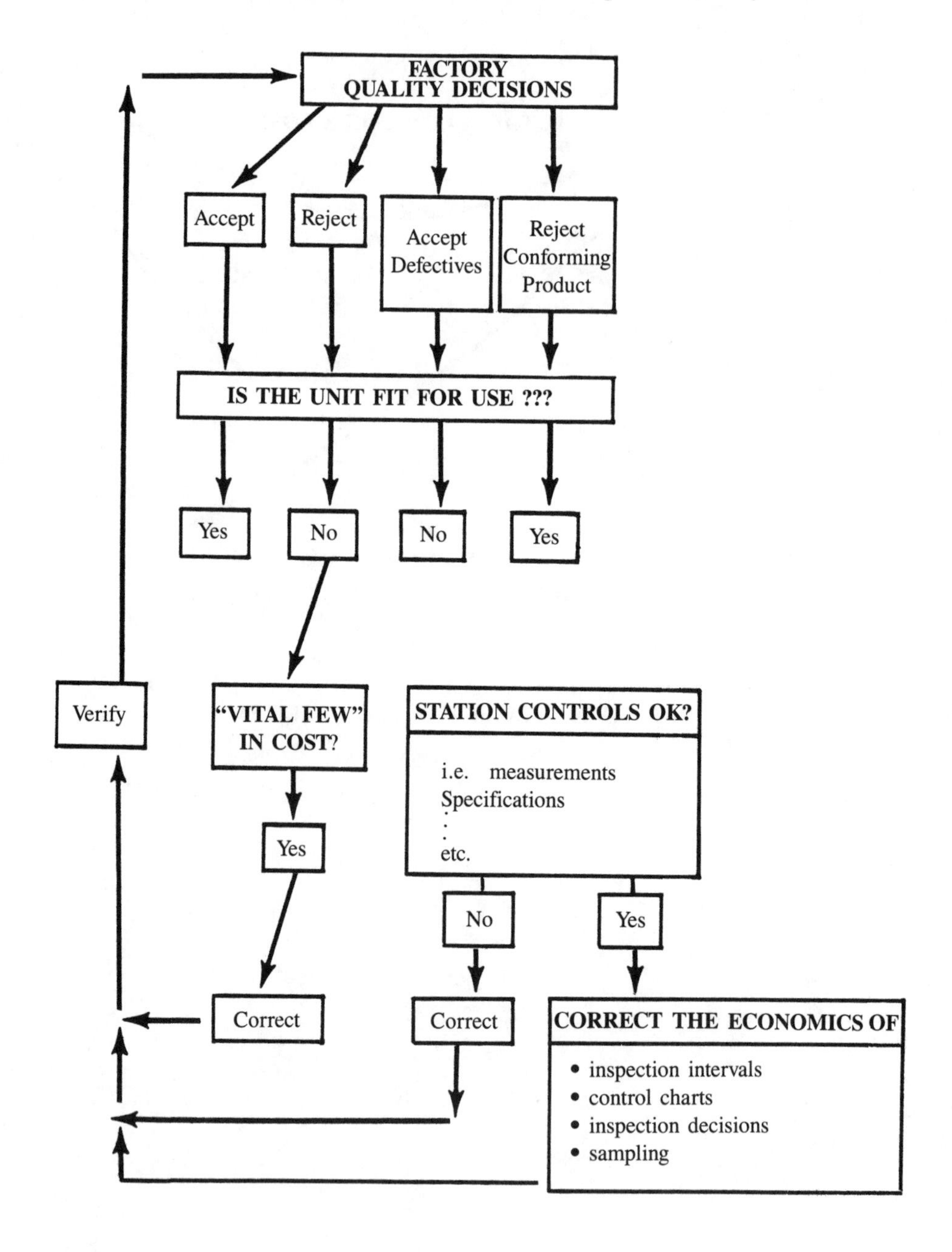

FIGURE 3 — Cost Traceability

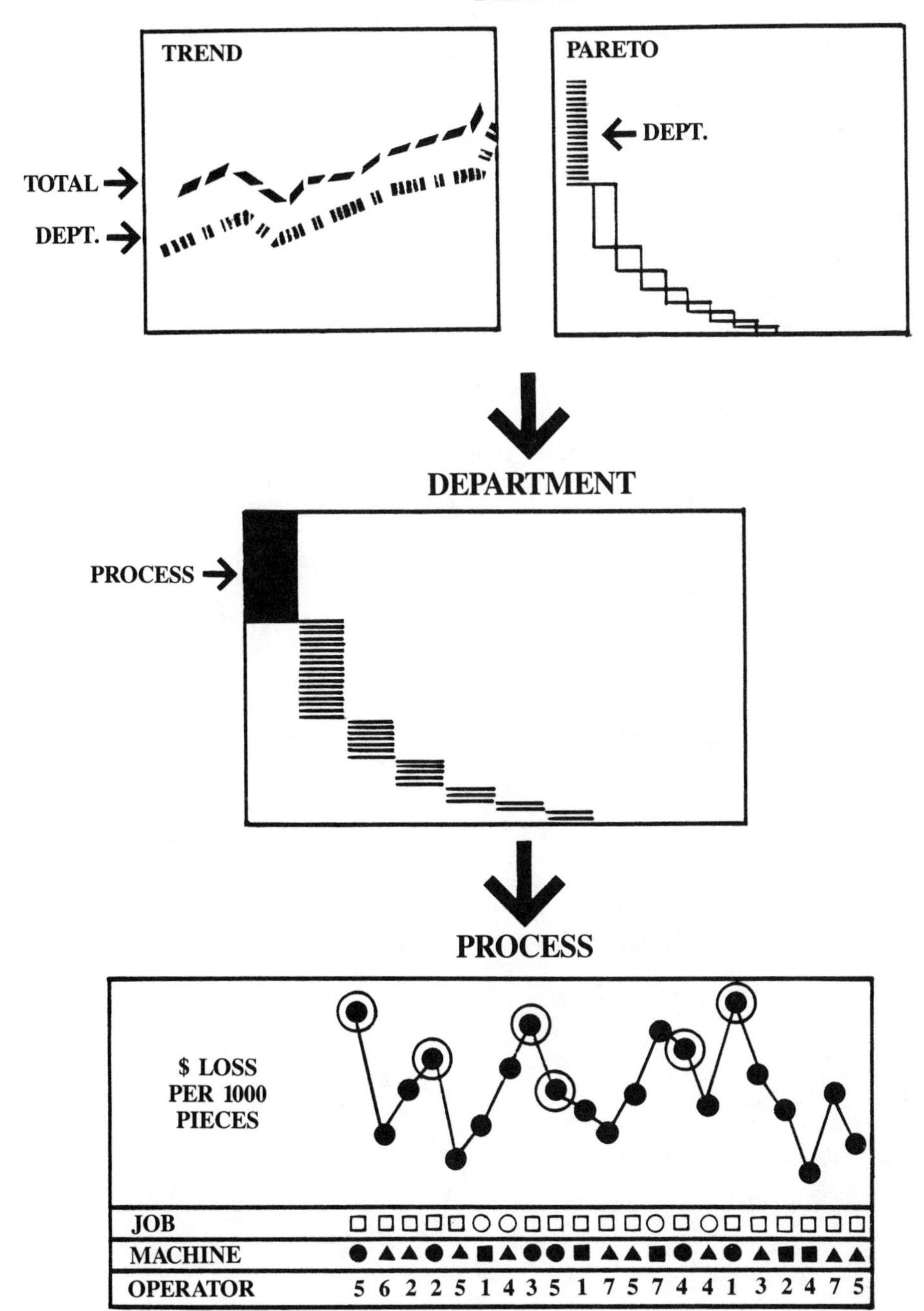

QUALITY COST — A PLACE FOR FINANCIAL IMPACT

O. G. Kolacek
Manager, Quality Assurance
Allis-Chalmers Canada Limited
Lachine, Quebec

Nations, continents and hemispheres have proved no barrier to the businessman's suspicion that Total Quality Assurance programs are too fancy and too expensive for him. Disproving this notion and at the same time charting a course any firm can follow, Allis-Chalmers Canada Limited has proved:

1) Total Quality Assurance uses its modest cash appropriations to build profits.
2) T.Q.A. keeps, attracts, and holds customers.
3) T.Q.A. almost eliminates field troubles.
4) T.Q.A. conserves interest costs by improving cash flow.
5) T.Q.A. turns Inspectors into quality prevention oriented coaches and Designers into quality planners.

The Total Quality Assurance concept involves an entire organization, top to bottom, in a mutual and never-ending effort to keep the benefits flowing. T.Q.A. is a system for making quality in plants' products the responsbility of almost everybody in the plant.

The basic goal is to improve quality and reduce cost. Therefore, the top management must be primarily concerned with maintaining the proper level of product quality at the lowest cost. Thus, the quality cost becomes the business face which the quality organization turns to top management in order to provide the most significant measurement and administrative tools in the business management of the Quality function. It provides a decision-making tool, a planning tool, a system of built-in checks and balances, and the media which translates many of the intangibles of quality into the great common denominator — dollars and cents.

Quality cost permits the management of the quality function to become true business management by contributing to the actual profit of the company. "Does Quality Assurance Department contribute to profit?" It would seem logical to take the position that since industrial production organizations exist to make a profit, then everyone on the payroll should contribute to that end. There should be no doubt about it. But doubts will continue in the minds of presidents and general managers as long as they are influenced by discriminatory accounting terminology such as 'indirect", "burden', "overhead", or in plain English, "parasite". Take the Accounting Department. Are they contributing to profit? You must be kidding! Surely accounting is a star example of a non-contributor to profit. Many of them have only the faintest idea what the product is, from which profit is supposed to be made. But accountants will tell you, and you better believe them, that until they sort out and arrange receipts and expenses and taxes and show a rational balance, nobody knows whether there is a profit.

So, the question about quality assurance contribution to profit should be not "whether", but 'how much". For that reason we need total quality cost reporting systems. You will find that nobody at first really has much interest in such a concept. Accounting hates it because it looks like one more job for them to do, and besides, it can't be of any use because it doesn't have to balance with normal reports.

We were fortunate to have in our organization a friendly accounting type who helped us identify and establish quality accounts and definitions to report quality cost. In order to make quality cost reports mandatory we needed a policy on product quality which states the responsibilities for quality function and includes the following statement regarding quality cost:

Quote:

Efforts relating to attaining and controlling standards of product quality shall be measured in terms of costs. These costs are broadly classified as:

Prevention Costs

The investment made through quality engineering and pre-planning efforts, basically problem prevention.

Appraisal Costs

The cost of examining products at any stage of manufacture to determine their compliance to specifications.

Failure Costs

These are simply the costs resulting from not meeting specifications. Unquote.

The Accounting instructions made the controller responsible for the cost assignments and the issuance of Quality Cost Reports by stating, and I quote: "In accordance with corporate policy 15.J on Product Quality, all operating divisions will be required to identify, classify and report quality costs, which are an integral part in the administration and management of company business, in accordance with the principles and definitions stated in this section." Such policies made the only person who can ultimately be held accountable for total quality costs the same person who is accountable for profits.

I have run into many Quality Assurance Managers who sincerely believe that total quality costs, including scrap and rework costs, are "their baby" and refer to them as "my costs". Well, I think they are wrong and they owe it to themselves and the company to try to keep the orientation correct. Perhaps their own management will not let them, but I do think a good Q.A. Manager will try, by various means, to educate his own management, not to hold him responsible for costs he does not create.

Let me review through this series of slides taken from one of our training courses, what the sequence of events and responsibilities for quality cost should be.

It is the Quality Assurance which must define the elements of quality cost.

Here is an example of a monthly quality manpower log whereby quality assurance is reporting to Accounting how the man hours of individuals were spent with regard to those elements of quality cost related to efforts of people. Accounting can convert these hours to a cost value. Other people contributing to quality costs from other departments would be identified in the same way.

It's the Controller's job to put a dollar value on these elements of quality costs. *Avoid* in every possible way, attempts by the Quality Assurance Department to put together quality costs or other financial figures, no matter how reliable your source is. If you do, you will very likely end up spending more time explaining and defending your numbers than you will using them.

Remember — it's the Controller's job to prepare, endorse and issue the Total Quality Cost Report — the Quality Assurance Manager is then only a recipient.

Now it is the Quality Assurance Manager's job to audit and evaluate that report and to use it not as a department manager but as the Administrator of the Quality Assurance Program. It is his job to interpret the report and advise top management on what it says.

As I said before and as is shown here, it is the top man who is ultimately accountable for total quality cost, but he will constantly be looking to you, the "specialist" for interpreting the results. It is really the best tool the Q.A. Manager has for initiating corrective action and to get others to do their quality functions. The question which remains to be answered is "What should be your total quality costs to sales ratio?"

The point I would like to make here is that the real job is to optimize quality costs — use it to show you the trend, and if comparisons must be made, then limit them to comparisons with your own past performance or with an objective that has been set for each product. You will have to keep reminding people that we cannot eliminate quality costs — perfection is too expensive, at least in the commercial world.

Something must be invested in product quality. Ten years ago our quality cost dollar was spent as follows: 65 cents on failures, both internal and external, 33 cents on appraisal, and only 2 cents on prevention, and that was a generous 2 cents.

Today, our company, by promoting the investment in preventive Quality Assurance Engineering, properly applied, pays for itself by reduced failure costs and even reduced appraisal costs adding to profit by net reductions in total quality costs.

How do we use these quality cost measurements? In our plan we have shown that if our business was yielding a profit of 3% of sales and our quality costs were 7% of sales, then we were in trouble. The point is, that any reduction of quality costs from this level of 7% can be directly converted to a contribution to profits by bringing the quality costs of 7% of sales down to 4% and transposing the difference over into the profit column and speak in terms of 6% profit and 4% quality costs.

We have shown our management what will happen as a result of implementation of our Total Quality Assurance Program by typical pie chart. The illustration shows that the allocation of the pie changes according to an increase in prevention and a decrease in failures, but also a decrease in the size of the pie, indicating less dollars being spent.

Here is the way we demonstrate the savings effect on improved quality cost ratios. If the quality cost level started at 3.2% of sales in 1968 and improved to 2.5% of sales in 1969, that difference in those two ratios applied to the sales in 1969 would yield a $486,00 contribution to profit, really a cost avoidance. Because if the ratio had not changed and we had remained at the 1968 level, that amount of money would have been spent, but it was not. We have calculated these cost avoidance dollars from year to year and since 1969, which was established as a base year, and we at Allis-Chalmers Canada Limited claim a contribution in 5 years of $2,378,000.00; in 1970, $291,000; in 1971, $344,000; in 1972, $582,000; in 1973, $510,000; in 1974, $651,00. The contribution for 1975 was $716,445. *Very good justification of the Total Quality Program to management.*

Let me show you a different view of total qualtiy cost in our company by examining the allocation of the quality dollar in terms of prevention, appraisal, internal and external failure costs, and the size of that dollar expressed as a per cent of sales.

ALLOCATION OF THE QUALITY DOLLAR

	EFFORT		RESULTS		QUALITY COST INDEX (Total Quality Costs As A % of Product Sales)	
	Prevention	Appraisal	Internal Failure	External Failure		
1967	5%	13%	36%	46%	1967	4.5%
1968	9%	20%	34%	37%	1968	3.9%
1969	18%	30%	32%	20%	1969	2.0%
1970	20%	30%	35%	15%	1970	1.9%
1971	24%	31%	35%	10%	1971	1.7%
1972	27%	32%	25%	16%	1972	1.8%
1973	27%	33%	29%	11%	1973	1.7%
1974	26%	35%	33%	6%	1974	1.5%
	Increased Investment		Trend-Line Lower Resultant Cost			Trend-Line Lower Total Quality Costs

A basic principle of Total Quality Assurance was demonstrated here. A 70 per cent reduction of total quality costs — equal to 3 per cent of sales — since development of the T.Q.A. System began. Total quality costs as a percentage of product sales have declined from 4.5 per cent in 1967 to 1.5 per cent in 1974. Allocation of the quality dollar also has swung heavily to prevention, and appraisal 5 per cent for prevention and 13 per cent for appraisal in 1967 to 26 per cent and 35 per cent respectively in 1974.

Our Total Quality Assurance System is monitored and audited by Corporate Auditors and we must report our quality program progress every year as follows.

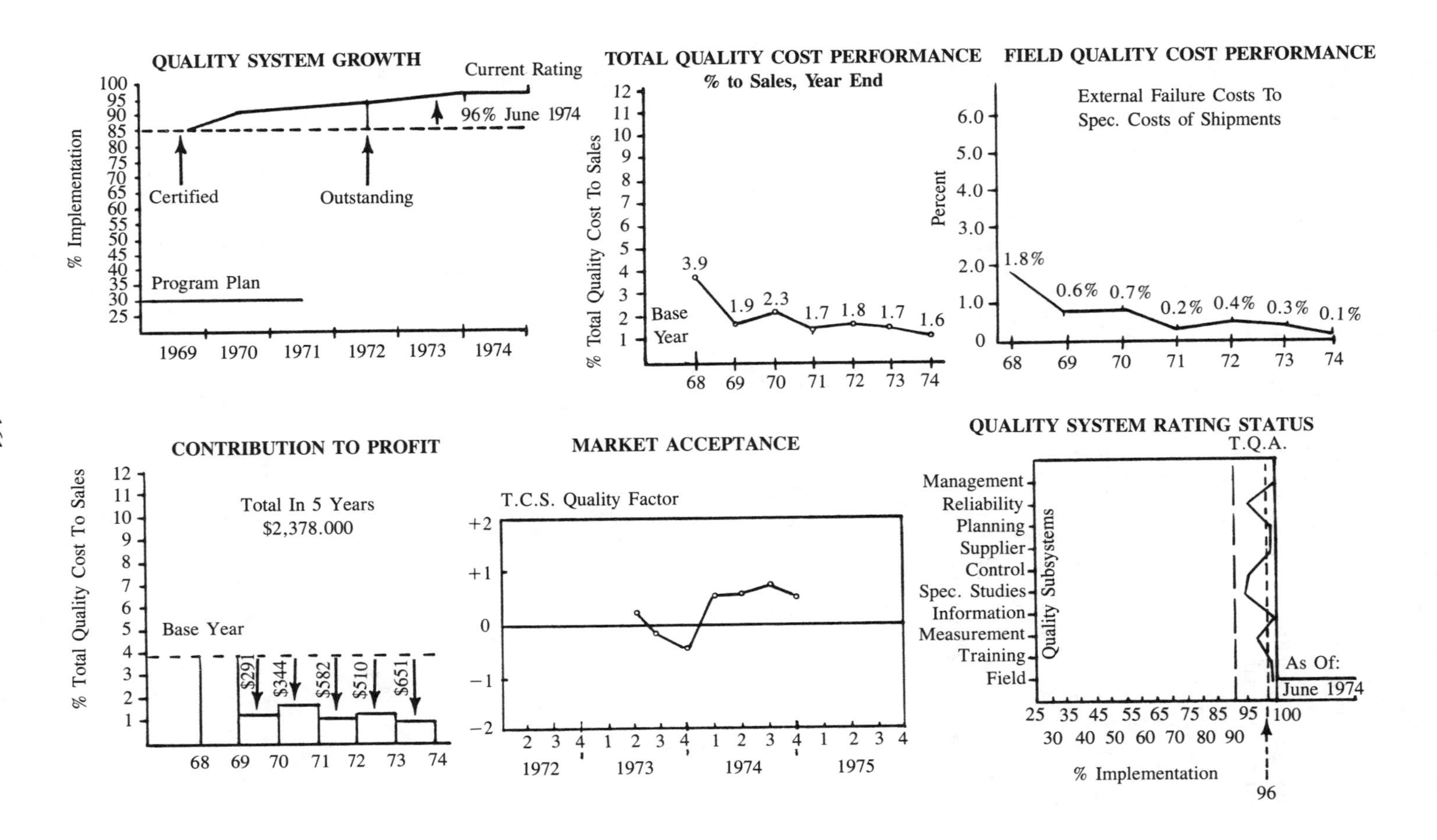
QUALITY SYSTEM GROWTH
Current Rating
96% June 1974
% Implementation
Certified
Outstanding
Program Plan
1969 1970 1971 1972 1973 1974
TOTAL QUALITY COST PERFORMANCE
% to Sales, Year End
% Total Quality Cost To Sales
Base Year
3.9 1.9 2.3 1.7 1.8 1.7 1.6
68 69 70 71 72 73 74
FIELD QUALITY COST PERFORMANCE
External Failure Costs To Spec. Costs of Shipments
Percent
1.8% 0.6% 0.7% 0.2% 0.4% 0.3% 0.1%
68 69 70 71 72 73 74
CONTRIBUTION TO PROFIT
% Total Quality Cost To Sales
Total In 5 Years
$2,378.000
Base Year
$291 $344 $582 $510 $651
68 69 70 71 72 73 74
MARKET ACCEPTANCE
T.C.S. Quality Factor
1972 1973 1974 1975
QUALITY SYSTEM RATING STATUS
T.Q.A.
Management
Reliability
Planning
Supplier
Control
Spec. Studies
Information
Measurement
Training
Field
Quality Subsystems
As Of:
June 1974
% Implementation
96

In the March-April 1974 issue of the "Harvard Business Review", it becomes clear, from a study of 57 corporations and 620 diverse businesses, that for the first time economists are now turning their attention to the effects of product quality on market share and return on investment.

We have at Allis-Chalmers Canada Limited understood the true business-oriented nature of Total Quality Assurance, because the improvements in product quality are quickly noted on the balance sheet.

Prevention and effective cost measurement are the keystones of our successful Quality Systems. The effective measurement of quality costs is a first step to more effective Quality Assurance and minimizing profit dissipation. There is a villain in our midst which is affecting most of our businesses. Get him identified and get him optimized and you will be doing your business a real service. Put the right tools to work through your Quality System. You cannot do today's quality job with yesterday's skills.

QUALITY COST — A PLACE FOR EVALUATION OF CUSTOMER SATISFACTION

H. James Harrington, Manager, Product Quality Assurance
IBM, San Jose

ABSTRACT

How do you spend the quality dollar to get maximum return on your investment? This is the perennial challenge that faces quality assurance managers.

An attempt to resolve this challenge was offered by a program called QUICO (Quality Improvement Cost Optimization). QUICO was initiated at Stanford University under a government grant in the early 1960's. It is a forerunner of the quality cost-analysis system as we know it today.

As its inception, quality cost analysis was a strong management tool. In many cases, quality assurance managers used it to expand their quality systems, but all too often the projected return from the slight increase in quality budget was so great that upper management questioned the data.

To add validity to the quality cost data, some managers have turned their cost-analysis programs over to the plant controller. This has proven to be an effective operating mode for, in most cases, quality assurance gains an ally. However, quality assurance gives the controller a double-edged axe that can be swung in both directions. In one direction it decreases product cost by increasing the quality budget; in the other direction it decreases product cost by decreasing the quality assurance budget.

Unfortunately, the controller is primarily oriented to ledger quality cost (direct quality cost) and has not been educated to understand and consider the harder-to-recognize indirect quality cost. Both direct and indirect quality costs (see table, below) will be treated here, but primary emphasis will be placed on the indirect quality costs.

Total Quality Costs

I. Direct Quality Costs
   A. Operating Quality Costs
      1. Controllable quality costs
         (a) Preventive costs
         (b) Appraisal costs
      2. Failure costs
         (a) Internal failure costs
         (b) External failure costs
   B. Equipment Quality Costs

II. Indirect Quality Costs
   A. Customer Incurred Quality Costs
   B. Customer Dissatisfaction Quality Costs
   C. Loss of Reputation

This presentation will primarily emphasize the indirect quality cost aspect and it will present how these indirect quality costs influence cost graphs.

Indirect costs form the shadow over the quality cost-analysis systems being implemented today. Fortunately, most systems operate to the left-hand side of the optimum operating point. If your system operates to the right-hand side of the optimum point of the combined direct cost curve, be sure that you fully consider the indirect cost before you cut controllable costs. A total quality cost system is as basic a quality tool as a micrometer. If you don't have a quality cost system, make this your goal for the coming year. If you have a quality cost system, use this effective tool with judgment. Trade in your opinion for some facts.

COST QUALITY CONTROL CHART FOR VARIABLES

A. D. Oak
Professor & Head of the Dept. of Statistics
Mulund College of Commerce
Mulund (West), BOMBAY 400 080 (India)

0.0 Abstract :—

The main purpose of statistical quality control is to devise statistical method for separating chance causes from assignable causes in a manufacturing process. However, this approach does not give any consideration to the cost of manufacturing a unit. The present practices of finding Control Limits and then drawing a control chart for variables does not consider 'the cost aspect' and also ignores some special features seen in 'variables'.

This paper provides :—

i) New approach to the concept of a 'variable';
ii) New design of Control Chart; and
iii) Consideration to both cost and quality simultaneously.

1.1 Present Approach:—

In the present set-up of statistical quality control as adopted in various industries the procedure for a control chart for variables, in general, is

i) to find upper and lower control limits (UCL and LCL) by using (a) $\overline{X}$-Chart, (b) R-Chart or (c) σ-Chart;
ii) to draw the control chart and plot the sample points;
iii) to read the trend of points and then draw inference about the efficiency of the machine and also the quality of the units produced.

This procedure, however, basically fails to cater for all the practical situations and requirements. The major drawbacks being :—

A) The only classification under 'variable' is not adequate.
B) This does not consider cost aspect at all.
C) In many cases the quality control of the machine is to be judged by the quantity control of the final product.

Let us consider each of these — A, B and C— separately.

1.2 Variable :—

Any measurable quality in terms of some units, is considered as a variable.

e.g. i) Dimensions of components, measured in units (length, height, thickness, diameter, capacity, area etc.).
ii) Physical or chemical properties in some units (resistance, shear strength, specific gravity, etc.).

However, the author feels that there is some special characteristic shown by some variables depending upon the nature of further use of the unit produced.

1.3 Illustrative Examples :–

Example : 1 – Suppose circular discs of certain metal are being produced on large scale on an automatic machine. Its diameter is to be X cmx. and $\overline{X}$ chart is being used in ASQC and further actual unit produced has diameter equal to y cms. then

i) $y > \overline{\overline{X}} + A_2\overline{R}$, if economically feasible rework is possible with which the unit may be put to the required size; and ii) $y < \overline{\overline{X}} - A_2\overline{R}$ then the question of rework does not arise and then in such a case total wastage of material is unavoidable.

Example : 2 – On an automatic filling machine (tea, toothpaste, soap powder, milk, wines etc.) suppose the containers are being filled in. The contents are supposed to be and are sold as × gms. and further if $\overline{X}$ chart is used. A particular unit is filled with y gms. by the machine.

i) If $y > \overline{\overline{X}} + A_2\overline{R}$ there is possibility of refilling the container or else to pass the unit even though it is over filled.

ii) If $y < \overline{\overline{X}} - A_2\overline{R}$ in this case also there is possibility of refilling or else to pass the unit even though under filled.

However, under both the cases no loss of material is involved although (i) will incur some additional cost of over filled material. Mostly in such cases the containers (aluminum foil, card board box, bottles, plastic bags etc.) are in a position to accomodate some additional material than the specified capacity. In these cases we find that the performance (quality control) of the machine is to be judged by the quantity control of the unit produced.

Example : 3 – In the case of medicinal injections sold in vials there is every possibility that while using

i) Some liquid is bound to stick to the inner surface of the container.
ii) The syringe needle may not be able to collect all the contents of the bottle.

This has direct bearing with the medicinal use of the dose prescribed and therefore the manufacturers do overfill, slightly, than the units specified.

This excess is also to be accounted for as the consumer is not charged directly for this.

Example : 4 – Some pharmaceutical companies follow a different mode. If 1000 c.c. of injection liquid is prepared and each vial is to be filled up with 1 c.c. of liquid. The approach shows that if, say, 950 vials are prepared out of 1000 c.c. then it signifies food performance of the machine and also the cost side is acceptable.

2.1 New Apporach :–

With these four examples (1), (2), (3) and (4) in mind the author proposes to classify variables into two classes :–

i) Variables involving *loss of material*; and

ii) Variables involving *no loss of material.*

Separate methods of drawing control charts are suggested further in 2.3.

2.2 Cost :–

The present practice of drawing control charts does not consider cost aspect at all. The recent trend in management demands cost consideration along with the performance of the machine. The author has proposed a new method of drawing control charts which satisfy all the present requirements and along with these also reveal the cost aspect.

2.3 New Method :–

A method suggested, to over come (A) and (B) both, is given below. The method is to be used in the case of variables which *do not involve* any loss of material. (e.g. Automatic filling tooth paste, powders, tea, coffee, etc.)

Suppose (i) Container is to be filled with M units of the product.

(ii) The price at which the product is sold to the customer is Rs.y per unit and that too for M units only.

(iii) Machine fills (M $\pm$ z) units of the product.

The producer sees it like this.

Loss = yz Rs. if the contents are (M + z)

Additional profit = yz Rs. if the contents are (m –z)

No loss no gains = 0 if the contents are M or z = 0

Units inspected = n.

Method :– M = 500 gms. p = 10 Ps./gms.

TABLE : I

Marked/Specified Quantity M	Actual Quantity filled m	Deviations d = (m – M) (+) ve in Units	Points +	Points –	(–) ve in Units	Cost Variation p(m – M) pd
500 gms.	501	1				+10
"	503	3				+30
"	496				–4	–40
"	498				–2	–20
"	500	0				– 0
"	502	2				+20
"	499				–1	–10
"	499				–1	–10
						Σp(m–M) = –20

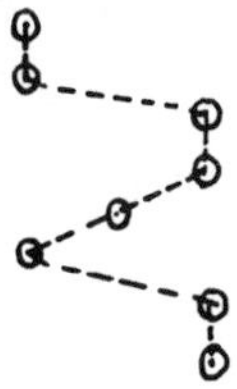

The constraint to be put in Σ p(m − M) $\leq$ 0. In other words 20 ps. is an additional profit. Here of course, one must give due consideration to the consumers' risk and to the producer's risk.

The trend of points plotted in the central two columns can give a clear idea about the performance of the machine and Σ p(m − M) could also give cost consideration.

The drawback being although the points show positive and negative deviations they fail to show the magnitude of deviations. However, this is over come by the two columns showing the magnitude of deviations. This chart should normally suffice for an over all picture of the performance of the machine.

This method can further be extended to R chart and σ chart.

3.1 Conclusion :–

If the variable does not involve loss of material then undue importance which is given presently by calculating $\bar{x}$, $\bar{\bar{x}}$, $A_2\bar{R}$ etc. can be reduced. This method is very simple to operate and understand.

This method does not show any violation of basic assumption of drawing a control chart.

This paper does not give any alternative method to control chart for variables involving loss of material.

The author does not claim that the method suggested above is fully exhaustive. However, even if it provides a new *line* approach to the concept of 'variables' and the control chart the basic aim is fulfilled.

REFERENCES

1. Cowden Dudley J., "Statistical Methods in Quality Control." Prentice Hall Publication.
2. Duncan Acheson J., "Quality Control and Industrial Statistics." Richard D. Irwin Inc. 1965.
3. Erricker B.C., "Advanced General Statistics." English University Press, London. 1971.
4. Grant E. L. and Leavenwarth R.S., Statistical Quality Control. McGraw Hill Kogakusha Ltd., 1972.
5. Juran J.M., "Quality Control Hand Book." McGraw Hill Book Co. 1962.
6. Oak A.D. "Cost and Quality Control Charts." Sept. 1974 issue of *Quality Progress,* American Society for Quality Control. Milwaukee, Wisconsin.
7. Wilke S.S., Mathematical Statistics. John Wiley and Sons. New York, 1962.

LIFE CYCLE COSTING — A NEW DIMENSION FOR RELIABILITY ENGINEERING CHALLENGE

T. David Kiang, Manager, Contract Reliability Engineering
Bell-Northern Research, Ottawa, Canada

ABSTRACT

Life Cycle Costing is the process of economic analysis to assess the total cost of system acquisition and ownership taking into consideration the operational constraints and performance requirements of the system. This paper provides insight into the current interest and pursuit of Life Cycle Costing by government and industry, and to those areas that have opened a new dimension in Reliability Engineering. The concept of Cost-Effectiveness-Performance is introduced, and some implications on Life Cycle Cost implementation are discussed.

INTRODUCTION

The advent of Life Cycle Costing has generated extensive interest in recent years among many countries who are battling inflation and calling for nationwide economic restraint. The scarcity of resources and the alarming escalation of material and labour costs have led to intense pursuit of Life Cycle Costing at various levels of the government and in some private sectors of the business enterprise. In this endeavour, they all seem to share a common goal: to achieve a proper balance of acquisition and ownership costs.

Over the past decade, Life Cycle Costing has evolved into an important costing discipline to aid management in their decision making process. The basic technique stamps from the concept of engineering economy [1]. In recent years, attempts have been made to link Life Cycle Cost to System Effectiveness [2]. It is in this latter emphasis that Life Cycle Costing has become a design trade-off tool. Because System Effectiveness is mainly in the realm of Reliability Engineering, it has expanded the horizon of Life Cycle Costing for active pursuit by Reliability Engineers.

WHAT IS LIFE CYCLE COSTING

Life Cycle Costing is the process of economic analysis to assess the total cost of system acquisition and ownership taking into consideration the operational constraints and performance requirements of the system. The term system here is defined as a major end item with all the components required for its operation and support to achieve its service objectives. A system may include both hardware and software, relevant support facilities, material, data, personnel and related services. In this respect, Life Cycle Costing is basically concerned with the total cost of the system over its full life.

To appreciate the extent and involvement of Life Cycle Costing, it is necessary to put the various phases of a system life cycle into proper perspective. Figure 1 depicts the six major phases of a system life cycle. The essence of Life Cycle

Costing associated with each phase of the system life cycle is summarized in Figure 2.

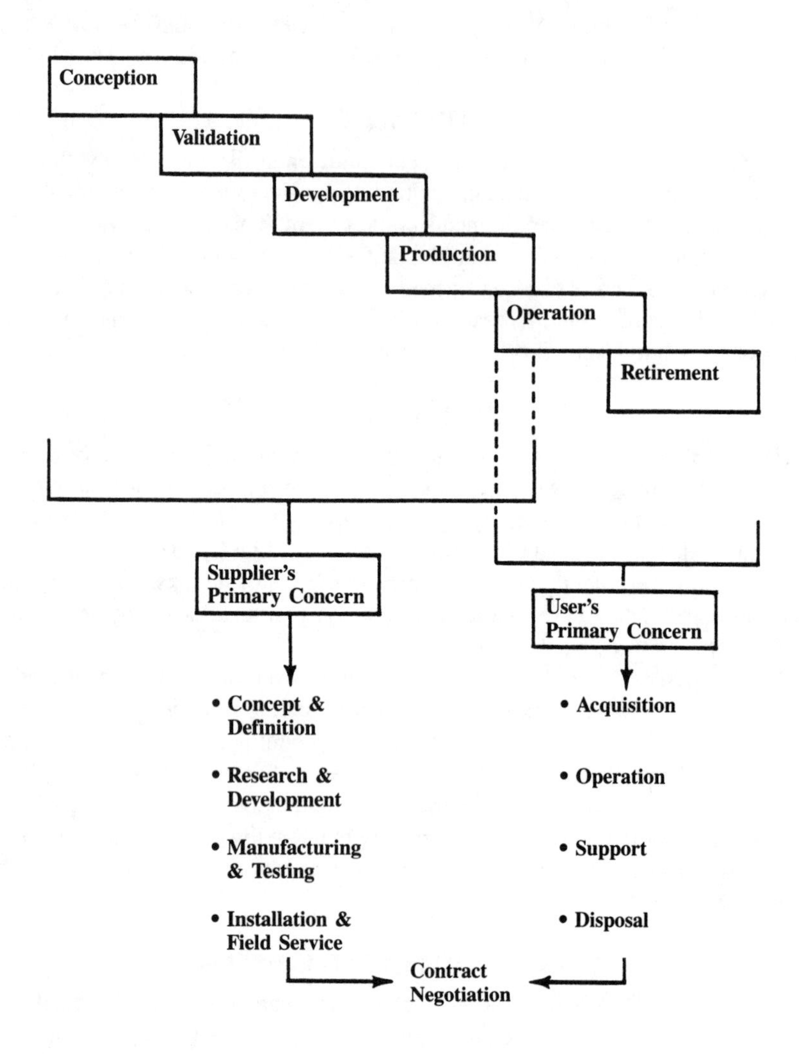

FIGURE 1 A System Life Cycle Profile

System Life Cycle Phase	Relation to Life Cycle Costing
1. Conception	Investigate different concepts and system usage scenerios and their impact on life cycle cost. Select preferred design configuration and estimate baseline life cycle cost.
2. Validation	Establish design-to-cost goals based on design specifications and operational requirements.
3. Development	Perform design trade-offs and evaluate impact of reliability and maintainability on operation and logistics support costs. Reassess baseline life cycle cost in relation to work breakdown structure.
4. Production	Assess cost impact on the extent of reliability testing and quality control to minimize warranty risk and unscheduled field service needs.
5. Operation	Balance cost of acquisition to cost of ownership. Update Life Cycle Cost Model based on actual field data collected to optimize cost of operation.
6. Retirement	Evaluate the economics of service life extension versus retirement from service. Analyse the effects of dispersed retirement.

FIGURE 2 Relation of Life Cycle Costing to System Life Cycle Phases

Figure 3 shows a typical Life Cycle Cost Program.

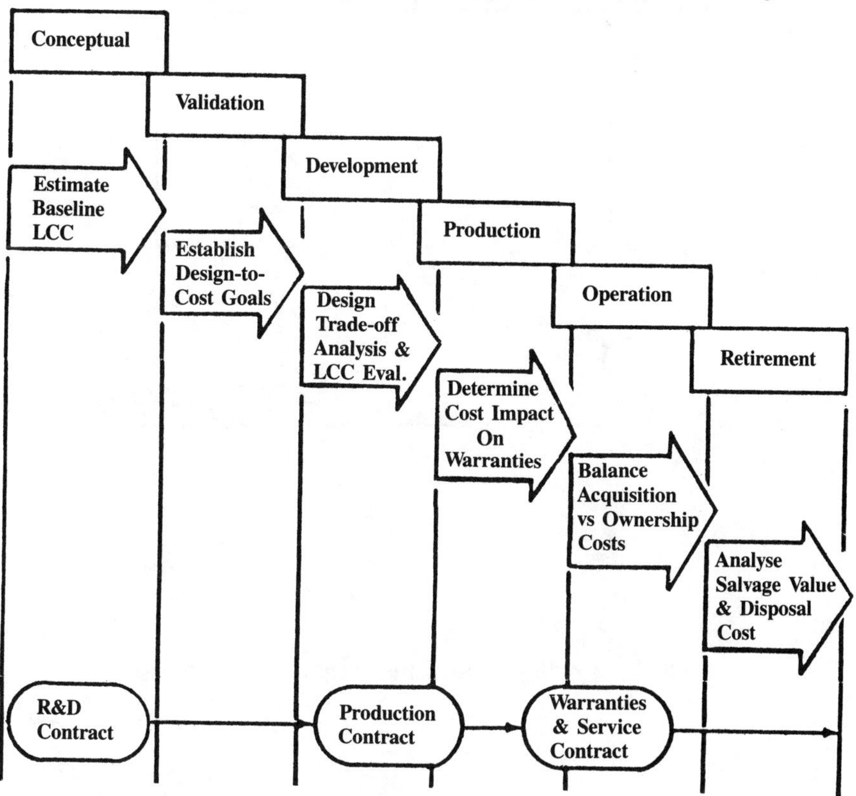

FIGURE 3 A Typical Life Cycle Cost Program

Figure 4 illustrates a simplified Life Cycle Costing Process.

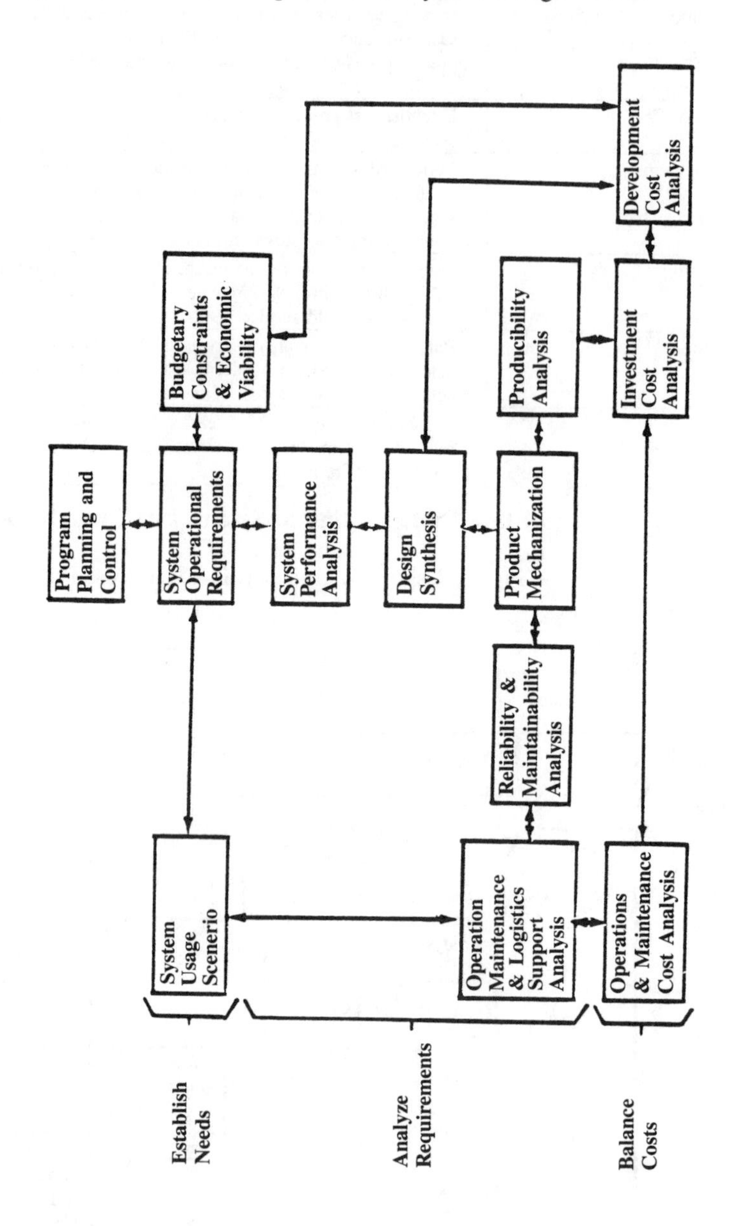

FIGURE 4 A Simplified Life Cycle Costing Process

THE COST-EFFECTIVENESS-PERFORMANCE CONCEPT

Reliability Engineers are directly involved in many facets of Life Cycle Costing associated with the various phases of the system life cycle. In an attempt to link life cycle cost to system effectiveness and performance requirements, the Cost-Effectiveness-Performance concept is proposed. This concept is shown in Figure 5. The relationship between System Effectiveness, Performance Requirements and Life Cycle Cost is depicted in Figure 6. This illustrates a surface area in 3-dimensional scale for trade-off analysis. The optimum Life Cycle Cost can be determined based on a set of parameters. In practice, this relationship may be much more complicated due to various exogenous constraints and other uncontrollable factors, which make the mathematical model difficult to formulate and compute. However, this concept may serve as a starting point in approaching a Life Cycle Cost problem. The value and accuracy of its solution will depend on the complexity of the situation and the assumptions made in establishing a realistic measure on Life Cycle Costing for comparative engineering studies.

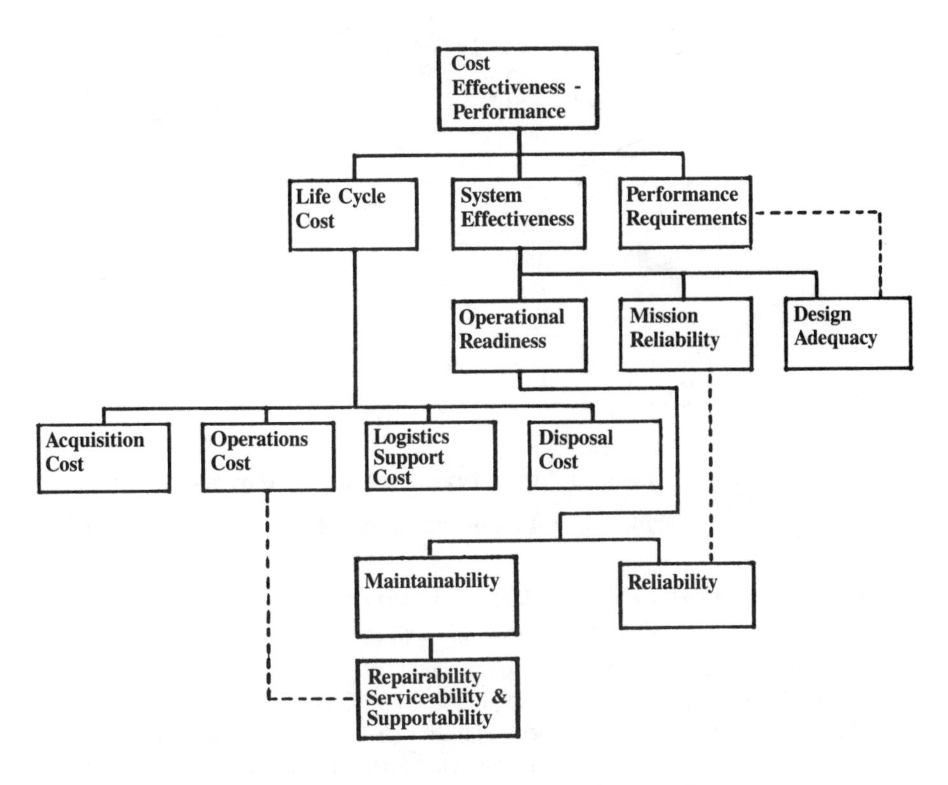

FIGURE 5 Cost-Effectiveness-Performance Concept

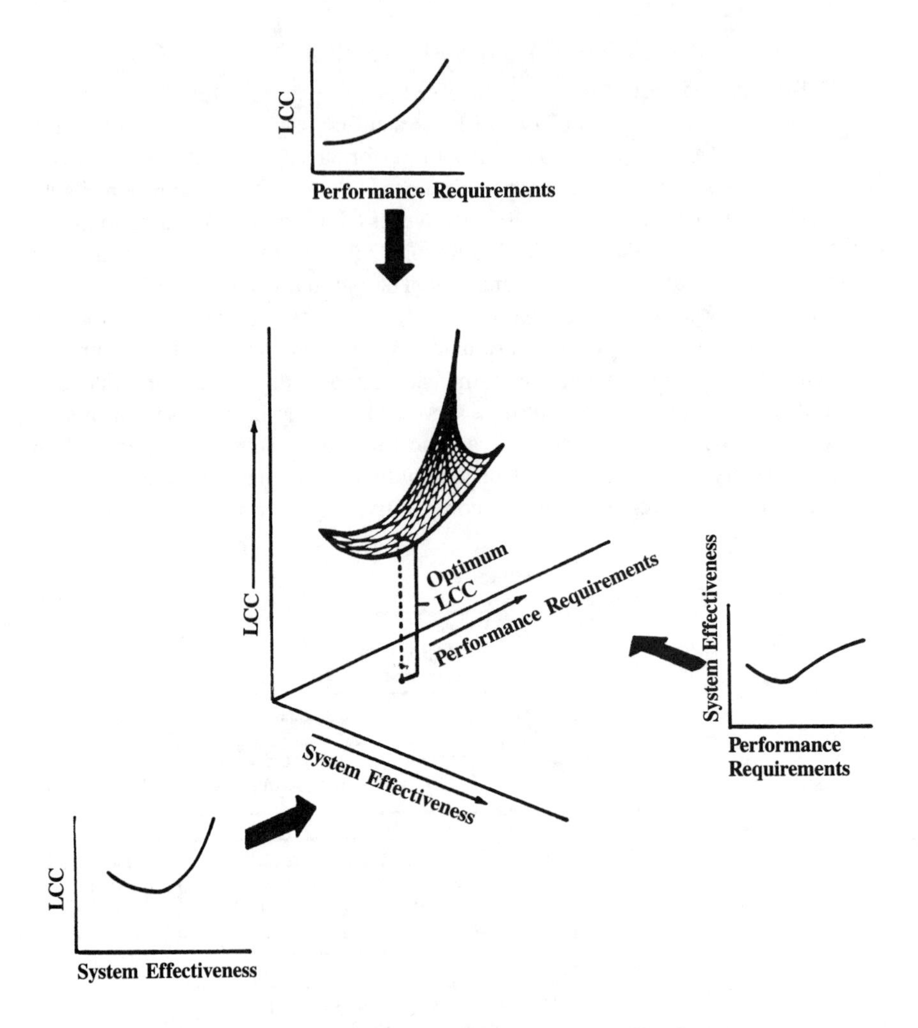

FIGURE 6 Relationship Between System Effectiveness, Performance Requirements and LCC

A METHODOLOGY FOR LIFE CYCLE COSTING AND APPLICATIONS

The methodology for Life Cycle Costing is basically concerned with the approach, formulation, and subsequent solution of a cost function relating the various cost contribution parameters to the total Life Cycle Cost. In this respect, a Life Cycle Cost model can be established. Figure 7 shows a typical Life Cycle Cost model. It serves as an objective cost function to be determined subject to the various constraints associated with the cost contributing parameters.

Depending on the amount of details sought, this model may be extended for general application in:

a) Life Cycle Cost comparison for new system or equipment procurement.

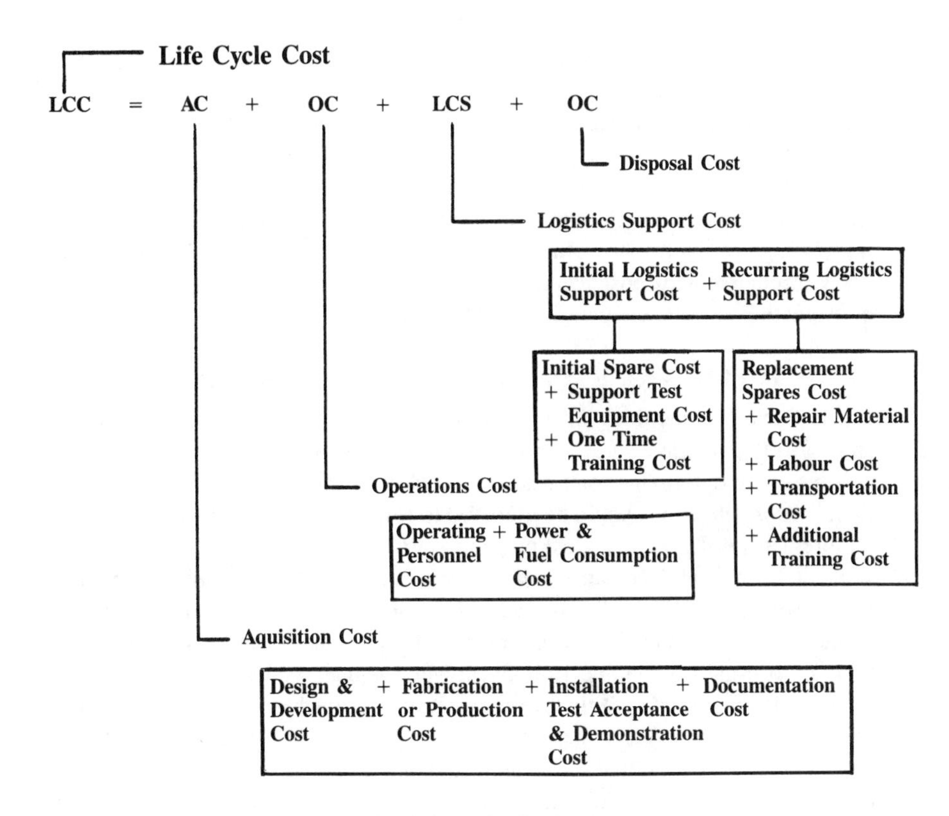

FIGURE 7 A Typical Life Cycle Cost Model

b) determination on the effects or changes of Life Cycle Cost on existing system or equipment modification.
c) trade-off analysis between initial acquisition cost versus recurring support cost for the desired degree of performance reliability and maintainability, and the levels of maintenance and logistics support concept.
d) economic analysis on the optimum retirement age of a system or equipment from inventory.
e) determination of system effectiveness associated with prescribed Life Cycle Cost.
g) sensitivity analysis and the associated cost impact resulting from the variation of system effectiveness parameters.
g) cost-effectiveness analysis on the level of maintenance, and the criteria for repair or throw away modules.

It should be emphasized at this point that the effectiveness of using this Life Cycle Cost model depends largely on the user's interpretation of the actual situation where such a model could be gainfully employed. The applicability and availability of data play a dominant role on the reality of the analysis results. Application of the Life Cycle Cost model is basically a mathematical analysis process relating system effectiveness parameters to cost. This quantitative approach is used to approximate a real life situation. The results obtained from such a process will not be absolute figures but of relative and comparative nature, and they should be treated accordingly.

The methodology employed here is not a cost accounting process, but rather an engineering tool for trading off system parameters against Life Cycle Cost. It gives engineering management an instant overview of Life Cycle Costing without going through lengthy cost accounting computations. The model would save time in evaluating various engineering options and design alternatives.

SOME IMPLICATIONS ON LIFE CYCLE COST IMPLEMENTATION

Life Cycle Costing does not happen overnight in a organization. It takes time and effort to put Life Cycle Costing to work. The concept of Life Cycle Costing as a means of optimizing total cost should be well accepted within an organization. The organization should be prepared to respond to technical and contractual requirements of Life Cycle Cost programs. Profit incentives were emphasized in some of the major maintenance and support programs for the commercial airlines [3], as well as in FFW/RIW contracts [4,5]. To enhance the effectiveness of implementing Life Cycle Cost programs, government and large service organizations must establish suitable Life Cycle Management Systems.

The functional group within the organization responsible for Life Cycle costing needs proper tools and technical support. One such tool is a suitable methodology for Life Cycle Cost evaluation. This would entail the use of an appropriate Life Cycle Cost model or models. Existing models [6] are generally available but it should be cautioned here that they were developed for specific applications. The decision here would be either to adapt an existing model that closely suits the application or to develop a model for the intended use. Before committing a model to full scale implementation, time and effort must be devoted to evaluating the validity of the model.

To support Life Cycle Cost analysis, relevant data are required. This calls for establishing a data base from which Life Cycle Cost data could be drawn upon. Such a data base relies heavily on actual reliability performance data of the product in service. A viable data base system should tie in closely with other systems that exist in the organization. In this way, inputs for cost information, maintenance and field service data, can be provided. The data base permits a basis for standardization of parametric costs, and it serves as reference points for Life Cycle Cost analysis. A data base is the focal point for data acquisition and retrieval in a Life Cycle Costing process.

To effectively control and manipulate the data required for Life Cycle Costing, there are obvious reasons to bring in some form of computer services. The extent of automation and computerization depends on the degree of sophistication sought in implementing Life Cycle Costing. If there are a large number of users, it may be necessary to design a computerized information system totally dedicated to the application of Life Cycle Costing. An example of such a computerized system scheme is shown in Figure 9.

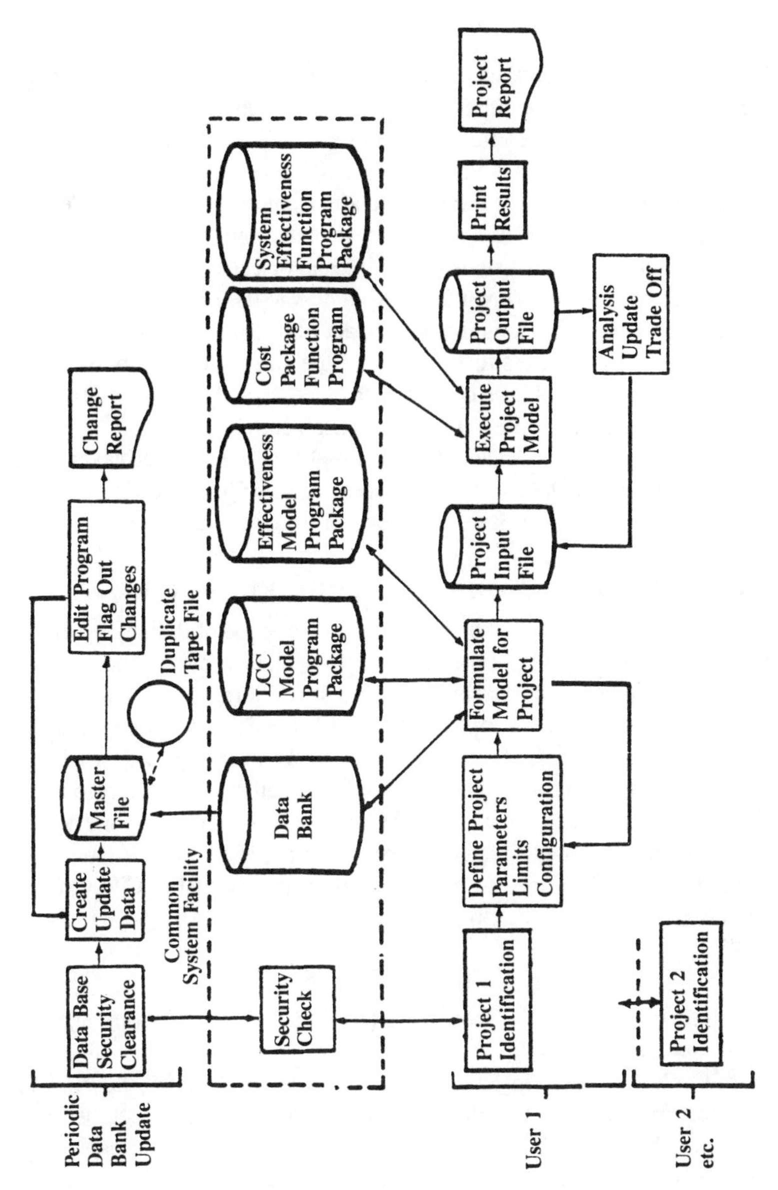

FIGURE 8 A Simplified Computer System Functional Flow Diagram for LCC

Implementation of Life Cycle Costing requires people involvement. It is essential that all the participants and contributors in a Life Cycle Cost program are tuned to the same frequency. In this respect, some form of indoctrination may be necessary at various levels of management and technical personnel involved in the Life Cycle Costing process. Guideline documents on prodedures and practices need to be established. Instruction manuals for Life Cycle Cost analysis need to be prepared. Training seminars and workshop sessions may serve as a vehicle to disseminate the knowledge of Life Cycle Costing. Depending on the number of people involved, it is envisaged that some co-ordination effort may be required to launch an effective indoctrination program in conjunction with the implementation of a full scale Life Cycle Cost program.

Life Cycle Costing is not a "cure-all" process for economic analysis and evaluation. There are limitations to the extent and applicability of the models developed, even the ones that are considered to be most sophisticated. Evaluation factors such as economic climate, political environment, technological change, and work disruptions would be extremely difficult to incorporate into the Life Cycle Cost models. Yet these factors could have significant impact on the outcome of the Life Cycle Costing process. These are situations where the value judgements of managers and decision makers would have come into the picture, and hopefully rational decisions could be made to achieve the intended objectives.

Life Cycle Costing provides an analytical tool for the decision makers. It does not make decisions for the users. The success of Life Cycle Costing depends on a realistic data base and on wise decisions in utilizing the results.

THE CHALLENGE TO RELIABILITY ENGINEERS

Life Cycle Costing has undoubtedly widened the scope of Reliability Engineering to appraise a system within the concept of its total life cycle. The exploitation of the relationship between System Effectiveness, Performance Requirements and Life Cycle Cost has indeed opened a new dimension for in-depth technical challenge. Besides the profit incentives associated with Life Cycle Costing, it tends to stimulate our thinking by relating reliability directly to cost. Confronted with the rapid changes of technology, Reliability Engineers must respond to the increasing demand of reliable designs, dependable processes, and improved product integrity. And these need to be done within the affordable Life Cycle Cost constraints in our current belt-tightened economy.

REFERENCES

1. "Engineering Economy", Second Edition, American Telephone and Telegraph Company, 1963.
2. "Life Cycle Cost/System Effectiveness Evaluation Criteria", G. A. Walker, The Boeing Company, Seattle, Washington, AD 916001, December 1975.
3. "Interface Between Maintainability & Commercial Aircraft Spares Support", J. E. Losee, Proceedings of the 1976 Annual Reliability & Maintainability Symposium, Las Vegas, January 1976.
4. "Failure Free Warranty/Reliability Improvement Warranty — Seller Viewpoints", S. T. Rosental, Transactions of the 1975 ASQC Technical Conference, San Diego, June 1975.
5. "Failure Free Warranty/Reliability Improvement Warranty — Buyer Viewpoints", O. Markowitz, Transactions of the 1975 ASQC Technical Conference, San Diego, June 1975.
6. "A Summary and Analysis of Selected Life Cycle Costing Techniques and Models", L. E. Dover, et al, Air Force Institute of Technology, Wright-Patterson AFB, Ohio, AD 787183, August 1974.

A BUSINESS PERFORMANCE MEASURE OF QUALITY MANAGEMENT

R. A. Cawsey, Director of Corporate Quality
Northern Electric Company, Limited
Montreal, Quebec

INTRODUCTION

The modern management approach in the control of quality is to consider this activity as a business venture with the goal of enhancing the profitability of the company on a continuing basis.

The prime objective of the company is, of course, to provide a good return on the capital employed. The product line or service sold is of little concern to most shareholders of the company so long as the operation shows a consistent pattern of profits and growth.

The shareholders' wishes are implemented by executive management and they establish what markets to enter and the "value of quality" needed to successfully penetrate the markets selected. It is the "value of quality" provided to the customer that largely determines the share of the market obtained which in turn will greatly affect product profitability.

A typical manufacturing company, having a yearly sales volume of $100 million, will earn in the order of $16 million before taxes and spend approximately $4 million on achieving the "value of quality" required by the marketplace. Using these figures as a model, the quality manager could be looked upon as the general manager of a $4 million enterprise within a $100 million operation.

The question now arises as to how quality management's business performance can be evaluated in relation to the objectives of the company.

The quality manager defines the "value of quality" requirement in terms of a measurable standard of outgoing quality that can be achieved within the cost constraints of remaining competitive. Therefore, his first objective is to establish a quality system that can deliver this level of quality. Techniques used for evaluating performance in this area, such as Statistical Analysis, Product Audit, Customer Feedback, etc., are the normal working tools of quality management. However, summaries of this type of information are not easy to translate into the working language of the business executive.

The second objective of the quality manager is to keep the costs of achieving his first goal to an absolute minimum and involves making trade-off decisions between the various quality cost elements. The manner in which progress toward this goal is measured and reported has a tremendous bearing on top management's recognition of the quality manager as a fully competent and respected member of the business team.

One of the most widely used indicators of performance in the business world is return on capital employed. Executive managements judge divisional profit centers and total company performance in comparison to business norms established on this base. Normally, the capital employed remains reasonably constant over the short term and earnings are taken as the key indicator of successful operation.

It is therefore essential for the quality manager to report on the overall effectiveness of his quality system in values that are readily understood by the company's executive management. In general this can be successfully accomplished only by evaluating the performance of the quality program in terms of its impact on profitability.

This paper describes a method of measuring performance that would appear to satisfy the latter objective.

CONCEPT DEVELOPMENT

The ideal product for manufacture would be one which does not require any form of quality control whatsoever and where the total manufacturing process can be considered as pure added value to the raw materials used. This would then allow the manufacturer to achieve his maximum profit potential within the limitations of competition in the marketplace.

With the majority of manufactured products, however, failures do occur. The manufacturer who relies on the customer to inspect his product will register his total quality costs in the external failure category. These costs could be large enough to completely wipe out any anticipated earnings.

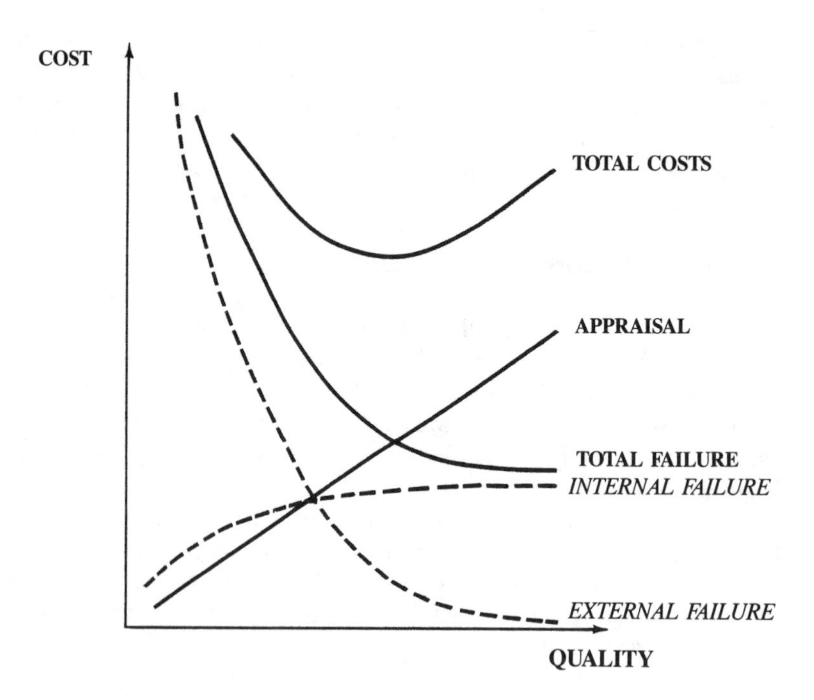

RELATIONSHIP OF MAJOR QUALITY COST COMPONENTS
(BASIC INSPECTION SYSTEM)

The introduction of a basic inspection activity will greatly reduce the costs of external failure by transferring most of the failures, at a much reduced cost, into the internal failure category. The cost of the screening operations and the costs of internal and external failures combined need to be balanced in order to provide a maximum return.

The basic inspection program, however, makes little or no attempt to determine the cause of failures. The creation of prevention and assurance functions, associated with a total quality program, are essential to further reduce failure costs and the cost of appraisal to a lower economically balanced position.

The problem of achieving an economic optimum now changes from one of balancing the cost elements of appraisal against failure to a condition where the interaction of the third component of quality assurance must be considered.

The components of the total quality cost structure illustrated below are arranged somewhat differently from that of traditional quality cost breakdowns and have been grouped to fit the development of this concept.

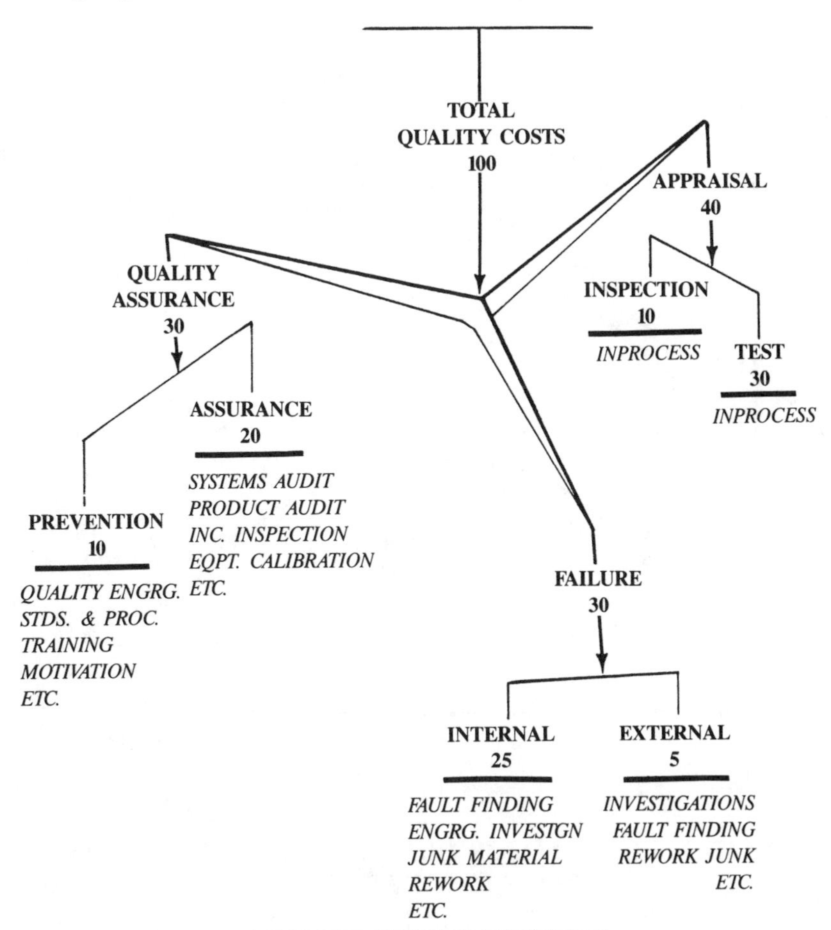

QUALITY COST STRUCTURE
(TYPICAL: ELECTRONICS INDUSTRY)

Although the work involved in Product Audit, Incoming Inspection and Equipment Calibration is an appraisal activity, it is felt that the principle purpose which these functions serve belongs in the quality assurance category. Incoming Inspection is essentially a product audit of the supplier and exists solely for the economic reason of preventing unsuitable parts being assembled prior to detection. System and Product Audit activities provide the feedback information required by Quality Management to plan any action found necessary to improve the system and product quality performance. Equipment Calibration assures that the accuracy of measurements can be maintained on a continuing basis.

A general overview of the relationships that exist between the three major quality cost components is illustrated below and the resultant presented in a form that looks upon the total cost of quality as a degradation to maximum earnings potential. This is based on the concept that all costs associated with maintaining an effective quality program come directly out of company earnings.

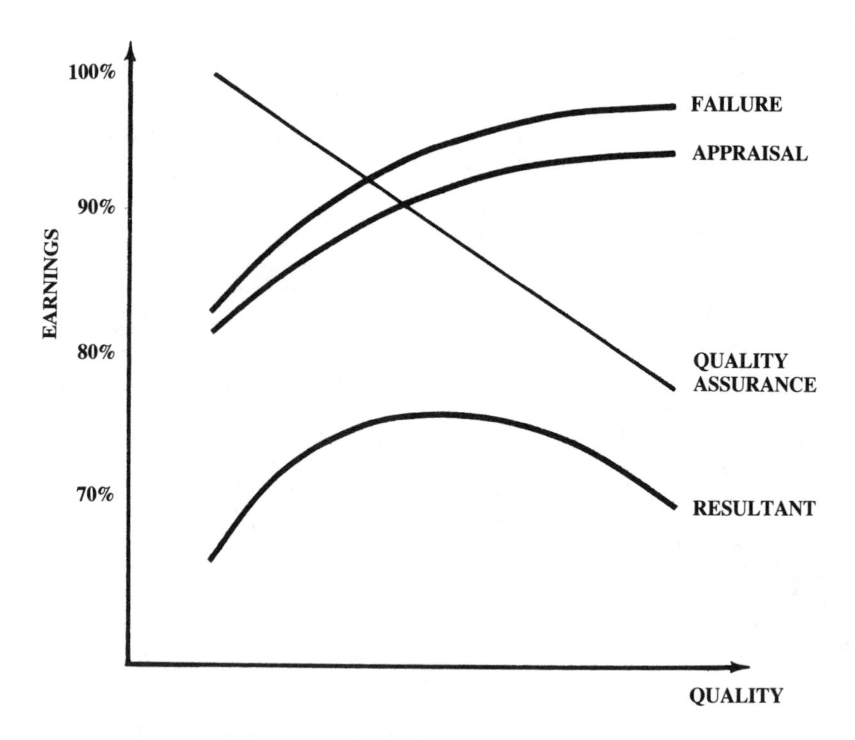

RELATIONSHIP OF MAJOR QUALITY COST COMPONENTS
(TOTAL QUALITY PROGRAM)

PERFORMANCE MEASURE

Since the common goal of management, including that of quality, is to maximize profit, any meaningful measure of the overall performance of a quality program must relate to the impact it has on the company's earnings. Based on the concept illustrated above, an expression of a performance rating suitable for this purpose would be:

$$100\left(1-\frac{T.Q.C.}{T.Q.C.+E}\right)$$

where T.Q.C. = Total Quality Costs per Month
E = Earnings before Taxes

and E+T.Q.C. represents the maximum earnings potential based on an ideal product.

Use of actual earnings before taxes as a base has a number of problems such as the time between the production activity and when the shipment is made, adjustments to earnings due to extraordinary items, price changes, etc. It is perhaps better to use a calculated earnings base such as Standard Cost of products produced during the month multiplied by an earnings ratio of the budgeted yearly forecast of earnings to cost of products produced. This gets over possible arguments that the performance index may be more greatly affected by inputs other than productive output or quality matters. Occasionally, time phase differences will still be evident should large preventive efforts be applied on new designs or an increased level of customer returns be experienced during the reporting period. These costs are easily isolated and explainable for short term reporting but must, however, be included as an integral part of any overall performance measurement.

Referring to the model initially presented in the introduction, a reasonable performance target could be set at 80% as shown in the following table. This table also depicts the expected shift in the distribution of quality costs that would occur during the build-up of a Total Quality program.

Distribution of Quality Costs	No Failures	Product With Failures		
	Ideal Product	**No Quality System**	**Basic Inspection System**	**Total Quality Program**
Sales	$ 100M	$ 100M	$ 100M	$ 100M
Earnings (Before Tax)	$ 20M	0	$ 12M	$ 16M
Prevention & Assurance	0	0	0	$ 1.2M
Appraisal	0	0	$ 4M	$ 1.6M
Internal Failure	0	0	$ 3M	$ 1.0M
External Failure	0	$ 20M	$ 1M	$ .2M
Total Quality	0	$ 20M	$ 8M	$ 4.0M
Business Performance Index	100%	0	60%	80%

CONCLUSION

By developing a deeper understanding of the interdependence between the various quality components and their effect on profitability, management will appreciate more fully the value in maintaining a healthy quality operation. It is felt that use of a performance measure such as described could be the vehicle for accomplishing this mission by providing the following benefits:

1. Allow top management to assess the profit contribution of quality management.
2. Provide a means for setting targets and tracking performance in the achievement of operation at a maximum earnings position.
3. Focus management attention on quality matters, particularly in highly competitive situations having slim profit margins.
4. Help the quality manager to determine the optimum balance between expenditures on quality assurance and appraisal activities relative to failure costs.
5. Ensure that quality management remains cognizant of the overall financial goals of the company.

THE HIDDEN ASPECT OF VENDOR QUALITY COST

William O. Winchell, Project Manager
Product Assurance Dept., Environmental Activities Staff
General Motors Technical Center, Warren, Michigan

The hidden aspect of vendor quality costs is an important subject to all of us because in most cases we not only buy from vendors but we are a vendor. For example, this area has a large concentration of manufacturing facilities. These facilities buy their raw material from vendors. Once the materials are processed into parts, much of the production is sold to other manufacturers. So the manufacturer is both a purchaser and a vendor at the same time.

For the purpose of this paper, we will look at hidden cost from just one perspective — that of the buyer. From the buyer's viewpoint, hidden vendor quality costs occur in three basic ways. By hidden, we mean that these costs are not really easily identified or perhaps not even seen as vendor related quality costs. The first type of cost is quality cost inside the vendor's facility. For the most part it is much the same as the buyer's quality cost. However, it is included as part of the piece price the buyer pays for the part. The vendor, perhaps rightly so, protects the identification of this investment for competitive reasons. The buyer is furnished the piece price, alone; and if this price is competitive, he's usually satisfied with this knowledge. The second type of hidden vendor quality cost is that which buyer spends, but does not usually segregate, to insure the quality of a product at the vendor's facility. This may be in the form of sending quality engineers to the vendor's plant to help him solve a crisis. The third type of hidden vendor quality costs is that the buyer spends in-house, again not usually segregated, specifically for vendors. An example of this may be process engineering required to repair or correct a purchased part not quite up-to-specification. Many of us call this fire-fighting to keep the line going. Of course there are also visible costs, that the buyer can and sometimes does segregate to specific vendors. An example of this type of cost is receiving inspection effort.

For a typical manufacturing firm these vendor related costs are a huge chunk of money-perhaps in the range of 10 to 20% of the selling price to your customer. Through proper control of these hidden and visible costs that are vendor related, we should achieve several results:

- — The first is to optimize our profits.
- — The second is to insure a product that is "fit-for-use" by our customers. "Fitness-for-use" of our products will insure our success in the marketplace. If we don't meet our customers' requirements, our competitors will; and we will not survive in the long run.

The basic purpose of this paper is to suggest to you methods by which you can control both the hidden and visible quality costs incurred in your vendor relationship. Very simply, two ways will be emphasized. The first way is to talk your vendors into adopting a quality cost program. The second way is to use your quality cost program to identify high magnitude vendor problems for resolution. Quality cost can be a valuable ingredient in your vendor relationship.

Before we discuss these methods in detail, let us first review the basic principles of quality costs and how they apply to our vendors. The American Society for Quality Control publishes a document entitled "Quality Costs — What and How," authored by the Quality Costs Committee. Included in this publication are the essential elements that make up a quality cost system, recommended definitions and methods of installing and gaining acceptance. In this publication the ASQC recognizes four basic categories of cost. They are identified as: *Prevention,* which primarily is the costs resulting from planning to prevent defects in products; *Appraisal* is costs relative to the conformance to quality standards. *Failure* costs are the results of poor planning and include the failure of products to meet the established or required quality standards. Failure costs are either *internal,* which are mainly manufacturing losses, or *external,* which are due to defective products shipped to the customer. Each category of prevention, appraisal, internal and external failure contains many elements. A few examples of elements in each category are listed below:

Prevention includes:
— Quality control administration
— Quality systems planning and measurement
— Vendor quality surveys
— Quality data analysis and feedback

Appraisal includes:
— All inspection and test costs
— Laboratory acceptance testing costs
— Vendor quality audits and surveillance

Internal Failure includes:
— Rework
— Scrap
— Retest and reinspection

External Failure includes:
— Returned material processing and handling
— Warranty replacement

The first step in a quality cost program, *that of measuring,* is the most difficult. Most businesses do not enjoy the sophisticated accounting system that permits interrogating for the cost of quality elements. A system for collection of this information has to be designed to fit each specific company. Once we have collected the information and categorized it we can now *analyze.* The most obvious is to examine those quality elements for which the greatest costs have occurred. It is here that the most effective and dramatic cost savings can be realized. But we must also note the quality elements having little or no costs, since insufficient effort may be occurring. For example, little action in vendor evaluations may result in high receiving inspection costs. *Action,* of course, is the result of analysis.

All the good cost reporting in the world is wasted unless someone takes corrective action. The logical next step is to take action in areas needing improvements. Specific persons must be assigned specific improvements against a time scale to assure effective results. Finally, we *control* the improvements in a continuing effort to do even better over the long term.

Now that we have reviewed quality costs in general, let us now discuss how they apply in the buyer-vendor relationship. Previously we discussed how the vendor related costs can be classified as either hidden or visible. Now I would like to discuss each of the categories in detail. Hidden quality costs, as you may recall, are in three parts:

— The first is those that are incurred by the vendor at his plant.
— The second is incurred by the buyer in fixing problems at the vendor plant.
— The third are those costs which are not usually allocated to vendors, incurred by the buyer as a result of potential or actual vendor problems.

Those quality costs incurred by the vendor at his plant are, for the most part, unknown to the buyer and, therefore, hidden. However, even though the magnitude is hidden, the type of cost is not. They are the same type quality cost as the buyer incurs. For example, the vendor certainly has prevention effort. If he manufacturers the product, he has expenses related to the quality engineering of the product. Even if the vendor is a small shop, this task must be done by someone and may very well be handled by the production supervisor if the plant lacks a quality engineering staff. Certainly effort is expended in the appraisal area even by the smallest vendors. Someone must inspect the product prior to shipping to the buyer. In a one man shop this is done by the man who made the part. Unfortunately, we all have failure costs whether we are a large shop or a small one. When the vendor makes a mistake in manufacturing, they must either rework the part or scrap it, causing an internal failure expense. If the vendor sends it to the buyer, it may be rejected with the resulting cost being considered as external failure.

The second type of hidden quality cost — that which is incurred by the buyer in fixing problems at the vendor's facility — is usually not specifically allocated to vendors. Except for an awareness of our troublesome vendors, there is usually no tabulation of the cost of the effort expended. Therefore, the actual expense is hidden. One example of this cost is the buyer sending a quality engineer to a vendor to resolve a crisis. Another example is the auditing a buyer may do at the vendor's plant to insure that he has the proper quality controls to build a product to the buyer's specifications.

The last type of hidden quality cost is that at the buyer's plant incurred on behalf of a vendor. Like the second type of hidden cost, we are certainly aware of it but we probably do not keep track of it. The magnitude of this expense is not known for specific vendors. Again, the actual expense is hidden. This type of cost may include in the prevention area the following:

— Preparing the specifications for packaging the product that is shipped by the vendor to the buyer.
— Specification and design of gages that must be used in the buyer's receiving inspection and, perhaps as well, by the vendor prior to shipping.

— Designation of appropriate specifications that the vendor must follow in the manufacture of the product.

For the hidden appraisal costs, we may incur expense for:

— Certain inspection operations and quality control effort in the buyer's production line related specifically to a vendor product.

— Review of test and inspection data on vendor parts to determine acceptability for processing in the buyer's plant.

— Calibration and maintenance of equipment necessary in the control of the quality of vendor parts.

In the internal failure area we may be troubleshooting or fire-fighting in order to use a vendor part that is not quite up to specification. Finally, regarding external failure, there may be field engineering required to analyze and correct a problem caused by a vendor. It must be remembered that this discussion of the types of vendor related hidden costs is by no means exhaustive. There are many more. Some of the types of cost not pointed out may be very significant depending on your individual situation.

There are, as discussed initially in this talk, vendor related quality costs that are apparent and perhaps much easier to identify in magnitude and assign to various vendors by the buyer. These are called visible costs. They are primarily in the appraisal and failure areas. Included in the visible quality cost category are the following:

— Receiving inspection

— Laboratory acceptance testing

— Scrap and rework that is vendor responsibility

— Warranty replacement on parts supplied by vendor

The visible costs, if tracked, are perhaps most significant because they can be good indicators of problem areas.

Now that we have looked at the various types of hidden and visible quality costs, let us discuss the possible methods through which we may optimize these expenditures.

The first step for the buyer in optimizing his vendor related quality costs is to determine what costs are important to him. It is suggested that comparing the relative magnitude of quality costs by category could be a good start. To do this, your own quality cost program could be invaluable. But, if you don't have a quality cost program, a special study could be initiated to determine this information. The ASQC publication "Quality Costs — What & How" would be a valuable guide in accomplishing this. For an example, let us assume a hypothetical situation in which vendor scrap and rework are the biggest problem areas for the buyer. If the buyer makes the assumption that through improvements in vendor caused scrap and rework warranty will be lowered, then for this company, vendor responsible scrap and rework are the important items in vendor relationship.

The next step is determining which vendors are causing the problem. Very likely you find that perhaps 20% of your vendors are causing 80% of your problems in scrap and rework. Now the buyer can narrow his effort to the vital few of the vendors and take appropriate action.

What is appropriate action? First and foremost, I would suggest that you promote that the "vital few" vendors start a quality cost program using the ASQC publication "Quality Costs — What & How." Through vendors launching such a program, the costs most visible to you will likely be reduced. If these are reduced, the hidden costs expended by both you and the "vital few" vendors should also be lowered. The result will be that the vendor's product and your product is more "fit-for-use." This should increase profits for both — a very desirable condition. What other action can be taken? Keeping track of vendor related costs, both hidden and visible, may be a gigantic undertaking for the buying company. However, some way must be found to insure that progress is being made. It is recommended that the buying company track the important visible costs for each of the "vital few" vendors. It is submitted to you that, in most cases, you will find that if progress in reducing the visible costs is demonstrated, hidden costs are also decreasing. What other action can be taken if we know the magnitude of the visible costs? It is possible that this information can be incorporated into a buyer's vendor rating system. The ASQC "Procurement Handbook" can supply valuable help to you in accomplishing this. A vendor relationship depends upon more than the traditional measures of meeting deliveries and receiving inspection rejections. It also must recognize the "fitness-of-use" of the vendor's products for which visible quality cost can be a valuable indicator.

Visible quality costs can also be used in other ways. Some companies debit vendors for the scrap and rework ocurring in the buyer's plant to put the responsibility for failures where it hurts most — in the pocketbook. However, in the long run this may be counterproductive in that some vendors may ask for a price increase to cover this situation. Another effective method is by reducing the amount of business given to the offender and rewarding the good performer with a greater share of the "order pie." A far more positive method is to use the visible costs to identify vendor quality improvements that are needed. The buying company would initiate projects jointly with vendors to resolve the problems that are the source of high quality costs. The problem may be solved through an action by the buying company. Perhaps the specifications are not correct or the vendor really doesn't know of the application of his product in the total package. On the other hand, it may be that the vendor's manufacturing process needs upgrading through perhaps better tooling. Through joint projects, using visible quality costs as facts, we can solve those problems and get better products. If quality costs are collected in a regular fashion, the results will document lower costs to both the vendor and buyer.

Let's now summarize what we have discussed. Basically, use quality costs in your vendor relationships. Through this tool, find out what costs and vendors are most important to focus upon. Suggest that these vendors adopt a quality cost program in order to obtain improvements in fitness for use of their products. Use your visible vendor quality costs as a basis for starting joint quality/reliability improvement projects with your suppliers.

Most important of all is that any quality cost program is not complete without an effective action program. The mere act of collecting quality costs alone will do nothing but add cost for you or your company. Only through pinpointing problems and solving these can we make progress.

MANAGEMENT BUDGET CONTROL: QUALITY LABOR STANDARDS

Ronald L. Pollard, Manager of Reliability and Quality Assurance
Addressograph Multigraph Corp., Mt. Prospect, Illinois

THE PROBLEM

Establishing the Quality Budget is a major management responsibility. When it comes to Quality Costs, it seems inevitable that the Quality Manager will have a different viewpoint than Executive Management, in particular, the Controller's office. The Controller views costs in a different manner than the Quality Manager.[1] Grimm shows how some of the communication gap between the Quality field and the Accounting field can be resolved. Of course, definitions of Quality Cost can be obtained from ASQC[2] and the procedures to tabulate Quality Costs and relate them to scrap, rework, warranty and other parameters are outlined in the literature.[3] What is not so clear is how these Quality Costs relate to the age old art of budgeting. Typically a budget is determined from a) a sales forecast and b) cost of living guidelines received from the Controller's office. The procedure probably followed is as follows: 1) Receive the sales forecast 2) Review what was spent last year 3) increase or decrease budget estimates based upon business expectations and 4) adjust for cost of living in accordance with the guidelines. 5) Present the ''budget'' to your superior and after some negotiations about ''cutting overhead'', arrive at a budget compromise. 6) A composite budget is submitted to corporate and perhaps new profit guidelines are established and 7) resubmitted to the division or plant for adjustments. There you are, back again, negotiating about ''overhead'' and cutting costs. Finally a budget is arrived at (not really agreed upon) which you have a ''gut'' feeling is marginal or unsatisfactory.

Where is Quality Cost in this environment? Where is scrap, warranty, prevention and all the other good things we talk about in our technical sessions? Is the appraisal function merely overhead? To be bandied about until the overhead is in line? Where is our break-even analysis? Our cost justification? One reason why the Quality Manager is in such a situation is that there are ''no standards'' for Quality labor and these standards, if they exist, are not related to Quality Costs in a convenient way for the executive office to make a rational decision. To the Executive — you, the Quality function, *cost money.* You are there because if you are not there, the business gets an unreasonable amount of customer complaints. The Executive then wishes to reduce these costs as much as possible without receiving customer complaints. This idea is of course too simple but it dramatizes the fact that the Quality Manager, has not presented his case in financial terms understandable to Executive Management. This problem of developing and approving a Quality budget is quite complex but there is a way to relate at least the appraisal activities to Quality Costs. This can be done by a) establishing labor standards b) comparing these to ''escape rates'' or complaints and

c) determining "loss on investment" and d) relate these three parameters to the budget. This paper will limit the discussion to the evaluation of appraisal activities and leave the subject of prevention costs for another time.

IDENTIFICATION OF QUALITY ACTIVITIES AND LABOR STANDARDS

For purposes of this discussion, it is important to realize that there are two types of standards — Labor Standards and Budget Standards. Quality labor standards are discussed by Juran[4] in which four methods are highlighted a) Historical Standards b) Engineered Standards c) Market Standards and d) Project Standards. These standards relate to inspection activities and will vary in accordance with 1) lot size 2) product mix and 3) flow of material. Labor standards, then, establish how much work an inspector can accomplish under a particular situation and may or may not be related to the allocation of funds made by the company for inspection purposes. In other words, how can work load be transferred to budget requirements? In addition, if you have a complex product line involving a variety of processes such as chemical, electrical, mechanical, and optical; how do you establish labor standards? The thought of utilizing traditional Industrial Engineering time studies for 100,000 different parts is mind boggling. Producing a good economical quality plan for all these parts, sub-assemblies, and machines requires enormous engineering efforts and money — much of which is not availalbe. While this problem is being confronted, another factor usually enters the picture — new products. The introduction of new products usually causes severe strains on the planning activity. As if the above problems were not enough, our company decided to install a "Direct Cost Accounting System" in one year's time. Direct Cost is the out of pocket cost on each product for material, labor, and overhead. This is in contrast to a fully absorbed cost where overhead is grouped into one account and redistributed over the entire product line in accordance with an allocation formula. This new Direct Cost system immediately requires specific costs or "time allocation" on each product. Obviously, the impact of this problem was enormous. In order to achieve budgeting standards, satisfactory labor standards had to be established. The most rational approach appeared to be to utilize the "Historical Standards" mentioned by Juran.[4] We had for instance, performed work sampling in Receiving Inspection and had established a base line from which to monitor the efficiency of the operation. The only information missing was the time allocated to each product line. A similar situation existed in the other parts of the plant, i.e., efficiency standards had been established but allocation of labor to specific product was missing. Although specific information on parts and machines was missing, we felt comfortable that the operation was standardized and the overall operation stablized. The strategy was then to gather specific information on products, relate time expended to production, (Historical Standards) established the historical standard effectiveness by comparing to escape rate and loss on investment, project this historical standard to new sales forecast and create a budget.

DEVELOPMENT OF DIRECT COST INFORMATION AND BUDGETING TECHNIQUES

In order to document inspection costs on each product, labor reporting cards had to be designed and the use of EDP for processing the information established. This task was accomplished with the usual de-bugging problems, reruns, training, etc. until we felt comfortable that the information was accurate. This information was utilized in budgeting.

Before we introduce budgeting, perhaps a short diversion is necessary to discuss two types of budgets that can be utilized in the Direct Cost system. There is a "Fixed Budget" and a "Variable Budget". Both types are used in industry and each has associated with it different management philosophies. In a fixed budget system, the budget is established at the beginning of the year and variances from the budget are "explained" every month. Reasons such as increased or decrease in business, technical problems, etc. are developed for explaining the variance. Usually the intensity of explanations is dependent on whether the variance is positive or negative. "Staying within budget" offers the easiest explanation task. Of course, during a recession or downturn, it is easier to "stay within budget" than it is during an expansion mode.

The "Flexing" or "Variable" budget is designed to overcome the shortcomings inherent in a fixed budget. Business conditions do change and it is more difficult to measure performance relative to a fixed budget than it is to a variable budget. In a fixed budget system, it is difficult to estimate what the costs should have been during a business change and, therefore, the overrun or underrun information does not show the costs related to efficiency and utilization. In other words, additional information is required to assess budget performance. In a "flexing" budget system, the throughput of the department is based upon utilization and efficiency and the budget standard utilized for flexing incorporates this information. Performance to budget is easier to determine. For instance, at Receiving Inspection, the number of lots received is a throughput parameter and is independent of "dollars purchased" or "direct labor" in the plant. If the area's efficiency and utilization are maintained at satisfactory levels, then it follows that a certain "throughput" will be accomplished. The acceptance or rejection of a lot plays no part in this budgeting scheme. The budgeting could "flex" then on "the number of lots received". Explanation of variances from budget are thus restricted to business problems rather than emotional problems concerning the inspector's abilities, attitudes and the like. In other areas of the plant, "Flexing parameters" established were: "Direct labor hours for machine shop, subassembly, and machine assembly work; "square yards produced" for such inspection activities involved in supplies manufacture. The point is that each area must decide on a factor that the budget can "flex" on and that the parameter should be related to inspection "throughput".

We felt that the "Flexing" budget offered the best control mechanism for us and eliminated the need to develop a "defensive posture" and provided an opportunity for our supervisors to concentrate on cost savings and maximizing efficiency. The disadvantage of the Flex Budget system is that it requires accurate information for a good standard.

The development of good standards was accomplished by gathering specific information on products via the labor cards previously mentioned. This information was then matched to historical production volumes, budget standards established, and a budget calculated from a new Sales Forecast. The relationship of Quality Costs in this procedure will be explained later. The calculations necessary to transfer product information to budget information is shown below:

Machine Quality Verification — Machines Produced

	Actual Labor Hours	Actual No. Hours Available	Actual No. Machines Built Last Year	Actual No. Hrs./Machine	Proposed Forecast No. Machines	Proposed Hours Needed
Machine Type						
X						
Y						
Z						

Receiving Verification — Lots Received — $/Lot

	Labor Hours	No. Hrs. Avail.	No. of Machines Produced	No. of Lots Rec'd.	Lots per Mach.	Hrs. per Lot	New Machine Forecast	Lots Expected Lots/Mach. X New Forecast	Expected Hrs. = Hrs./Lot X Expected Lots
Machine Type									
X									
Y									
Z									

Supplies Verification — Square Yards Produced

	Labor/Hrs.	No. Hrs. Available	Yds. of Product Made	Actual No. Hrs./Yd.	Proposed Yards	Proposed Hours Needed
Product						
X						
Y						
Z						

As an example of the calculations, let's review the Machine Quality Verification sequence. Approximately 2 months data were available at budgeting time so this information was transformed into a percentage of time worked on a product model (Column 1). This percentage was used to calculate the number of hours worked on a model for a whole year (Column 2). In Column 3, the number of machines built during the year was listed.

A budget standard was then established by dividing the number of hours worked by the machines built (Column 4). The new forecast of machines to be built was listed in Column 5. A new budget was then calculated from the new forecast and budget standard (Column 6). Similar type calculations were made for Receiving and Supplies Verification areas. It appears, at this point, a valid forecasting technique is available but some adjustments are necessary. In the case where no historical data are available, which happens on the introduction of a new model; some estimating must occur. One of the better ways of estimating is to compare the new product with existing machines and select a machine similar in complexity and use the budget standard for that machine on the new product.

It appears that we are ready to submit our proposed budget for approval and substantiate our request. Before we do this, the budget must be related to Quality Costs. How can we do this?

QUALITY COSTS AND BUDGET RELATIONSHIPS

If we are going to compare Quality Cost information with the budget, we must first answer the basic question concerning the effectiveness of the program. In our case, the effectiveness of the Quality Program is measured by compiling data from an independent assessment of machine quality. When our machines are shipped, they are installed by another division (Service) and the results of this installation forwarded to the manufacturing division. The "Percent of Machines which Function Properly" when installed by Service is the parameter followed to show the effectiveness of the Quality Program. The results are shown in Figure 1.

FIGURE 1
ESCAPE RATE
PERCENT OF MACHINES
WHICH FUNCTION PROPERLY
WHEN INSTALLED BY SERVICE

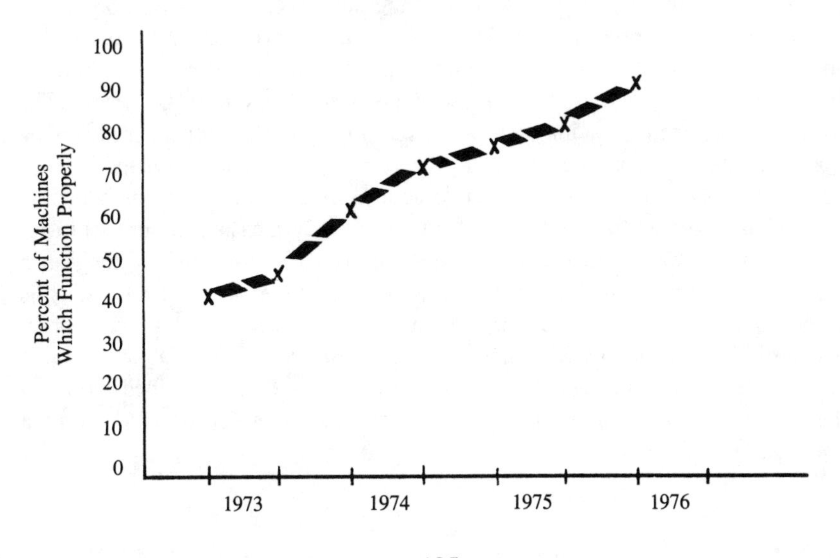

	1971	'72	'73	'74	'75	'76
Quality Engineering As a % of QualityBudget	-0-	7.5%	15.6%	14.1%	13.3%	13.4%
Capital Investment Inspection Equip. & Failure Analysis		$50,000	$218,000	$51,000	$18,000	$25,000
Warranty % Sales					0.98	1.12
R&QA Budget (%Sales)					1.0	1.0

As can be seen from Figure 1, an excellent improvement has been attained since the inception of a "Total Quality Control" program. Investment in capital and Quality Engineering has yielded positive results. It is unfortunate that during the early years of the program, Quality Cost information was just not available. The collection of these costs is a tremendous undertaking and certainly could be the subject of another paper. As you can see, warranty cost as a percent of sales is also tabulated in Figure 1. This important segment of Quality Cost is the result of mistakes in design, manufacture and possibly Service. Is this cost too high? How can we relate it to budgeting for appraisal activities? Where are we on the curves in Figure 2? The fact that we know the external failure costs does not pinpoint our location on the curve. One helpful approach in answering this question is provided by Juran where he shows ratios of Quality Costs categories[4] (Figure 3). Use of this information may assist in determining whether Quality Costs are near optimums. This estimate is qualitative in nature and one does not know which side of optimum he is on. In one case, increasing inspection activities may help and in the other case, decreasing inspection would be the direction to go. In addition, the cost magnitude of reducing failure costs by 1% may be significant and the guide presented by Juran has zones differing by 10% to 20%! It appears that it would be quite an undertaking to find out where you are on the curve. One has to decrease or increase inspection and monitor the results. The wrong decision can bring undesirable results. In our business field, feedback on warranty extends in some cases over 1 year's time. Failure Analysis helps to determine the area of interest for corrective action, that is; design, manufacturing, or purchased goods. It is quite complex to accurately measure the impact of inspection operations on field quality and therefore, one has to turn to long-term monitoring compared with changes in action taken at the factory level. I see this drawback as the major reason why Quality Costs are not considered as a practical tool by some Quality Managers. I believe, however, that if labor standards and budgets are established by some valid reasoning such as described early and this information compared with Quality Costs, then these costs will become a practical tool.

FIG. 2

QUALITY COST RELATIONSHIPS (4)

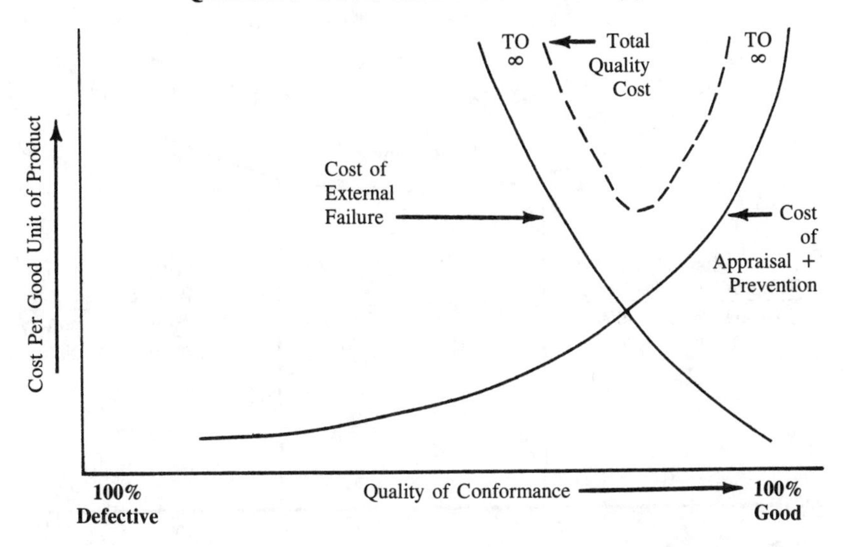

RELATIONSHIP OF MAJOR QUALITY COST COMPONENTS (5)

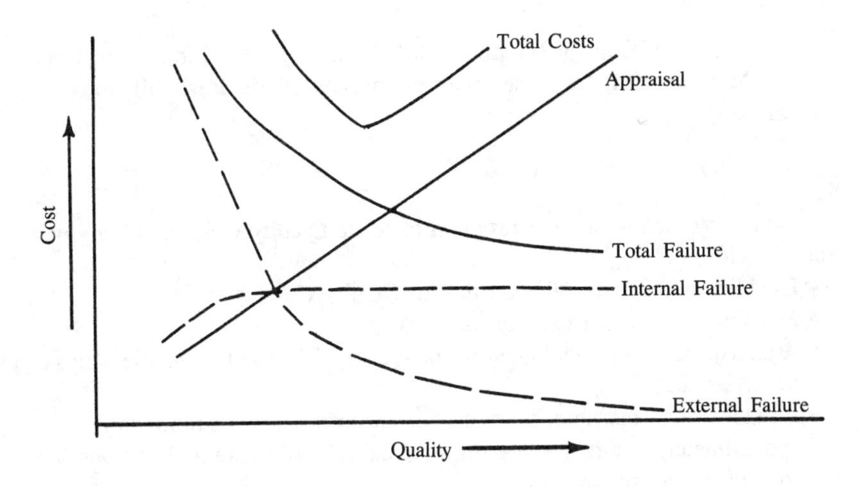

(4) "From Quality Control Handbook-3rd Edition by J. M. Juran. Copyright 1974 McGraw-Hill, Inc. Used with permission by McGraw-Hill Book Company".

(5) "From A Business Performance Measure of Quality Management by R. A. Cawsey *Transactions* 1976. Copyright© 1976 American Society for Quality Control, Inc. Reprinted by permission".

FIGURE 3
RATIO FOR QULAITY COSTS CATEGORIES (4)

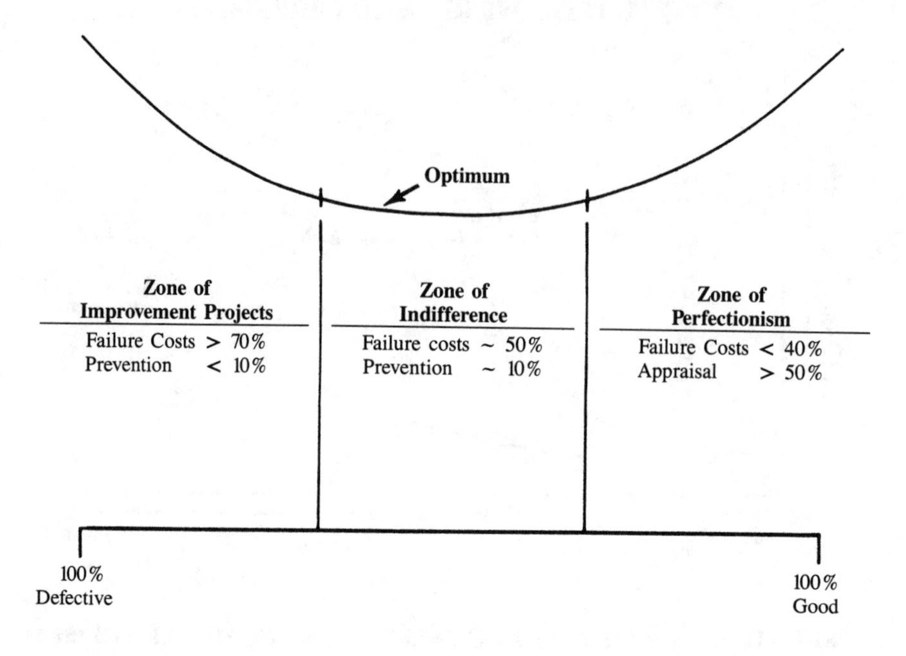

(4) "From Quality Control Handbook-3rd Edition by J.M. Juran. Copyright 1974 McGraw-Hill, Inc. Used with permission of McGraw-Hill Book Company".

In summary then — a valid program to relate Quality Costs and the Quality Budget is to:

- Establish good financial reporting of Quality Costs.
- Establish good Field Quality information.
- Establish trends on Field Quality and Quality Costs so that relationships can be developed.
- As a temporary measure, relate the current costs to Juran's ratios so you know approximately where you are on the Quality Cost curve and can develop a strategy for a proposed budget.
- Be able to justify efficiency of your operation through labor standards.
- Present your budget every year with a Quality Cost curve supplement — eventually the use of Quality Costs will be established.

REFERENCES

(1) *Quality Costs: Where Are They In The Accounting Process?* A. F. Grimm — 1974 ASQC Technical Conference Transactions — Boston.

(2) Quality Cost-Cost Effectiveness Committee, ASQC, *Quality Cost — What & How* 2nd Edition — American Society for Quality Control, 310 W. Wisconsin Ave., Milwaukee, Wisc. 1971.
(3) Feigenbaum A. V., *Total Quality Control: Engineering and Management* McGraw-Hill Book Co., New York, New York 1961.
(4) *Quality Control Handbook*-3rd Edition 1974-J. M. Juran, Editor-in-Chief — McGraw Hill.
(5) *A Business Performance Measure of Quality Management* by R. A. Cawsey — 1976 ASQC Technical Conference Transactions.

EXTENDING EFFECTIVENESS OF QUALITY COST PROGRAMS

Richard K. Dobbins, Principal Quality Assurance Staff Engineer
Honeywell, Inc. — Process Control Division/Fort Washington, Pa.

EVOLUTION OF QUALITY COST PROGRAMS

Does your company have a quality cost program? If so, would it be classified as *active* or *passive*? What primary purpose did the program originally serve? What purpose does it serve now? And, most importantly, what reaction does it generate within your executive offices?

The answers to questions like these would prove very interesting, if it were possible to obtain factual replies from all American business concerns with more than 500 employees. While no valid studies are available, the ASQC Quality Costs Technical Committee has reason to believe the following relationships would probably apply:

1. *The majority of companies still do not use any formalized quality cost program.*
   - Most manufacturing companies regularly monitor certain key quality cost elements, such as scrap, rework, inspection budget costs, etc.
   - Other management techniques are frequently employed, which may bring considerable temporary cost consciousness to particular products or departments. Examples include:
     A. Budget controls and restrictions.
     B. Value Engineering or cost reduction studies.
     C. Cost improvement goals within Management by objectives.
     D. Employee suggestion systems.
     E. Management bonuses, incentive systems, and profit sharing.
   - Formal quality cost programs are quite rare within service industries, or companies having high service-to-product ratios.
2. *Use of formalized quality cost programs is steadily growing within Manufacturing Industries, and there is a noticeable quickening in growth rate.*
   - In 1967, when the first edition of *"Quality Costs — What & How"* was published, quality cost analysis was mainly an unknown academic curiosity within technical circles of the quality profession.
   - In 1971, when the expanded second edition of this manual was published, these techniques had been accepted in principle by only the most progressive of Manufacturing Industries, and experimental programs had begun.
   - By 1977, when *"Guide For Reducing Quality Costs"* was published, routine use of quality cost analysis was an accepted practice for a significant minority of manufacturing companies. Still others were just instituting trial programs.
   - Within the past five years, interest levels have been progressively increasing, as indicated by attendance at the tutorial and management sessions

at the Annual Technical Conferences, regional, divisional and section symposiums.

3. *Initial applications of formal quality cost programs are usually directed towards specific pre-existing malignancies of major proportions.*
   - Quality cost techniques are used to identify and focus attention on these known aggravated problems, (generally product lines or manufacturing processes).
   - After quantification, priority managerial approval is obtained to mount a coordinated attack by all concerned parties.
   - Cost analysis reports document the early economic results of this campaign, until all proposed corrective actions have been implemented, (or other crises require reduction in attack effort).
   - Subsequent reports document the longer term impact of the earlier task team corrective actions, and soon reports become very routine until a new crisis causes the cycle to be repeated.
4. *The normal use of a quality cost program may, therefore, deteriorate into a "scorecard" posting, with little sustaining executive influence.*
   - Reports serve chiefly as a reminder of previous "glorious accomplishments", plus a general reassurance that present day quality is still "in control".
   - Complacency is further promoted through favorable comparison with:
     A. Non-exact general industry approximations.
     B. Past company history before any quality cost program.
     C. Conservative or "safe" company goals.
   - Most quality cost programs are maintained only on a composite company or divisional basis, without ready segregation of specific product lines or manufacturing processes.
   - Unless violent unfavorable events occur, little effect may be indicated in the overall "scorecard" rating.
   - When used only in a passive manner, the analysis indicates results which have already occurred, with no opportunity to change what is now past history. This further increases the "scorecard syndrome" on the part of executive reaction!

QUALITY COST ANALYSIS — SO WHAT?

Unless dynamic response is introduced into your quality cost program, it may be found that your company's reaction will have rotated full circle from *"Quality Cost Analysis — What's That?"* to *"Quality Cost Analysis — SO WHAT?"* Companies should not use a quality cost program simply for the sake of using one, or because they believe they have a better quality cost rating than the average company in their line. And they certainly shouldn't start one because best Q.A. theory indicates it is the *right* way to do things. It is much more important to "do the right things" than it is to "do things right"!

DOING THE RIGHT THINGS

From an executive management viewpoint, what type of "right things" need to be done in which Quality Assurance can exert a positive influence? Examples include:

- Early identification of problems, with assessment of magnitude and trend patterns.
- Establishment of priorities for efficient utilization of available limited resources, and decision making.
- Administrative structuring to have middle management exercise *active* responsibility, within their authority, for *total* company interests.
- Combat complacency and inertia through actions to improve the status quo.

A well designed quality cost program can be an effective management control tool, instead of simply a "scorecard" of composite quality cost attainment. But, how can this happy state be achieved?

An active quality cost program will assist and permit the *management* of total quality costs consistent with company policy. But, to "manage", we must *control.* To "control", we must *measure.* To "measure", we must *define.* In "defining", we must *quantify.* Keeping these elementary requirements in mind, the following guidelines are presented for construction of a powerful quality cost program.

ANATOMY OF A DYNAMIC QUALITY COST PROGRAM

1. *Provide for detailed cost identification and segregation down to the lowest level where control is desired.*

   Remember, in combined form, the individual elements are not capable of measurement, and therefore can't be controlled. It may not be necessary to segregate each individual product model, but general product lines or families should be separated from other costs. Within manufacturing facilities, consider whether "machine shop" might be too broad a listing, and yet "machines lathes", "turret lathes", "multi-spindle", and "automatic screw machines" might all be combined into "lathe family".

2. *Reflect all identified factory or field quality costs back to the vertical lines responsible for the products and services involved. This includes allocating the entire Quality Department budget!*

   Don't lump any quality cost elements into big, nondescript collections, such as "field service costs", or "general scrap". Should the Field Service Manager be held accountable for the costs associated with servicing field products which become defective during the warranty period, or should the responsible Vertical Line Manager? Should not that responsibility include *all* factory and field costs identified with that product or service? If so, then the Q.A. and Inspection costs attributed to the support of that product (both in the factory and in the field) should also be allocated back to that Vertical Product Line. Generally speaking, all identified or allocated costs should reflect back to the same organizational groups which received credit for the units of production, and net sales billed.

3. *Establish bold cost improvement objective goals, which are believed to be achievable, but only through sustained, hard effort.*

   Don't establish next year's goals based upon actual costs for this year, plus an allowance for inflation. This keeps increasing management's "threshold of pain" to the point of insensitivity to all but major calamities. *Long established* costs do not necessarily mean *unavoidable costs.* challenge them!

4. *The cost improvement objectives established should be incorporated as personal goals of the lowest managers who have overall discretionary authority for the product or service involved.*

   Individual objectives for the product lines or services should then be combined as composite objectives upwards within the chain of command, in the same manner as line responsibilities are combined upwards. How else can you require these managers to control their quality costs in the same manner they now control their direct and indirect costs, and manage their production schedules?

5. *Consider all high quality cost elements as being potential cost breakthrough opportunities.*

   Use Pareto analysis and trend analysis techniques to identify such opportunities for concentrated attack through design, material, process, methods and equipment challenges. Initiate Value Engineering studies, and invite suggestions from the people who work on the product line or in the processing area involved. Compare how your competitors handle that aspect within their products. Consider new technology which has been introduced since your design or process method was first implemented.

6. *Incorporate specific quality cost objectives into all new or improved product design efforts, or services offered.*

   Quality cost objectives should be just as important as formal estimates of direct production costs, which *are* normally a specified criteria of every new release. (Also, don't forget that MTBF goals should be included in all Sales-Engineering project specifications for new releases.) In the same manner as the direct cost estimates, quality cost objectives should be agreed to by the Vertical Line Manager who will finally be responsible for the new release. Until achieved or renegotiated, Engineering should be held jointly responsible with the Vertical Line Manager for excessive direct or quality costs associated with the new product release.

7. *Require rapid collection of current data, and perform timely trend analysis using appropriate multiple bases.*

   The object of this exercise is the creation of an "early warning system", to permit necessary interim, adaptive and corrective actions. Rapid actions will prevent or limit additional avoidable losses from being sustained. Doing this will provide the dynamic response needed to overcome the historical "scorecard syndrome".

8. *Use the quality cost program to manage the available Quality Assurance budget and resources where they will be most effective.*

Whether your departmental budget and resources are generous or meager, *stop doing other people's work for them,* and concentrate on your own major quality responsibilities. As Quality Manager, you are *not* responsible to cure every problem discovered! But you *are* responsible to see that each such problem is properly addressed by the appropriate action parties, and then satisfactorily resolved!

BENEFITS OF A DYNAMIC QUALITY COST PROGRAM

Quality Assurance should intensify efforts towards the identification of existing or potential problems, assessing their magnitude and impact, referral to the proper parties for resolution, and monitoring corrective action efforts and effectiveness of same. In all these endeavors, a dynamic quality cost program could be the vehicle to manage the majority of these efforts.

Such a quality cost program would then be truly an effective management control tool, and not just a historical "scorecard" grade. Just as a machine tool provides a distinctive mechanical advantage, so this diagnostic control tool provides a distinctive "managerial advantage". This advantage can then be directed towards more effective management decisions, increased profitability, better return on investment, improved utilization of Quality Department resources, and elimination or improvement of product lines or services with marginal future economic worth. Isn't this a giant-sized improvement over former intuitive or stereotyped management methods? It can assure a progressive company of increased profitability and cost effectiveness, and permit it to engage in keener competition with smaller risks. That, in turn, can mean greater market share, whether times are good or bad.

THE QUALITY ASSURANCE CHALLENGE

Have you offered the executive management of *your* company a dynamic quality cost program yet? Consider how your competitive position might now have been improved, *if* such a program had been implemented two years ago! Can you afford to wait until *after your major competitor has one in operation?*

BIBLIOGRAPHY

"Cost Effectiveness of Corrective Action", R. K. Dobbins, *30th Annual Technical Conference Transactions,* ASQC, June 7-9, 1976, pgs 115-122.

Guide for Reducing Quality Costs, Quality Costs Technical Committee, ASQC, Milwaukee, 1977.

"Quality Cost Management for Profit", R. K. Dobbins, *29th Annual Technical Conference Transactions,* ASQC, May 12-14, 1975, pgs 358-367.

Quality Costs — What & How, Quality Costs Technical Committee, ASQC, Milwaukee. 1971.

"Systems Analysis in Management", D. Burchfield, *Quality Progress,* January, 1970, Page 18.

"A METHOD FOR PREDICTING WARRANTY COSTS"

Andrew F. Grimm, Manager, Quality Assurance
Clark Equipment Company — Axle Division, Buchanan, Michigan

Data generated in a reliability evaluation of a product can be an important starting point for predicting potential costs involved with product failure at any future point in time. A conversion of failure rates into cost data can prove to be a valuable information contribution to business managers in forming decisions about continuation of a product program or at least provision of initial and ongoing capital budgeting plans for a potential product's production and introduction to its market.

The objective of this paper is to illustrate a method for translating a product's reliability profile and design alternative considerations into a cost structure that can provide business management with information they can understand. The extra steps of this method could prove worthwhile in simplifying illustration of the underlying potential warranty problems and associated costs that might not be truly understandable in its original technical form.

Basically, the three parts of the method are the original reliability evaluation of a design and its alternatives, the application of Markov Chain Theory in extending failure rate data into the future, and the analysis of applied costs to this point by the net present value method in order to determine project profitability potentials between the original design and its alternatives. Markov Chain construction can provide information about summation of product failures at future points in time as production continues and the population of product in use increases. This profile provides information about product design failure rate information can be applied in predicting the probable quantities rate of return of failed goods. Of course, return of goods may be covered under both implied product warranty as well as stated warranty policy.

In order to explain the method, an actual example has been selected to illustrate the steps taken in arriving at project cost information that will be in a simplified form that business managers can find to be usable in forming decisions about a project's viability and continuation. Although this project example is real, certain elements have been changed so as to prevent recognition of the actual product. This author believes that the changes that have been made to prevent recognition of the product do not materially affect the outcomes contained in the example's illustrations. The example selected is a small analog computer that has a fixed program for controlling a mechanical function of some machinery. Knowing the actual controlling outputs of this device are not necessary to understanding the evaluation of the reliability of the product. However, let us end this product's application with the note that it is an integral part of a higher order system. The product part we will use in our illustration is not maintainable, but is replacable in the overall system.

The computer is composed of a variety of electronic components mounted on three stacked PC boards. These electronic components consist of the usual varieties including transistors, pots, capacitors, integrated circuits, diodes, resistors, etc.

Our study will be confined to the transistors, ICs and diodes involved in the design. The original design contains the following quantities of each component under study: IC's -2, Transistors - 14, silicon diodes - 25.

Reliability information about these components is drawn from MIL-HDBK-217B, "Reliability Prediction of Electronic Equipment". Table 1. Contains a summary of this information as used in this study. Data was generated for components operating at 25° C ambient and at two levels of quality, i.e: commercial vs military, (Jan). Table 1. contains the pertinent information concerning the design alternatives under consideration. This information includes the combinations of commercial versus military quality components for each design alternative which in turn influence the total unit failure rate. Of course, the total unit incorporates other components which for simplicity in this example are considered held constant. Additional information contained in Table 1, is the cost to build per unit which takes into account the added cost of "burn in" and sort testing of military quality components. Comparing the failure rate column with cost to build does not clearly define a relationship as to which combination delivers the optimal reliability to cost relation. Thus the business decision becomes quite difficult unless additional means of analysis is found to overcome the indecisiveness of this problem. The method we will propose is one means of bridging the current state of our problem.

TABLE 1

Design Alternative	Component Quality Levels IC	Diode	Transistor	Failure Rate %	Cost to Build/unit
1	c	c	c	12.6	$30.00
2	m	c	c	5.9	30.14
3	m	m	c	3.8	31.89
4	m	c	m	4.6	31.54
5	c	m	c	10.5	31.75
6	c	m	m	9.3	33.15
7	c	c	m	11.2	31.40
8	m	m	m	2.5	33.29

m - military, (Jan) c - commercial

Let us assume that we must generate a returned goods rate based on the failure rates indicated in Table 1, for each component design quality level. The method used is the Markov chain. Application of this technique will develop the number of units expected to fail within each time segment of the warranty period. Our example will utilize a five year warranty period. Instead of developing Markov chains for each of our eight design alternatives, we shall only develop an example of a Markov chain for a 10% failure rate over a 5 year warranty period. Figure 1 is the graphical illustration of this example Markov chain.

FIGURE 1

Markov Chain for: Failure Rate, 10%; Warranty Period, 5 years.

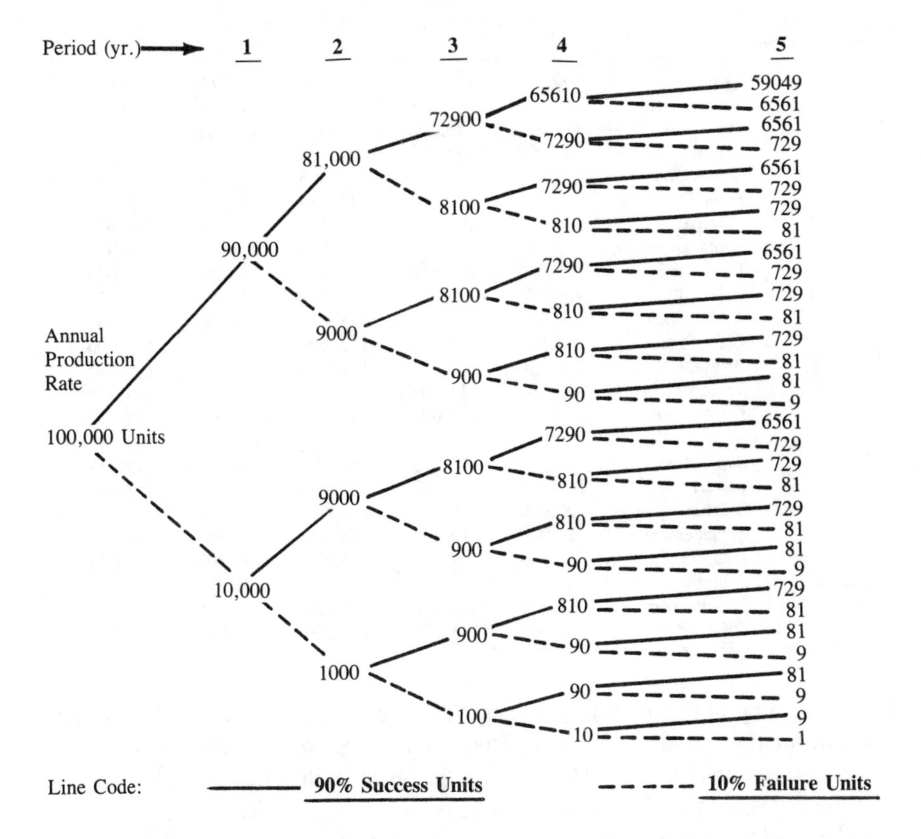

Note that failures are replaced with additional production in the following time segment and that those replacement units are subject to the same failure rate in each time segment. The Markov chain is an excellent tool to handle this condition for production replacement. Table 2 is a summation of the Markov chains for each design alternative in the example.

TABLE 2

Design Alternative	Sale Type	Warranty Year 1	2	3	4	5
1.	Original	100,000	111,012	122,024	133,036	144,048
	Replacement	12,600	14,188	15,776	17,364	18,952
	Total Prod.	112,600	125,200	137,800	150,400	163,000
2.	Original	100,000	105,552	111,104	116,656	122,208
	Replacement	5,900	6,248	6,546	6,944	7,292
	Total Prod.	105,900	111,800	117,650	123,600	129,500
3.	Original	100,000	103,656	107,311	110,967	114,622
	Replacement	3,800	3,944	4,089	4,233	4,378
	Total Prod.	103,800	107,600	111,400	115,200	119,000
4.	Original	100,000	104,388	108,777	113,165	117,554
	Replacement	4,600	4,812	5,023	5,235	5,446
	Total Prod.	104,600	109,200	113,800	118,400	123,000
5.	Original	100,000	109,398	118,795	128,193	137,590
	Replacement	10,500	11,603	12,705	13,808	14,710
	Total Prod.	110,500	121,001	131,500	142,001	152,500
6.	Original	100,000	108,435	116,870	125,305	133,740
	Replacement	9,300	10,165	11,030	11,895	12,760
	Total Prod.	109,300	118,600	127,900	137,200	146,500
7.	Original	100,000	109,946	119,891	129,837	139,782
	Replacement	11,200	12,454	13,709	14,963	16,218
	Total Prod.	111,200	122,400	133,600	144,800	156,000
8.	Original	100,000	102,437	104,875	107,313	109,750
	Replacement	2,500	2,563	2,625	2,688	2,750
	Total Prod.	102,500	105,000	107,500	110,001	112,500

Combining information from tables 1 and 2 and figure 1 allows development of information for tables 3 and 4. (Note: Due to the size of tables 3 and 4, they are contained in Appendix B). Table 3 is the business information matrix leading to the NPV of table 4. A discussion of each element in Table 3 will allow an understanding of each idea. The design alternative is taken from Table 1 and indicates the mix of commercial quality to military quality in the three components. The year column signifies each year in the warranty period. Units for sale is taken from Original Sale Type in Table 2. Cash In Flow at $50 is the total income of units for sale at a price of $50 per unit. Units Built is the sum of original production and replacement production based on Markov chain outcomes at each design alternative failure rate. Cost to Build is the product of Units Built times the cost of manufacture. Again, unit cost of manufacture is based on the mix of military to commercial quality of the three basic electronic components we are studying in our example. Units Returned Warranty is the sum per period of failures calculated in the Markov chain. Additional Cost @ $25 Warranty is the product of Units Returned times the cost of handling warranty that is additional to the production cost shown in the Cost to Build column. Cash Outflow is the difference between the column, Cash Inflow @ $50, and the summation of the columns, Cost to Build plus Additional Costs @ $25 Warranty.

Table 4 represents the steps in calculating the net present value for each of the design alternatives. This is the business decision matrix. The first column is Cash Outflow. This column is the ending column in Table 3. The next column is Depreciation of capital assets. For simplicity, we have selected the straight line depreciation method for a five year period. Net Income B_4 Taxes is the difference between Cash Inflow and the summation of Cash Outflow plus Depreciation. The next column, Net Income After Taxes is based on a corporate tax rate of 48%. (Note: The corporate tax rate may vary according to changes in tax law and may be subject to change. This means that the reader should check the current corporate tax rate before using the method.) Net Cash Flow shows the first figure as a negative number. This figure is the cost of capital for the project term, in our case 5 years. The negative figure is also the sum of the Depreciation column for the project term. The remaining figures in Net Cash Flow are the difference between Net Income B_4 Taxes and Net Income After Taxes. The second from the last column is the rate of discount. This is the rate available to borrowers of capital on the money market at the time the project is being considered. The final column, NPV is the product of the Net Cash Flow times the value of the discount rate in that year of the project. Discount rate values are usually obtainable from discount tables contained in many texts in Accounting, Finance, Managerial Economics, etc. The project with the highest project NPV is the most desirable business alternative provided assumptions we have made hold true for the project term. Limiting assumptions are; cost of capital is correctly defined; corporate tax rates do not change; discount rates will hold fairly steady; etc.

TABLE 5

Design Alt.	Failure Rate (%)	Production Cost Per Unit	5 Yr. Total Net Income After Taxes	5 Yr. Total Sales	5 Yr. Period Net Present Value
1	12.6	$30.00	$4,004,000	$30,506,000	$3,044,128
2	5.9	30.14	4,688,770	27,776,000	3,596,890
3	3.8	31.89	4,344,064	26,827,810	3,338,260
4	4.6	31.54	4,378,421	27,194,200	3,361,391
5	10.5	31.75	3,658,014	29,698,800	2,783,744
6	9.3	33.15	3,348,448	29,217,500	2,547,167
7	11.2	31.40	3,683,680	29,972,800	2,802,026
8	2.5	33.29	4,054,540	26,218,750	3,118,676

Table 5 summarizes pertinent information from the myriad of calculations in previous tables. The first three columns have already been explained, but are included to define individual project identities. Column, 5 Yr. Total Net Income is the sum of the project term figures from the Net Income column in Table 4. 5 Yr. Total Sales is the sum of the Cash Inflow column in Table 3, and 5 Yr. Period Net Present Value is Total NPV in Table 4.

Analysing the results indicates that from a business decision viewpoint, design alternative 2 is the most feasible, relating to NPV. It is also most feasible when considering Net Income After Taxes. However, it is not the most powerful when solely considering total sales. It isn't too bad when evaluating unit production cost.

From the Engineering viewpoint, design alternative 8 would be most feasible since it indicates the lowest failure rate. If the product in fact does require high reliability considerations in use, note that the NVP is fourth best. The balance between technological and business evaluation is fairly good. The value of this method gives information that can allow a less uncertain environment in which a decision can be made.

The method outlined attempts to apply business decision techniques to technological evaluation in order to create a less risky environment in which project selection must be conducted. Cost information applied to technological reasoning rounds out the manager's decision making reponsibility toward a higher level of certainty. If the limiting assumptions do in fact hold true for the period under study, we find that this method can actually allow for control of the amount of warranty returns and associated costs in each period by selecting the design alternative most appropriate to the business and technical environment conditions and criteria.

APPENDIX A

BIBLIOGRAPHY

1. Goodman, A. H. and J. S. Ratti, ''Finite Mathematics With Applications'', New York, The MacMillan Company, 1971
2. McGuigan, James R. and R. Charles Moyer, ''Managerial Economics: Private and Public Sector Decision Analysis'', Hinsdale, Illinois, The Dryden Press, 1975
3. Van Horne, James C., ''Financial Management and Policy'', 4th Edition, Englewood Cliffs, N.J. Prentice-Hall, Inc., 1977
4. MIL-HDBK-217B, Reliability Prediction of Electronic Equipment''.

APPENDIX B
TABLE 3

Des. Alt.	Yr.	Units For Sale	Cash In Flow At $50	Units Built	Cost To Build	Warranty Units Returned	Warranty Costs At $25/Unit	Cash Out Flow
					@ $30.00			
1	1	100000	$5000000	112600	$3378000	12600	$315000	$3693000
	2	111012	5550600	125200	3756000	14188	354700	4110700
	3	122024	6101200	137800	4134000	15776	394400	4528400
	4	133036	6651800	150400	4512000	17364	434100	4946100
	5	144048	7202400	163000	4890000	18952	473800	5363800
					@ $30.14			
2	1	100000	5000000	105900	3191826	5900	147500	3339326
	2	105552	5277600	111800	3369652	6248	156200	3525852
	3	111104	5555200	117650	3545971	6546	163650	3709621
	4	116656	5832800	123600	3725304	6944	173600	3898904
	5	122208	6110400	129500	3903130	7292	182300	4085430
					@ $31.89			
3	1	100000	5000000	103800	3310182	3800	95000	3405182
	2	103656	5182800	107600	3431364	3944	98600	3529964
	3	107311	5365550	111400	3552546	4089	102225	3654771
	4	110967	5548350	115200	3673728	4233	105825	3779553
	5	114622	5731110	119000	3794910	4378	109450	3904360
					@ $31.54			
4	1	100000	5000000	104600	3299084	4600	115000	3414084
	2	104388	5219400	109200	3444168	4812	120300	3564468
	3	108777	5438850	113800	3589252	5023	125575	3714827
	4	113165	5658250	118400	3734336	5235	130875	3865211
	5	117554	5877700	123000	3879420	5446	136150	4015570
					@ $31.75			
5	1	100000	5000000	110500	3508375	10500	262500	3770875
	2	109398	5469900	121001	3841782	11603	290075	4131856
	3	118795	5939750	131500	4175125	12705	317625	4492750
	4	128193	6409650	142001	4508849	13808	345200	4854049
	5	137590	6879500	152500	4841875	14910	372750	5214625
					@ $33.15			
6	1	100000	5000000	109300	3623295	9300	232500	3855795
	2	108435	5421750	118600	3931590	10165	254125	4185715
	3	116870	5843500	127900	4239885	11030	275750	4515635
	4	125305	6265250	137200	4548180	11895	297375	4845555
	5	133740	6687000	146500	4856475	12760	319000	5175475
					@ $31.40			
7	1	100000	5000000	111200	3491680	11200	280000	3771680
	2	109946	5497300	122400	3843360	12454	311350	4154710
	3	119891	5994550	133600	4195040	13709	342725	4537765
	4	129837	6491850	144800	4546720	14963	374075	4920795
	5	139782	6989100	156000	4898400	16218	405450	5303850
					@ $33.29			
8	1	100000	5000000	102500	3412225	2500	62500	3474725
	2	102437	5121850	105000	3495450	2563	64074	3559525
	3	104875	5243750	107500	3578675	2625	65625	3644300
	4	107313	5365650	110001	3661933	2688	67200	3729133
	5	109750	5487500	112500	3745125	2750	68750	3813875

APPENDIX B
TABLE 4

Des. Alt.	Yr.	Cash Out Flow	Straight Line Dep.	Net Income Before Tax	Net Income After Tax @ 48%	Net Cash Flow	Discount @ 8.75%	NPV
1	0					−$200000	1.0000	−$200000
	1	$3693000	$40000	$1267000	$658840	698840	.9195	642583
	2	4110700	40000	1399900	727948	767948	.8456	649377
	3	4528400	40000	1532800	797056	837056	.7775	650811
	4	4946100	40000	1665700	866164	906164	.7150	647907
	5	5363800	40000	1834600	953992	993992	.6574	653450
							Total NVP	=3,044,128
2	0					−$200000	1.0000	−$200000
	1	3339326	40000	1620674	842750	882750	.9195	811689
	2	3525852	40000	1711748	890109	930109	.8456	786500
	3	3709621	40000	1805579	938901	978901	.7775	761096
	4	3898904	40000	1893896	984826	1024826	.7150	732751
	5	4085430	40000	1984970	1032184	1072184	.6574	704854
							Total NPV	=3,596,890
3	0					−$200000	1.0000	−$200000
	1	3405182	40000	1554818	808505	848505	.9195	780201
	2	3529964	40000	1612836	838675	878675	.8456	743007
	3	3654771	40000	1670779	868805	908805	.7775	706596
	4	3779553	40000	1728797	898974	938974	.7150	671367
	5	3904360	40000	1786740	929105	969105	.6574	637089
							Total NPV	=3,338,260
4	0					−$200000	1.0000	−$200000
	1	3414084	40000	1545916	803876	843876	.9195	775944
	2	3564468	40000	1614932	839765	979764	.8456	743929
	3	3714827	40000	1684023	875692	915692	.7775	711950
	4	3865211	40000	1753039	911580	951580	.7150	680380
	5	4015570	40000	1822130	947508	987508	.6574	649187
							Total NPV	=3,361,391
5	0					−$200000	1.0000	−$200000
	1	3770875	40000	1189125	618345	658345	.9195	605348
	2	4131856	40000	1298043	674982	714982	.8456	604589
	3	4492750	40000	1407000	731640	771640	.7775	599950
	4	4854049	40000	1515601	788112	828112	.7150	592100
	5	5214625	40000	1624875	844935	884935	.6574	581756
							Total NPV	=2,783,744
6	0					−$200000	1.0000	−$200000
	1	3855795	40000	1104205	574187	614187	.9195	564745
	2	4185715	40000	1196035	621938	661938	.8456	559735
	3	4515635	40000	1287865	669690	709690	.7775	551784
	4	4845555	40000	1379695	717441	757441	.7150	541571
	5	5175475	40000	1471525	765193	805193	.6574	529334
							Total NPV	=2,547,169
7	0					−$200000	1.0000	−$200000
	1	3771680	40000	1188320	617926	657926	.9195	604963
	2	4154710	40000	1302590	677347	717347	.8456	606588
	3	4537765	40000	1416785	736728	776728	.7775	603906
	4	4920795	40000	1531055	796149	836149	.7150	597846
	5	5303850	40000	1645250	855530	895530	.6574	588721
							Total NPV	=2,802,026
8	0					−$200000	1.0000	−$200000
	1	3474725	40000	1485275	772343	812343	.9195	746949
	2	3559525	40000	1522325	791609	831609	.8456	703209
	3	3644300	40000	1559450	810914	850914	.7775	661586
	4	3729133	40000	1596516	830189	870189	.7150	622185
	5	3813875	40000	1633625	849485	889485	.6574	584747
							Total NPV	=3,118,676

QUALITY COST AND PROFIT PERFORMANCE

Francis X. Brown, Staff Assistant
Roger W. Kane, Staff Assistant
Westinghouse Electric Corporation
Corporate Product Integrity
Pittsburgh, Pennsylvania 15235

ABSTRACT

Use of failure cost (rather than total quality cost) as the principal financial measure of quality performance opens the door to more extensive and effective applications of this management tool. The measure is better correlated with other indicators of quality performance; the demonstrated correlation with profit performance forms a rational basis for quality cost performance norms; and the projected effect of a specific quality improvement on overall business performance can be forecast with better confidence.

BACKGROUND IN PIMS

From its earliest beginnings, the quality profession has engaged in the pursuit of profitable excellence. The list of searchers is as long as it is illustrious — Shewhart, Feigenbaum, Juran, and Schoeffler. . .Schoeffler? — While not a member of the quality community, he and his colleagues at Harvard broke the barrier that had prevented a direct coupling of quality costs to measures of business performance.[1] Prior to the landmark Profit Impact of Market Strategies (PIMS) study, quality cost methods, in general, utilized targets and emphasized prevention expenditures to reduce failure costs thereby bringing total quality costs into balance. In 1951, Feigenbaum wrote: "An unprofitable cycle is at work that operates something like this: the more defects produced, the higher the failure costs. The traditional answer to higher failure costs has been more inspection. This, of course, means a higher appraisal cost. . .and the higher they go, the higher they are likely to go without preventive activity".[2] In the same book Feigenbaum says ". . .the product can be provided with those qualities which motivate purchase by the consumer and thus increase salability." As recently as 1970 Juran and Gryna recognized the problem of being unable to measure the effect of quality on business performance with these words: "The facts on cost of quality are often precisely ascertainable. However, facts on value of quality are more nebulous. In particular, while the factors of quality reputation and customer goodwill are conceded to be of great importance, the present methods for evaluating them are quite primitive.[3]

And there matters stood until the PIMS work was published in 1974. PIMS concluded, at least qualitatively, that the influence of Quality Reputation on Return on Investment was very potent indeed and second only to Market Share. Here, at last, was an analytic tool to couple quality to profitability.

The work reported in this paper was begun in 1975, and resulted from efforts to utilize the PIMS methodology to quantify the impact of quality on profitability. In order to achieve the quantification several criteria were established:

(1) Provide a financial measure of quality performance that is consistent with other measures of quality performance;

(2) Provide a measure which responds in a timely manner to changes in actual quality performance;

(3) Provide a measure that is credible to Management, acceptable to the Controller organization, and suitable for inclusion in the routine financial reports of the business.

The primary data sources used for this study are the quality cost data reported by Westinghouse divisions in response to a Corporate requirement, and financial statements of these same divisions, (coded in this paper).

SEARCH FOR A MEASURE

Initially our attention was put on *total* quality costs (TQC), with the expectation that the best performing divisions would have the lowest TQC. However, a ranking of divisions by TQC did not agree with rankings based on other measures — such as the division's reputation for quality, or the findings of Corporate Quality Program Audits. A better measure was found when attention was limited to failure (F) costs alone. Divisions with low failure costs were also found to have better reputation for quality performance.

Figure 1 illustrates the performance of 3 similar divisions between 1975 and 1977. All three divisions introduced quality program improvements during 1975 which contributed to lower failure costs in 1976 and 1977 — lower in dollars and lower as a percentage of Value-Added Sales.

Failure cost relative to value-added sales (F% VAS) was found to be a better indicator of quality performance trends than the more common measure of F% total sales billed, since VAS is a better measure of *opportunity* for failure costs to occur. (Value-Added Sales is the difference between gross sales billed and the cost of purchased material or services.)

Figure 1 shows not only that these divisions achieved a reduction in F% VAS, but also (as predicted by the PIMS study) that the profit performance of these divisions improved when failure costs were reduced.

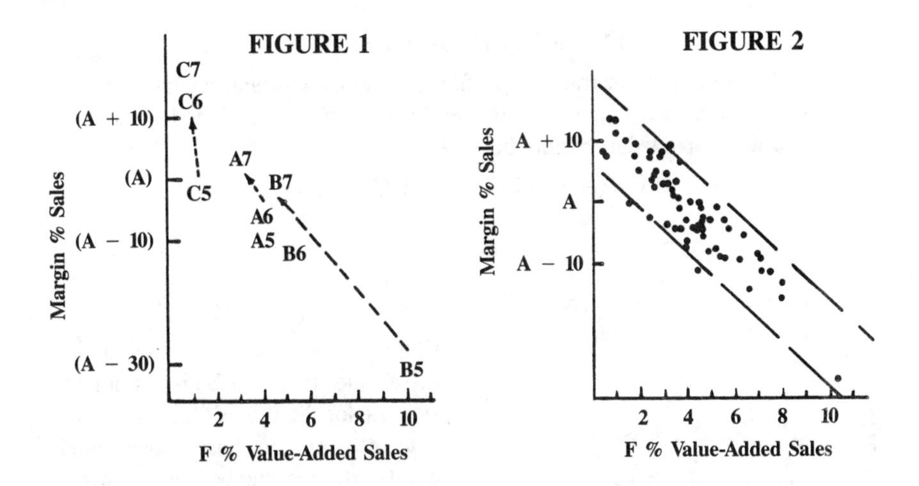

Figure 2 shows the relationship when data for other divisions is added to the nine data points of Figure 1. The letter "A" on the vertical scale represents what was defined to be an acceptable level of profit performance. The data show good linear correlation (r= −0.90) between profit margin (M % sales) and failure costs (F % VAS).

From this data the least-squares regression line was found to be:

$$M = (A+13) - 3.6\ F_{vas}$$

The observation that failure costs and profit performance are highly correlated does not in itself prove that either is a direct cause of the other. External factors such as an industry-wide change in demand for the product can cause both profit margin and F % VAS in a particular division to shift — even though no change may have occurred in quality performance itself.

Whether or not a cause-and-effect relationship is present, the correlation of margin to failure costs can be highly useful — for example, to test the credibility of reported failure cost data. For the 64 data points shown in Figure 2, the difference between actual profit margin and the margin predicted by the regression mode has a standard deviation of 3.5 percentage points. The two parallel lines in this figure lie two standard deviations (seven points of profit margin) above and below the least-squares regression line, and therefore represent upper and lower boundaries of a 95% confidence band for margin as a function of failure cost.

When data points are found to lie outside the appropriate confidence band, they should be investigated. The few exceptions observed were investigated, and in nearly every case this investigation revealed errors in the reporting of failure cost data — for example, use of estimates or accrual rates rather than actual failure costs.

As mentioned earlier, some of the correlation between margin and failure costs may be the result of both factors responding simultaneously to a third variable. Nontheless, there are also indications that a strong cause-and-effect relationship is present, and that a sustained program for reducing failure costs can make a major contribution to improved profit performance.

THE MULTIPLIER EFFECT

Figure 3 helps to demonstrate the profit impact of a one-dollar change in failure cost. (The same measurement base is used for both variables.) The least squares regression line for 64 data points becomes

$$M = (A + 12.6) - 5.45\ (F\%\ \text{Sales}).$$

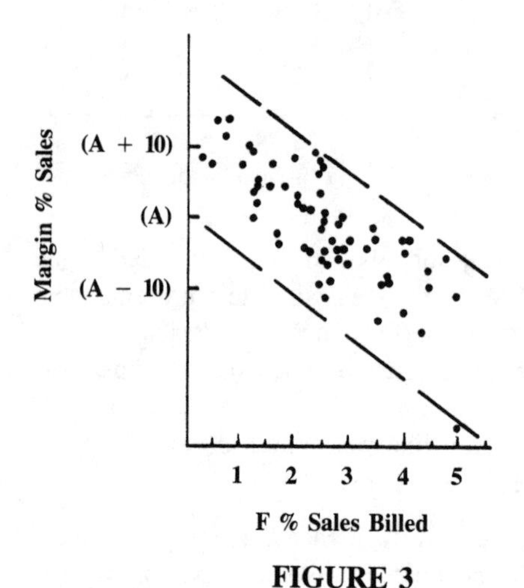

FIGURE 3

Although the linear correlation for M:F% sales (r = −0.79) is not quite as good as for M:F% VAS., use of the same measurement base (sales billed) reveals the existence of a strong "multiplier effect". The slope of 5.45 means that profit margin (before taxes) improves an average of $5.45 for every $1.00 reduction in reported failure costs.

Most managers are aware that some costs of quality failure are not easily identified as failure costs, but that they affect profit nonetheless*:

- in-process quality failures may cause parts shortages which create production downtime at subsequent operations;
- some in-section rework is absorbed in productivity and reported as a labor variance;
- chronic rework or excessive "shrinkage" may necessitate scheduling of overtime or purchase of additional production facilities;
- completed product which fails in final test may result in increased work-in-process inventory and reduced billings for the month;
- product failures in the field can, at least in the long run, contribute to reduced market share or poor price realization. They may also contribute to past-due or uncollectable receivables.

Although most managers recognize failure effects such as these (sometimes called "symptoms of failure disease"), it has not been common practice to quantify the expected benefit in these areas when attempting to justify capital expenditures for quality improvement. Some of the observed slope of 5.45 may be only coincidental, but our experience indicates that a multiplier effect of at least three or four is directly related to such hidden effects of quality failure.

FINDING OPTIMUM "BALANCE"

This multiplier effect of failure costs on profits also explains why *minimum* TQC is hardly ever *optimum.* Reported quality costs are typically at their lowest when the voluntary costs of quality (prevention plus appraisal) are about equal to the reported failure costs — or 40% to 60% of TQC, as shown by the solid lines in Figure 4. If, however, the "true" cost of failure is found to be more nearly four times the *reported* costs, the point of true minimum cost would move further to the right, as shown by the broken lines of Figure 4.

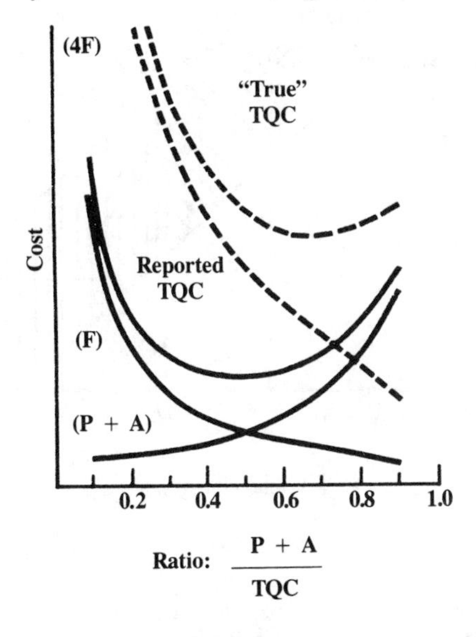

FIGURE 4

Our data tend to confirm this. Reported TQC is at a minimum when prevention plus appraisal cost (P+A) constitute about half of the reported total quality cost (TQC), but *profits* continue to improve until (P+A) contributes 70% to 80% of the reported TQC.

EVALUATING PROGRAM EFFECTIVENESS

Since a particular level of profit margin (A) was defined as "acceptable", and since our data confirms the PIMS finding that divisions with better quality performance also earn more profits, we used this relationship to define an acceptable level of failure costs. The level of failure costs at which the M:F regression line crosses M = A % sales is described as F_{par}, and the level of failure costs at which the regression line crosses M = (A-8) % sales is defined as F_{tol}.

For the divisions included in Figure 2 data these points were found to be F_{par} = 3.6% VAS and F_{tol} = 5.8% VAS (See Figure 5). Divisions whose reported failure costs are less than or equal to F_{par} are considered "green" — shown as zone "G" in Figure 5.

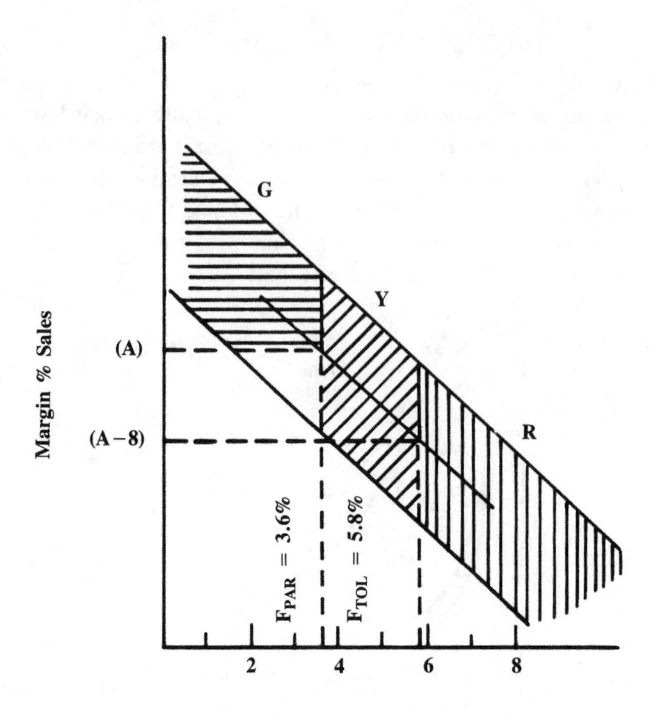

FIGURE 5

On the other hand, when reported failure costs exceed F_{tol} the status is considered "Red", and the division is suspected to have some form of "failure disease". We contact these "Red" divisions to determine which specific symptoms of failure disease might be present (low productivity, missed billings, loss of market share, etc.) and begin to work with them in establishing a program for improvement.

As you might suspect, when reported failure costs lie between F_{par} and F_{tol} the performance is considered "yellow". If left alone, yellow will usually change to red.

FAMILY OF PROFILE TYPES

Finally, we should say a word about the selection of divisions included in Figure 2. Not all Westinghouse divisions fit this profile type, although it is the most common.

Some divisions continue to report margin and failure costs which fall above and to the right of the data plotted in Figure 2. Typically these divisions also tend to be in higher risk divisions, with higher technology requirements for their products or services. Such divisions fit the profile types III or IV shown in Figure 6.

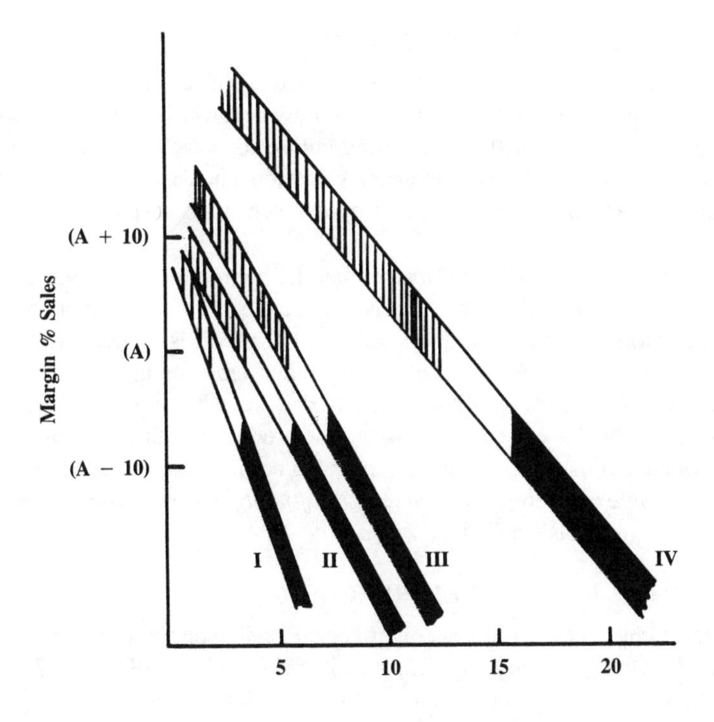

FIGURE 6

Other divisions report margin and failure costs which fall below and to the left of the Type II profile divisions. In many cases, these divisions are engaged in a more mature business, with reduced technological requirements. However, the steeper slope of these Type I divisions also implies that fewer of the real costs related to failure are identified and reported. (So long as the reporting is consistent from month-to-month, and the portion of failure costs which is identified and reported is a valid indicator of performance trends, this "under-reporting" need not cause concern.)

Performance standards will of course differ for the various profile types. At present, we are using the same level of profit margin (M = A % Sales) to define F_{par}. As shown in Figure 6, our observed values of F_{par} are as follows:

Type I:	2.0 % VAS
Type II:	3.6 % VAS
Type III:	5.3 % VAS
Type IV:	12.4 % VAS

As more years of margin and failure cost data are collected, it will become less important to group divisions by profile type.

(SUMMARY)

The observed correlation between failure costs and profit performance has many uses as a tool of quality management. A particular division which shows no margin-to-failure correlation is likely to be reporting failure costs which are not responsive indicators of actual performance trends. A division who data *does* fit a profile-type has most need for quality program improvement if its status is ''yellow'' or ''red''.

Perhaps of greatest interest is the *slope* of the M:F regression line; every dollar of improvement in *reported* failure costs is likely to result in several dollars improvement in profits. In many instances, this kind of payoff is more fruitful than the more traditional efforts to reduce direct product labor or material costs.

In a nutshell, Shewhart[4] said it first and best: ''. . . when *assignable causes* of *variation* in quality have been *eliminated*. . . the product may then be considered to be controlled. . . this state of control appears to be, in general, a kind of limit to which we may expect to go *economically* in *finding* and *removing* causes of *variability*. . .'' (emphasis added.)

REFERENCES

1. Schoeffer, Sidney, et. al., ''Impact of Strategic Planning on Profit Performance. Harvard Business Review, Vol. 52, No. 2., March-April 1974 p 137.
2. Feigenbaum, A. V., ''Total Quality Control'', McGraw-Hill, 1961.
3. Juran, J. M., Gryna, Jr., E. M., ''Quality Planning and Analysis'', McGraw-Hill, 1970.
4. Shewhart, W.A., ''Economic Control of Quality of Manufactured Product'', D. Van Nostrand, 1931.

QUALITY COSTS — WHAT DOES MANAGEMENT EXPECT?

Dr. Frank M. Gryna, Jr.
Professor of Industrial Engineering
Bradley University, Peoria, Illinois 61625

INTRODUCTION

The measurement of quality cost has proven its importance in the quality control field[1]. However, my experience indicates that only about 40% of the American companies have a system that goes beyond measuring scrap and rework dollars. Furthermore, some companies do not even dollarize scrap and rework. Thus, for many companies quality costs still represent an opportunity.

One of the factors influencing the introduction of a quality cost system in a company is the attitude of top management. This paper will look at quality costs from the viewpoint of upper management levels — plant manager to president. This will be done by describing ten actual situations ("scenarios") and offering some conclusions for each.

SCENARIO # 1 — WHAT ARE QUALITY COSTS?

A quality control manager decides to measure quality costs. He works with the accounting department and proposes that quality costs be compiled for four basic categories: prevention, appraisal, internal failure and external failures. The accounting people declare that they certainly want to cooperate with him but "but books aren't kept that way." They also point out that scrap and rework dollars have long been known and are rather stable at about 2% of direct labor and direct material costs. As the 2% is included in the standard cost system, the accounting people feel that scrap and rework costs are covered and there is no need for further measurement.

Finally, the data is collected and presented to the plant manager. His initial reaction is "these are the costs of the quality control department and they certainly are too high." This misunderstanding horrifies the QC manager. Next, the manufacturing manager refutes the data because he claims that part of the "scrap" is unique to the type of manufacturing process and clearly unavoidable. Meanwhile, the accounting people continue to emphasize the inclusion of scrap and rework dollars in standard costs. The plant manager is confused.

Conclusions. Top management expects that any presentation of quality costs will cleary identify the meaning of the costs. Further, they expect that any arguments about including certain elements of costs such as unavoidable manufacturing scrap will be settled before presenting the final report. The misunderstanding that quality costs are the same as the costs of the quality control department is an offshoot of the idea that the quality program in a company is the same as the work of the quality control department. We know that this is not the case but, like it or not, many people believe it. It's probably too late to change the term quality cost but you may want to consider using another term in your own company.

We need to say what quality cost is and what quality cost is not. Quality cost is a collection of costs throughout the entire company associated with the prevention, generation, measurement, and correction of quality problems. Quality costs are not equal to the costs of the quality control department just as federal expenses are not equal to the expenses of the Treasury Department. Quality costs go beyond the standard amount for scrap and rework included in standard costs. However, the standard cost system assumes that such costs are inevitable — the QC system views these costs as cost reduction opportunities.

Avoidable and unavoidable costs in connection with quality should be clearly labeled. In addition, careful, careful explanation is needed for any "hidden costs." For example, suppose a packaged product must meet a minimum weight requirement. If the process variability is wide then the process average must be set sufficiently above the minimum weight to have a high probability of meeting the minimum on individual packages. Although the product meets the specification, the high process average represents a giveaway which is a hidden cost of quality.

SCENARIO # 2 — PAR

A QC manager presents his first quality cost study and shows that quality costs are equal to 11% of sales dollars. There is general agreement by upper management that the figure is high but top management asks "what should it be?" The QC manager replies that it is difficult to have a specific numerical answer but the industry average is about 6% — he read this in a magazine article. Now the arguments begin. Someone points out that the 6% probably varies by type of industry. Another argument is that the amount of quality cost depends on the quality philosophy of the company — is the company aiming to be the top quality leader or just average with competition? Another question is how different companies define qulaity costs — is overhead included, is unavoidable manufacturing scrap included, are periodic losses of new contracts due to poor quality image included? These questions are a mixture of wind, snow, ice, and trouble.

Conclusions. Management not only wants to know the total quality cost but needs to know how to interpret the total. Comparison to other companies or industry in general can be dangerous for the reasons cited.

A better tact is this: select a few major projects to improve, dollarize the size of the quality costs on these projects, set a cost reduction goal, estimate the investment required and calculate a measure of effectiveness such as return on investment or payback period. Of course, the complete participation of responsible departments in all this is a must.

Finally, trends may be more important than the absolute value of quality cost. They can serve as an early warning device and help to set specific goals for the future.

SCENARIO # 3 — WHAT SHOULD I DO?

The QC manager of a consumer product presents a quality cost report including the four basic categories. The executive vice-president reviews it and agrees that the overall costs are high and that the categories make good reading but then nothing happens. Finally, he asks "what should I do with this information?"

In another plant making a product that is a rental product, the quality cost report to management is extremely detailed. Many ratios highly useful to QC professionals are presented and the emphasis is both on investment costs and losses. Again, the quality cost report simply is not getting enough action from upper management.

In two other cases, upper management express a desire to relate quality costs to objectives for functional managers, e.g., through a tie-in to bonus plans or a management by objectives program.

Conclusions. Management expects to be shown how to interpret a quality cost report and the benefits and costs of alternative actions. QC measures quality. This scorekeeping is necessary but management expects somebody to show them how to win the pennant. Specific improvement projects with numerical objectives must be defined as part of the proposed action to be taken. This is in contrast to just dramatizing the high total and asking for general support across the board on quality.

The improvement projects can be on products or on tasks within a quality program. For example, objectives for next year on specific products might be "quality costs for the company shall be reduced by 20%" or the "average leak rate for product x shall be reduced to 4%." Examples of objectives for tasks rather than products might be "quality costs shall be determined for at least one product" or "numerical reliability and maintainability objectives shall be defined for at least one product."

SCENARIO # 4 — THE TOTAL

A QC manager calculates quality costs as a total of $1.5 million for a plant with about 1,000 production workers. The cost is primarily scrap, rework, and warranty charges. He presents this information to justify his request for an increased budget next year.

The plant manager reacts by agreeing that it is useful to collect all of the costs concerning quality across all departments and show it in one total. He agrees that the total is high but says that their product is relatively expensive and perhaps the absolute figure is not completely meaningful. After reviewing the literature, the QC manager restates the quality costs as 9.2% of the sales income and shows a breakdown of the total into the usual four basic quality cost categories. There is much argument about the usefulness of all this for the top management people. (The president later told me that he did not recall ever hearing about the four categories — zero impact.)

Conclusions. The total quality cost figure is often dramatically high. However, presenting this figure by itself to top management people generally is not sufficient. The figure must be related to sales or some other base and some breakdown of the figure should be made. The problem is that the nature of the breakdown often depends on the industry and on the specific management person involved (see Scenarios 5, 6, and 7). Useful as the prevention, appraisal and failure categories are they may not be the right categories for your case.

It may help to distinguish between fixed and variable quality costs. Other possible breakdowns are by product, vendor, department defect type or defect cause depending on the level of detail desired. Pareto scores again.

Finally, quality cost information may need to be presented as part of a larger "instrument panel" which shows quality costs plus other quality indicators that can help management understand the significance of the dollar figures. The quality indicators used for top management of an equipment manufacturer are shown in Table 1:

TABLE I. Example of Quality Indicators

Internal quality indicators

1. Machine shop scrap and rework hours
2. Assembly scrap and rework hours
3. % of finished product which is incomplete at end of line
4. Dollars of purchased material rejected

External quality indicators

1. Defects found on delivery to dealers
2. Failure incidents per unit time
3. User rating of product based on opinion poll
4. Dollars of warranty expense

Each of these indicators is related to an appropriate base. Note that only 2 of the 8 indicators are in dollars. The primary reason is that a large share of the quality costs are in labor hours and this governs the emphasis on quality improvement.

SCENARIO # 5 — UNITS, UNITS, UNITS

A group of QC managers are comparing notes on their experiences with quality cost. Bill comes from a chemical company and mentions how his division president calculated the cost of spoilage as 13¢ per share of stock outstanding. This was instrumental in convincing the president to direct corporate QC to set up a quality cost system throughout the division.

Tom comes from a mechanically oriented company where the quality cost for all divisions is $76 million per year. He comments about the value of converting this dollar figure into equivalent physical terms. For his case, the $76 million converts to 5,900 extra personnel. 1.1 million square feet of space, and $6 million of extra inventory being carried. These figures colletively are equivalent to the entire yearly production of one plant of his company. This physical analogy had great impact on his management.

Ed works for a food processing plant where rework alone is three-quarters of a million dollars a year. This figure was instrumental in convincing the plant manager to set up the training program for foremen in process control methods. In this case the effect on delivery promises to customers was also an important factor in dramatizing the importance of the rework problem.

Conclusions. Management expects us to present quality costs in a language suited to their needs. Often that language is dollars but not always. Sometimes productivity and delivery schedules or other matters are equally or more important. The need is to discover what language fits the territory.

SCENARIO # 6 — BASE, BASE, BASE

The QC manager of a company that designs and manufacturers mechanical assemblies prepares a quality cost study and explains the results to engineering, manufacturing, purchasing, and other major function areas.

Everyone is impressed that the total quality cost is high but there is considerable difference of opinion on how to interpret the figure. A further discussion breaks down the total by plants but now the arguments really become heated. The different plants make different products and the only agreement reached is that the absolute quality cost dollar figure is not sufficient to make comparison between plants.

A number of different bases are proposed such as expressing quality cost as percent of sales, quality cost as percent of factory costs, and so forth and the discussion continues into the night.

Conclusions. Management expects the QC manager (in cooperation with others) to determine the most useful way to present quality cost figures. The literature emphasizes quality cost as percent of sales. However, other bases include direct labor hours, direct labor dollars, standard manufacturing cost dollars, value added dollars, burden, and product units.

It is difficult to logically determine beforehand which base or bases to use. In my experience, several bases should be selected and tried for a period of time and then a judgement made on which bases are best.

Figure 1 shows total quality costs expressed in absolute dollars and two different bases. Note how the conclusions vary with the base used. The data for this figure was taken from Reference 2 which includes a further discussion of bases.

SCENARIO # 7 — MY PLANT IS DIFFERENT

In my experience, the reaction of plant managers and company presidents concerning quality cost is quite varied. The president of an electrical component manufacturer is oriented to meeting the needs of the appliance industry and focuses on quality and sales income rather than quality and cost reduction. On the other hand, the president of a pharmaceutical company wants to control the cost of quality but is more concerned about the seriousness of a major recall on any of his products. The division manager of an appliance firm is more concerned about the customer getting value than he is about reducing quality costs. The president of a small firm producing lawn equipment echos the same belief. The plant manager of a chemical company explains that a directive from corporate headquarters set the stage for him: reduce the cost of nonconforming material. A vice-president of a high technology company is cost/benefit oriented. In his words: "I don't mind spending money for quality but I sure wish I could be more certain of my investment rather than spend it in a defensive manner."

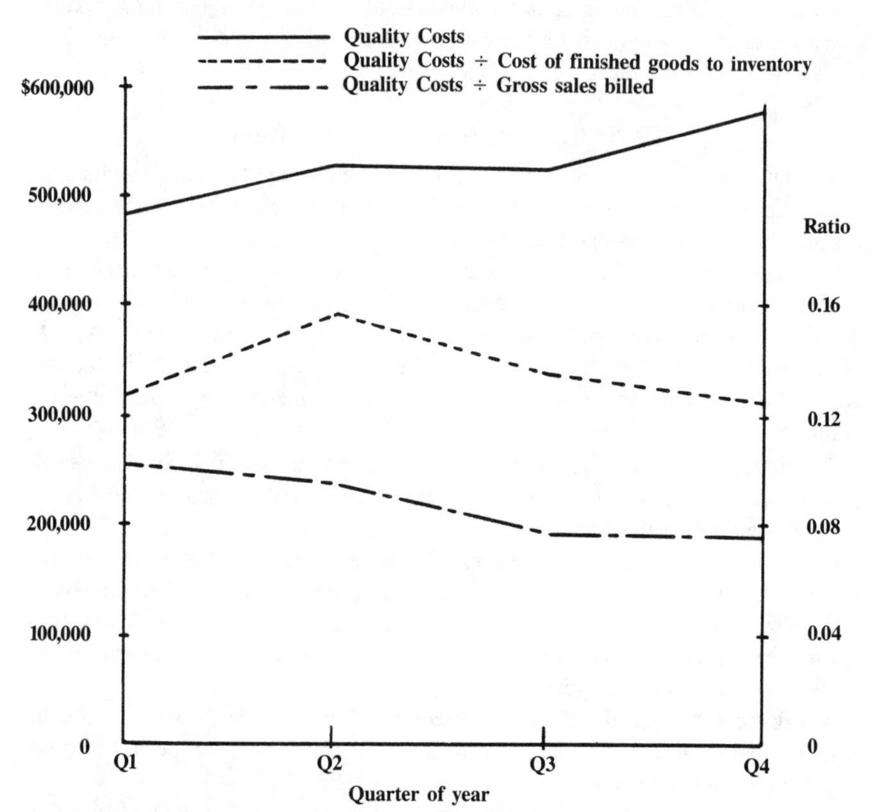

FIGURE 1. Quality Costs Using Several Bases.

Conclusions. Management expects us to present quality costs in a way that is unique for their situation rather than emphasizing the conventional methodology of quality costs. If the orientation is toward reducing the cost of nonconformance, the conventional methodology will meet the test. Other priorities require a different tact.

The priorities of the executive depend on the industry but also the scope of the responsibility. For example, a plant manager responsible only for manufacturing may stress conformance to specifications. A division or company president with broader responsibilities must weigh the cost of nonconformance with design, sales and other matters influencing profit.

SCENARIO # 8 — VALUE

A discussion with a division manager for a major product line in the appliance industry makes it clear that his road to increased profits is to increase sales rather than just reduce quality costs. He feels that product value is the key to increased sales. For example, his company can provide a digital readout of temperature in the main compartment of a refrigerator for about $40 additional in the price. However, a study concludes this would be poor value from the customer's viewpoint and therefore the idea is dropped.

The same manager also makes a distinction between the real and preceived performance of the product. The real performance of the product is what the products can actually do. The perceived performance is the performance as the customer sees it. For example, refrigerators are designed to have a certain temperature in the refrigeration compartment. If hostile environments are encountered (high humidity, high usage due to children) then excessive temperatures occur. It is possible to design the refrigerator to handle these temperatures but it would not be economical — too few customers are involved. Now he relates this to quality cost. If such a refrigertor is marketed then the cost of service calls during warranty for the customers with extreme temperatures is viewed as a quality cost that is planned.

This division manager reacts to questions about quality cost by directly orienting to value provided to the customer. This is not the intent of quality cost systems but it is his orientation.

Conclusions. The quality cost concept as we QC professionals view it is too narrow for some management people. The management level that is responsible for manufacturing and design and sales expects customer talk and not just prevention, appraisal and failure costs.

Table 2 shows an example of a method for comparing alternative designs of a product taking into account both performance and cost (i.e., value). For each performance factor a fraction is calculated. The numerator is a product of a weighting constant for the factor and the score achieved by the specific design for that factor. The denominator is the unit cost of achieving the factor by the specific design. Design B shows more value than Design G. Such an analysis could compare alternative internal designs or an internal design with competition.

TABLE II. Value Comparison of Food Waste Disposer

Performance Factor	Design B	Design G
Grinding time	9/0.87 = 10.3	6/2.40 = 2.8
Fineness of grind	4/7.82 = 0.5	4/11.88 = 0.3
Frequency of jamming	9/1.98 = 4.6	9/2.46 = 3.7
Noise	4/0.45 = 8.9	4/0.52 = 7.7
Self-cleaning	2/0.49 = 4.1	4/0.58 = 6.9
Elec. safety	16/0.52 = 30.8	16/0.43 = 37.2
Particle protection	6/0.30 = 20.0	2/0.37 = 5.4
Ease of servicing	4/0.52 = 7.7	6/0.98 = 6.1
Cutter life	9/0.83 = 10.8	9/1.32 = 6.8
Ease of installation	9/0.33 = 27.3	0/0.70 = 11.8
Total	72/14.11 = 5.1	69/21.44 = 3.2

Conventional quality costs measure costs incurred by the manufacturer. This is separate from costs incurred by the user of the product (see Reference 3) which in turn is related to income. Although the quality cost concept has proven its use as a cost reduction measure, management may increasingly look to the QC manager to relate quality and sales income. For example, in both the automotive and appliance industries, the share of the market given to suppliers for certain companies directly depends on a quantitative quality rating which is computed periodically. Suppliers whose quality rating is higher than competition are given a higher share of the market for the next purchasing period.

One chemical company reported that there may be more potential for profit improvement (through increased market share) by improving the products supplied to the customer as contrasted to reducing scrap and rework costs.

SCENARIO # 9 — LET'S NOT GET BURNED AGAIN

A manufacturer of major mechanical assemblies looks back on several new product developments. Classically each new product is carefully tested and evaluated before marketing (wouldn't it be nice if this really happened?). However, on some new products the warranty costs were not only high but the company image suffered by product weaknesses that were not uncovered during the evaluation. This clearly hurt sales. In addition, the company has not been sufficiently prepared to handle the problems of servicing and provisions of spare parts. Again, the image suffered — and this clearly hurt sales.

Management said "let's not get burned again." Some new reliability efforts were instituted including the prediction of warranty costs as early as possible upon the introduction of a new product.

Ironically, management now notes two payoffs: potential field problems are anticipated and the prevention efforts have resulted in cost reduction. The original emphasis was to "get the customer off our back" but (pleasant surprise) this has been accomplished and a cost reduction also achieved.

Conclusions. Particularly for companies making a diversity of new products, upper management expects the quality function to provide early warning devices on warranty costs and other quality related matters. These early warning devices must provide sufficient reaction time to prevent major field quality problems from occurring and to do this early enough to eliminate the need for compromising either costs or delivery schedules. Weibull Probability Paper for warranty cost analysis can be helpful.

SCENARIO # 10 — THE CORPORATE ROLE

The president of major division of a chemical company decides on the need for a quality cost system throughout his division and instructs the division QC people to proceed. The division QC people develop the system and are currently running a pilot run of the system in one plant. So far, the results look encouraging — the necessary data can be collected in a reasonable time. A meeting on quality costs is held at another plant of the division and that discussion raises all kinds of questions. For example, the meaning of manufacturing scrap, avoidable vs unavoidable costs, who is responsible for publishing the data, exactly what format will the data be in, what use will be made of the data by plants, and more.

Conclusions. Management often looks to corporate QC people to develop and institute an overall quality cost system for the company. However, management knows that reductions in quality costs come from plant cafeterias rather than corporate dining rooms. Thus, the corporate people must work closely with the individual plants to develop the quality cost system. Before procedures are issued, sufficient time should be allotted for extensive discussions with the plants to make sure that the plants can provide the data, and that the resulting reports will be useful both for comparing the different plants and for conducting the quality improvement programs. Along with this should be the realization that the issuance of the reports alone will not secure a lasting reduction in quality costs. Large quality costs should be used to justify and provide to the plants the additional manpower needed to determine the causes of the qualtiy problems and institute corrective action.

SUMMARY

The determination of quality costs is still unknown in many companies. This presents an opportunity for the QC manager. However, the priorities of top management are quite varied even though the profit objective is common. The challenge for the QC manager is this: view quality as a business problem and adapt conventional quality cost methodology to meet overall company needs.

REFERENCES

1. Juran, J. M., "Quality Control Handbook," Third Edition, McGraw-Hill Book Company, New York, N.Y., 1974, Chapter 5.
2. "Guide for Reducing Quality Costs," American Society for Quality Control, 1977.

3. Gryna, F. M. Jr., "Quality Costs: User vs. Manufacturer," *Quality Progress,* June 1977, pp. 10-13.

BIBLIOGRAPHY

Groocock, J. M., "The Cost of Quality," Pitman Publishing Company, New York, N.Y., 1974.

Juran, J. M. and F. M. Gryna, Jr., "Quality Planning and Analysis," McGraw-Hill Book Co., New York, N.Y., 1970.

"Quality Costs — What and How," Second Edition, American Society for Quality Control, 1971.

THE PHILOSOPHY AND USEFULNESS OF QUALITY COSTS

William N. Moore, Manager Manufacturing Operations
Westinghouse Electric Corporation
Process Equipment Department, Sykesville, Md. 21784

DEVELOPMENT OF THE GUIDE

The "Guide for Reducing Quality Costs" was developed following a realization by members of the ASQC Quality Costs Committee that very little guidance existed on what to do with quality costs once they were defined, categorized and collected as described in "Quality Costs — What and How". Several of the members of the committee had experience in setting up and implementing successful quality cost reduction programs. Several of the companies represented had internal guides and literature on the subject. It was felt that quality management and general management would benefit from a how-to-book dealing with what uses could be made of quality costs and the structure of programs to reduce the costs. The "Guide for Reducing Quality Costs" is an attempt to tell management:

- How to analyze their quality programs using quality costs.
- How to structure programs for reducing costs.
- How to find areas needing improvement.
- How to reduce failure costs.
- How to reduce appraisal costs.
- How to prevent quality costs.
- How to measure improvement.

The guide represents many hours of work on the part of those people listed in the front of the guide. My presentation in this session will closely follow the content of the first five sections of the guide:

1. Purpose and Quality Definitions
2. The Quality Cost Improvement Philosophy
3. Usefulness of Quality Costs
4. The Quality System
5. Finding the Problems Areas

Following sections of the paper summarize the content of each section of the guide, but do not repeat the content or contain detailed explanations. The reader is referred to the guide for such detail.

PURPOSE

The purpose of the publication is to provide guidance to general management and quality program management to enable them to structure and manage programs for Quality Cost Reduction. The guide provides the procedures for taking total quality costs and using them for identifying problem areas and for reducing these costs. Quality costs covered include those in the design, manufacture, inspection, test, and product service phases. Improvement programs must encompass all phases of product life, from design through use by the customer.

THE QUALITY COST IMPROVEMENT PHILOSOPHY

The guide is about a quality program not confined to the control of quality in the manufacturing stage. Most people recognize that product quality is determined by many factors outside this stage, but many quality programs do not concern themselves with these factors. In some cases, quality program efforts have been attempts at not allowing things to get any worse (control) instead of striving to get better (improvement). As a result, things have gotten worse in many places, simply because controls are not 100% effective and can never be. Improving quality is much like improving product costs. It is everybody's job and everybody is for the idea, but, *until, there is a management commitment to improve and a formal program for forcing improvement, it just doesn's happen.*

The guide describes what each function must do to satisfy the customer's needs and reduce your quality costs. Also described are ways to prevent the production of defectives through involvement of people in Marketing, Design, Purchasing, Accounting, Manufacturing, and Quality Assurance. It describes ways to find problems and correct their causes. It tells you how to use the costs associated with quality and how to reduce those costs.

Quality improvement results in cost improvement. Designing and building a product right the first time always costs less. Solving problems with existing products by finding their causes and eliminating them results in measurable savings. To cash in on these savings requires that the quality performance of the past be improved and the guide describes ways to do that.

The chart in the guide shows how quality cost analysis bridges the gap between the elements of a prevention oriented quality program and the means used by general management to measure performance — the profit and loss statement.

USEFULNESS OF QUALITY COSTS

Quality costs are useful for strategic planning and programing quality improvement.

STRATEGIC PLANNING

Strategic quality program planning is vital to the continuing profitability of many segments of American Industry. The pressures for safer, cleaner, and more reliable products are becoming stronger each year. We must find ways to meet these increasing demands and still remain competitive. The key to doing this is to improve quality using the methods described in this guide, and reduce costs as a result.

To improve quality and reduce quality costs there has to be a trigger for making changes in the status quo. The firm's *Strategic Plan* is an ideal device to force changes; and the inclusion of quality and quality cost improvement plans in the overall strategic plan is recommended. This gives the quality and quality cost situation the management visibility too often lacking. Because quality and quality cost improvements are set forth as business objectives (along with the more conventional business objectives) against which management performance is evaluated, there is effective motivation for action.

Relationship Between Quality Costs and Strategic Planning

Quality costs for a profit center are made up of costs incurred in several activities. The chart in the guide shows the buildup of costs from all functional departments into an overall quality cost analysis for the entire profit center.

Quality costs are incurred by all major functions in an organization, so problem areas can exist anywhere. Careful analysis must be done to find the most costly problems and programs must be developed to attack them. Many times a strategic program is needed. When this need exists, a strategic quality program should be developed using inputs from all functions and it should become a part of the profit center's overall strategic program.

The Planning Process

Strategic planning should be done in a step by step process. The basis steps are:

- Review past performance and present position
- Appraise the environment
- Set objectives
- Select a strategy
- Implement the strategic program
- Report and evaluate the plan

The final step in the cycle provides needed inputs for the first step so planning becomes a continuous process. Following are general considerations for accomplishing each step in the cycle.

Review Past Performance and Present Position — A thorough review provides a realistic assessment of past performance, current conditions, and future potential.

Appraise the Environment — There are numerous environmental factors which may significantly interact with the quality programs. Typical examples are:

- The activities of other departments.
- Changes in customer demands.
- New safety and liability regulations.
- Actions of competitors.

Set Objectives — From the knowledge and understanding achieved in the status review and the environmental information, the strengths and weaknesses of the quality program should be known. So, specific objectives with target completion dates can be established.

Select a Strategy — Once clear objectives have been established, a definite strategy should be formulated and clearly stated.

Implement the Strategic Program — Planning for the implementation of the strategy is the most important step of the planning process. All prior effort leads up to the implementation or action step. There are many cases of "that was a sound strategy but nothing happened as a result". This situation is due to a failure to follow up the strategy with appropriate action programs.

Report and Evaluate the Plan — The final step in the planning cycle is one of integrating the strategic quality plan into the total strategic plan and evaluating the costs and benefits of the plan.

Through better planning, quality performance can be improved. Continued achievement of good performance can identify the quality program as a key contributor to the success and assure that Quality will play an enlarged role in future plans and activities.

PROGRAMMING IMPROVEMENT

The Strategic Quality Plan describes a management commitment to quality and quality cost improvement. The quality cost data indicate the areas that are candidates for improvement. When the highest cost areas are analyzed in greater detail many improvement projects become apparent. For example, high warranty costs are a trigger to rank customer failure problems for detailed investigation, with the aim of looking into product design, process control, or inspection planning for the cures to the highest cost problems. Regardless of what the high quality cost category may be, the mere act of identifying it should lead to actions to reduce it.

To effectively program quality improvement efforts, it is necessary to:

- Recognize and organize quality related costs to gain knowledge of magnitude, contributing elements and trends.
- Analyze quality performance, identify major problem areas and measure product line and/or manufacturing section performance.
- Implement effective corrective action and cost improvement programs.
- Evaluate effect of action to assure intended results.
- Program activities for maximum dollar pay off and maximum effective manpower utilization.
- Budget quality work to meet objectives.

THE QUALITY SYSTEM

Perhaps the most important result of the collection, analysis, and use of quality costs is the exposure given to the total quality system as it really exists in the organization. The collection of costs forces definition of all activities contributing to the quality of the product. Analysis of quality cost data forces evaluation of the effectiveness of the contribution of each activity, the relationships among the many activities, and the all-important communications links that tie activities together.

Each organization will define its quality system differently, but the overriding requirement is that whatever definition is used must be comprehensive, That is, *it must include all efforts that affect product quality,* wherever the efforts are accomplished in the organization.

If there is a weakness in the quality system, it is usually a deficiency in the integration of the elements and their sub-elements into a working whole. Almost every manager can look at his system elements and convince himself that his organization has something going in each area — perhaps he even has a shelf full of procedures manuals to prove it. However, much of what is actually going on in the quality system might be superficial, or might not be well integrated into the operations and traditions of the total organization.

The concept of quality costs is a potent tool for management precisely because it can be used to force the integration of all the separate quality activities into the mainstream of the product cycle; that is, into a total quality system. It forces the entire organization to examine each cost element (and each quality-related activity) in the context of the total quality cost (and total quality system).

FINDING THE PROBLEM AREAS

When quality costs are displayed to managers who have not been exposed to the concept, the initial question is likely to be "how much should they be?" or "how does this compare with other organizations or products?" Unfortunately, it is *not* practical to establish any meaningful absolute standards for such cost comparisons. A quality cost system should be "tailored" to a particular company's needs, so as to perceive trends of significance and furnish objective evidence for management decisions as to where assurance efforts should be placed for optimum return. The search for "industry guidelines" or other standards of comparison, while natural, is quite dangerous, since it leads to quality costs emphasis for "score-carding" instead of utilization as a management tool for improving the status-quo.

The futility of establishing meaningful absolute quality cost guidelines is more apparent if you just consider:

1. Inherent Key variations in how companies interpret and capture quality cost data;
2. Critical differences in product complexity, process methods and stability, production volume, market characteristics, management needs and objectives, customer reactions, etc;
3. The awkwardness or inappropriateness for many companies of the most prevalent form of quality cost measure (% of net sales billed), considering effect of time differences between time of sales billing and incurrence of actual quality costs.

This last factor is particularly important for periods involving an expanding or contracting product volume or mix, unstable market pricing, shifting sales/leasing revenue ratios, or changing competitive performance criteria. Accordingly, it is much more productive abandon efforts to compare your quality cost measurements with other companies, in favor of meaningful analysis of the problem areas contributing most significantly to *your* quality costs, so that suitable corrective actions can be initiated.

Analysis techniques for quality costs are as varied as those used for any other quality problems in industry. They range from simple charting techinques to complicated mathematical models of the program. The most common techniques are:

1. Trend Analysis
2. Pareto Analysis
    A. By Element Group
    B. By Department
    C. By Product
    D. Other Groupings

Trend Analysis is simply comparing present cost levels to past levels. It is suggested that costs be collected for at least one year before attempting to draw conclusions or plan action programs. The data from this one year (minimum) period should be plotted in several ways.

The Pareto Analysis technique involves listing the factors that contribute to the problem and ranking them according to the magnitude of their contribution. In most situations, a relatively small number of causes or sources will contribute a relatively large percentage of the total costs. To produce the greatest improvement, effort should be spent on reducing costs coming from the largest contributors.

REDUCING APPRAISAL COSTS[1]

Edgar W. Dawes
Corporate Quality Assurance Manager
Veeder-Root Company
Division of Western Pacific Industries
Hartford, Connecticut 06102

INTRODUCTION

The costs of appraisal sometimes approach half of the total quality costs. Although most quality cost improvement programs properly concentrate on reducing failure costs first, programs for appraisal cost improvement can also have a significant impact. We will discuss several techniques for improving these costs.

- Inspection and Test Planning
- Equipment and Methods Improvement
- Statistical Quality Control
- Appraisal Accuracy Studies
- Decision Analysis
- Work Sampling

INSPECTION AND TEST PLANNING

Getting the most out of the available appraisal resources requires careful planning. Determining where control points should be and the amount of inspection and test to do should be the job of professionals — not left to the judgment of the individual doing the inspection or test job.

In-process controls are a vital part of a prevention oriented quality system. They provide a powerful means for reducing the incidence of defective finished product and for reducing quality costs. In addition, an effective in-process inspection system often makes it possible to reduce the amount of final inspection required. In-process inspection control involves inspections or verifications performed at significant stages of the manufacturing process. If defective parts or subassemblies are being produced, the trouble can be detected early and corrected before it affects the quality of the finished product. *The system must be efficiently designed so that every inspection will serve an essential purpose.*

The requirements that the finished product has to meet, along with finished product inspection and test specifications, should be thoroughly reviewed during the development of the in-process inspections and controls. This review provides the Quality Engineer with the information needed to determine the type and degree of in-process inspection required at various stages of the manufacturing process or assembly operations. It also helps him select the kinds of in-process controls that will prevent the manufacture of defective product and yield the best economic return. A periodic review of the planned inspection and tests should be made to assure that the levels remain economical in light of quality history. We will briefly review five types of inspection and test controls and the advantages and disadvantages of each:

- Operator Inspection
- 100% In-line Inspection
- First Piece Inspection
- Patrol Inspection
- In-process Acceptance Inspection

Operator Inspection

In-process controls can be enhanced by requiring an operator, stationed at a machine or at a processing station, to inspect his own work. The operator must be provided with the proper gages and be instructed in their proper use. He should be trained to recognize when an item is unacceptable in appearance. Enough time should be provided in his work standards to allow him to perform the inspection with reasonable care.

These are the advantages of operator inspection:

- The operator usually handles every piece coming off the line.
- He is thoroughly familiar with the item he is making.
- He is in a position to spot defects quickly and call for help to correct problems as soon as they appear.

One caution is that special care may be necessary for the operator to keep the records normally required for an effective inspection procedure.

100% In-Line Inspection

Inspection or testing may be carried out on a 100% basis at designated points in the manufacturing line. This type of inspection or testing is normally performed by inspection or test personnel in the Manufacturing Department. Its purpose is to screen out items that either do not conform to quality workmanship standards or are not likely to pass the finished product inspection.

The following are some of the advantages and disadvantages of 100% in-line inspection.

Advantages

- It saves the cost of further processing of a product that is likely to fail final inspection and test.
- It provides data on quality performance that can be used to take corrective action.

Disadvantages

- The in-line inspection function may become a routine step in the manufacturing line and because the rejects are being screened out, there may be less emphasis on the prevention of defects.
- It tends to duplicate inspection and increase inspection costs.
- It is not 100% effective because performing 100% inspection does not guarantee that all of the defective items will be detected.

First-Piece Inspection

In first-piece inspection, several pieces at the beginning of every new run are inspected to determine whether the set-up has been properly made and whether the tooling is adequate. The sample should provide a complete check of the machine or operation set-up. If the machine has nine spindles, for instance, samples should be taken from each spindle. Usually, the first five pieces produced by the new set-up constitute a large enough sample.

The advantage of first-piece inspection is that since the items turned out by a process or operation are evaluated at the beginning of the run, any necessary correction can be made before the run is started.

Patrol Inspection

The inspector patrols the operations at periodic intervals and inspects the items being produced. Since inspection is performed concurrently with the operation, patrol inspection can provide faster response than inspection after the item has been completed.

It is advantageous to use patrol inspection under the following conditions:

- When a process turns out a high percentage of defective products and requires frequent inspection.
- When a process is erratic and the operator is unable to do a thorough job of inspection.
- When there is a need to collect special detailed data on the performance of the process.
- When an audit of the process is required.

In-Process Acceptance Inspection

This is the classic type of inspection. All the items made at an operation in a given period of time are inspected together as a lot. They must be inspected before they are approved for release to the next operation.

In-process acceptance inspection provides several advantages:

- It makes it possible to control the quality level at each successive stage of the manufacturing process.
- It provides data to use in preparing performance reports to help pinpoint problem areas.

On the other hand, it also has some limitations:

- It does not prevent defects since the inspection is performed after the items have been completed.
- It delays the movement of parts from one operation to the next.
- It is not easily applied to continuous processes.

Of the five types of inspection, none is completely effective by itself. An efficiently designed in-process inspection system requires several of them in combination. How they can best be combined to serve particular needs depends on an evaluation of the following factors:

- The cost of each type of inspection.
- The type of manpower each type requires.

- The history of quality performance. Has the process been in control in the past?
- The type of process. Is it continuous, or can the items produced be collected and inspected in batches?
- The stability and the capability of the process.
- The nature of the product characteristics being controlled. Are the characteristics critical or minor?

IMPROVING EQUIPMENT AND METHODS

Many of the most profitable areas for savings of inspection and test costs lie in improved equipment and methods used to do the job. Since inspection and test are not usually measured and controlled to the extent production jobs are, they are not usually too efficient. Improvements can often be made by:

1. Providing equipment which can perform inspection and test tasks faster or without operators.
2. Building inspection or test devices into production equipment.
3. Designing improved record and reporting systems which require less time and effort.
4. Applying industrial engineering techniques to improve inspection and test station layouts and methods.

STATISTICAL QUALITY CONTROL

Powerful tools that can be used to help achieve in-process control are capability studies, control charts and sampling inspection.

A capability study shows whether a machine or process is inherently capable of turning out items that conform to specification. The results of the study may even indicate that it is possible to reduce the amount of inspection without adversely affecting quality. On the other hand, a capability study may reveal that a certain percentage of the items will always exceed the tolerances of the specification. In this case, it may be necessary either to relax the tolerances or to acquire a machine that is capable of meeting them.

Control charts are another excellent tool for increasing the efficiency of the in-process control techniques. In any series of measurements, there is variation. Sometimes the variation is only the natural outcome of "constant causes" inherent in the process. In other situations, there is also variation due to what are called "assignable causes". In the first case, the variation is normal and the process should be left alone. In the second, the variation indicates that something has gone wrong with the process and action should be taken to correct it. The problem is to decide which type of variation is present.

Acceptance sampling techniques provide a means of measuring and controlling quality without the necessity of checking all the units produced. By using a sampling plan, it is sometimes possible to significantly reduce the costs associated with appraisal while still maintaining adequate control.

ACCURACY STUDIES

There are failure costs associated with incorrect quality decisions on the part of inspectors, testers, or operators. These can be in the following areas:

1. Falsely rejecting acceptance material.
2. Falsely accepting rejectable material.

There are many plans which rate appraisal personnel in relation to these two errors. Perhaps the easiest to apply is one involving accuracy as a percent of defects correctly identified. This involves submitting a known number of good and bad units to an individual and rating his ability to correctly separate the units. The number of incorrect decisions is then multiplied by the cost of each wrong decision and an extension is made showing the cost implications.

DECISION ANALYSIS

In the early manufacturing of a new product (and in spite of good quality planning), the need for adjustment of measurement and test controls is generally revealed. This creates a need to analyze the effectiveness of decisions made on components, sub-assemblies and final product in terms of the earliest possible detection of defects. A technique called decision analysis is sometimes helpful in such a determination. This involves an analysis of accept-reject decisions of inspectors and testers, and identifies the point in the process where such decisions are made. Further, the model or part on which such decisions are made is also specified.

Summaries of such results frequently show trends for individual inspectors, especially where inspection planning, visual standards or training is less than adequate. These trends show up in terms of two inspectors servicing the same area and rejecting significantly different amounts of material. Also, decision times (i.e., the time required to make an accept-reject decision) may be significantly different for two inspectors in the same area.

Obviously, the planning engineer must address to these differences and provide improved control (and therefore improved costs) by more effectively utilizing the appraisal personnel.

WORK SAMPLING

The technique of work sampling consists of sampling work elements of an individual or group and using probabilistic theory to estimate the total time spent on a given activity. When applied to appraisal personnel, who very often do not have repetitive work elements, it can be used to more effectively structure the work routines. For example, if work sampling determines that 10% of an inspector's time is spent walking from one end of the department to another, then obviously some change in his geographic assignment or work station could be made to minimize such a cost.

Generally, work sampling has been found to be a better tool for measuring indirect labor rather than direct labor, inasmuch as repetitive work elements can be studied by either time study or predetermined time standard systems.

REDUCING APPRAISAL COSTS

In summary, we can (and should) ask some pertinent questions to determine the effectiveness of our appraisal efforts:

1. Are inspection points located to maximize the return on dollars spent for inspection?
2. Are inspection stations and methods engineered for the most efficient work accomplishment?
3. Could inspection and test operations be economically automated by using special purpose instrumentation, or tape or computer controlled equipment?
4. Could inspection and test record and data reporting functions be more efficiently performed using the computer or other modern data handling devices?
5. Is it possible to control processes sufficiently to prevent production of defectives and eliminate product inspection?
6. Could statistical quality control techniques be profitably used?
7. Could tests now being performed by outside laboratories be performed at less cost in-house or vice versa?
8. Are some tasks now being performed by highly paid inspectors or testers which could be performed by lower classification employees?

REFERENCES

(1) "Guideline for Reducing Quality Costs", American Society for Quality Control, Milwaukee, Wisconsin, 1977.

REDUCING FAILURE COSTS AND MEASURING IMPROVEMENT

William O. Winchell, Project Manager
Product Assurance Department
Environmental Activities Staff
General Motors Technical Center
Warren, Michigan 48090

Today I would like to discuss with you two topics — reducing failure costs and measuring improvement — from the ASQC publication "Guide for Reducing Quality Costs." Reducing failure costs is a tough thing to do and I am sure you will agree that once it is reduced, it is even tougher to maintain the new level. Eliminating failures is something like an old leaky tube of toothpaste. You can fix the leak and force the toothpaste in the right direction — through the nozzle; but not for long — soon it leaks in another spot. But there is a reason for this. It is because in most cases the real cause of the problem has not been identified. Without knowing what the real problem is, we surely can't fix it.

So often the quality manager sees his job only as informing management of symptoms of problems such as high rejects and excessive returns from the field. His fix, considering his perceived scope of activity, is to put more inspection on to keep the bad parts from escaping until that day comes when the problem is corrected. But without effective corrective action that day never comes. Management looks to the quality organization for more than this and rightly so. The quality organization must not only report problem symptoms, it must involve itself in the mainstream of defining the real problem and proposing the best possible solution. The quality organization must develop an effective corrective action system that gets at the root cause of problems in order to permanently reduce failure costs.

The ASQC publication "Guide for Reducing Quality Costs" recommends four key ways of reducing the cost of failures. These are to:

— Communicate the problem area to all affected persons.
— Create a desire in others to mutually solve the problem.
— Provide leadership in the planning and carrying out of a logical investigation by those in the company best able to resolve the problem.
— Follow up to insure that the problem was permanently fixed.

Quality performance reports are a good tool for communicating problem areas to those that can solve the problem and to make top management aware of the progress. However, they must be understandable, clear and to the point. They should summarize information and not overwhelm the user with needless detail, while pointing out and emphasizing the significant trends. Certainly, the reports should be designed to meet the needs of those using it for taking corrective action, as well as for those tracking the progress. Obviously, there is no one best format; the reports must be individually designed to meet the needs of your organization. They must be user oriented. If they are not they will be useless documents on the desks of those people that the quality manager needs most — those that can find and fix the problem.

Creating a desire in others to mutually solve a problem is critical for most quality managers. It is directly reflected in the quality manager's ability to solve the chronic quality problems existing in his company. This is because the causes of the chronic problems are not normally the ones that the quality manager can correct by himself. A chronic problem may be caused by bad designs, improper tooling, improper manufacturing sequences or many other things which, of course, must be corrected by someone else. To get someone else to do something about the problems is a selling job, to a large extent. For the quality manager this involves selling himself, his corrective action program and, most importantly, the fact that the problems are serious enough to justify spending the time and money necessary to solve them.

It is important to remember that in order to effectively sell your programs, good justification must be the corner stone. In the early stage of investigation the cost of the problem, the amount of effort needed for solution and the tangible and intangible benefits should be estimated as best you can and presented in your corrective action reports. Adequate justification will gain support from your management when the problem is stated in terms they understand, For example, the engineering department can often be stimulated to correct a design problem when it is shown that it affects product reliability and that a specific amount of excessive warranty cost is being generated. The production supervisor must be convinced that a solution to a rework problem will help him in his daily battle with the efficiency of his operation. Above all, don't give up if your best selling job falls on deaf ears the first time around. If you are still convinced that the solution to the problem will contribute positively to your company, look for ways around the road blocks. Remember, the quality manager must provide the leadership for failure reduction.

Of course, a key to your failure reduction program is a systematic corrective action system through which logical investigations and resolutions of problems can be processed. The ASQC publication "Guide for Reducing Quality Costs" suggests several forms that can be used as a framework around which your system can be designed. Experience has shown that certain aspects need to be determined and documented early to insure the effective operation of your system. They are:

— A step-by-step plan for investigating and solving the problem. Each step should have the responsible individual listed along with a target date for completion. It is vital that the plan be kept up to date, reflecting changing conditions, and that periodic status be reported.

— Both tangible and intangible benefits expected to be achieved in terms management can understand.

— The projected cost to achieve the desired results.

By having a plan, those involved will be confident that progress will be made and that benefits at the end of the road outweigh the costs that will be incurred along the way. In addition, priorities can be systematically assigned to problems so that those having the greatest benefit to your company will be attacked first, while those having smaller payoffs wait on the back burner.

Following up to insure corrective permanent action boils down to a periodic measurement of the results. The initial measurement will determine if the plan is achieving the projected results at the estimated cost. If it did not, additional investigations to obtain the needed results should be launched. Periodic remeasurements, either as part of your formal quality reporting system or by special studies, will help insure that the gains are permanent. Holding the improved quality levels is vital and often overlooked. It is common to see improvement trends reverse themselves and head toward former levels. By proper reevaluation, these downward trends can be caught early and corrected before they become serious.

In summary, a program to reduce failure costs has to consider:

— Communication of problem areas to all those affected.

— Creating a desire in others to mutually solve the problem.

— Providing leadership in the planning and carrying out a logical investigation of the problem with those best able to resolve the problem.

— Following up to insure that the problem was permanently fixed.

Now let us discuss our next topic, which is included in the ASQC publication "Guide for Reducing Quality Costs," how can we measure improvements in a company's quality system? Being a quality cost session, you might think that we would stress that quality cost is the best way of measuring improvements, but this is not always true. Only sometimes can quality cost provide the answer and, when it is used, it must be applied with a great deal of care and discretion. The ASQC publication "Quality Costs — What and How" is a good reference to help you stay clear of the pitfalls inherent in comparing costs.

For example, don't be misled in comparing percentages. One might assume that a failure cost decrease from 70% of total quality cost to 60% is a big improvement. But if total quality cost, on a unit basis, rose between the two periods from $2.00/unit to $2.50/unit due to added inspection, failure cost on a unit basis actually increased. Indeed this company is still in trouble. Quality costs are usually good for determining the magnitude of in-plant improvements that require the measurement of inspection, scrap or rework costs. But again, be careful. Don't make conclusions on records alone — use personal observations too. Take a walk to the production line and see if things are what the records indicate. So often an innocent mistake in recording information or a major shift in the product mix to something easier to build will cause results on paper to misrepresent what is really happening. There is really no substitute for a first hand look at the situation.

Perhaps the greatest pitfall to guard against in using quality costs is to measure the effect of product improvements relative to field failure or warranty. To illustrate this, let's suppose that we have a reliability problem that only shows up after six months in the customer's hand. Assume that we fixed that problem in January, 1978. Even if the product goes directly to customer, it is July of that year before the effect on warranty cost is felt. Obviously, it is absurd to use warranty expense for the month of May to track the difference. This inherent lag makes it tough to apply quality costs to finding out if field performance is better, especially if your boss is breathing down your neck for an answer. Something else must be used to obtain timely answers such as life cycle testing or possibly outgoing quality audits of completed products that incorporate durability provisions. Such tests can discover system or product discrepancies far more quickly and at less expense than would be associated with waiting for failures in the customer's hands.

Other sources of data besides warranty failure costs that are good indicators of field performance, depending upon each individual situation, are:

- — Field trouble reports originated by perhaps your field service engineers. These provide more detail regarding your quality problems allowing a more accurate comparison between present and past performance.
- — Market research surveys that determine what the customer thinks about your product. This indeed may be an appropriate way of getting a handle on potential lost sales due to quality problems.
- — Spare part sales that supplement warranty failure cost information may better identify product performance trends.
- — Customer complaints, properly tabulated and summarized, are another means of tracking progress. However, caution must be exercised in that this data may not represent your normal customer — they usually are from the most irate of customers.
- — Installation phase reporting of quality problems found in the erection and commissioning activities of large and complex projects certainly is a valuable source for tracking progress for these type of products.

In summary, there are many ways besides quality cost to track improvements in a company's quality system. The choice of which method or methods to use depends upon your individual situation. One method that is inadequate for one situation may be the best in another. Most importantly look before you leap. Carefully look at all the alternatives before selecting those that you will use for your immediate needs.

WHAT ARE QUALITY COSTS

Joan H. Alliger, Product Control Engineer
Eastman Kodak Company
Rochester, New York 14650

ABSTRACT

In the area of quality control, the most efficient route to meet industries' challenges of improving product quality and lowering quality costs is to make the product (or provide the service) correctly the first time. Departures from this route incur unnecessary expense.

With the proper tools for measuring quality cost data and a program for presenting quality cost information, trends can be established as guides for management in pursuing these challenges.

The purpose of this presentation is to introduce a system to organize certain quality related costs enabling management to effectively measure and optimize quality costs as well as product quality.

The system is described in detail in the ASQC publication, "Quality Costs — What and How", prepared by the Quality Costs Technical Committee, ASQC.

DEVELOPING A COST EFFECTIVENESS PROGRAM — HOW TO START

Victor J. Goetz, Manager
Quality Planning and Development
Warner-Lambert Company
Morris Plains, N.J. 07950

ABSTRACT

Can we answer the question "are we getting our money's worth for the dollars we are putting into the quality program"? The purpose of this paper is to provide assistance in "How to Start" a program to measure the cost effectiveness of the quality function at your location. Problem areas, pitfalls, and key points to be considered are discussed. The steps to be taken are outlined and answers are supplied to the typical questions that will be raised by your management. The function of and the results to be expected from the program are detailed. The emphasis is placed on how to utilize available data for effective cost improvement.

INTRODUCTION

Can we answer the question that may be put to us by top management "are we getting our money's worth for the dollars we are putting into our quality program?"

Obviously, the question is an extremely difficult one to respond to since, if all concerned in Engineering, Purchasing and Manufacturing were to do their job, there would be no need for Quality Assurance. While we are certain that the investment in Quality Assurance is worthwhile, it is very difficult to determine if you are realizing your money's worth — one subjective measure, of course, is the quality acceptance of the product by the ultimate consumer; another is the acceptance of the quality of the product, and the conditions under which it is manufactured, by regulatory agencies, or our customers and a third is: Are we making money?

We recently responded to this question as put to us by the president of one of our divisions. Today, I'd like to discuss some of the things which we explored in preparing a response.

The purpose of this paper is to provide assistance in starting a program at your location. We will identify some of the problem areas, some pitfalls and some key points to be considered. (See Appendix I). This is not an easy project. It's complicated; it involves much digging out of information and difficult decisions to be made since there are few clear-cut rules as to what exactly is a "quality-related" cost or how to start your own particular program. So let's start at the beginning.

PROJECT STEPS

The first step is to review some of the basic literature that has been published in the area of quality costs (See Bibliography). This review is useful for learning about or refreshing your knowledge of quality cost concepts, recommended methods of analysis and report presentations. You will find that most of this material is very complete in describing how to analyze and present the quality cost data that identify problem areas, monitor quality improvement activities and provide a management tool for long-term trend analysis. They are, however, seriously lacking in the specifics of how to start the project and how to develop the original data for a Quality Cost Analysis or Performance Measurement Report.

The next activity is to discuss, within your organization, some of these quality cost ideas. This may also involve discussions with your accountants and with your top management to get their thinking and interests in the area of quality cost and to define your quality cost effectiveness program objectives.

After you have had these discussions with your management and have decided to develop a cost effectiveness program, prepare a brief outline of the methods you are going to use and a list of activities to be done. (See Appendix II). Next, select a location or an area of concern. It is better to start small. Quality costs are more useful at the department or location level then they are at the division or the group level.

MANAGEMENT PRESENTATION

The next step is to make a formal presentation to the top manager and the major staff heads at the location, i.e. Engineering, Production, Quality, Accounting, Purchasing, Technical Services, etc. At this meeting you will define the objectives of the program, demonstrate that you have top management support for the study and clear the air of any misconception of the goals of the program, or how the information is going to be used.

The major topics for discussion and the points to be made at this meeting can be stated as follows:

1. This is a "one-shot" study to justify starting a program of quality improvement and cost reduction.
2. The emphasis of the program will be on *Operations.*
3. The role of the Quality Department will be deemphasized.
4. The result of the program is a top management tool that:
   a. Demonstrates the effectiveness of the total quality program working within a location.
   b. Measures the efficiency of the Quality Department itself.
   c. Provides a means of analyzing quality problems and measuring cost to benefit ratio of correcting them.
5. The "one-shot" study"
   a. Develops a reporting format
   b. Shows where you are now
   c. Develops a par for your location
   d. Indicates areas of concern worth investigating
   e. Quantifies what we may already know intuitively
   f. Highlights and quantifies high cost areas

g. Establishes the largest single cost element

h. Shows that some major cost elements may not be being evaluated

During this meeting you must be prepared to answer the following typical questions:

1. Of what value, will be this information?
2. What will it tell me that I don't already know?
3. Where am I going to get the resources to work on these newly identified areas of concern when I cannot get the resources for the problems that I already know about?

Some of the answers that can be provided are that the study:

1. Provides new insights into your accounting and quality data systems.
2. Tells executive management what is happening.
3. Provides back up data and information to do failure analysis more effectively.
4. Shows that the present system is not responsive to changes in quality or production.
5. Provides a trailing indictor that shows the results of cost reduction efforts or the lack thereof.

It is important to emphasize that you will not know exactly what the study will provide until you do it the first time. This is a "one-shot" study with the decisions to be made later concerning an ongoing program.

REVIEW OF RECORDS

The next step is to have the top management assign someone in the Accounting or Financial Control Department to assist you in reviewing the accounting system utilized in your organization.

Here you may meet some resistance. You will be digging deep into someones private domain. Emphasize that you are only interested in the meaning of the data that you find in the financial reports. You are planning to view the information from a "horizontal" view across cost centers rather that a "vertical" view within each cost center as does the present accounting system. This across cost center view point, i.e., prevention, appraisal and failure classification tends to highlight high total cost areas that might be too small within a given cost center to be noticeable.

A major fallout from this study will be to learn that many company decisions are made from data that is highly suspect. Once you start to review the financial records you will be interested in the budget control statements and the various types of variance and expense reports. Obtain a general feeling of how they keep the books, what kind of accounts they use, and what kind of interim reports they prepared to convert the raw data into the final accounting reports. You will generally find that the organization probably uses a "variance" accounting system. This means that the accounts report the differences between the actual costs and the established standards. There may be both favorable and unfavorable variances combined within any given account, resulting in only the net variance being reported. Therefore, you will have to review the source of the data since you will be more concerned with the unfavorable variances.

When looking at the accounting records for scrap, rework, recovery, inspection or quality information, you will find that the accounts may include alot of superfluous information, as far as quality is concerned, or may not even include any of the quality information you think that you need. The point is to get started, make some preliminary decisions about the accounts and then go to the accounting records and see what you can find. The investigator must piece things together, using the books where possible, resorting to analysis, estimates and even creating new data if necessary.

This first and most difficult phase of the cost effectiveness program may require a period of 3 to 6 months to complete and prepare a management presentation.

SUMMARY

In summary, this paper has supplied information to assist the user to start a quality cost program and, through this, develop some information for a cost improvement effort. This disciplined approach to evaluation of available data at your location will identify and quantify many areas of concern.

No amount of writing can fully describe this activity. You must do it yourself at your location, not just to know about it intelligently, but in order to really understand a quality cost program.

APPENDIX I
SOME KEY POINTS TO BE CONSIDERED

1. Just the discipline of collecting and analyzing this cost information will identify and quantify many areas of concern.
2. Use the standard approach — failure, appraisal, prevention — for measuring the total annual quality cost and the impact it has on profits.
3. Use a two-man team — quality and finance — to collect these costs.
4. Utilize your current accounting systems and procedures to the maximum extent possible.
5. Utilize your accounting and manufacturing personnel for obtaining data and estimates. Utilize both actual and estimated cost values.
6. Don't spend $10 to find $1.
7. You will never capture every dollar related to quality costs.
8. Avoid using burden cost — too easy to show large dollar improvement with little effort and no real change in the location's costs.
9. Use these quality costs to determine the need for a quality improvement program and to select the specific quality cost reduction projects which will yield the greatest payoff.
10. Quality costs serve as a benchmark for judging the effectiveness of the quality cost reduction program.
11. The total costs of quality will only show improvements as the major areas of concern are reduced. This is why the report is a trailing indication.
12. Don't make direct *between* location comparisons — trends *within* a location are what is important. Every location is different.
13. Develop meaningful indices for measuring performances.

14. Tailor your quality cost report to your audience. Talk quality to company management in terms of $— the universal language.
15. Don't computerize the report, but use data from the computer, if available.
16. Don't report the total quality cost periodically. Set up to monitor on a regular-weekly or monthly- basis only the specific high quality cost areas that are being worked on.
17. Repeat total quality cost survey 12-18 months later to measure success of program and to determine the next steps to take.

APPENDIX II
SOME TYPICAL COST EFFECTIVENESS PROGRAM ACTIVITIES

1. Review literature
2. Define objectives and develop and outline program
3. Select location
4. Make management presentation
5. Review accounting system
6. Define classification and cost elements to be included
7. Identify areas of concern
8. Collect data
9. Develop report format and allocation method
10. Prepare quality cost analysis report
11. Identify potential areas for improvement
12. Present base line data and recommendations to management
13. Select areas for action
14. Develop action program
15. Demonstrate usefulness of system
16. Identify refinements and improvement to quality cost system
17. Transfer responsibility for report to accounting function

APPENDIX III
BIBLIOGRAPHY

Feigenbaum, A. V. "Total Quality Control", McGraw-Hill Book Company, New York, 1961 — Chapter 5.

"Guide for Reducing Quality Costs", American Society for Quality Control, Milwaukee, Wisconsin, 1977.

Juran, J. M. (ed) "Quality Control Handbook", 3rd Edition, McGraw-Hill Book Company, New York, 1974 — Section 5.

Juran, J. M. and F. M. Gryna, Jr., "Quality Planning and Analysis", McGraw-Hill Book Company, New York, 1970 — Chapters 4 & 5.

"Industrial Quality Cost for General Management", — STAT-A-MATRIX Institute, Edison, N.J., 1975 — Chapters 6 thru 11.

"Quality Costs — What and How" 2nd Edition — American Society for Quality Control, Milwaukee, 1971.

CAN QUALITY COST PRINCIPLES BE APPLIED TO PRODUCT LIABILITY?

Dorian Shainin, Consultant
Dorian Shainin Consultants, Inc.

Paul D. Krensky, Consultant
Paul D. Krensky, Associates, Inc.

Edgar W. Dawes, Corporation Quality Manager
Veeder-Root Company

ABSTRACT

INTRODUCTION

Let's assume that product liability litigation is underway against a company. The nature of the problem would indicate that the company is faced with the following:

- Manufacturing Test Records will probably be reviewed under the Rights of Disclosure Law.
- Shipped Product Locations must be listed for purposes of a recall.
- The Product involves an original equipment manufacturer. Distributors and installers are also being sued.

What actions must be taken and what will be the cost of the required responses?

- In the case of records, examples can be cited where a company's key Managers have spent weeks with a Legal Counsel in reconstructing the tests made on a specific unit of product.
- Recent studies made by the Consumer Products Safety Commission show how product recall effectiveness depends on several factors including Distribution Channels, Initial Product Cost, End Use, and Product Life. The complexity of these factors can be shown to have a dramatic effect on both recall costs and effectiveness.
- The Plaintiff's Attorneys will generally involve as many parties as possible in a suit. This makes it especially difficult for any original equipment manufacturer as his distributors and installers usually become involved. Under such circumstances, defense costs usually skyrocket.

The old axiom that "An Ounce of Prevention is Worth a Pound of Cure" prevails in these situations. This discussion develops the relative costs of preventing litigation versus responding to its awesome challenge.

COST CATEGORIES

Costs go far beyond those cited above. The Panel discussion will develop the following typical cost elements:

Quality Category	Prevention Costs	Litigation Response Costs
Design Quality	**New Product Safety Activities** Costs associated with reviews and changes to guard against safety failure or customer foreseeable misuse.	**Design Failure Mode Analysis** Costs associated with determining the reason for the customer encountered design failure and what changes must be made on shipped product.
Manufactured Quality	**Manufacturing Plan** Costs of predetermining the product's safety characteristics and the planning and implementation of total conformance on these in manufacturing.	**Manufacturing Failure** Costs of determining "what went wrong" and the changes and costs necessary to prevent recurrence.
Customer Quality	**Customer Safety Planning** The prevention costs to assure delivery and use (throughout life) of a safe product.	**Customer Failure Costs** Litigation loss cost settlements and associated costs of elimination further customer and company risks including product recalls, field retrofits, etc.

QUALITY COSTS — A NEW PERSPECTIVE

Robert C. Kroeger
Manager, Reliability & Quality Assurance
General Electric Company, AESD
Utica, New York 13503

ABSTRACT

At General Electric the definition of product quality is sufficiently broad to include product performance as perceived by the user after delivery. If one examines quality costs in this broader perspective, some unique insights are developed. Case histories were selected from programs over the last 15 years to illustrate cost effectiveness in this broader definition of quality. They were then used in the development of an in-house data system which relates the quality data base (failures at various levels of assembly) to a cost data base (the cost of finding and removing failures at each level of assembly) to optimize appraisal planning for in-house product flow. This cost-optimized appraisal system allows the quality engineer to make balanced decisions on how much to test and at what level of assembly, such that the overall cost of quality is minimized.

INTRODUCTION

Within the General Electric Company, quality is defined as "the relative excellence of the composite of all product attributes in fulfilling the needs and reasonable expectations of those whom the product serves, as they perceive such fulfillment from time of offering throughout product life." Since this broad definition of quality includes long-range performance of the product, it embraces reliability as well. This paper will deal with quality costs in this broader sense of "quality."

In January of 1971, Robert Seamans, Jr., who was then Secretary of the Air Force, addressed the IEEE Reliability Symposium in Washington with these words: "Save now and pay later concepts have been evident in several Government programs. A failure that can be eliminated for one dollar during concept formulation can cost $10 during engineering development, $100 in production, and $1000 in the operational phase." Initially, it was presumed that these words were intended to be interpreted qualitatively. Subsequent studies and data received from the Air Force, however, revealed that in fact this was the actual performance history of some ten Air Force programs that were then in work. Accordingly, we at GE's Aerospace Electronic Systems Department (AESD) in Utica, New York, began to examine the costs of finding and eliminating failures in-house through an orderly factory program which included environmental screening as part of the product flow. The case histories described in this paper were derived from efforts during the manufacture of aerospace and defense electronics.

CASE HISTORIES

Figure 1 reveals the failures experienced per system at each level of assembly during the manufacture of F-111 Attack Radar Systems in the late 1960's. Subassembly, in Figure 1, refers to the first level of assembly at which piece parts are interconnected, which for AESD at that time was a printed wiring assembly. PWA's were assembled into Line Replaceable Units (LRU's) and several LRU's made up a Radar System. During environmental burn-in, LRU's were checked to a performance specification for the first time. Systems containing several LRU's were then integrated, acceptance tested, and submitted to a demonstration of reliability. Shown below each of those levels of assembly is the cost of locating, removing, and repairing a failure and restoring the system or black box to acceptable performance under the specification. At the lowest level of assembly this cost was $53, at the unit level it was $154, and at the system level of assembly it was about $300. For reliability demonstration the cost was slightly less because engineers, rather than factory hourly personnel, were monitoring and running the reliability demonstration test. Noting the 6 to 1 cost leverage in finding failures and removing them at the lowest level of assembly, we very quickly adopted a quality program that was based on "driving the failures upstream"; that is, finding them at the lowest possible level of assembly — even at the supplier's plant, if that were possible. Note that failures at system level could be removed at a cost of roughly $300. Using Dr. Seamans' data, the removal of these failures should be worth 10 times $300 (or $3000) to the Air Force. At this rate it wouldn't take many hidden failures that could escape to the field to make this entire screening program very cost-effective.

F-111 Attack Radar System
IN-PROCESS PRIMARY ELECTRICAL FAILURE DISTRIBUTION

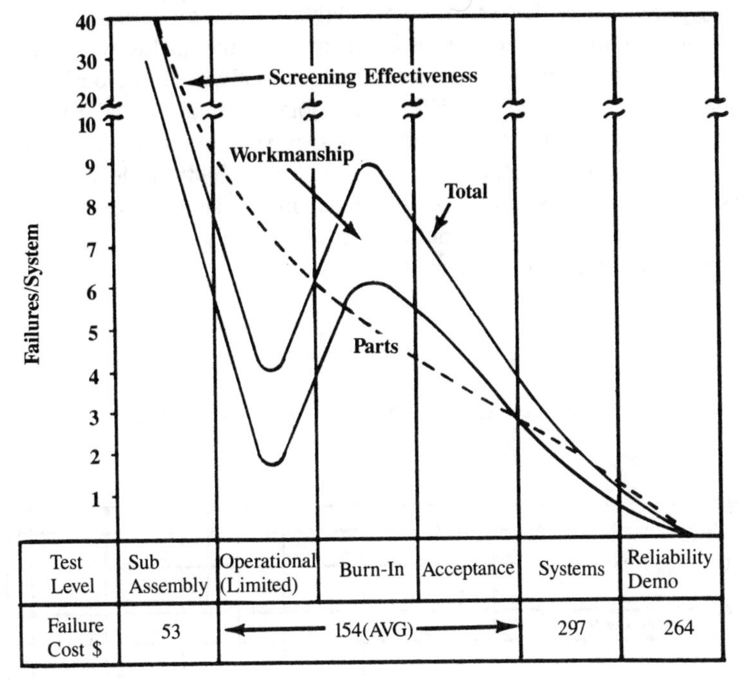

Test Level	Sub Assembly	Operational (Limited)	Burn-In	Acceptance	Systems	Reliability Demo
Failure Cost $	53	← 154(AVG) →			297	264

FIGURE 1

Figure 2 depicts the reliability performance of a missile system over several years of procurement. The customer purchased from AESD about 4,000 of these vehicles each year, so it was a very large production run. To verify continuing reliability on this program, a portion of each month's shipment was exposed to a reliability demonstration test in accordance with MIL-STD-781B, Test Plan 20. Demonstration of 150 hours MTBF was required, and evidently it was possible because reliability prediction showed that the MTBF of this equipment could be as high as 300 hours. Figure 2 shows that the equipment did not meet the required MTBF over the first two years of the program; yet the need for this device in the field caused the customer to direct shipments of what, from a reliability/quality standpoint, was substantial hardware. As a result of an effective reliability growth program, equipments were eventually manufactured and delivered that substantially exceeded the reliability requirements. In fact, the learning curve, or reliability growth curve if you prefer, showed an improvement or growth rate of approximately $\propto = .5$.

In Figure 3, a depiction of different data for the same missile, the solid line is a common statistical measure of quality — defects per unit — and the dotted line is the kill probability actually demonstrated by sample rounds fired from each shipped lot. The on-receipt defects per unit decreased by a factor of five over the life of the program, an improvement attributable to the attention to quality and reliability performance that was shown in Figure 2 and to learning experience over five buys on this product. At the same time that the quality was improving and reliability growth was being demonstrated internally, the kill probability rose from about 80% to very close to 100%. The story in Figures 2 and 3 is simple: As we solved the problems in-house, the customer benefited from reduced defects on receipt and from substantially increased kill probability of devices selected for final use. At the same time, the cost of supporting such devices in the field was reduced in proportion to the number of failures removed in the factory and thereby prevented from escaping to the field.

RPM MODEL ARRAY MISSILE EXAMPLE

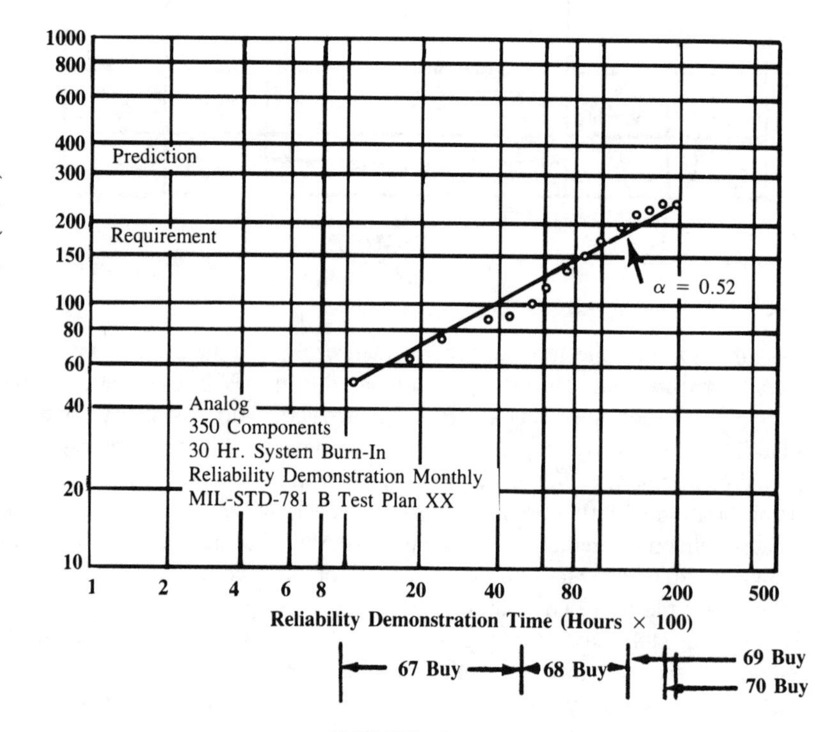

FIGURE 2

PRODUCT DEFECTS VS FIELD PERFORMANCE

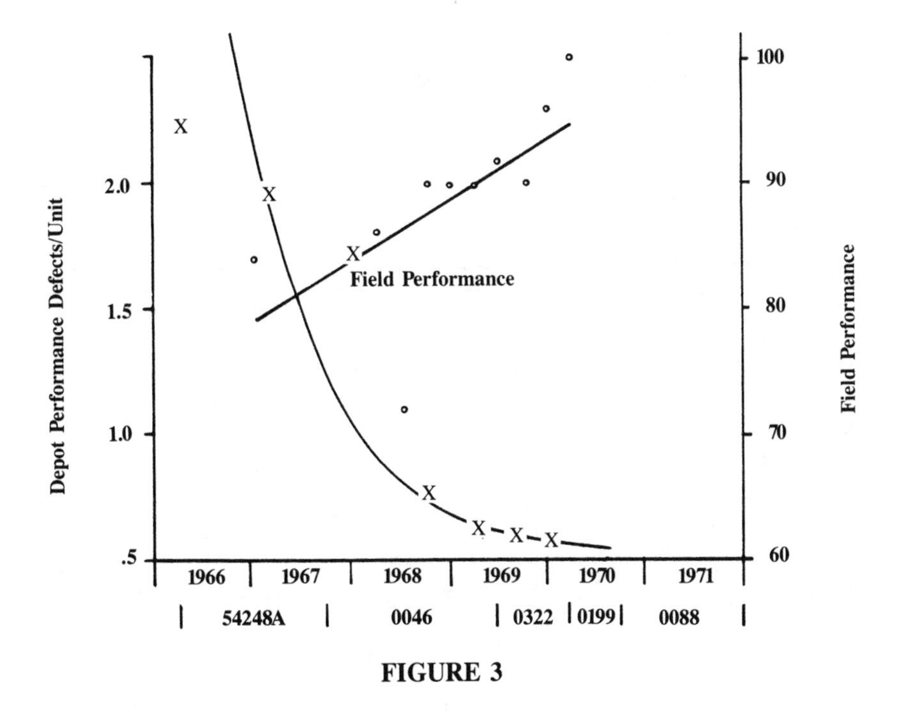

FIGURE 3

Figure 4 summarizes another cost story. Portrayed is the equipment failure rate vs. equipment on-time for family of electromechanical equipments, called Magnetic Tape Transports, used on the P3C airplane. Delivery of this equipment under contract was predicated on passing a "Shakedown Test" with sinusoidal vibration and cyclic temperature applied while electrical performance was checked. It was required that the last 50 hours of such testing be failure-free. The intent of the test was to remove early mortality or incipient failures and thereby deliver high quality material to the field. As you might imagine, some equipment took several hundred hours of operation before achieving 50 failure-free hours; and some required little more than 50 hours. The data were segregated into two groups: equipments which were operated for more than 100 hours, and those which were operated less than 100 hours to get 50 hours failure-free. The subsequent field performance of these two populations is shown in Figure 4. Note the dramatic difference in field performance between the two equipment populations, with the equipments subjected to the longer operation (>100 hours) during shakedown experiencing at least double the field MTBF. Evidently the shorter shakedown period left residual failures in the equipment that subsequently were manifest in field performance. When such data were shown to the customer, requirements for additional shakedown time were added to subsequent contracts as a priori requirements for delivery.

P-3C MTT F-3 Models

FAILURE RATE VS EQUIPMENT AGE

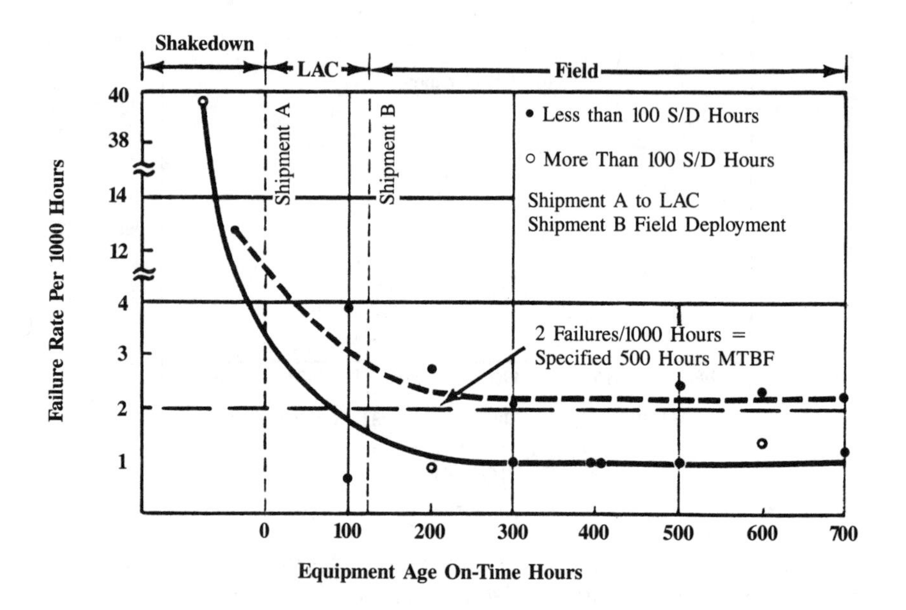

FIGURE 4

Figure 5 shows the same shakedown performance for a family of digital logic equipments on the same program. Notice that in this case there is very little difference between equipments that had more than 86 hours of shakedown and those that had less than 86 hours. The conclusion from these data was that, unlike the electro-mechanical analog MTT, a 50-hour failure-free shakedown requirement was sufficient to remove incipient failures in digital logic designs. We might also point out that the field MTBF of the logic units was well above the specified 800 hours, and that this achieved performance was related to the effectiveness of the 50-hour failure-free shakedown period. It is certainly easy to calculate the effect on life-cycle-costs, especially field maintenance costs, for an equipment whose field MTBF far exceeds its specified reliability. Certainly, head-end program investment in screening, design conservatism, and reliability disciplines had a payoff in reduced costs of operation and ownership for this equipment.

P-3C DPS Logic Units CRL Model
EQUIPMENT AGE VS FAILURE RATE

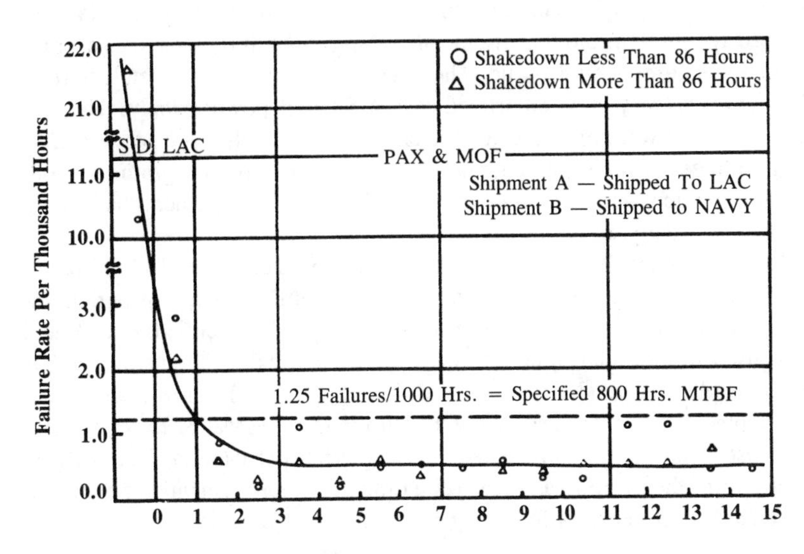

Equipment Age In Hundreds Of Hours

FIGURE 5

F-111 Attack Radar System
HVPS FAILURE RATE AT LRU BURN-IN

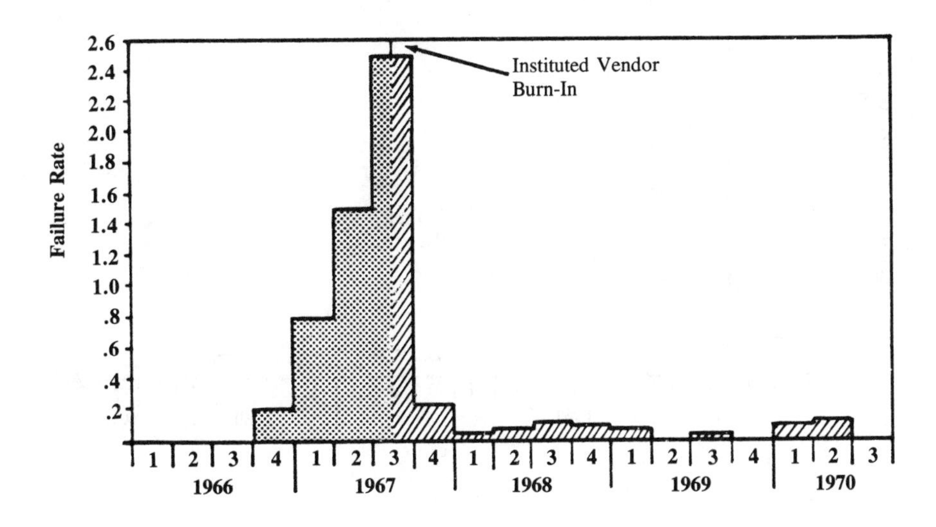

FIGURE 6

Figure 6 portrays the story of a component used in the F-111 Attack Radar System. A high voltage power supply (HVPS) was procured from a small vendor in the Northeast. In 1967 GE/AESD was building and shipping one radar system per day — thirty radars a month. The failure rate shown — 2.5 failures per shipped system — means that in a month's shipment of thirty radars, an average of 75 HVPS failures would occur. The vendor was directed to install burn-in facilities at his factory at an amortized cost of $85 per power supply. The cost of the failures that were prevented on the basis of the data that you see was $245 per high voltage power supply, for a net unit saving to the program of $160. This is a dramatic manifestation of the type of quality management that should be addressed in daily activities.

Figure 7 depicts actual data for what, at that time, was considered to be a rather amazing result. The question was asked, "Could a similar burn-in program improve the performance of a balky *mechanical* device?" (specifically, a camera that was used in the radar system to take scope pictures). Anticipating similar favorable results, the vendor was directed, in late 1967, to burn-in each camera prior to shipment. The amortized cost of the burn-in facilities was $225 per camera, and the cost of the failures prevented was also $225 per camera. The net cost to the program, therefore, was zero; yet a dramatic improvement was achieved in the reliability of cameras following the addition of burn-in.

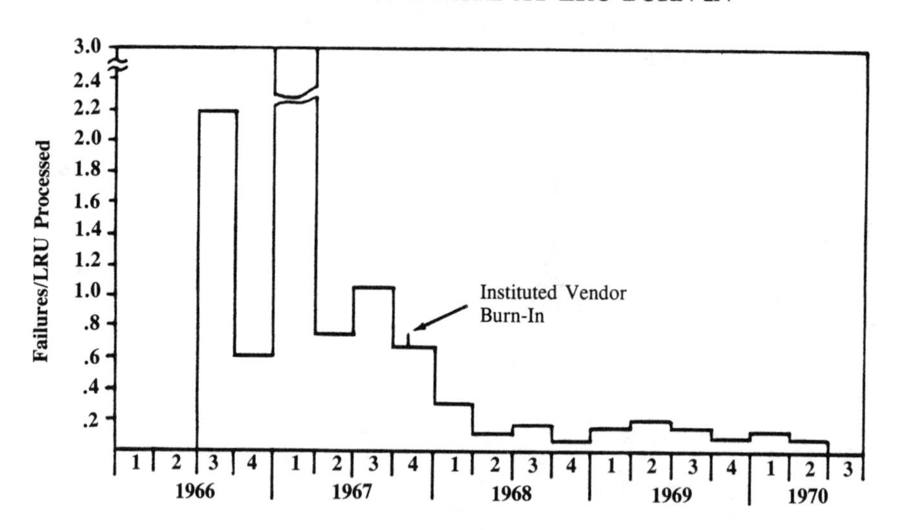

FIGURE 7

Figure 8 portrays similar results, but in this case with rotational machinery: electromechanical devices, end-instruments, synchros, pick-offs, and instrument motors. In this instance, intuition would tell one that burn-in would only prematurely wear such parts and thereby reduce equipment life. But, as is often the case, intuition was wrong. The data show clearly that screening in cyclic temperature burn-in for as little as 48 hours — 2 days — would remove 93% of the defects that were manifest or inherent in these devices without any damage to the life characteristics of the rotational machinery. The effect on the program? A $400 saving per device!

These last three examples have to do with "quality" interpreted in the broader sense of life performance of equipment as delivered to the customer, and "quality costs" in the broader business sense of how many dollars and other resources can be devoted to the goal of performance excellence in delivering such hardware.

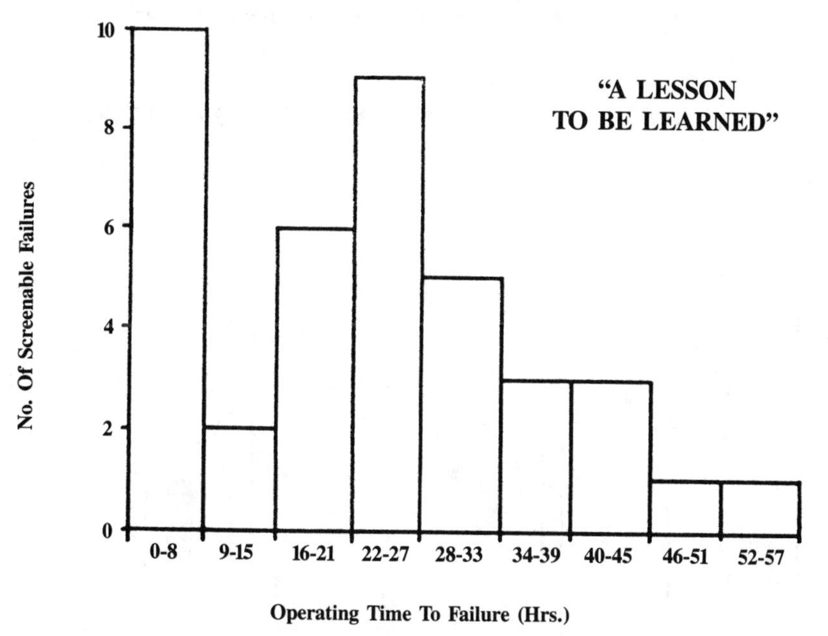

FIGURE 8

Figure 9 is more specifically related to quality costs in the classic sense. Total quality costs are separated into prevention, appraisal, internal failure, and external failure for businesses selected for their product and process similarity to GE/AESD. Costs are shown as a percentage to net sales billed. Within each business, the amount spent on quality and its apportionment to the above categories are discretionary. One might imagine the objectives to be realized by such expenditure as: #1) to minimize external failures, i.e., those failures that affect customer relationships, #2) to minimize internal failures, #3) to minimize the cost of the sum thereof. Internal investments in prevention and appraisal toward those goals would be iterated and managed according to the needs of the individual business, such that the total is an acceptable portion of net sales billed.

As you can see, some of the businesses spend a great deal more on quality costs than others and yet must have unhappy customers due to external failures. Some businesses experience serious difficulties on the factory floor as measured by the cost of internal failures. In many of these businesses, *the mix* of expenditures is wrong. Failures *can* be eliminated; failures *can* be found and analyzed and ultimately prevented. For proper quality cost management, not only should the total cost be minimized, but the external and internal failures should be kept to a minimum, in that order. Moreover, as you can see from Figures 1 through 8, examining quality costs in the classic sense as portrayed in Figure 9 is only part of the total business picture. Clearly, the performance of products as perceived by the user over product life has a significant effect on external failures and therefore on quality costs.

OPERATING QUALITY COSTS TO NET SALES BILLED

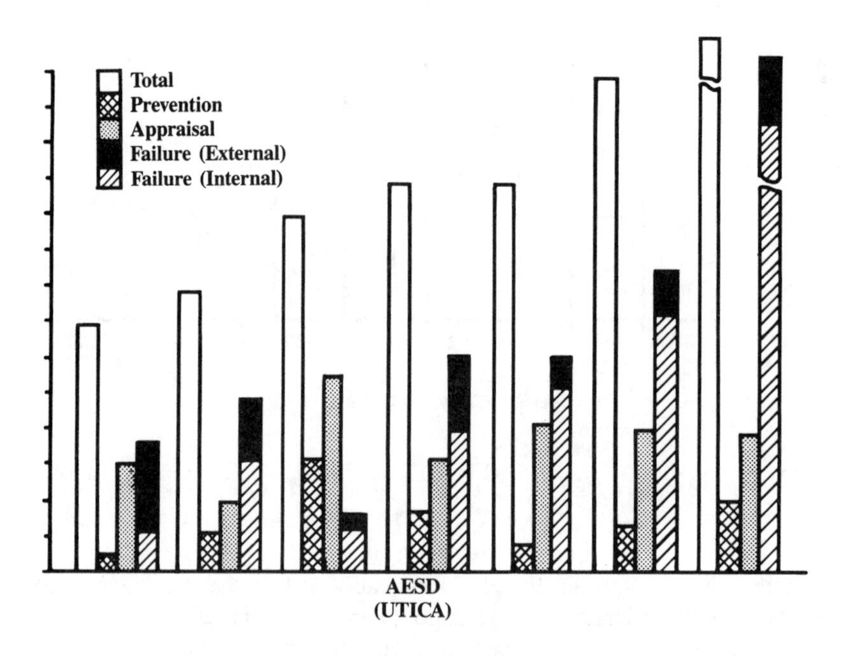

FIGURE 9

DISCUSSION OF DATA/RESULTS

The above illustrations make a common point for practitioners in both the Reliability and the Quality disciplines — that to control product costs and performance, quality problems should be identified as early in the product cycle as possible. In terms of the allocation of quality cost dollars, recognition of this fact is generally translated into a relatively high expenditure of prevention and appraisal dollars early in the program, with a gradual reduction as quality levels become known and as corrective actions result in quality improvements.

There are two basic reasons for this approach. First, the advent of automatic test and inspection equipment has dramatically altered the relationship between the prevention and the appraisal cost drivers. Automatic testing has reduced the recurring costs of testing each unit compared to the potential cost of part failures such that 100% incoming testing throughout a program is often practical. When more complex parts (such as microprocessors) are tested, recurring test costs may be low, but planning and programming costs per device are dramatically escalated. This situation may necessitate making a decision on how much to test for the entire program before the first item has been tested, because the big cost driver in automation is in programming the test set. Once the programming has been done, there is little reason (except where very high volume is involved) to test less than 100% for the duration of the program. The testing problem is further complicated by Large Scale Integrated Circuits (LSI's), where planning (programming) time is high, test running time is high, and, in some cases, complete exercise of the product is impractical. The quality/cost control system must contend with increasing costs to locate and correct defective components and modules at each state of manufacture — a situation wherein knowing incoming quality is of increasing importance, but where such knowledge costs much more to acquire.

The second reason for addressing the issue of the relationship of program phase to the allocation of appraisal effort is that even if we accept the idea that extra expenditures for appraisal early in the program will yield a cost return, the basic question of just how much appraisal will be cost effective still remains; and even if we agree that we should generally reduce appraisal effort as the program or product matures, how do we decide on the proper amount of reduction? When does such a reduction cease to be cost-effective?

COST MODEL

In addressing these questions at AESD, we decided that if we were to make appropriate decisions on our appraisal planning we would need a system that could account for a broad variety of products, production volumes, and hardware complexities, utilizing a wide range of measurement equipment. Our challenge was to develop a model or a series of *models that would integrate the quality data base with the cost data base* in such a manner as to permit a direct dollar tradeoff among the prevention, appraisal and failure costs associated with:

1. testing or not testing
2. assembly levels at which to perform tests
3. amount of testing to do (e.g., 100% or sample)
4. parameters to be tested at each hardware level

The model would have to be applicable at all hardware levels from incoming purchased material through final test of the end product. We also decided that our system should be capable of providing this information very early in a program — during the initial costing phase whenever possible. Information to support the kinds of decisions mentioned above should be available at a level that will permit the quality engineer to make detailed appraisal decisions; and it should be available at a summary level for management to determine the effectiveness of the total appraisal program and each segment thereof.

The cost model which we developed for our incoming test and inspection of purchased material in response to the above criteria includes the fixed cost per lot and variable cost per item for incoming test and inspection and the cost of each probable failure after assembly into one of our products. The model uses these data to identify the most economical quantity to test and inspect in order to minimize the total incoming test and inspection costs plus the costs of probable losses after assembly. The model utilizes various assumed levels of test/inspection at incoming ranging all the way from 100% to a complete elimination of incoming appraisal effort. The level shown to result in the smallest net cost is used by the Quality Assurance Engineer in determining the apropriate appraisal plan for each purchased item.

In addition to supporting decisions regarding the level of incoming test and inspection, our system provides visibility of characteristics checked at incoming on which no defects have been found and characteristics checked at incoming on which defects have been found at higher levels of assembly so that the incoming test and inspection plan can be iterated to increase both the efficiency and the effectiveness of the incoming screen.

The cost performance shown in Figure 9 has been the result of ten years of attention, control, and improvement; but we are reaching an asymptote on failure costs — a point of diminishing returns on any *major* effort to reduce them further. This is why our quality cost optimization emphasis has been shifted to the appraisal area.

Through implementation of the cost/quality data system described above, we hope to reduce our appraisal costs while making additional, but less dramatic, reductions in our failure costs. If we can continue to reduce the cost of quality while improving our product quality and reliability, we will be making the kind of contribution to the business that management and customers have a right to expect.

BAYESIAN COST ANALYSIS OF RECTIFYING REJECTED LOTS

David L. Kimble, Mathematical Statistician
Ecological Analysts, Inc.*
1910 Olympic Boulevard
Walnut Creek, California 94596

ABSTRACT

This paper gives a technique of equating the cost of correcting a defective item to the cost of sampling inspection. The purpose of this technique is to provide a useful tool for a manager whose job is to decide whether a quality control acceptance sampling plan should include rectification of rejected lots. This technique gives the manager a way of relating the cost of rectifying to the actual returns of rectifying.

INTRODUCTION

I am assuming that the quality control system is based on drawing a random sample from a lot to determine if the lot is above or below an acceptable level of quality. If the lot is estimated to be of unacceptable quality, it is rejected. The problem in many settings is what to do next. For example, if the lot is a quantity of work completed by a single worker in a production setting, there are four choices of action: (a) the rejected lot can be corrected by reworking the lot, (b) the worker can be given a "demerit" for doing a poor quality job (if the worker receives too many demerits he/she is taken off production), (c) the quality can be estimated and documented (the system can be used strictly as a feedback system), or (d) any combination of the three can be used. When a rejected lot is reworked (rectified), the quality of this rework is often as poor as the original work. Occasionally, the rectification miss rate** on some coding operations is as great as 95 percent, which leaves great doubts as to the gains of rectifying. This paper only addresses the decision of whether to rework rejected lots or not to rework rejected lots.

The technique assumes that the quality control decision rates are already determined by conventional means and the sample size and rejection number are fixed. For the sake of presentation, this paper deals with attribute sampling plans only, but this method can be generalized to variable plans.

METHOD

The general method of this procedure is simply to use Bayesian Decision Theory as described in many texts (Hogg & Craig, 1970, DeGroot, 1970, Lindgren, 1968) and to solve the Bayes risk for the unknown costs. The analysis starts by defining that the decision to be made is to choose either a QC system that does not rectify rejected lots or one that does rectify rejected lots. The associated costs of these decisions are used as the loss function. The three components of the costs are the cost to inspect the QC sample, the cost to rectify a rejected work unit, and the cost of having defective items in the outgoing lot. The cost function for both decisions is:

Cost Function

Decision One: No Rectification	Decision Two: Rectification
$nC_1 + NpC_4$	If work unit is accepted $nC_1 + NpC_4$
	If work unit is rejected $nC_1 + NC_2 + NprC_4$

* This paper was written while Mr. Kimble was employed by Statistical Methods Division, U.S. Bureau of the Census, Washington, D.C. 20233

** Rectification miss rate is the proportion of the defectives in a lot either not corrected by the rectifier or introduced by the rectifier.

Where

n = Quality control sample size

N = Lot size

C_1 = Cost to inspect one item in quality control sample.

C_2 = Cost to inspect one item during rectification.

C_4 = "Cost" of having one defective item left in the lot

p = proportion of the items defective in the lot; this is a random variable.

r = rectifier miss rate*

The next step of this method is to determine the risk of each decision. If the "no-rectification" decision is chosen, regardless what the quality control sample shows, the lot will not be rectified; thus the risk is the cost function. If the rectification decision is selected, whenever the number of items-in-error found in the quality control sample exceeds a specified number, the lot is rejected and rectified. To compute the risk for this rule, we look at the probability distribution associated with the number of defectives found in the quality control sample. The risks for both of these decisions are:

Risk

Decision One: No Rectification	Decision Two: Rectification
$nC_1 + NpC_4$	$[nC_1 + NpC_4]\, G(p)$ $+[nC_1 + NC_2 + NprC_4]\,[1 - G(p)]$

where

G(p) is the probability of accepting the lot if p proportion of the items are defective

The Bayes risk for each of the decisions can be determined by taking the expected value of the risk with respect to the prior distribution f(p) of p the incoming quality; i.e.,

$$\int_0^1 \text{Risk}\ f(p)\, dp$$

This implies:

Bayes Risk

Decision One: No Rectification	Decision Two: Rectification
$nC_1 + NC_4\, E(p)$	$NC_4 \int_0^1 p\, G(p)\, f(p)\, dp\, [1 - r]$ $+ nC_1 + NC_2\, [1 - \int_0^1 G(p)\, f(p)\, dp]$ $+ NrC_4\, E(p)$

$$*r = \frac{\text{number of defectives not corrected by the rectifier} + \text{number of correctly produced items that were changed by the rectifier to incorrect items}}{\text{(total number of defectives in the lots before rectification)}}$$

where

$E(p) = \int_0^1 p\, f(p)\, dp.$

If we use estimates for the cost of inspecting one item (C_1), the cost of rectifying one item (C_2), the rectifier miss rate (r), and the parameters of the prior distribution, then the Bayes risks are linear functions of C_4 only.

If we choose the decision which will minimize the Bayes risk, then by equating the Bayes risks at the two decisions, we can easily solve for the break-even value of C_4 (call it C_4').

$$C_4' = \frac{C_2}{1 - r} \cdot \frac{1 - \int_0^1 G(p)\, f(p)\, dp}{E(p) - \int_0^1 p\, G(p)\, f(p)\, dp} = \frac{1}{1 - r}\, A\, C_2$$

$$\text{where } A = \frac{1 - \int_0^1 G(p)\, f(p)\, dp}{E(p) - \int_0^1 p\, G(p)\, f(p)\, dp}$$

See Figure 1.

The decision rule with minimum Bayes risk can then be characterized as follows: If a manager subjectively values the cost of leaving one item in error more than the cost C_4', then the procedure of rectification should be chosen. If a manager subjectively values C_4 less than C_4', then the procedure with no rectification should be chosen.

In order to view this solution in a way current to the coding rectification study, if we assume r to be an independent variable and the break-even cost C_4' to be the dependent variable, the function would be $C_4 = \frac{1}{1 - r} A\, C_2$ where A is the constant computed from the above formula. (See Figure 2).

For the range where C_4 is greater than the C_4' as a function of r, the decision would be to have a rectification procedure. For the area below the C_4' function, the decision would be to not have a rectification procedure.

Example: The General Coding Operation for the 1980 Census

The operation where the coding of most items on Bureau Census questionnaires is performed (called General Coding) has used a quality control plan that rectifies rejected work units. It is believed that rectification corrects only a small proportion of the defective items. By applying the technique presented here, the cost of correcting a single item can be determined as a function of the rectifier miss rate. (Later through an experimental study, the rectifier miss rate will be estimated.)

By using the data of the General Coding of the 1977 Oakland Precensus, the constant A used in the curve of Figure 2 can be calculated in the following manner: Assuming that the work units' proportion defective p is distributed as a beta random variable, then the parameters of the prior distribution f(p) can be estimated by standard methods (Johnson & Kotz, 1969). (These were estimated from the Oakland Coding data to be $\alpha = .09$ and $\beta = 19.91$.) The average error rate for all the work units (.0045) can also be tenably used as an estimate of E(p). Because the sample sizes are approximately one-tenth of the lot size and the lot sizes are large (on the average 1160), G(p) can be easily justified to be the binomial probability distribution. Because of the one-in-ten sample rate, n is approximately equal to 116. By using the assumption that the cost of inspecting an item in a quality control sample is equal to the cost of inspecting an item in the rectification phase, then C_2 is equal to C_1. These quite justifiable assumptions imply the break-even cost to be

$$C_4' = 26.68 \frac{1}{1 - r} C_1$$

See Figure 3.

This function means that if the estimated rectifier miss rate is $r = .83$ (this is the estimate of r), then the break-even cost C_4' would be 156.94 C_1. In other words, a manager must be willing to pay at least $\$7.04^+$ to correct one item if he/she chooses to rectify rejected lots. If a manager is not willing to pay that amount, then he/she should *not* choose to rectify rejected lots.

CONCLUSION

This technique can be used to decide if a quality control system should include rectification of rejected lots. The technique relates the rectification miss rate to the Bayes risk cost of leaving a defective item in the outgoing lot. Once the rectifier miss rate is estimated, a manager can then decide if the cost of correcting one item is worthwhile. If it is, then the system should rectify rejected lots; if it is not, then the system should not rectify rejected lots.

REFERENCES

DeGroot, Morris H., 1970, *Optimal Statistical Decisions,* McGraw-Hill, New York

Hogg, Robert V., and Allen T. Craig, 1970, *Introduction to Mathematical Statistics,* Macmillan, London

Johnson, Norman L., and Samuel Kotz, 1969, *Discrete Distribution, Distribution in Statistics,* Houghton-Mifflin, Boston

Lindgren, B.W., 1968, *Statistical Theory,* Macmillan, London

---

+ Based on a verifier production rate of 140.7[++] items per hour estimates from the Oakland Precensus, $44.19 per 7-hour work day wage and overhead cost per rectifier, and r = .83.

++ The 95% confidence interval of the verifier production rate was estimated by random groups method to be [124.440, 156.96]. The 95% confidence interval of the $3.26 break-even cost would then be [$6.31, $7.96].

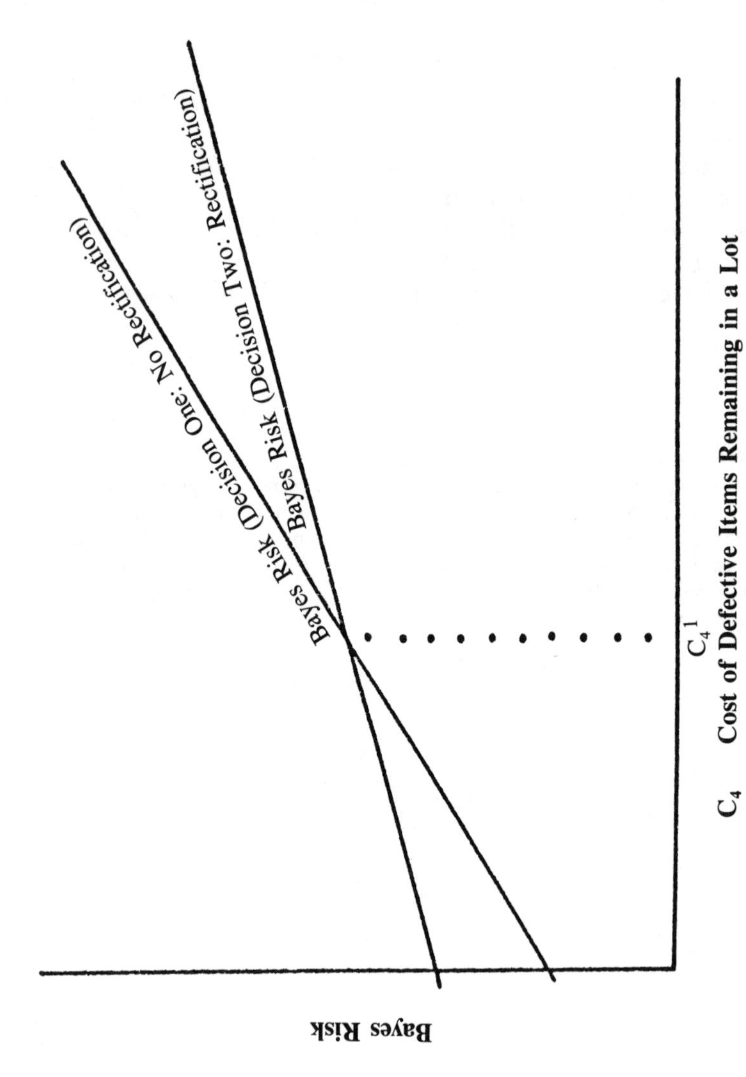

FIGURE 1 The Bayes risk of the two decisions.

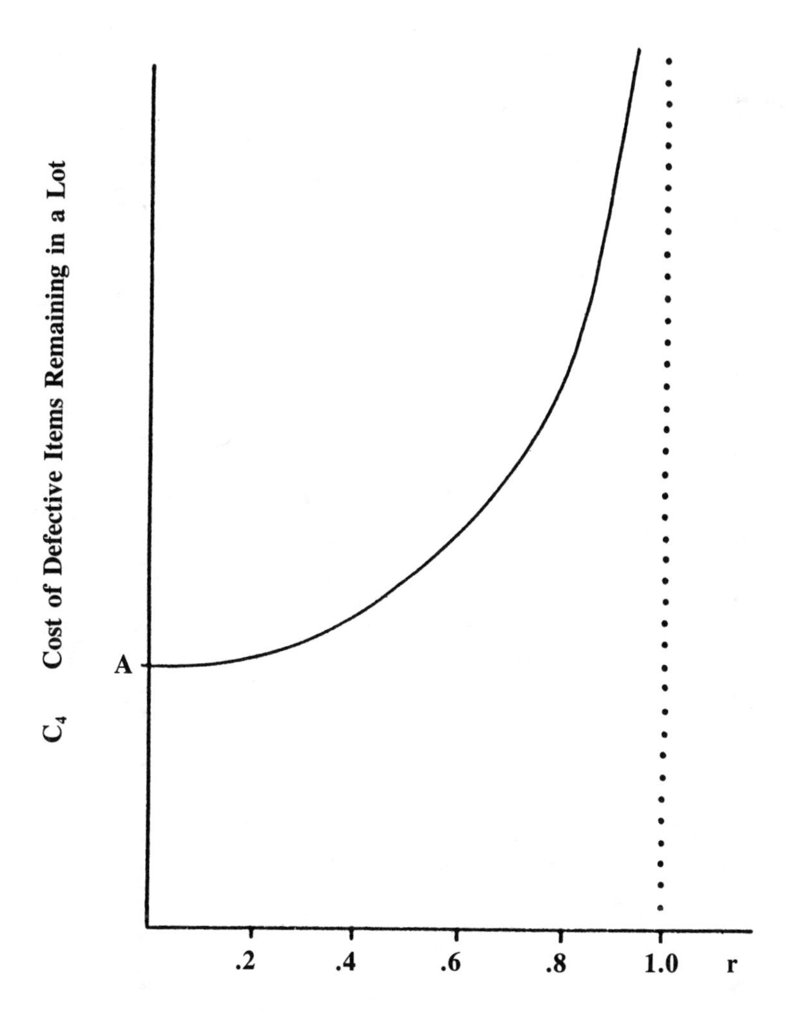

FIGURE 2 The break-even cost of the two decisions as a function of r (rectifier miss rate) for "A" not calculated.

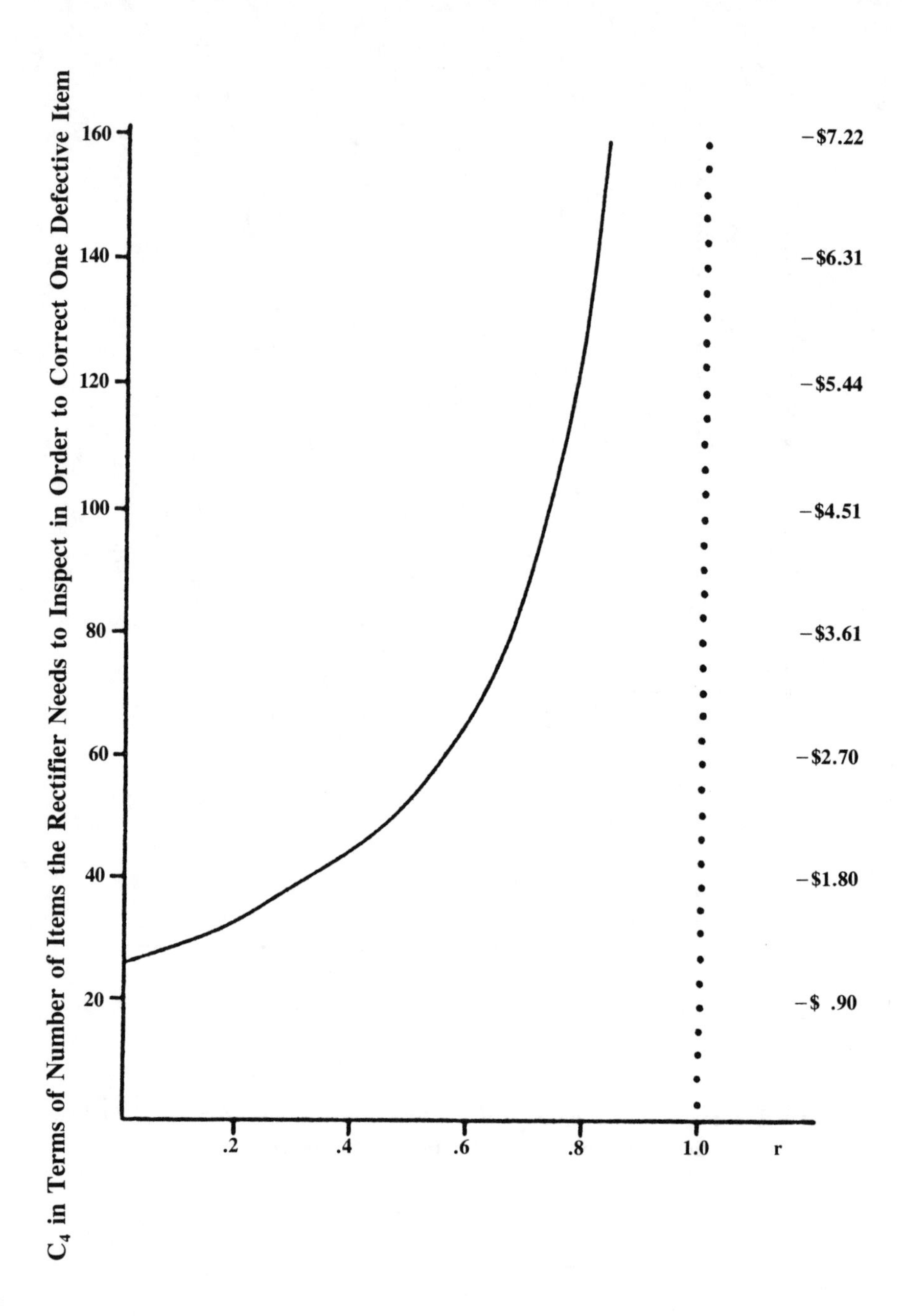

The break-even cost of the two decisions as a function of r (rectifier miss rate) for "A" calculated from Oakland General Coding Data.

FIGURE 3

"GUIDE FOR REDUCING QUALITY COSTS"

James F. Zerfas
Manager, Quality Control
Jeffrey Mining Machinery Division
Dresser Industries, Inc.

Lance Arrington
President, Philip Crosby Associates, Inc.

The following text is extracted from the ASQC publication *"Guide for Reducing Quality Costs"* and represents the combined efforts of the members of the Quality Costs Technical Committee. Each speaker in the tutorial will elaborate on the subject matter from his experience and opinions and will provide appropriate examples to illustrate the key points.

THE QUALITY COST IMPROVEMENT PHILOSOPHY

It is a fact, too often not recognized, that every dollar saved in the total cost of quality is directly translatable into a dollar of pre-tax earnings. It is also a fact that quality improvements and quality cost reductions cannot be legislated by Management demand — they have to be earned by the hard process of problem-solving. The first step in the process is the identification of problems; and a problem in this context is defined as an area of high quality cost. Truly, every problem identified by quality cost is an opportunity for profit improvements.

USEFULNESS OF QUALITY COSTS

Quality costs are useful for strategic planning and programming quality improvement.

Strategic Planning —

To improve quality and reduce quality costs there has to be a trigger for making changes in the status quo. The firm's Strategic Plan is an ideal device to force changes; and the inclusion of quality and quality cost improvement plans in the overall strategic plan is recommended. This gives the quality and quality cost situation the management visibility too often is lacking. Because quality and quality cost improvements are set forth as business objectives (along with the more conventional business objectives) against which management performance is evaluated, there is effective motivation for action.

Today's quality program must be directed toward improved product quality and reduced quality costs. The Quality Manager must develop strategic quality plans which will assure continuous quality improvement. This program must be integrated into the total company planning activities. Strategic quality planning is a continuous process of evaluation, decisions, and actions. All activities, customer demands, competitor's activities, the history of the quality role, and the future quality role must be considered in this planning.

Programming Improvement

The Strategic Quality Plan describes a management commitment to quality and quality cost improvement. The quality cost data indicate the areas that are candidates for improvement. When the highest cost areas are analyzed in greater detail many improvement projects become apparent. For example, high warranty costs are a trigger to rank customer failure problems for detailed investigation, with the aim of looking into product design, process control, or inspection planning for the cures to the highest cost problems. Regardless of what the high quality cost category may be, the mere act of identifying it should lead to actions to reduce it.

In summary, to effectively program quality improvement efforts, it is necessary to:

- Recognize and organize quality related costs to gain knowledge of magnitude, contributing elements, and trends.
- Analyze quality performance, identify major problem areas and measure product line and/or manufacturing section performance.
- Implement effective corrective action and cost improvement programs.
- Evaluate effect of action to assure intended results.
- Program activities for maximum dollar pay off and maximum effective manpower utilization.
- Budget quality work to meet objectives.

THE QUALITY SYSTEM

Perhaps the most important result of the collection, analysis, and use of quality costs is the exposure given to the total quality system as it really exists in the organization. The collection of costs forces definition of all activities contributing to the quality of the product. Analysis of quality cost data forces evaluation of the effectiveness of the contribution of each activity, the relationships among the many activities, and the all-important communications links that tie activities together.

FINDING THE PROBLEM AREAS

When quality costs are displayed to managers who have not been exposed to the concept, the initial question is likely to be "how much should they be?" or "how does this compare with other organizations or products?". Unfortunately, it is *not* practical to establish any meaningful absolute standards for such cost comparisons. A quality cost system should be "tailored" to a particular company's needs, so as to perceive trends of significance and furnish objective evidence for management decisions as to where assurance efforts should be placed for optimum return.

Analysis techniques for quality costs are as varied as those used for any other quality problems in industry. They range from simple charting techniques to complicated mathematical models of the program.

The most common techniques are:

1. Trend Analysis
2. Pareto Analysis
   A. By Element Group
   B. By Department
   C. By Product
   D. Other Groupings

THE TEAM APPROACH

Once a problem has been identified and reported and the involved personnel are committed for action, the job is started but far from complete. The efforts of people involved must be planned, coordinated, scheduled, implemented, and followed-up. Problems can normally be thought of as one of two types; those which one individual or department can correct with little or no outside help and those requiring coordinated action from several activities in the organization. Examples of the first type of problem are operator-controllable defects, design errors, and inspection errors. Examples of the second type are product performance problems for which a cause is not known, defects caused by a combination of factors not under the control of one department, and field failures of unknown cause.

To attack and solve problems of the first type, an elaborate system is not required. Most can and should be resolved at the working level with the foreman, the engineer, or other responsible parties. Usually, the working personnel of these departments have sufficient authority to enact corrective action, within defined limits, without specific approval of their superiors. If documentation of such problems is desired, the use of a "Quality Improvement Project" is recommended.

Unfortunately, problems of the second type are normally the most costly and are not as easily solved. Causes of such problems may be numerous and unknown. Solution may require action from several sources. The investigation of the problem and the planning of its solution must be coordinated and scheduled to assure that effective action is taken. One of the best devices for doing this is the quality improvement committee. Working with the data and problem analysis reports and headed by an individual who is interested in solving the problems, this committee develops the plan, and then coordinates and schedules the investigation and action.

REDUCING FAILURE COSTS

This section discusses four steps to an effective, corrective action program and describes some techniques which have been found to work in accomplishing each step. The four are:

1. Make all persons concerned aware of the problem and its possible causes.
2. Create a desire in others to solve the mutual problem.
3. Plan and carry out a logical investigation of the problem with others involved.
4. Follow-up on action taken.

PREVENTION OF QUALITY COSTS

Prevention activities are of two basic types — those related to employee attitudes and resulting approach to their jobs, and those formal techniques which identify potential problems early in the product cycle and eliminate them before they become expensive.

Employee attitudes toward quality are determined in large part by their beliefs about what is really wanted by their boss and higher management of the business and by the degree to which they are personally involved in the improvement program. Management's commitment to quality improvement must be visible to all employees. The first line supervisors play key roles in a successful quality improvement program. They should thoroughly understand and be involved with the elements affecting their people.

Formal programs for preventing deficiencies in the product have been used by some companies for many years but, unfortunately, are not in general use. Some examples of these programs are:

- New product verification programs which require that new products be thoroughly reviewed, inspected, tested, and proven prior to release for quantity production.
- Design review programs requiring a thorough and detailed review of new or significantly changed designs by a group representing all segments of the organization.
- Supplier selection programs which require an evaluation of the potential supplier's ability to supply the required quality before an order is placed with him.
- Reliability testing to discover problems before they cause high field failure costs.
- Thorough training and testing of employees placed on critical jobs.

REDUCING APPRAISAL COSTS

The costs of appraisal sometimes approach half of the total quality costs. Although most quality cost improvement programs properly concentrate on reducing failure costs first, programs for appraisal cost improvement can also have a significant impact.

MEASURING IMPROVEMENT

There are many sources of information on product performance no one of which usually give sufficient data to determine whether the customer's needs are being met. It usually takes a combination of sources to get needed data and the optimum combination varies with the product and market. For example, different data sources would be used for a consumer product than for an industrial product. Some of these sources are:

- Change in quality costs
- Field failure and repair reports
- Installation phase reporting
- Personal observation by company personnel

- Life testing of company and competitor's finished product
- Market research on customer opinion and user costs
- Data from spare parts sales
- Customer complaints
- Outgoing product audits

SOURCE SURVEILLANCE AND VENDOR EVALUATION PLAN

Narendra S. Patel
Digital Equipment Corporation
Hudson, MA 01749

ABSTRACT

A. Purpose of this Paper

1. Why such a plan is necessary in the new decade.
2. How such a plan can be sold to upper management. Cost of such a program, three segments of quality cost will be addressed.
3. Who and how "who" will benefit from this plan.

B. Vendor Rating — Product Rating

How vendor and its product can be rated and evaluated. Various methods of rating vendor will be discussed at great length.

C. Maintainability of Quality Levels With Current Good Vendors

1. How to motivate a good vendor to keep producing an acceptable quality level.
2. How to encourage a new or poor vendor to produce an acceptable quality level.

QUALITY IN THE NEW DECADE

Since the theme of this conference is "Quality in the New Decade," I would like to present some ideas which should be very useful to the incoming Quality Assurance Department of most industries in this competitive world. The purpose of this program is outlined below:

A. Savings in operating cost to supplier quality assurance.
B. To assure better quality and vendor service.
C. Reduction of reject rates.
D. To build confidence in the end product.
E. Reduction of in-house and field costs.
F. Increasing customer satisfaction.

Like all living species, the human being is concerned with quality. He has to determine what to eat, what clothes to wear, etc. Today, in the developing countries, where there are few mass production industries, people pay high prices for their household furnishings and their custom made clothing. These people select the fabric of their choice, go to the tailor, and have their clothing custom made. What about quality? Since the clothing was designed to the clients specified requirements, he is satisfied, but at a high price.

Now let us talk about more industrially developed countries, where there are many industries, lots of companies, and mass production. How can one company put out a profitable product in competition with others? The first thought is to sell the product at a lower price than the competitor. If the features are the same as those of the competitor, this is fantastic. A very simple answer, isn't it?

When it comes to reality, and the question is raised as to "How?", it becomes difficult to answer. In view of today's inflationary costs, how can one sell his product at a reduced price, and still show a profit?

The Industrial Revolution, which made possible an enormous expansion of manufactured products, has created some problems for which present day solutions are still inadequate. Take this example: The president of a large organization cannot solve quality problems like the head of a small company. He must use every managerial tool and skill at his disposal. In a small company, the head of the business can usually go directly to the area in question, talk to the people involved, and come up with an answer.

Mass production and mass usage has created problems of feedback, and because of lack of feedback, quality improvement sometimes becomes difficult for toy firms, computer plants, and the automotive industry, as well as others. Each of these industries receive thousands of parts in their incoming inspection department. It is expensive to inspect and test all these parts. Upper management often looks in this direction to economize.

There is one way such a problem can be solved, and a company can produce a profitable product on the competitive market. In most organizations, the majority of problems arise due to the poor quality of incoming material. Taking a preventive approach, such as evaluation of vendor design before he starts production can save thousands of dollars for the client company in the long run. It is an absolute necessity to develop good communications between the vendor and the vendee's purchasing department. After this line of communication has been established, it becomes easier for a buyer to enter a vendor's facility for the purpose of rating and evaluation.

Before I start talking about the rating system, let me say something about costs, and the view of upper management in this area. In any organization, Quality Control is an overhead expense, and the only way such an expense can be justified is if Quality Assurance shows the effectiveness of its existence by saving money for the company. This can be achieved if appropriate planning has been involved.

Following are some typical errors in the distribution of quality costs in many organizations:

TYPICAL		DESIRED
25%	Cost of Quality Program	65%
75%	Cost of Losses Caused By Vendor Non-Conformance	35%
	35% a.Internal Scrap Rework 24%	
	40% b. External Warranty & Downgrading 11%	

The goal is to spend most of the cost of quality money for running programs by changing the distribution of these costs through closed loop effective, formal, demand types of corrective action systems that make problems go away forever: the true test of effectiveness.

It has been shown that adding money at the front end can save a substantial amount down on the line, on such items as internal and external failures. Customer satisfaction will be increased, because of the lower failure rate, which will, in the long run, assure a good product. Both the producer and the consumer cannot help but benefit from such a plan.

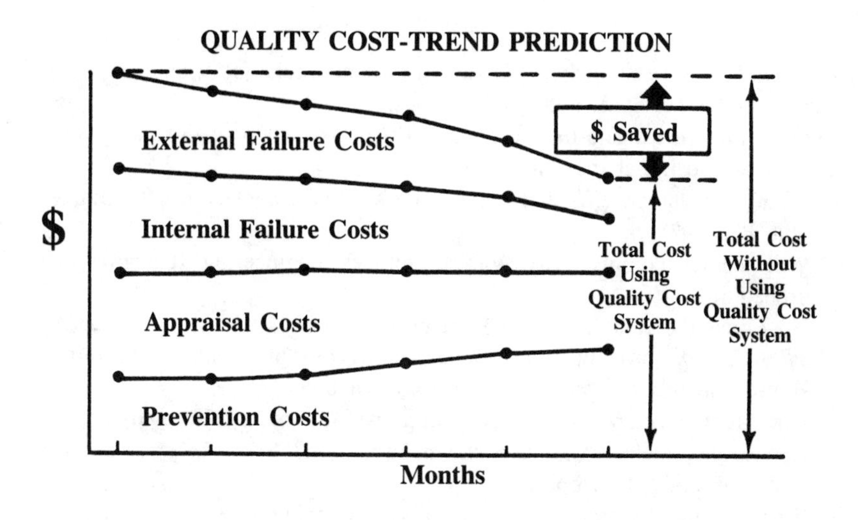

FIGURE 1

Factors for analysis and evaluation of commodity vendors are limited mainly to price, quality, and service. This is indicated in the breakdown given below:

Prices: This will be a competitive factor for equivalent quality lines. However, pricing structures will vary with respect to quantity discounts.

Quality: This is also a competitive item. On similar lines of off-the--shelf items, there may be advantages to a vendee's specific requirement of analyzing the range in quality being offered.

Service: This is a most significant factor, which includes the vendor's willingness and ability to fill the vendee's requirements reliably.

SURVEYING SUPPLIERS

New suppliers should be surveyed in the following manner: Prior to placing an order, qualification shall be defined in terms of the following capabilities:

1. Technical
2. Manufacturing
3. Quality Control
4. Financial
5. Management

In these situations, a survey shall be used to determine the extent of Development, Test Manufacturing, and Quality Control Engineering. Plant and equipment controls over facilities, manpower, and materials, in addition to the vendor's ability to plan and/or organize, should be determined at this time.

A check list of the following questions should be included for a vendor capability survey:

1. *Personnel:* Key personnel in the vendor's plant, including titles, functions, background experience, and qualifications.
2. *Engineering:* Investigate their design and development procedures.
   a. How are design changes incorporated?
   b. What procedures are used to integrate tooling and manufacturing techniques with R & D?
   c. Will vendor comply fully with customer's engineering standards and procedures?
   d. How does the vendor control Engineering Changes?
3. *Quality Control:* Investigate the vendor's inspection and control procedures.
   a. How frequently are procedures checked for update?
   b. How frequently are tools, gauges, and test equipment calibrated?
   c. How is the vendor's Q.C. department organized?
   d. To whom does Q.C. report?
   e. What is vendor's procedure for handling incoming material?
   f. Does vendor require a Certificate of Compliance?
   g. What records are kept on incoming inspection? Are they files?
4. *Production:* Investigate the vendor's planning and inventory systems.
   a. Determine the manner and frequency in which the production performance is reported.
   b. Who is responsible for machine and manpower loading.
   c. Is there a system for comparing current loading against the forecast?
   d. What is the vendor's system for stock and material control?
   e. How does the vendor dispose of scrap, surplus, and obsolete material?
5. *Finance:*
   a. Determine the vendor's current financial status as determined on his most recent balance sheet.
   b. Check vendor's profit and loss statement.
   c. Investigate his cash flow forecast for the remainder of the current accounting period.
   d. Determine available types and sources of additional capital, should it be required.
   e. What type of accounting system is used? (Standard cost, job cost, etc.)
   f. How does the vendor estimate initial production cost?
   g. How does the vendor treat processing costs, Engineering, R&D, and factory overhead?
   h. How does the vendor view selling and administrative costs?

6. *Labor:* What has been the labor history of the vendor, and are his employees unionized?
7. *Marketing:*
   a. What is the past business record of the vendor?
   b. What is the current volume of the vendor's business?
   c. How does the vendor's business break down?
      (1) Military (2) Subcontract (3) Other ________
   d. What is the projected volume of the vendor's business?
8. *Others:*
   a. How has the vendor performed for other customers?
   b. How much cooperation can the vendor provide in furnishing information about the following items when needed?
      (1) Engineering (2) Quality (3) Process/Product (4) Financial Organization (5) Stability.

VENDOR'S RESPONSIBILITY

A. After a vendor has been selected according to the foregoing plan, he shall have total responsibility for the effective control and maintenance of product quality. It will be his responsibility to ship only those products which conform to the specification which is agreed between the vendor and the buyer.

B. The vendor shall maintain and provide test and inspection records to the buyer upon request. The buyer will be allowed to perform quality and process audits as required at the vendor facility subject to reasonable notice.

C. The vendor will notify the buyer of any discrepancies discovered which could have serious effect on products shipped in the past. In addition, the vendor shall recommend an appropriate action.

D. The vendor shall provide information on any changes he makes in his process — people. He shall also fill out necessary forms such as certificates of compliance, product release form, product qualification form, etc.

TABLE I
CERTIFICATE OF COMPLIANCE

TO:
BUYER

PURCHASE ORDER NO. ____________________

DATE OF SHIPMENT ____________________

QUANTITY ____________________

Catalog #	Material	Batch Lot #	Guaranteed Per Buyer Specification

STATEMENT

THIS IS TO CERTIFY THAT THIS SHIPMENT IS IN CONFORMANCE WITH THE INSPECTION DETAIL SHEETS, SPECIFICATIONS, AND/OR DRAWINGS AS LISTED ON THIS ORDER; AND THAT THE REQUIRED PHYSICAL, ELECTRICAL OR CHEMICAL TEST REPORTS OF SAID MATERIAL ARE ON FILE AND ARE AVAILABLE UPON REQUEST.

AUTHORIZED SIGNATURE ____________________ DATE __________

BUYER'S RESPONSIBILITY

A. Quality engineer from buyer's plant will provide technical guidance to the vendor's technical people.

B. Will perform a complete quality survey of the supplier at least once a year and provide report to vendor.

C. Buyer will make sure that vendor understands quality requirements of the end users.

D. Buyer will allow vendor sufficient time to change in case buyer makes any changes in his requirements.

Evaluation of Vendor Performance — After the source has been selected, it is necessary to evaluate the vendor performance to determine if the selected vendor provides the same quality and service agreed upon.

Data Banks for Evaluation — Data from several departments or divisions shall be pulled. Field failure, and failure return rate shall be analyzed. In each case, the cost of correcting the failure shall be calculated.

Value Analysis — The purchasing department shall be asked to look into the total cost for keeping these products in service over their designated life, rather than looking into the original purchase price.

Life Cycle Costing — Cost after warranty is an important factor to most users. The following chart shows Life Cycle Costing:

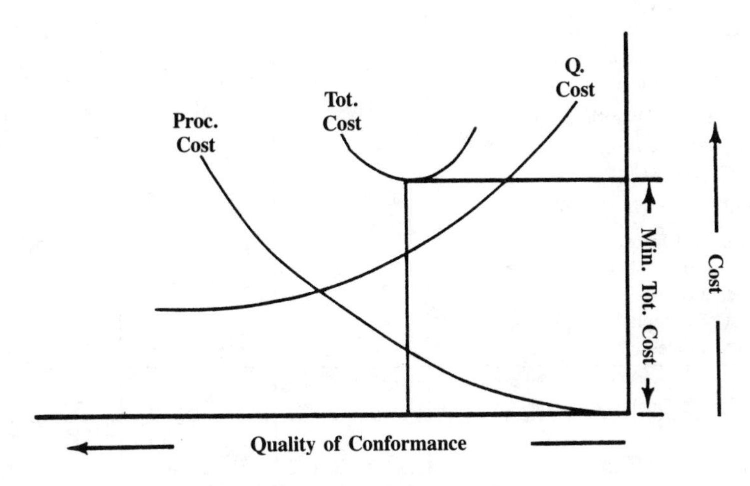

FIGURE 2 Life Cycle Costing

These include: A. The original cost. B. Cost of Down Time. C. Maintenance, Labor and Parts Cost. D. Cost of Carrying Inventories. E. Other Costs.

TABLE II
SUPPLIER QUALITY RATING

FROM: __________ TO:________

Quality Cost Factors	Vendor "A"	Vendor "B"	Vendor "C"
Cost of Defect Prevention:			
Vendor Survey			
Cost for Qualifications			
Approval of Samples			
Spec. Revision			
Cost of Defect Detection:			
Incoming Inspection			
Lab Test Costs			
Cost of Defect Correction:			
Additional Inspection			
Processing of Rejects			
Mfg., Losses			
Cost of Complaints			
Loss of Sales			
Total Cost of Quality Assurance			
Total Value of Purchases			
Quality Cost Ratio*			

$$\text{*Quality Cost Ratio} = \frac{\text{Total Quality Cost}}{\text{Total Purchasing Value}} \times 100$$

TABLE III
SUPPLIER DELIVERY RATING

FROM: __________ TO:________

Availability/Acquisition Cost Factors	Vendor "A"	Vendor "B"	Vendor "C"
Cost of Acquisition/Availability:			
Follow-up & Expediting Expenses			
Telephone & Telegraph Expenses			
Plant Visitation Expenses			
Mfg. Losses due to Vendor Delinquency			
Total Acquisition/Availability Cost:	$ 4,000	$ 2,500	$ 2,600
Total Value of Purchases:	$80,000	$65,000	$91,000
Acquisition/Availability Cost Ratio*	5%	3.8%	2.86%

*Acquisition/Availability Cost Ratio is based upon Total Acquisition Costs and the Total Value of Purchases.

TABLE IV
VENDOR SERVICE RATING

Characteristics Of Vendor	Vendor		
Excellent = 5 points; Average = 4 points; Below Average = 2 points	A	B	C
1. **Personnel Capabilities:** a. Caliber & availability of sales & technical personnel. b. Is Management Progressive? c. Technical knowledge of supervision. d. Cooperation on changes & problems. e. Technical field service availability. f. Labor relations. 2. **Facilities Capabilities:** a. Capability for anticipated volume. b. Latest technology & equipment. c. Excess production capacity. d. Geographical location. e. Financial capacity to stand behind product failures f. Investing capital in the organization? 3. **R & D Capabilities:** a. New product development. b. Alerting for future needs? c. Does the vendor update the buyer with the latest techniques? 4. **Product Service Capability:** a. Offers emergency assistance? b. Does vendor provide consultation for potential troubles? c. What type of warranty is furnished? d. Is the vendor willing to accept responsibility? e. What is the vendor's record for reliability in past dealings?			
Total Service Points			

WEIGHTED POINT PLAN

This plan is designed to develop a close working relationship between Purchasing and Quality Control. The plan provides a tool for evaluating vendors of specific items on an individual basis. S.Q.A. will assist Purchasing in two important areas: 1. Making the vendor aware of quality requirements. Quality levels shall be maintained on each of the characteristics involved. Details of test and inspection shall be outlined and attached to the purchase order for quotation. This will eliminate any misunderstanding resulting from oral instructions. 2. Evaluating the vendor's ability to meet the requirements by reviewing the vendor's plant, Q.A. procedures, etc.

In conjunction with the itemized list, a weight scale is developed for use in determining a rating figure for each item. For example:

QUALITY	=	40 points
PRICE	=	35 points
SERVICE	=	25 points
TOTAL	=	100 points

It is important to be aware that the weighing of the Quality, Price, and Service factors will vary, depending on the market conditions.

TABLE V

QUALITY

Part No.	Lots Received	Lots Accepted	Lots Rejected	% Accepted	X Factor	Q.C. Rating
Vendor A	30	27	3	90.0	.40	36.0
Vendor B	30	28	2	93.3	.40	37.3
Vendor C	10	8	2	80.0	.40	32.0

PRICE

Part "A"	Unit Price	Discount	Discounted Price	+ Transportation Charge	Net Price
Supplier A	2.00	10%	1.80	.06	$1.86
Supplier B	2.50	15%	2.12	.12	2.24
Supplier C	3.00	20%	2.40	.06	2.46

Part "B"	Lowest Price	Net Price	= Percentage	X Factor	Price Rating
Supplier A	1.86	1.86	100%	35	35.0
Supplier B	1.86	2.24	83%	35	29.1
Supplier C	1.86	2.46	76%	35	26.6

SERVICE

	Promises Kept	X	Service Factor	=	Service Rating
A	90%		25		22.5
B	95%		25		23.8
C	100%		25		25.0

COMPOSITE RATING

Rating	Supplier A	Supplier B	Supplier C
Quality (40)	36.0	37.3	32.0
Price (35)	35.0	29.1	26.6
Service (25)	22.5	23.8	25.0
Total Rating	93.5	90.2	83.6

The rating system outlined below was developed and used for Incoming Inspection of Semi-Conductor Grade Chemicals:

1. A rating of 1.0 is given for each lot which passes incoming AQL.
2. A rating of 0.9 is given if the lot is defective due to the lack of support documentation, for example, the Certificate of Compliance.
3. A rating of 0.5 will be given to any lot that fails the AQL, but on which the IQR disposition is "USE AS IS."
4. A rating of 0.0 will be given to any lot that fails the AQL, and is either scrapped, screened 100%, or returned to the vendor. The result of each lot inspected should be recorded in the daily log, and the ratings will be compiled every three months, using the following formulas:

$$\text{RATING} = \Sigma \frac{\text{lot ratings}}{\eta} \text{ X } 100$$

The lot rating equals the sum of all lots inspected in the three month period, while "η" equals the number of lots inspected.

"A" through "D" below are applicable to 6 months time increments, to the same part types, and same revision suffixes.

VENDOR QUALIFICATION CRITERIA FOR PERFORMANCE EVALUATION

A. *Fully Approved Supplier:* Accepted for three consecutive shipments.

B. *Not Fully Approved Supplier:* Accepted on 1 or 2 consecutive shipments.

C. *Conditional Supplier:* Two consecutive shipments rejected for the same discrepancy, or rejected 3 consecutive shipments for different reasons.

C.* *Conditional Supplier:* Only 1 shipment received, and subsequently rejected.

C.** *Conditional Supplier:* Rejected one shipment, accepted one shipment, rejected one shipment. (No two consecutive shipments both rejected or accepted.)

C.*** *Conditional Supplier:* Rejected two consecutive shipments for different discrepancies.

D. *Disapproved Supplier:* Rejected three consecutive shipments for the same discrepancy, or rejected five consecutive shipments for different reasons.

The following formulas have been developed to rate quality alone:

$$\text{Quality Rating} = 70 + \left(\frac{\Sigma \text{ LR}}{\text{N}} - 70\right) \sqrt{\text{N}}$$

N = The number of lots submitted during a given period.

$$\text{LR} = \text{Lot Rating} = 70 - 10 \frac{\text{P} - \text{P}^1}{\sqrt{\text{p}^1 (100\text{-p}^1) / \text{N}}}$$

p^1 = A.Q.L. Percent
P = Percent defective of sample
N = Sample size

Employing this equation, each vendor starts out with a score of 70, so he is considered average. Quality ratings are interpreted as follows:

ABOVE 80 — Quality average better than A.Q.L.
51 to 80 — Acceptable quality
BELOW 51 — Quality definitely worse than A.Q.L.

Maintainability of Quality Levels With Fully Approved Suppliers by Positive Communication — It is most important to establish a good relationship with the supplier. It is absolutely necessary to develop two-way communications, the purpose of which is to supply essential information, performance data, and to identify troubles which may arise. This procedure stimulates corrective action, and improves the ability of the buyer and the supplier to work closely.

Good performance on the part of a supplier should be recognized, and letters of commendation should be sent to him. Presenting an award will go far to encourage the supplier to keep up his good performance level.

As the communications are developed, there is every chance that the supplier will move from his purely defensive position to one of making constructive proposals and suggestions.

SUMMARY

In the new decade, most industries will reduce incoming inspection. Most of the hardware and mechanical parts will be replaced by electronic or solid state technology.

More and more complex systems are going to be developed. Such systems will require inspection and check-out by sophisticated electronic instruments. It will be difficult for a small buyer or manufacturer to justify such a system as might be necessary to properly inspect his incoming material. For this reason, it will become necessary for him to depend on the supplier for this service. As a result, a "Supplier Rating System" will be developed, most likely, by the end of the decade.

Many medium and large companies will support "Field Residents," whose sole job will be auditing the product, and the processes at the suppliers facility. Such a system, as described above will result in a tremendous reduction in the in-house failure rate.

BIBLIOGRAPHY

J.M. Juran & Robert G. Fitzgibbons. "Vendor Relations." Quality Control Handbook.

Thomas G. Shahna Zarian. "Vendor Certification." Industrial Quality Control.

QUALITY COSTS AND STRATEGIC PLANNING

F.X. Brown, Consultant
Westinghouse Electric Corporation

ABSTRACT

Modern Planning Activities focus on integrating functional activities — Marketing, Finance, Design, and Production — into a coherent Business Strategy. This paper discusses a method for including the Quality Function in the Strategic Planning Process and presents guidelines, directions, methods for Quality Planners to participate in Business Strategy through the use of Quality Cost information.

I. INTRODUCTION

Strategic Planning, in its simplest sense, is the integration of the resources of a business to achieve, effectively, a specified business objective.

In a general sense, this paper will define the Planning Process from the Vantage Point of the business planner and will demonstrate a role for the Quality function, starting from the broad conceptual framework of a strategic plan and proceeding to an increasingly detailed Quality Cost view of the Planning Function.

II. CONCEPTUAL NATURE OF THE STRATEGIC PLAN

From the Vantage point of the Business Manager and his planner, the resources available are:

- Capital
- Time
- Human talent

These resources are employed to achieve the Business Objective, Profit, in a manner which:

- Increases Revenue
- Decreases Investment
- Increases Cash Flow
- Decreases Costs

The inter-relationship among these four primary variables is shown, schematically, in Figure 1.

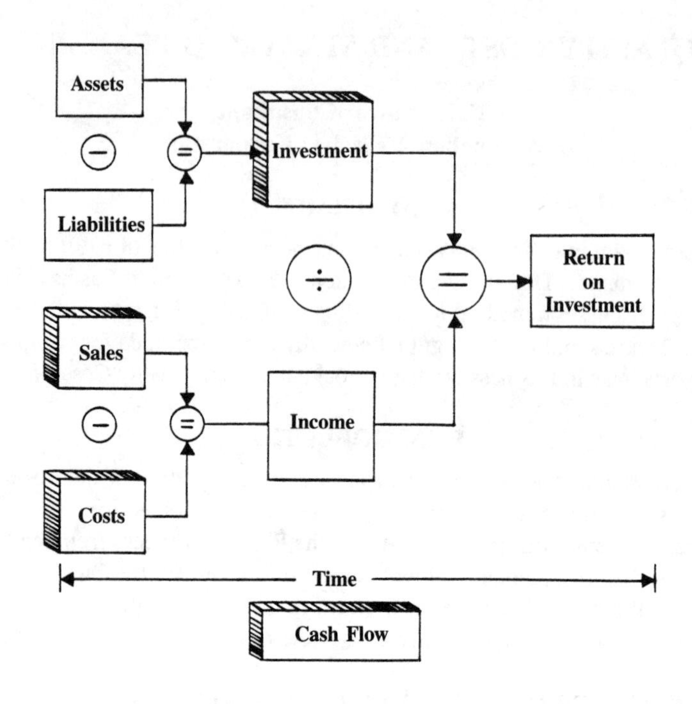

FIGURE 1

In a conceptual sense, primary responsibility for developing action plans is then assigned to various functional areas of the organization with secondary and tertiary responsibilities as illustrated in Figure 2.

Variable Responsibility	Revenue	Cost	Investment	Cash Flow
Primary	**Marketing**	Manufacturing	C.E.O.	Finance
Secondary	Engineering	Engineering	Manufacturing	**Marketing**
Tertiary	N/A	Distribution	**Marketing**	Manufacturing
n				

FIGURE 2

From this assignment of responsibility naturally flow functional plans to achieve the objectives of the business. As an example, a Marketing Strategic Plan might include the following sub-objectives in its areas of responsibility. (Figure 3)

MARKETING STRATEGY

Variable	Sub-Objectives
Revenue	1. Increase Market Share-0.7% a) Promotion-5,000,000 b) Selected Price Reduction-1.0% 2. Increase Service Sales-2.0%
Cost	N/A
Investment	1. Reduce Distributor Inventory-4% a) Trim Product Lines-A&B b) Install Micro-computer
Cash Flow	1. Improve Receivables-5 days a) Increase Cash Sales-4% b) Revise Incentives 2. Install Computer Billing and follow system

FIGURE 3

Each of the Functional Plans are then integrated into the overall Business Plan. A not-so-casual reader might make three observations at this point. First: The strategies which optimize one of the four variables are not likely to optimize overall business performance. Second: The Functional Strategies may not optimize even one variable. Third, and most importantly from our view: Not a word has been said about Quality either as a Function or as a Financial Input to the Plan.

The first two observations reach the crux of the Strategic Planner's job — to integrate the resources of the business to achieve an overall objective. This necessarily implies that some Functional Strategies and the Business Performance of some variables will individually be sub-optimized in the interests of overall Optimization.

The third observation reaches the crux of the Quality Manager's job and raises the question — What is (should be) the role of quality in the Strategic Plan?

III. A ROLE FOR QUALITY

In general, in the strategic sense, quality is not a primary function as Marketing, Engineering, and Manufacturing. It is only in those instances of business catastrophe that the Quality Function is called upon to "develop a plan for recalls" or some other short term plan to recover from the disaster of the moment.

In the ordinary course of events, the role of Quality is to support the primary functions in the achievement of the business goals. From a Strategic Planning vantage point, the role for quality is:

A ROLE FOR QUALITY

The • Collection

and

• Analysis

of

• Relevant (Financial) Data

for

• Action (by OTHERS)

FIGURE 4

in a functional sense and we further accept the role of quality as defined above, then we can proceed to integrate Quality into the Strategic Planning Process.

Such an integration for the sample Marketing Strategy of Figure 3 might look like Figure 5.

MARKETING-QUALITY STRATEGY

Variable	Marketing	Quality
Revenue	1. Market Share-0.7%	1. Customer Survey On Quality 2. Warranty Analysis For Cause
Cash Flow	1. Improve Receivables-5 Days	1. Analyze Receivables for Quality Causes

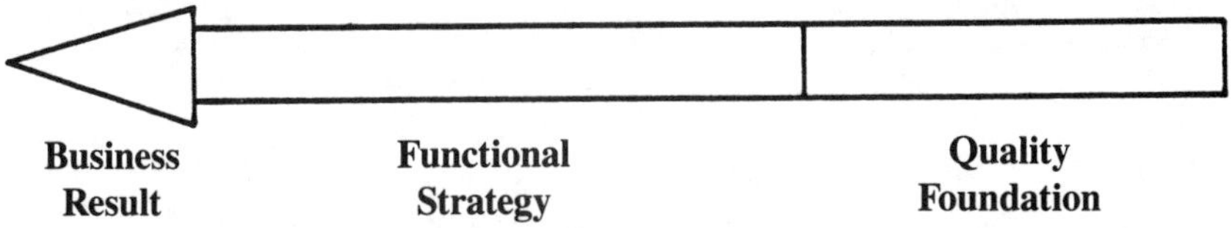

FIGURE 5

In similar fashion, the Quality integration in each of the Primary Functional areas can be outlined in terms of Financial Results. If, for example, Manufacturing strategies are aimed at Productivity Improvement and Inventory reductions, the quality plan might well call for Internal Failure cost reductions to support both strategies. Planning Models such as PIMS[(1)] and Portfolio Analysis[(2)] take specific account of Inventories, Market Share, Capital Investment and so forth on the Financial Results of the business. The use of Regression Analysis on the influence of Failure Costs on Profit Margins has been documented[(3)].

The application of classic quality techniques to Receivables, Inventories, Plant and Equipment, Product Design and other areas of Business Strategy await only our imagination.

V. A FEW WORDS IN CONCLUSION

The part played by Quality Costs in Strategic Planning as outlined in this paper is one of second-order support of the primary strategies employed in producing business results. It is pro-active in the sense that it actively seeks opportunities to support others, rather than reactive in the sense that it tries to restore lost Market Share or reputation. In the longer term, a pro-active approach has the merit of ultimately persuading business managers of what we already know: "Product quality comes close to being a panacea."[4]

REFERENCES

1. Schoeffler, S., et.al., "Impact of Strategic Planning on Profit Performance", Harvard Business Review, Vol. 52, No. 3, March-April 1974, p. 137.
2. ____, 1979 SPI Conference Transactions, The Strategic Planning Institute, Cambridge Ma., November 1979.
3. Brown, F. X. and Kane, R. W., "Quality Cost and Profit Performance", 32nd. Annual Technical Conference Transactions, American Society for Quality Control, May 1978.
4. Heaney, D. F., 1979 SPI Conference.

COST REDUCTION THROUGH QUALITY MANAGEMENT

Frank Scanlon
Secretary — Director Quality
The Hartford Insurance Group

ABSTRACT

For several years, quality professionals have been attempting to develop and implement administrative quality costs. The Hartford has incorporated many of these ideas and concepts into The Hartford's Quality Program. This presentation gives a step-by-step approach to selling to management, developing and implementing the program, and management reporting.

INTRODUCTION

Today will be the first day of your development of an ongoing custom-made administrative Quality Program for your company!

The success of an administrative Quality Program is measured in the same manner as your manufacturing Quality Program, that is, it adds dollars to your profits and can protect or enhance the reputation of your company.

For several years, quality professionals have been attempting to develop and implement administrative type of Quality Programs. Over the past three years, we have utilized many of the ideas and concepts presented by quality professionals and have successfully incorporated them into Hartford's own Quality Program. I am proposing to consolidate all of the various ideas and concepts and put together the basic ingredients of one administrative Quality Program.

Administrative Quality Programs, for the purpose of this paper, are identified as programs for those employees that are neither management nor directly involved in the manufacturing process.

For any Quality Program to be effective, it must be built in order that it can:

- Be measured
- Identify repetitive type of errors
- Identify major costly errors
- Provide statistics

The measurement — In the past, we have tried to become too sophisticated in the measurement of administrative quality costs. Unlike manufacturing, where it is reasonable and practical to show a separate line for actual cost of quality, the same degree of sophistication cannot be applied to adminstrative work processing. This does not mean we should abandon the development of quality costs, but only use a different approach. I am proposing the utilization of existing credible reports and statistics along with standard costs to develop a quality cost base for administrative functions, for example:

— Develop a predetermined standard average cost to fix an error multiplied by the number of errors detected will give you failure costs.
— Number of hours spent checking work multiplied by the standard cost hourly rate will give you the appraisal costs.
— Number of hours spent taking corrective action multiplied by the standard cost hourly rate will give you prevention costs.

If this approach makes sense to you, then you can build a viable quality improvement program for the administrative functions. A suggested approach is outlined in the following steps.

METHODOLOGY

Step I. Sell the Program to Management. — In order for this program to be effective, the need and purpose must be clearly identified and agreed upon by all levels of management.

1. *Need.* We must improve the quality of our administrative areas for the following reasons:
   (*NOTE*: Quantify as many of the following items as possible prior to your presentation.)
   a. Cost to Fix Errors.
      — 17 to 25% of all the paperwork is being reprocessed due to errors.
      — Errors in Accounts Receivable can impact profits, e.g., undercharging errors are seldom brought to the company's attention whereas overcharges usually are.
   b. Employee frustration.
   c. Reputation of the company.
   d. Negative impact in meeting schedules.
   e. Customer complaints

2. *Purpose.* The purpose of this program will be:
   a. Minimize the needs.
   b. Provide first line management with the proper tools of identifying problems and developing corrective action plans.
   c. Provide management with a report which will measure the level of quality and display improvement or slippage.

Step II. — With management's commitment and participation, you can proceed and put together your custom-made administrative Quality Improvement Program. This Quality Improvement Program has six ingredients:

1. Commitment/Leadership. Commitment and Leadership for a Quality Program must come from the top down. Quality cannot be effective without the full commitment and participation of top management.
2. Sampling. Once you have determined what credibility and tolerance will be acceptable and you know the existing error rate, published sampling tables can be utilized. Sampling in lieu of 100% checking is recommended as:

— Sampling will fulfill the requirements of the program.
— 100% checking is costly in terms of dollars and timeliness.
— 100% checking is only 85% effective.
— There is a tendency for individuals to commit more errors when they know 100% checking program is in effect.

3. Checking. The sampled items are compared to the written instructions to see if they conform. We define error as "An error is an act or omission that will require a document to be reprocessed." Individuals other than the person who did the job can perform the document check as long as they are competent in that job function.
4. Recording. The purpose of recording data is to provide the data necessary to perform an analysis and error trending. Proper recording will permit you to determine who is making errors and whether or not a specific unit is improving or deteriorating. This process includes those errors identified through: Sampling (processing), other units within the company (internal), and by outside sources such as customers, vendors, etc. (external).
5. Analyze. Analyzing the recorded data will enable a manager/supervisor to categorize the errors to identify a unit's major problem.
6. Corrective Action. With major problem identified, the manager/supervisor can develop a practical corrective action plan addressing the cause of error. By correcting the cause of the errors, it is logical that this type of error will not recur as frequently in the future, thereby achieving quality improvement.

Step III. The Detail.

1. *Key Elements to be Accomplished Prior to Implementation.*
   — Identify administrative functions in the work process.
   — Identify current administration quality costs.
   — Identify appraisal method.
   — Develop the procedures necessary to record, analyze and report.
   a. *Identify clerical work process functions.*
      — Do not spend your time checking one of a kind functions, but concentrate on those which support the manufacturing process, such as, receiving, shipping, billing, engineering, inspection and marketing.
   b. *Identify current administration quality costs.*
      — Utilize the comptroller to develop the Quality base measurement and standard costs to make certain all costs are included, that is, appraisal, failure and prevention. Make certain different failure costs are developed for processing, internal and external errors.
   c. *Identify appraisal method.*
      — Determine what is the most effective and economical appraisal method within your company. There may be times when 100% checking should be utilized; however, generally speaking, published sampling table can be used. The method of appraisal should be developed to be included into the workflow process. This will per-

mit the most economical method to detect errors as close to the source as practical.

d. *Develop the procedures necessary to record, analyze and report.*
  — The instructions or procedures for your Quality Program should be similar to the other instructions within your company. If these instructions are geared for your first line management, then they should be written accordingly.

2. *Implementation.*
  — Train supervisors on how to use their Quality Program.
  a. Mechanics. (How to) Utilizing workshop sessions, train supervisors on the proper completion of worksheets and reports. Then the training should emphasize how to analyze the error data and the management techniques required to develop practical corrective action plans.
  b. Attitude. (Why the Quality Program?) It is their Quality Program to detect errors in order that they can develop practical corrective action plans.

3. *Management Reports Contain.*
  — Cost of Quality
  — Error Trend
  — Corrective Action Plan
  a. *Cost of Quality*
    — Should be shown in both dollars and as a percentage of either sales or budget.
  b. *Error Trend*
    — Should display an error trend comparing actual to an objective to show improvement or deterioration.
  c. *Corrective Action Plan*
    — This report should contain a narrative defining the problems and corrective action activities. This section should also include information as to whether or not previous corrective action plans were completed and successful.

Note: As Quality Manager, it is important for you to sell to senior management that these reports are the by-product of the Quality System — not its prime purpose.

4. *Responsibilities.*
  a. Comptroller.
    — Development and verification of standard costs.
    — Evaluate quality reports to validate the actual percentage of improvement or deterioration.

Note: It is to your advantage to involve the comptroller for support and reinforcement as much as possible in the development and verification of standard costs.

  b. Quality Manager
    — Evaluate and analyze the comptroller's standards.
    — Discuss the problems with functional managers asking questions such as:

Are they taking effective corrective action?

Are they addressing the cause of error or just fixing the same errors again and again?

— Periodic review. The administrative process can change more frequently than the manufacturing process. Without updating the quality system to meet these changes, the Quality Program quickly becomes obsolete.

— Audit the quality system itself. Is what you think is happening actually taking place? If you are a single location company, you can personally audit the system at regular intervals. In multiple location companies, use your internal auditing system or other corporate review teams.

SUMMARY

1. An administrative type of Quality Improvement Program does require the quality professional to be a true manager, that is, to plan, organize and control.
2. The major difference between manufacturing and administrative quality control system is in the manufacturing process. The end products are consistent, whereas, in administrative areas the process is consistent, but it does not yield a consistent end product.
3. You are only a catalyst. You can only get this job done through the efforts of your first line managers and supervisors.
4. An administrative quality system cannot be as scientific as the manufacturing system, but through proficient management, the system can be as effective and less expensive.
5. In the '40's, the quality professionals had the courage to try something new in the manufacturing process — now is the time to stand up and be counted!

GUIDE FOR MANAGING VENDOR QUALITY COSTS

W. O. Winchell, Manager, Customer Satisfaction Activities
Consumer Relations & Service Staff
General Motors Technical Center
Warren, Michigan 48090

ABSTRACT

In this part of the tutorial, key portions of the new ASQC publication "Guide For Managing Vendor Quality Costs" will be reviewed. Covered first will be the fundamentals regarding quality costs found in vendor relationships. Following this will be suggestions as how to effectively manage these quality costs. Through starting joint programs with key vendors, significant quality problems can be solved quickly and permanently. Also, if conditions permit, encouraging vendors to start their own quality cost program can ultimately result in valuable improvements. Most important of all is that no quality cost program is effective without prompt corrective action.

The subject of our talk today is managing vendor quality costs. This is an important subject to all of us because in most cases we not only buy from vendors but we are a vendor to someone else. Key to first understanding and then managing vendor quality costs is recognizing that much of these costs are hidden. By hidden, we mean that these costs are not really easily identified or perhaps not even seen as vendor related quality costs.

For the purpose of this talk, we will look at these hidden costs from just one perspective — that of the buyer. From the buyer's viewpoint, hidden vendor quality costs occur in three basic ways.

The first type of cost is quality cost inside the vendor's facility. For the most part it is much the same as the buyer's quality cost. However, it is included as part of the piece price the buyer pays for the part. The vendor, perhaps rightly so, protects the identification of this investment for competitive reasons. The buyer is furnished the piece price alone and, if this price is competitive, he's usually satisfied with this knowledge.

The second type of hidden vendor quality cost is that which buyer spends, but does not usually segregate, to insure the quality of a product at the vendor's facility. This may be in the form of sending quality engineers to the vendor's plant to help him solve a crisis.

The third type of hidden vendor quality costs is that which the buyer spends in-house, again not usually segregated, specifically for vendors. An example of this may be process engineering required to repair or correct a purchased part not quite up-to-specification. Many of us call this fire-fighting to keep the line going.

Of course, there are also visible costs that the buyer can and sometimes does segregate to specific vendors. An example of this type of cost is receiving inspection effort.

For a typical manufacturing firm, these vendor related costs are a huge chunk of money — perhaps in the range of 10-20% of the selling price to your customer.

Through proper control of these hidden and visible costs that are vendor related, we should achieve several results:

- — The first is to optimize our profits.
- — The second is to insure a product that is "fit-for-use" by our customer. "Fitness-for-use" of our products will insure our success in the marketplace. If we don't meet our customers' requirements, our competitors will, and we will not survive in the long run.

The basic purpose of my talk is to suggest to you methods by which you can control both the hidden and visible quality costs incurred in your vendor relationships.

Very simply, two ways will be emphasized. The first way is to talk your vendors into adopting a quality cost program. However, discretion must be used since all vendors cannot support a quality cost program. They may be too small or special circumstances may prevent successfully accomplishing it.

The second way is to use your quality cost program to identify high magnitude vendor problems for resolution. Before we discuss these methods in detail, let us first review the basic principles of quality costs and how they apply to our vendors.

The American Society for Quality Control publishes a document titled "Quality Costs — What & How," authored by the Quality Cost Committee. Included in this publication are the essential elements that make up a quality cost system, recommended definitions and methods of installing and gaining acceptance.

In this publication, the ASQC recognizes four basic categories of cost. They are identified as:

- — *Prevention* — which primarily are the costs resulting from planning to prevent defects in products.
- — *Appraisal* — costs relative to the conformance to quality standards.
- — *Failure* — costs are the results of poor planning and include the failure of products to meet the established or required quality standards.
  - — failure costs are either *internal,* which are mainly manufacturing losses, or *external,* which are due to defective products shipped to the customer.

Each category of prevention, appraisal, internal and external failure contains many elements. Let's look at a few examples of elements in each category.

Prevention includes:

- — Quality control administration
- — Quality systems planning and measurement
- — Vendor quality surveys
- — Quality data analysis and feedback

Appraisal includes:

- — All inspection and test costs
- — Laboratory acceptance testing costs
- — Vendor quality audits and surveillance

Internal Failure includes:
- Rework
- Scrap
- Retest and reinspection

External Failure includes:
- Returned material processing and handling
- Warranty replacement

The first part of the quality cost program, *that of measuring,* is the most difficult. Most businesses do not enjoy the sophisticated accounting system that permits interrogating for the cost of quality elements.

A system for collection of this information has to be designed to fit each specific company.

Once we have collected the information and categorized it, we can now *analyze.* The most obvious is to examine those quality elements for which the greatest costs have occurred. It is here that the most effective and dramatic cost savings can be realized. But we must also note the quality elements having little or no costs, since insufficient effort may be occurring. For example, little action in vendor evaluations may result in high receiving inspection costs.

Action, of course, is the result of analysis. All the good cost reporting in the world is wasted unless someone takes corrective action. The logical next step is to take action in areas needing improvements. Specific persons must be assigned specific improvements against a time scale to assure effective results.

Finally, we *control* the improvements in a continuing effort to do even better over the long term.

Now that we have reviewed quality costs in general, let us now discuss how they apply in the buyer-vendor relationship. Previously we discussed how the vendor related costs can be classified as either hidden or visible.

Now I would like to discuss each of the categories in detail. Hidden quality costs, as you may recall, are in three parts:
- The first are those which are incurred by the vendor at his plant.
- The second are incurred by the buyer in fixing problems at the vendor plant.
- The third are those costs — which are not usually allocated to vendors — incurred by the buyer as a result of potential or actual vendor problems.

Those quality costs incurred by the vendor at his plant are, for the most part, unknown to the buyer and, therefore, hidden. However, even though the magnitude is hidden, the type of cost is not. They are the same type quality costs as the buyer incurs. For example, the vendor certainly has prevention effort. If he manufactures the product, he has expenses related to the quality engineering of the product. Even if the vendor is a small shop, this task must be done by someone and may very well be handled by the Production Supervisor if the plant lacks a Quality Engineering Staff. Certainly effort is expended in the appraisal area even by the smallest vendors. Someone must inspect the product prior to shipping to the buyer. In a one man shop this is done by the man who made the part. Unfortunately, we all have failure costs whether we are a large shop or a small one.

When the vendor makes a mistake in manufacturing they must either rework the part or scrap it causing an internal failure expense. If the vendor sends it to the buyer, it may be rejected with the resulting cost being considered as external failure.

The second type of hidden quality cost — that which is incurred by the buyer in fixing problems at the vendor's facility — is usually not specifically allocated to vendors. Except for an awareness of our troublesome vendors, there is usually no tabulation of the cost of the effort expended. Therefore, the actual expense is hidden. One example of this cost is the buyer sending a Quality Engineer to a vendor to resolve a crisis. Another example is the auditing a buyer may do at the vendor's plant to insure that he has the proper quality controls to build a product to the buyer's specifications.

The last type of hidden quality cost is that at the buyer's plant incurred on behalf of a vendor. Like the second type of hidden cost, we are certainly aware of it, but we probably do not keep track of it. The magnitude of this expense is not known for specific vendors. Again, the actual expense is hidden. This type of cost may include, in the prevention area, the following:

- Preparing the specifications for packaging the product that is shipped by the vendor to the buyer.
- Specification and design of gauges that must be used in the buyer's receiving inspection and perhaps as well by the vendor prior to shipping.
- Designation of appropriate specifications that the vendor must follow in the manufacture of the product.

For the hidden appraisal costs, we may incur expense for:

- Certain inspection operations and quality control effort in the buyer's production line related specifically to a vendor product.
- Review of test and inspection data on vendor parts to determine acceptability for processing in the buyer's plant.
- Calibration and maintenance of equipment necessary in the control of the quality of vendor parts.

In the internal failure area we may do troubleshooting or fire-fighting in order to use a vendor part that is not quite up to specification.

Finally, regarding external failure, there may be field engineering required to analyze and correct a problem caused by a vendor.

It must be remembered that this discussion of the types of vendor related hidden costs is by no means exhaustive. There are many more. Some of the types of costs not pointed out may be very significant depending on your individual situation.

There are, as discussed initially in this talk, vendor related quality costs that are apparent and perhaps much easier to identify in magnitude and assign to various vendors by the buyer.

These are called visible costs. They are primarily in the appraisal and failure areas.

Included in the visible quality cost category are the following:
— Receiving inspection.
— Laboratory acceptance testing.
— Scrap and rework that is vendor responsibility.
— Warranty replacement on parts supplied by vendor.

The visible costs, if tracked, are perhaps most significant because they can be good indicators of problem area.

Now that we have looked at the various types of hidden and visible quality costs, let us discuss the possible methods through which we may optimize these expenditures.

The first step for the buyer in optimizing his vendor related quality costs is to determine what costs are important to him. It is suggested that comparing the relative magnitude of quality costs by category could be a good start. To do this, your own quality cost program could be invaluable. But, if you don't have a quality cost program, a special study could be initiated to determine this information. The ASQC publication "Quality Costs — What & How" would be a valuable guide in accomplishing this.

This slide illustrates a hypothetical situation in which vendor scrap and rework are the biggest problem areas for the buyer. If the buyer makes the assumption that through improvements in vendor caused scrap and rework, warranty will be lowered, then for this company vendor responsible scrap and rework are the important items in the vendor relationship.

The next step is determining which vendors are causing the problem. Very likely, you will find that 20% of your vendors are causing 80% of your problems in scrap and rework. Now the buyer can narrow his effort to the vital few of the vendors and take appropriate action.

What is appropriate action? First and foremost, I would suggest that you promote that the "vital few" vendors start a quality cost program using ASQC publication "Quality Costs — What & How," if appropriate. Discretion must be exercised before insisting on this. Some companies may be too small to support a quality cost program or to accomplish special studies. Special circumstances may exist in other companies that would prohibit this action. However, if a vendor finds that launching such a program or study is feasible, the costs most visible to the buying company will most likely be reduced by doing so. If these are reduced, the hidden costs expended by both the buying company and the "vital few" vendors should also be lowered. The result will be that the vendor's product and the buying company's product are more "fit for use". This should increase profits for both which is very desirable. Also, improved profitability for the vendor may eventually result in lower prices for the buyer in a competitive market.

What other action can be taken? Keeping track of vendor related costs — both hidden and visible — may be a gigantic undertaking for the buying company. However, some way must be found to insure that progress is being made. It is recommended that the buying company track the important visible costs for each of the "vital few" vendors. It is submitted to you that, in most cases, you will find that if progress in reducing the visible costs is demonstrated, hidden costs are also decreasing.

What other action can be taken if we know the magnitude of the visible costs? It is possible that this information can be incorporated into a buyer's vendor rating system. The *Guide for Managing Vendor Quality Cost* describes a system utilizing the cost of the purchased part and the identifiable quality cost. Although not theoretically perfect, this system is simple and has achieved significant improvements in quality. It is well worth your consideration.

The *ASQC Procurement Quality Control Handbook* can also supply valuable help to you in accomplishing this. A vendor relationship depends upon more than the traditional measures of meeting deliveries and receiving inspection rejections. It also must recognize the "fitness for use" of the vendor's products for which visible quality cost can be a valuable indicator.

Visible quality costs can also be used in other ways. Some companies debit vendors for the scrap and rework occurring in the buyer's plant to put the responsibility for failure where it hurts most — in the pocketbook. However, in the long run this may be counterproductive in that some vendors may ask for a price increase to cover this situation. On the other hand, many buyers find it very effective to reduce the amount of business given to the offender and reward the good performer with a greater share of the "order pie".

Perhaps a far more positive method is to use the visible costs to identify vendor quality improvements that are needed, the buying company would initiate projects jointly with vendors to resolve the problems that are the source of high quality costs. The problem may be solved through an action by the buying company — perhaps the specifications are not correct or the vendor really doesn't know of the application of his product in the total package.

On the other hand, it may be that the vendor's manufacturing process needs upgrading through perhaps better tooling and the vendor must take some action. Through joint projects, using visible quality costs as facts, we can solve those problems and get better products. If quality costs are collected in a regular fashion, the results will document lower costs to both the vendor and buyer.

Let's now summarize what we have discussed. Basically, use quality costs in your vendor relationships. Through this tool, find out what costs and vendors are most important to focus upon. Suggest that these vendors, if appropriate, adopt a quality cost program in order to obtain improvements in "fitness for use" of their products. Use your visible vendor quality costs as a basis for starting joint quality/reliability improvement projects with your suppliers.

Most important of all is that any quality costs program is not complete without an effective corrective action program. The mere act of collecting quality costs alone will do nothing but add cost for you or your company. Only through pinpointing the important problems and solving these can we make progress.

QUALITY COSTS — A REVIEW AND PREVIEW

Clyde W. Brewer, Director — Quality
Otis Engineering Corporation

ABSTRACT

Two purposes of this paper are: (1) to present a short history of the activities of the American Society for Quality Control (ASQC) Quality Costs Technical Committee, (2) to present some non-traditional areas which may be considered in your Quality Costs System.

In order for the output of a quality system to be effective, meaningful, quantitative and measurable, consideration should be given to non-traditional quality costs such as effectiveness of sales and marketing operations, performance of the financial organization, industrial relations concerning human functional requirements, and pilferage losses. This type of cost analysis is more difficult to isolate and quantify; normally, measurement is associated with success or failure of the undertaking.

Evaluation of these expenses provides management with additional significant information to assist in their decision making process.

HISTORICAL REVIEW

This paper has been developed from the "Historical Report" compiled by the ASQC Quality Costs Technical Committee. In 1961, the National Chairman of ASQC, A. V. Feigenbaum, appointed the Economic Survey Committee with Rocco Fiaschetti as chairman to determine the national quality costs. The broad objective was to develop a survey of American industry for a determination of quality costs as a percentage of net sales billed. A secondary purpose was to dramatize the magnitude and importance of product quality, and to gain recognition for ASQC as a reliable and up-to-date source for quality costs information. Results of this survey were never published. It was perceived that differences in accounting systems and methods of collecting costs made comparisons among companies inequitable although trends of a company's costs would be beneficial. The study pointed out the need for standardization of quality costs terminology and methodology.
Under the leadership of Daniel Lundvall, the committee was officially renamed the Quality Costs Technical Committee. The basic task during this period was to produce a document to categorize the contents of a quality costs program and define the elements. This was the beginning of the handbook, *Quality Costs — What and How.* The first issue was distributed at the Annual Technical Conference in 1967. "The Bibliography of Articles" regarding quality costs was started in 1967, published in 1973 and updated annually thru 1979 by Edgar W. Dawes, thereafter by Clyde W. Brewer.

Led by Chairman D. L. Field (1970) and Robert Cornell (1971), a new charter was written. A quality costs course for the ASQC Education and Training Institute (ETI) was developed and standardized in 1971. Additional activity included development of a speakers list, interface with the accounting profession and suggested curriculum for colleges and universities. *Quality Costs — What and How* was revised in 1969 to incorporate new improvements. Leading contributor was Hal Freeman.

Under the guidance of Chairman Joseph Farnam (1972-1975), eight task groups were formed. Task Group Seven's purpose was to develop and publicize methods used to reduce the cost of quality, and actions resulted in the 1977 publication, "Guide for Reducing Quality Costs." This task group was led by Chairman Nathan Moore.

John T. Hagen, Chairman (1975-1977), formulated a public relations subcommittee to develop national publicity. Lou Pasteelnick was chairman of this subcommittee which encouraged the presentation of quality costs panels at regional and divisional conferences. Rules and regulations of the committee were amended and reissued in April, 1976. Of special importance is Task Group Fourteen which was dedicated to the definition of vendor quality cost.

Richard Dobbins assumed the chairmanship in 1977 and emphasized the broadening of interest. Task groups with organized purposes became standing committees leaving task groups for special projects. Committee activity at the Annual Technical Conference was expanded to three sessions in Chicago in 1978, five sessions in Houston in 1979, with William Cabral as Program Chairman and six sessions in Atlanta in 1980 with Clayton C. Brewer as Program Chairman. Task Group Fourteen, under the leadership of William Winchell, authored the new publication, "Vendor Quality Cost" in 1979 and released for sale in 1980. Richard Dobbins has also established quality cost teams of ASQC Quality Costs Technical Committee members which will conduct Quality Costs Seminars at the request of ASQC Sections or Divisions.

MANAGEMENT CONSIDERATION

Productivity decay appears to be a management dilemma. As quality professionals, we can elect, if we desire, to accept a truly significant management challenge. One method can be to expand the traditional scope of quality responsibility. Operational audits of effectiveness of non-traditional areas can provide some interesting observations. You have a choice to expand the scope of your Quality Audit or the Internal Financial Audit function. This decision must be influenced by professional creditability. This can also be a co-op effort. The objective is to visibly present ineffective or defective performance and introduce proven closed-loop corrective action techniques. Although we have not developed our audit process at Otis to encompass all elements of the following suggestions, each has been identified as a future goal to measure.

NON-TRADITIONAL THOUGHTS

Before we visit some new considerations, we need to establish some boundaries which are described in some of today's business and management textbooks. Present day accounting techniques are primarily developed for the manufacturing activity. Inputs and outputs can be easily identified and measured. Input/output relationships become more difficult with non-manufacturing responsibilities.

Traditionally, work and performance measurement is delegated to the immediate supervisor and to the employee. This laxity in formal work performance measurement is diametrically opposed to the cardinal management fundamental that permanent improvement in any performance is impossible unless every objective is measurable. Every defined task has some measurable characteristics. Accounts receivable clerks could well be measured by posting entries completed per hour, day, week, etc. Normalization to loss elements are as readily assessable as the time charged to the machinist for rework and scrap.

Costs of these types are customarily defined as discretionary. Managed or programmed costs, resultant from management allocations with maximized limits, do not have measurable direct relationships to compare to inputs, as indicated by costs or outputs as related to sales, services or product.

Sales and Marketing Operations

Consider using some of the following to report effectiveness of this business function:

Cost attributed to overstaffing and related expenditures resulting from inflated sales predictions.

Sales and marketing costs associated with defect complaints, investigations, and soothing irate customers.

Excessive marketing expenditures for low quality products.

Finance

Quality of performance of a financial organization could enhance some of the following:

How effectively do they procure money (time value of money — loans — line of credit)?

Measure accounts receivable delay in collections from unsatisfied customers. This would include interest expenses for interim financing to provide adequate cash flow.

Are all travel and business expenses cost justified by measurement or results?

Industrial Relations

Establish and audit this function with some or all of the following:

How successful is Personnel or Industrial Relations in procuring human technology (people) and the related atmosphere and environment for motivation?

Educational services procure the skills and technology improvements. Are the measurable results of training and educational expenses commensurate with the resource allocations?

Are safety expenditures correlated with actual results?

Do cost justified capital expenditures actually produce the predicted results?

Are costs associated with inadequate or inferior equipment or improper usage measurable? This could include all office machines such as typewriters, copy machines, communications equipment, and certainly the computer.

Pilferage Losses

Dollar losses associated with employee theft are just as costly as dollars lost to scrap and rework. Are they significant? You be the judge. Wayne Hopkins, the U. S. Chamber of Commerce's Senior Associate for Crime Prevention and Control has stated, "Businessmen are woefully uninformed about crimes against their own companies. They're so busy with their management duties they don't take time to establish policies and procedures to fight crime that is doing them great harm. Businesses don't know who is committing the crime or what to do about it." U.S. Department of Commerce estimated the cost of crimes against business in 1975 to be $23.6 billion. Unofficial corporate security information estimates reach $100 billion. The Law Enforcement Assistance Administration (LEAA) awarded a $300,000 grant to the American Management Association (AMA) to study and attempt to define how large the problem actually is.

An alarming fact is that society is becoming oriented to accept pilferage losses as a fringe benefit for employees. Unfortunately, organizations appear willing to accept and ignore the profit drain.

CONCLUSIONS

Based on the differential cost of labor and material associated with the clerical and professional element of an organization, loss could likely outdistance the traditional production worker.

Hopefully, let me leave you with a question. Would not one of the items presented be worthy of investigation?

BIBLIOGRAPHY

Booth, William E., "Financial Reporting of Quality Performance," *Quality Progress* February 1976, pp. 14-15.

Brewer, C. W., Frongillo, T. D., "Assurance Costing for Profit," University of Dallas, 1977.

Crosby, Phillip B., "Quality is Free," McGraw-Hill, 1979.

Dawes, Edgar W., "Quality Cost — A Tool for Improving Profit," *Quality Progress,* Vol. III, No. 9, September 1975, pp. 11-13.

Dobbins, Richard K., "Quality Costs — A Place for Decision Making and Corrective Action," *30th Annual Technical Conference Transactions,* Toronto, Canada, 1976, pp. 115-122.

Harrington, H. James, "Quality Costs — The Whole and Its Parts," *Quality,* May 1976, pp. 35-36. June 1976, pp. 148-149.

Hendrickson, R. S., "Let's Forget About Quality — Let's Go For Profit," *Quality,* March 1971, pp. 148-149.

Griffen, Dick, "Crimes Against Business," *Chicago Daily News,* February 1977, Six part series.

Juran, J. M., Gryna, Dr. Frank M., Bingham, R. S., Jr., "Quality Control Handbook," McGraw-Hill, Inc., Third Edition, 1974.

ABSTRACT

QUALITY COST PRINCIPLES — A PREVIEW

Frank J. Corcoran, Q.A. Specialist, Training and Management Services, Singer Kearfott Division
Little Falls, New Jersey 07424

Task Group #12 of the Quality Cost Technical Committee has been assigned the responsibility of developing a Standard Text for Quality Cost. This will be a completely new guide to expand on "Quality Costs — What and How" etc. incorporating current information and establishing a set of principles that will provide the fundamental base for developing, implementing and teaching "Quality Costs Systems". Progress schedules and compilation of subjects covered in this text will be presented. Contents presently planned will be discussed with references, description of material and other details. Authors/contributors with background information and submittal information will be highlighted. It is hoped this second generation text can consolidate, provide direction and eliminate differences in concepts and opinions presently being presented in many papers, articles, periodicals and seminars.

From an original assignment of review for update of one of the best selling documents published by ASQC to the "Gargantuan" task of compiling, editing and publishing a basic text for Quality Costs is the history of Task Group No. 12 of the Quality Costs Technical Committee. Events can be described most concisely by the following TWX from the Chairman of the task group to the Chairman of the Committee.

TWX

Your request of Task Group #12 at meeting of 11/4/77.

Concensus of the task group is to replace "Quality Costs — What and How" with a completely new guide. Therefore, our purpose is "To develop a publication as the standard Quality Costs Document which will establish principles, definitions, applications, examples and methods of Quality Cost Management". The strategy is starting with "Quality Costs — What and How" and to develop current information to be included in the document. Current publications, articles and practices indicate a more direct interface with accounting discipline and service industry and therefore, subsequent terminology, theory, etc. must be established.

Expansion of information in "Quality Costs — What and How" is required. Actual Case Histories must be obtained and all information molded into a firm set of Principles and Problems for the Quality Profession.

Title will be "Quality Costs — Principles and Problems" and/or "The Principles of Quality Cost Management".

The obvious question is why not just update "Quality Costs — What and How". The committee felt that although first published in 1967 with a second edition in 1977 that this was an excellent basic document and provided an excellent starting point for Quality Costs. It provides definitions, concepts and procedures and with the "Guide for Reducing Quality Costs" published in 1977, analysis methods and techniques. These documents provide a firm base. However, a need for one complete text containing above information and incorporating all of the new philosophy, applications, uses and Quality Costs committee work should be written.

The Quality Cost Committee Task Group #14 will publish the "Guide for Managing Vendor Quality Cost" this year. This too is a very excellent document and expands on the Quality Cost Concept. From the "Bibliography of Articles Relating to Quality Costs Concepts and Improvement" you can see the increased interest in writing, presentation of papers and numerous examples, and theories on the subject. All of this activity is tied together with a common thread, dollars and the label Quality Cost. In order to prevent misconceptions and keep the subject under control, a standard is required. The "Principles of Quality Costs" is proposed to help fill this gap. This is a long term project and much research and many changes are anticipated.

In 1979 a proposed outline was made by the Task Group and presented and accepted by the Quality Cost Technical Committee last September. Members of the Task Group have been assigned sections for their contribution to the document. This outline follows on pages 3 and 4.

Naturally, the Task Group solicits and will consider contributions made by members of the society and interested parties. We do not consider the above outline as cast in bronze or final in any way. Our present schedule calls for publication in 1982.

My original motivation for this undertaking dates back to the ASQC Akron-Canton Fall Workshop of November 5, 1976. At this workshop, Frank X. Brown of Westinghouse Electric Corporation presented the Quality-Accounting Relationship. This was a very provocative presentation and definitely something new to me in Quality Costs. I had been accustomed to the usual definition session on Prevention, Appraisal and Failure, and had experience with the Beech Report and Aerospace Industries Association (AIA) "Quality Resources Study". This discussion showed me there was a great deal more to the subject than the basics. Subsequent articles on ROI, Committee discussions on Service Industries, Product Liability and Quality Costs A One Day Seminar and Workshop sponsored by ASQC Quality Costs Technical Committee has justified this.

Our goal is to maintain the fine record of "What and How" and "Reducing Quality Costs" and make an outstanding contribution. Your help will be appreciated.

SUMMARY

Task Group #12 of the Quality Cost Technical Committee has been assigned the responsibility of developing a Standard Text for Quality Cost. This will be a completely new guide to expand on "Quality Costs — What and How" etc. incorporating current information and establishing a set of principles that will provide the fundamental base for developing, implementing and teaching "Quality

Costs Systems". Publication is scheduled for 1982. Contents with description of material and other details is given. This second generation text will consolidate, provide direction and eliminate differences in concepts and opinions presently being presented in many papers, articles, periodicals and seminars.

BIBLIOGRAPHY

Code **Author, Title, Publication, Date, LCS Code**

T-184 Brown, F. X., "The Quality-Accounting Relationship", Fall Workshop, Akron-Canton Section, ASQC, Nov. 5, 1976 LCS 010:30:000 FAM

T-129 ASQC, "Quality Costs — What and How", ASQC 1967 — Publication — Available from American Society for Quality Control, 310 West Wisconsin Ave., Milwaukee, Wisconsin 53203

Winchell, William O., "Guide for Managing Vendor Quality Costs", ASQC Publication — see T-129

Moore, W. N., "Guide for Reducing Quality Costs", ASQC Publication — see T-129

PRINCIPLES OF QUALITY COSTS

— OUTLINE —

I. INTRODUCTION
- A. Background and History (of book)
- B. Purpose and Scope
- C. Contributors

II. QUALITY COST CONCEPT
- A. History (of concept)
- B. Applicable Specifications
- C. Philosophy
  1. Economics of Quality.
     - a) Cost of Quality versus Value of Quality.
     - b) Optimum Point diagram.
  2. Goal (Reduction of Total Quality Cost to a minimum).
  3. Management language and appeal (dollars).
  4. Quality/Accounting interface.
- D. Theory
  1. Management of Quality Costs.
  2. Classification (Definitions and categorization of Quality Cost elements).
     - a) Prevention Costs.
     - b) Appraisal Costs.
     - c) Internal Failure Costs.
     - d) External Failure Costs.
  3. Measurement Bases (Selection and Use of).
  4. Optimization (See Economics of Quality).
  5. Saturation Point.

6. Trend Analysis.
   a) Cause and effect lag.
   b) Breakthrough.
7. Pareto Analysis.
   a) Maldistribution of Quality Losses.
   b) Vital Few versus Trivial Many.

III. QUALITY COST PROGRAM IMPLEMENTATION
A. How to Start.
1. Trial Program.
2. Selling Top Management.
3. Obtaining Accounting Cooperation.

PRINCIPLES OF QUALITY COSTS
— OUTLINE —

I. INTRODUCTION
A. Background and History (of book)
B. Purpose and Scope
C. Contributors

II. QUALITY COST CONCEPT
A. History (of concept)
B. Applicable Specifications
C. Philosophy
1. Economics of Quality.
   a) Cost of Quality versus Value of Quality.
   b) Optimum Point diagram.
2. Goal (Reduction of Total Quality Cost to a minimum).
3. Management language and appeal (dollars).
4. Quality/Accounting interface.

D. Theory
1. Management of Quality Costs.
2. Classification (Definitions and categorization of Quality Cost elements).
   a) Prevention Costs.
   b) Appraisal Costs.
   c) Internal Failure Costs.
   d) External Failure Costs.
3. Measurement Bases (Selection and Use of).
4. Optimization (See Economics of Quality).
5. Saturation Point.
6. Trend Analysis.
   a) Cause and effect lag.
   b) Breakthrough.
7. Pareto Analysis.
   a) Maldistribution of Quality Losses.
   b) Vital Few versus Trivial Many.

III. QUALITY COST PROGRAM IMPLEMENTATION

A. How to Start.

1. Trial Program.
2. Selling Top Management.
3. Obtaining Accounting Cooperation.

INNOVATIONS IN QUALITY COSTS IN THE NEW DECADE

Clayton C. Brewer
Aerospace Electronics Systems Department
General Electric Company
Quality Cost Technical Committee Member

ABSTRACT

The purpose of this presentation is to identify trends in the new decade in quality cost measurement and control. Trends will occur when the need arises, since necessity is the mother of invention. Management attention will be focused on quality costs improvement as the need is recognized for quality and productivity improvements as affected by consumerism, inflation, competition and technology.

These major factors affecting quality costs will be described and the concepts and applications will be identified in which trends exist or are likely to develop.

The following sources have been especially selected for their potential contributions in quality costs in the new decade:

Applicable papers from the Quality Cost Bibliography
Quality Cost presentations in this 34th Annual Technical Conference
Observations, experience, and deductions of the writer and some of his associates on the Quality Cost Committee

INTRODUCTION

Four major factors affecting quality cost measurement and control in the new decade are consumerism, inflation, competition, and technology.

Consumerism

Consumerism is a reaction to poor quality by consumers. As a consumer we can relate readily to consumerism. My wife had used a well-known cookbook for the twenty-nine years of our marriage. It had become dog-eared from use and I bought a new copy to replace it. The metal rivets holding the loose-leaf binder together were still as good as ever on the old book, but one of the plastic substitutes to replace the rivets was broken on the new product. As a quality professional, I was angered. Now, can you imagine how I felt? Have you or a member of your family returned defective or poor workmanship products during the past year or otherwise made a complaint?

Consumers are angered at poor quality and feel that they should receive better quality for their money. A movement starting in the 1960's has led to lobbying, boycotting, and legislation against poor quality.[1] The most effective action, however, is when consumers do not receive the quality *they feel is right* at the price *they feel is right,* and they switch to the product that offers the value they demand. The demand for improved quality by consumers is felt not only in the consumer goods and service industries, but is carried over into all industries. Marketing and quality management must become aware when there is the need for quality improvement through communications with the customer, or inevitably, they will through complaints and lost business as competitors become more sensitive to the consumers' demands. Awakening will occur when management, responsible for the business plan, and quality management recognize that the quality requirements are really set by the consumer, and when they try to see their product through the consumers' eyes.

The strategic planning process and the relationship to quality costs for two major corporations, General Electric Company[2] and Westinghouse Corporation[3], are presented in this 34th Annual Technical Conference. Both Corporations report failure costs at the corporate level, and establish market share as a business objective using quality improvement as the strategy for achieving the objective. Both speakers presented a method for selectively reporting potential problem areas to Corporate. The General Electric Company speaker, L. J. Utzig, pinpointed a major change in focus in quality control from the traditional "Improvement in internal effectiveness — *To get the same or more quality for less dollars* — by eliminating the negatives of quality" to "Use quality as a strategic element to counter competitive threats and meet higher demands of customers — *To get more dollars for better quality* — through higher market share or higher prices or both." The reader is recommended to review the ASQC publication, *Guide for Reducing Quality Costs,* which covers strategic quality planning.[4]

During the next decade, we can expect to see a growing awareness of the need for quality improvement and of the implementation of quality programs sensitive to consumer requirements as follows:

- Strategic quality planning to improve market share
- Product Service policies to assure customer satisfaction
- Implementation of planned quality improvement programs
- Marketing surveys to determine consumer opinion of the quality of the products
- Improved analyses of external failure costs and complaints to determine causes, whether specification misinterpretation, design, workmanship, process control, or vendor responsibility
- Quality standards responsive to consumer requirements
- Quality systems designed to prevent or minimize the repetition of quality problems

Inflation

Inflation is a national and/or international problem, affected by a number of interrelated causes in which the experts do not always agree. The consumer price index, a customary measurement of inflation, is the aggregate price of the items people buy determined by a controlled sample indexed to a standard year. It represents the changes in prices we pay. The chart labeled Productivity Gap (Figure 1) shows a great need for productivity improvement in the United States.[5]

The Productivity Gap is defined as compensation per hour minus output per hour. The chart (Figure 1) labeled Productivity Gap shows the relationship between the Consumer Price Index and the Productivity Gap. The scale is in percent change. Compensation and Output have been plotted, and the difference between the two lines, Compensation and Output, shaded on this chart, is the Productivity Gap.

When this difference is plotted on the same scale and compared to the Consumer Price Index, the correlation is almost perfect. Notice as the Productivity Gap changes, the Consumer Price Index changes with it. Productivity must increase to offset inflation.

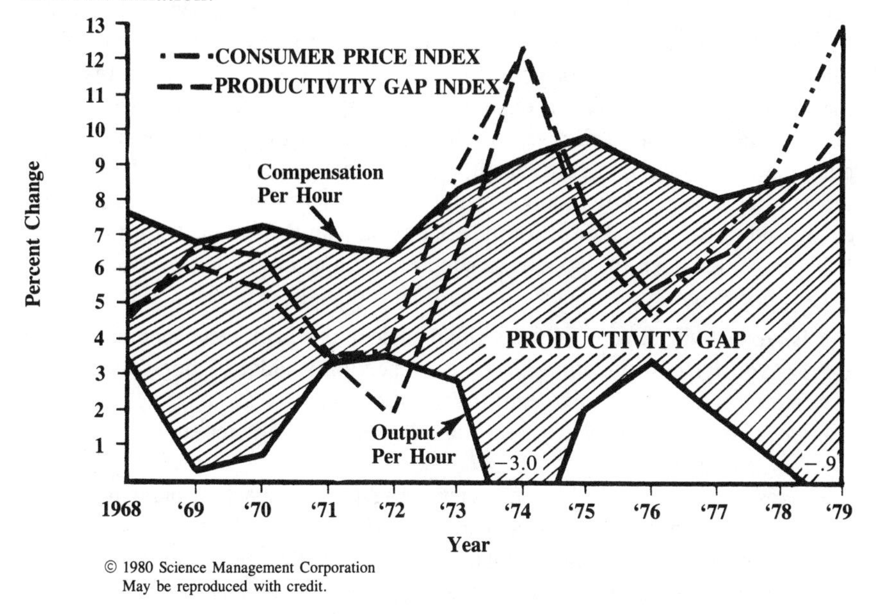

FIGURE 1. PRODUCTIVITY GAP

It may be interesting to conjecture on the causes for the rise in the cost of living in 1973 and attribute it to the increase in oil prices by OPEC nations, and to attribute the increase in the Productivity Gap in the winter of 1974 to severe weather, coupled with the oil embargo causing plants to cut back production. The speculation may be right; however, it is wasted effort, because there is not much that the quality professionals can do to affect either suggested cause.

Productivity has increased six times since 1900, which has caused a six-fold increase in the standard of living, and with the exception of the 1974 recession and the most recent quarters, productivity has increased every year since 1959. The productivity rate increased at an annual rate of 3.2 percent from 1947 to 1967, but has dropped to 1.5 percent during the last decade. Productivity must increase to offset inflation.

The great need for productivity improvement has been shown, and quality professionals and quality costs will have a great part in satisfying this need.

Productivity can be measured in many ways, but it is always a rate of output divided by input. Many persons consider productivity improvement applicable to manufacturing only, and look only to the improvement of the effectiveness of direct labor or to reduce direct labor by technology and automation. The manufacturing workforce as a percent of the total workforce has gone from 30 percent in 1947 to 22 percent in 1980.[6] Management needs to look to other areas also for productivity improvements. The use of quality cost measurements can contribute to improved productivity.

A review of the Quality Cost Bibliography[7] shows that many of the papers were written on quality cost control without getting further than the control of scrap and rework. Why not as much attention to the other major quality costs? Scrap and rework is highly visible, and it is easy to measure in both the defect reporting system and the finance systems. Perhaps a more significant reason is that nearly all quality control engineers feel responsible for quality assurance, and therefore, quality defectives and the corrective action system receives their attention. Although the quality engineers may feel responsible for planning for test and inspection labor, many feel much less responsible for the productivity of test and inspection labor or for the effectiveness of the test plan. Furthermore, few companies have the system to support test and inspection productivity.

How does your system stack up regarding productivity of test and inspection?

- Is productivity covered in the position guide of the quality engineer or planner?
- Do you have an incoming test and inspection standard labor system?
- Same for in-process test and inspection?
- Does your system relate the test and inspection characteristics inspected to the defectives found?
- Does your system call for a review of test and inspection plans to reduce inspection on characteristics with no history of rejections or when quality has improved and to add unplanned characteristics being rejected?
- Does your system have reports showing productivity as affected by test and inspection labor efficiency and by improvements in methods for both the quality control engineer and management?
- Does your organization require test and inspection productivity goals for test and inspection management and for the quality planner?
- Is management as concerned over the productivity of test and inspection labor as they are over the manufacturing direct labor?

If, after reading the above, the reader recognizes inadequacies in his system to control test and inspection productivity, and sets up a system to do so, the savings in improved quality and productivity will probably be equal to the savings in scrap and rework.

Other than test and inspection, there are other areas of productivity affected by the quality professional.

The new ASQC publication, *Guide for Reducing Vendor Quality Costs,*[8] on sale for the first time at this 34th ATC meeting, includes the control of hidden and visible vendor quality costs and the use of vendor quality performance index, which is actually a productivity measurement.

$$\text{Vendor Quality Performance} = \frac{\text{Vendor Quality Costs} + \text{Material Cost}}{\text{Material Costs}}$$

It is suggested that the reader review the two papers in the 34th ATC Transactions, which introduce this publication.[9]

Quality Control of the service functions within a manufacturing organization and of the service industries[10] are becoming recognized. The last three years an entire quality cost session has been devoted to quality costs in the service industries at the ASQC Annual Technical Conferences. The quality assurance and quality cost principles apply in the white collar and support functions. It is suggested that the reader review the two papers on service quality costs appearing in the 34th ATC.[11] A speaker is currently preparing a paper on software quality costs for presentation in the 35th ATC in San Francisco in 1981. The quality cost committee would welcome a companion presentation on software quality costs.

If the quality professional buys the need for productivity improvement as a business objective and the quality cost system as the method for achieving it, and if he is convinced the quality cost principles apply to service functions and service industries also, then it should be easy to accept the use of the quality organization and the quality cost system for control of non-traditional costs to improve profitability as was proposed by Clyde Brewer[12] in the companion paper to this presentation at this 34th ATC.

In the new decade, there will be a growing recognition by all levels of management to improve productivity, and there will be an increasing feeling of responsibility for quality professionals to improve productivity of quality labor, and use the quality and quality cost systems to improve productivity of non-traditional costs. During the new decade, there will be an increasing trend for quality professionals to develop and apply systems for productivity improvement as follows:

- To develop test and inspection standard labor systems
- To improve productivity of test and inspection labor
- To develop systems for improving the effectiveness of test and inspection planning
- To establish systems for improved effectiveness of indirect labor and of processing costs
- To apply quality cost principles to new industries
- To accept responsibility for improving productivity of non-traditional costs

Competition

Japanese industry is competing with American industry and is winning increased market share in many fields. Japanese industry with government assistance has made inroads into many markets. According to *Quality* magazine the Japanese will spend one billion dollars to break into the computer and semiconductor markets.[13]

The spectacular success of the Japanese to break into a market indicates that they are strong in strategic planning or development of a market or product. Strategies appear to be to gain market share by productivity improvement and quality improvement. The reader is recommended to review a paper presented by the General Manager of a Japanese electronics calculator firm who referred to himself as "Quality Control" specialist and management of a company.[14]

Excerpts from the paper are quoted. The italic portions are mine.

". . .judging from the products actually marketed, it may be concluded that in the past 25 years, the *speedy growth of business enterprises have been brought about by a succession of timely developed, manufactured and sold new and at-*

tractive quality products. . .The opinion of Japanese economists is that the *greater part of the profit of an enterprise has been attained through attractive new products timely supplied on markets.*

"To maintain proper profits every year, *manufacturers of electronic products have to develop a succession of attractive new products*. . .

"Looking back upon the history of Japan's electronic industry over the past 20 years it should be noted that all enterprise members, i.e., management engineers, sales personnel, factory workers, etc., had been exerting their full efforts in various fields such as investment in engineering study and development, *exploitation of new demand, encroaching on other firms' shares of the market by offering a new product and development of profitable, quality products so that the enterprise survives in the industry.* As a result, our country has reached such a level that high-quality electronic products typified by the transistor radio and TV can be distributed on the markets in various parts of the world.

"It is already 25 years since quality control techniques were introduced from the US into our country. In order to understand it correctly and put it to actual use, Japan's management group has been making its utmost efforts. Thus the so-called QC Circle activities have widely spread even among factory workers in the past 10 years."

The above paper includes techniques used in model planning which the author claims should have more widespread use.

Through the writings of Dr. Joseph M. Juran[15], the Quality Revolution in Japan has been defined. The serious reader should review Dr. Juran's papers for a detailed analysis of the differences between American and Japanese quality, of which an abbreviated list of the main points are listed below.

Development — Japanese a do more thorough job of debug.
Investment — Japanese take greater risks for longer time periods. Government policies help Japanese to accumulate investment.
Management commitment to quality improvement
Standards of quality are high and government assistance is given by the export standards required for exporting a product.
Training of all personnel from the top down in the most extensive training program in the world.
Participation — Japanese workers participate actively in solving quality and productivity problems.
Quality Improvement Program with goals and measurements.
Recognition — The Japanese system solicits and recognizes participation.

The last six items are characteristics which the Japanese system has to a much greater degree than American industry. American industry needs to tighten up in these areas.

Technology

The demand for improved quality and productivity hastens technological improvements such as advances in materials, increased design sophistication, automatic production processes, miniaturization of components and assemblies, use of robots, and improved computer equipment and software applications, resulting in greater requirements on product testing and test equipment.

Although the demands are greater, the same technology has provided improved capability to meet the requirements. Automatic production processes, computer aided design/manufacture/test, and microprocessor controlled or automatic test equipment minimizes much of the workmanship and test validity problems. Quality costs can be useful in the justification for these facilities.

The penalty for failure is greater in time and money placing greater requirements on quality feedback and cost systems. At the beginning of the past decade, almost all financial systems had the capability to provide scrap and rework costs for quality cost control as a by-product of the accounting system, but they lacked the capability to provide an integrated quality cost control system. Changes to computer equipment and software have been so great that most companies are probably using greatly improved computerized systems today because the cost of doing business is reducing in this area, opening opportunities for improvement justified on the savings in operating costs.

With the new computer capability, there will be a trend for more integrated systems to provide quality cost reporting at the profit center level and by product line and functional area. Measurements will be more accurate as the validity of input data is needed and is assured.

Quality Costs, being predominately labor costs, are input costs; however, most systems relate quality costs to sales. Sales is an output figure with variances in season, materials and inventories (materials, labor, and finished goods) that amount to several times the expected changes in the quality costs being measured. The sales as a measurement base is only effective as a long term indicator using rolled data; however, it was convenient and available when quality cost measurements made its debut. There will be a trend to greater use of direct labor or a value added (contributed value) base adjusted for inventory changes as measurement bases. When this happens, trend reporting will be more meaningful, and cause and effect relationships more apparent.

Quality costs reporting at the product line and functional area will be the performance measurements of the quality cost improvement program for the respective product line or functional area. The elemental costs comprising quality costs are measured, and the minimization of the elemental costs become the goals. When costs are interactive, optimization becomes the goal.

During the new decade there will be a trend towards integrated finance and quality measurement systems to include integrated data bases, real time systems, test and inspection standard cost systems, and the use of models for optimizing costs.

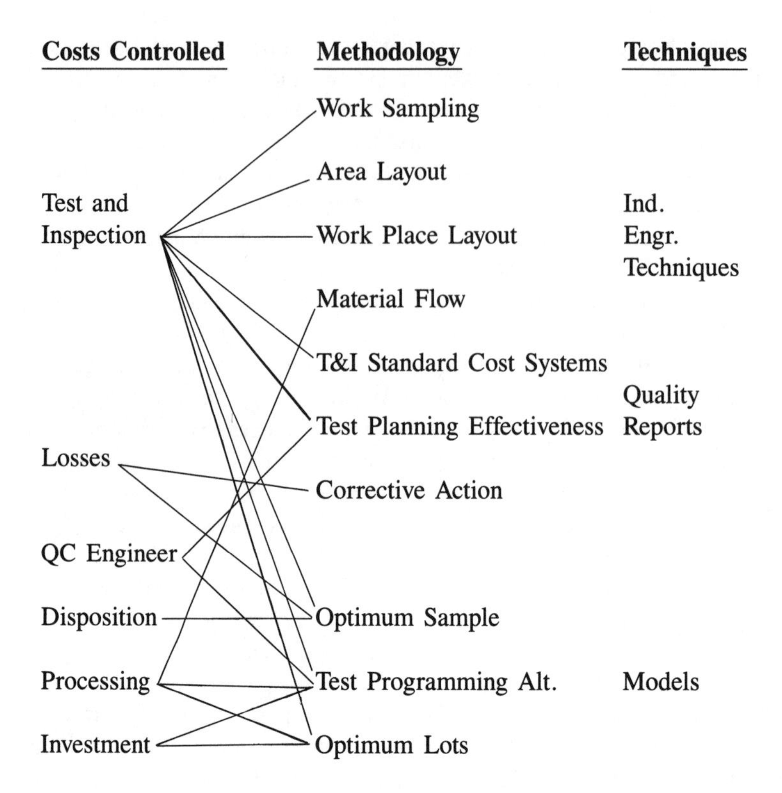

FIGURE 2 Integrated Finance and Quality Systems

An example of what can be done with such a system is shown in Figure 2. The system optimized the costs affected by or affecting incoming materials. The test and inspection, QC Engineering, losses on the floor for vendor responsibility and disposition costs are quality costs. The processing costs of materials from dock to stock and investment for materials in the materials cycle are included since they affect or are affected by the incoming materials system.

The methods used for optimizing the costs are:

- Industrial Engineering Techniques: Work sampling, area layout, work place layout, and materials flow.
- The T&I Standard Cost System is used for productivity reports for management, and is necessary for the models.
- The Corrective Action Report provides for corrective action at commodity, vendor, and drawing level.
- The models are:

  Optimum Sample which provides the sample size for the minimum total costs of test and inspection, losses and disposition costs.

  Test Programming Alternatives provides the breakeven point at which it would be more economical not to test rather than to program a part for the automatic test equipment.

Optimum lots model provides the number of lots for purchasing to bring in the material in order to optimize the processing costs, test and inspection costs, and the investment in materials in the materials cycle.

FOOTNOTES & BIBLIOGRAPHY

1. See Juran, J. M. *Quality Control Handbook,* Third Edition, Section 4, "Quality and Income," PP. 4-30 to 4-33. McGraw-Hill Book Company, 1974.
2. See Utzig, L. J., "Quality Reputation — A Precious Asset". *34th Annual Technical Conference Transactions — ASQC.* May, 1980.
3. See Brown F. X., "Quality Costs and Strategic Planning". *34th Annual Technical Conference Transactions — ASQC.* May, 1980. Also, Brown, F. X., "Quality Cost and Profit Performance". *32nd Annual Technical Conference Transactions — ASQC.* May, 1978.
4. ASQC Quality Cost Technical Committee, *Guide for Reducing Quality Costs — ASQC.* 1977 publication — available from American Society for Quality Control, 310 West Wisconsin Ave., Milwaukee, Wisconsin 53203.
5. By permission of Science Management Corporation, Moorestown, NJ.
6. See Geier, James A. D., "Improving Productivity" *Tooling* and *Production.* December 1979.
7. ASQC Quality Cost Technical Committee, *Quality Cost Bibliography.* ASQC, Rev. Nov., 1980 — available from American Society for Quality Control, 310 West Wisconsin Ave., Milwaukee, Wisconsin 53202.
8. Op. Cit. ASQC Quality Cost Technical Committee, *Guide for Reducing Vendor Quality Costs.*
9. See Winchell, William O., "Guide for Managing Vendor Quality Costs", *34th Annual Technical Conference — ASQC.* May, 1980. Also Dobbins, Richard K., "Procurement Rating System", *34th Annual Technical Conference — ASQC.* May, 1980.
10. See Bramblet, J. — Sodosky, T. and Wadsworth, H., "Control Charts Based on Cost for Use in Service Industries", *Administrative Application Division — ASQC* Yearbook. 1974.
    Also Clancy, Wendell, "Liability Takes Aim at the Insurance Agency", Quality Management & Engineering, April 1975.
    Also Latzko, W. D., Vice President, Irving Trust Co., 1 Wall Street, N.Y. "Process Capability in Clerical Operations".
    Also Latzko, W. D., "Reducing Clerical Quality Costs", *28th Annual. Technical Conference Transactions — ASQC.* May, 1976.
11. See Hershaver, James C., Ruch, William A., and Adam, Everett E. "Quality Measurement for Personnel Services", *34th Annual Technical Conference — ASQC.* May, 1980.
    Also Scanlon, Frank, "Cost Improvement Through Quality Management", *34th Annual Technical Conference — ASQC.* May, 1980.

12. See Brewer, Clyde W., "Quality Costs — A Review and Preview", *34th Annual Technical Conference — ASQC.* May, 1980.
13. Walsh, Loren C., "Editorial", *Quality* Magazine. February, 1979.
14. Harada, Akira, "Quality Control of New Product Development in Japanese Consumer Electronics", *34th Annual Technical Conference — ASQC.* May, 1972.
15. Juran, J. M., "Japanese and Western Quality — A Contrast, Part I", *Quality* — January, 1979.
    Juran, J. M., "Japanese and Western Quality — A Contrast Part II", *Quality* — February, 1979.
    Juran, J. M., "American Manufactures Strive for Quality — Japanese Style". Reprint From *Business Week*, March 12, 1979.

QUALITY PROGRAM MODELING FOR COST EFFECTIVE TAILORING

W. C. Wilhelm, Manager
Quality Assurance Systems
General Dynamics, Pomona Division
Pomona, California

ABSTRACT

A popular topic discussed among government and defense industry Quality people concerns the "tailoring and scrubbing" of specifications and requirements.

At General Dynamics, Pomona Division, we have implemented an approach for tailoring Quality Assurance requirements based on a modeling concept that is identifiable with various program phases using MIL-Q-9858A and Department of Defense Directive 5000.1, Life Cycle for Systems Acquisition.

To date, over a dozen Quality Program Plans, covering four different contract phases, involving a wide range of product lines, have been published. This has resulted in substantial savings in Quality program planning costs.

In the early nineteen seventies, there was a growing concern by the Department of Defense for the increasingly high cost of defense procurement. Special studies and committees were funded in government agencies and industry to determine the basic reasons for cost escalation and to make recommendations.

Most significant among these studies were the findings of a special industry and government task force chaired by Dr. Joseph Shea, Vice President of Raytheon Corporation. The Shea Report, as the findings are commonly referred to, presented a significant conclusion: There are over 40,000 specifications and standards which can be invoked on government contracts. The specifications and standards themselves are not the problem. The implementation and interpretation by either the government or the contractor is the prime reason for the increased costs.

The Shea Report pointed out that the 40,000 specifications and standards were really the most logical way to describe how to best perform specific tasks. The report's recommendations were to standardize common families of specifications to reduce potential conflicts on the same subject and to tailor these documents so they specifically identify the needs of the item to be procured. Traditionally, contractors have had the tendency to over-interpret the specifications while the Government was prone to over-enforce them.

A list of frequently used specifications were identified as "cost-drivers." On this list was the military specification MIL-Q-9859A Quality Program Requirements. For those of you not in the defense related business, MIL-Q-9858A delineates the requirement for the contractor to have a Quality program that provides a systematic method for defect prevention, detection, and appraisal. This specification for defense contractors is basically the same as the Quality requirements imposed by H.E.W. — Food and Drug Administration for the manufacture of medical devices.

One of the recommendations from the Shea Report was the tailoring of requirements with the intent of reducing cost escalation rate. The challenge was to assure that the tailored requirements complied with MIL-Q-9858A and were cost effective. In this regard General Dynamics, Ponoma Division has developed and implemented what is a rather unique response to this challenge.

For over twenty years General Dynamics, Pomona, was basically a single product house. This made compliance with MIL-Q-9858A a relatively simple task. There was only one Quality system at one level of maturity... production.

In the early seventies, General Dynamics, Pomona evolved from a single product house to a multi-program environment with various product commitments at all levels of maturity... conceptual, engineering development, initial production, as well as full production. Our single program approach to meeting the Quality requirements of MIL-Q-9858A was no longer viable and cost effective. Attention began to be directed toward specifications tailoring with emphasis on controlling cost drivers.

Our first task was to determine the character of General Dynamics' environment from a Quality point of view. It became obvious that there was a wide range of Quality requirements due to the multiplicity of programs and that these programs were in many phases of maturity. The many types of programs at varying maturity meant great variety in hardware, processes, and controlling systems. Another key factor was that General Dynamics was and is organized to provide autonomy for each program. This had resulted in a wide disparity of Quality policies.

Recognition of the enormity of the impact of a multi-program environment resulted in the establishment of specific goals:

- Provide individual Quality program plans tailored to each program's needs.
- Assure standardization to division Quality policies to make maximum use of the Division's Quality resources.
- Utilize the Quality program plans as working documents.
- Minimize the need for subordinate detailed Quality Assurance procedures.
- Satisfy the requirements of MIL-Q-9858A for each program.
- Assure optimum Quality program cost effectiveness.

Once goals were defined, the next major task became one of defining the means for achieving our goals. With programs in-house that ranged from conceptual design to production we decided to use the Department of Defense Life Cycle Acquisition definition of program phases and associated milestones within each phase. This program phasing arrangement described in D.o.D. Directive 5000.1 provided for four basic life cycle phases (conceptual or advance design, engineering development, full scale development or pilot production, and production). In simplified format, Figure 1 depicts the phases, their individual milestones, and the key review points prior to phase transition.

Dod 5000.1 Provides Definition Of Life Cycle Phases In Milestone Achievement Terms

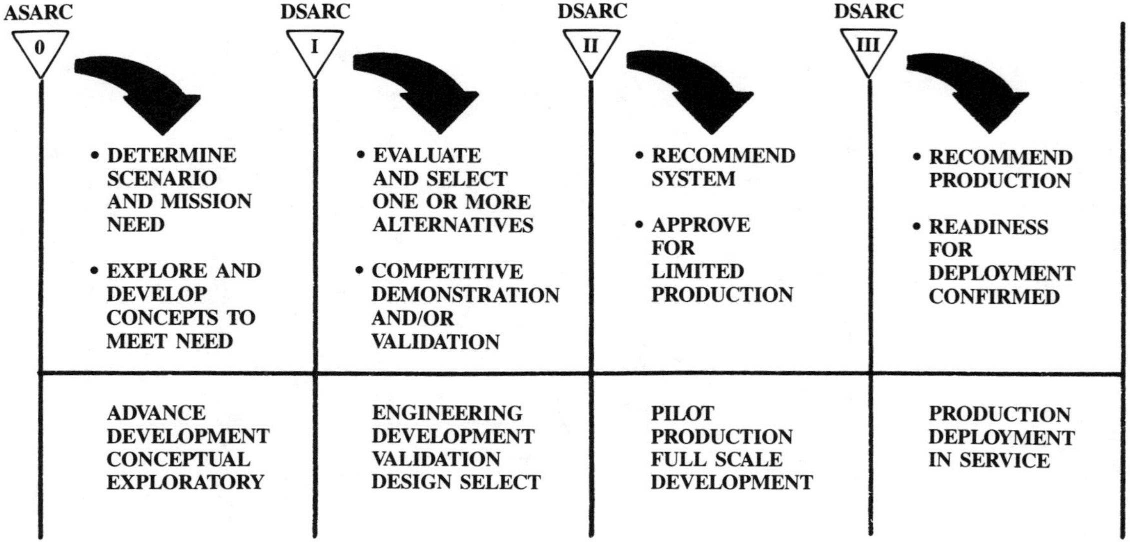

ASARC 0	DSARC I	DSARC II	DSARC III
• DETERMINE SCENARIO AND MISSION NEED • EXPLORE AND DEVELOP CONCEPTS TO MEET NEED	• EVALUATE AND SELECT ONE OR MORE ALTERNATIVES • COMPETITIVE DEMONSTRATION AND/OR VALIDATION	• RECOMMEND SYSTEM • APPROVE FOR LIMITED PRODUCTION	• RECOMMEND PRODUCTION • READINESS FOR DEPLOYMENT CONFIRMED
ADVANCE DEVELOPMENT CONCEPTUAL EXPLORATORY	ENGINEERING DEVELOPMENT VALIDATION DESIGN SELECT	PILOT PRODUCTION FULL SCALE DEVELOPMENT	PRODUCTION DEPLOYMENT IN SERVICE

F186675 /38

FIGURE 1. Basis for Phase Definition

It soon became obvious that D.o.D. Directive 5000.1 provided an ideal base for constructing four Quality Program Models to the requirements of MIL-Q-9858A with the objectives of the specific phase addressed. This is depicted in Figure 2.

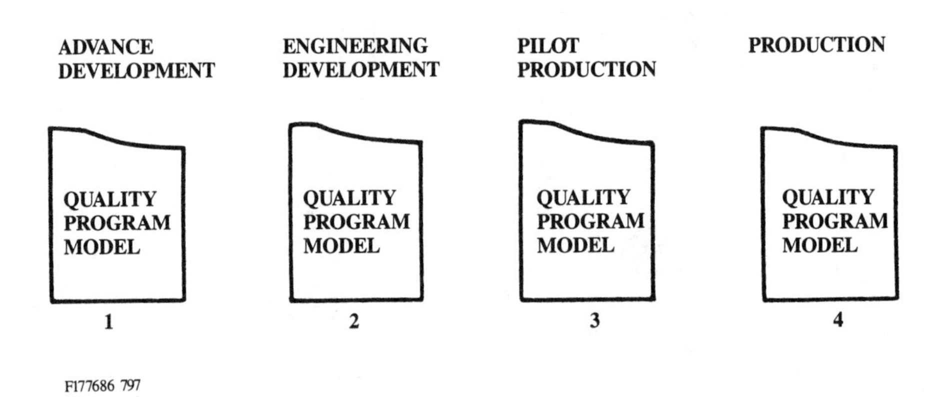

FIGURE 2. Formulating the Quality Program

Next we tackled the job of determining model format, ingredients, and the level of procedural content required to satisfy MIL-Q-9858A in context with the phase of maturity being addressed. The basic approach used in each model is shown in Figure 3. It consisted of:

— A Quality requirement matrix which correlated the Program Instructions with MIL-Q-9858A on a subject-by-subject basis to assure complete response.
— Milestone charts for showing how Quality tasks will be scheduled for implementation.
— Individual Quality instructions each addressing a specific MIL-Q-9858A subject.

Result: A Quality Assurance Program Plan Designed To Phase, Tailored To Program, and Has Management Utility for the Life of The Contract

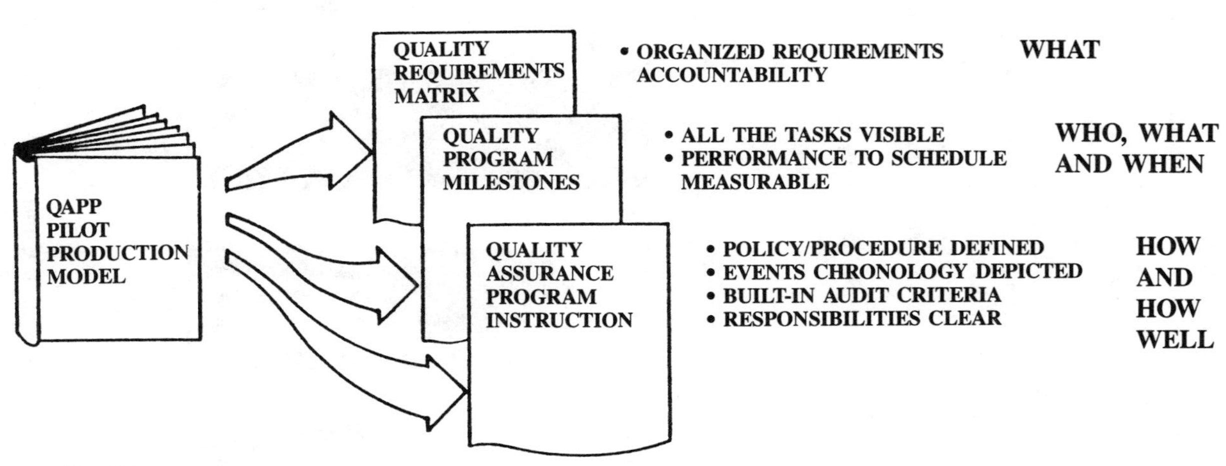

FIGURE 3. Procedural Aspects of a Model

Each Quality Assurance Program Instruction provided a narrative in response to a specific MIL-Q-9858A subject. For further clarity we provided a flow chart depicting each major action in chronological order including all feedback loops. As an aid to the user, audit criteria was added to provide a meaningful "self check" for compliance with each Program Instruction.

As we began developing the procedures for each of the four models, several interesting points emerged.

1. Procedures dealing with the same subject but at a more mature phase showed a marked increase in complexity and control with one exception. The pilot production model procedures rather than the expected production level procedures exhibited the greatest degree of control and complexity. This was due to the proofing and verification aspects of assuring full production readiness. This is illustrated in Figure 4.
2. Certain types of verbs dominated specific phases. This characterization of phase verbiage is shown in Figure 5.

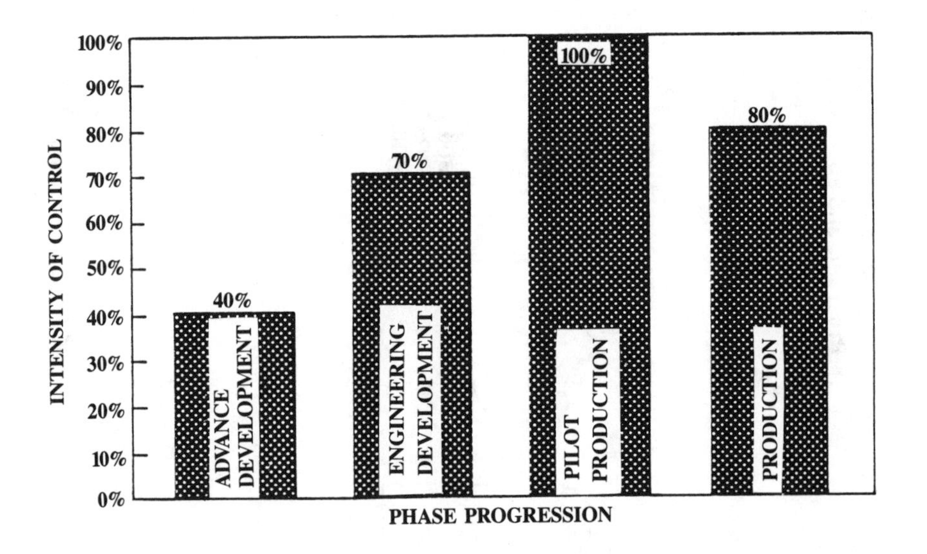

FIGURE 4. Percent of Change in Control According to Model Phase

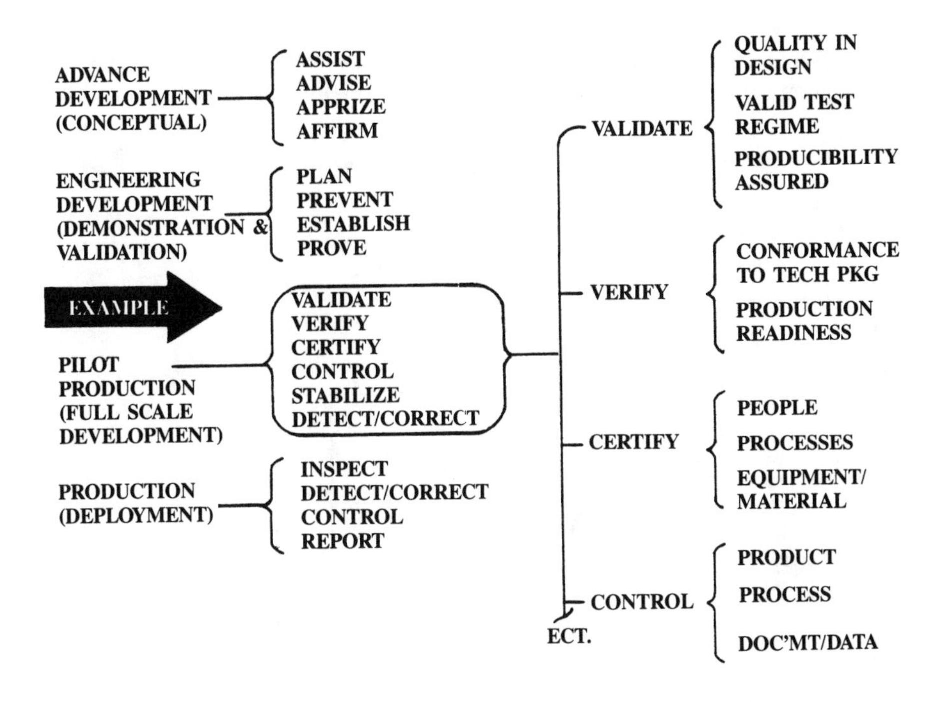

FIGURE 5. Phase Characterization of Verbiage; Quality Program

Once the Quality Assurance Program Instructions, milestone charts, and quality requirements matricies were finished for each phase, the appropriate table of contents, cover and back were added.

General Dynamics/Pomona was now ready for the real test. . .tailoring the appropriate model to the needs of each program. The task appeared formidable with over a dozen major active programs and every one of the four phases represented.

Each existing contract was thoroughly reviewed to identify all quality related requirements, specifications, standards, clause references, work statement line items or contract data requirement list items. The detailed review of each contract was documented on a special form called a Quality Assurance Requirements Profile (QARP). This form provided a ready reference of the contract requirements and served as a check list of the most common requirements occurring in contract provisions.

Some of the types of special contract quality requirements that frequently appear in military product contracts concern configuration control, component serialization, First Article Inspection, personnel certification in key processes, etc. . . .These requirements are generally superimposed on the requirements of MIL-

Q-9858A. Sometimes by contract language or clause, MIL-Q-9858A is altered. This approach to identification of unique Quality requirements is illustrated in Figure 6.

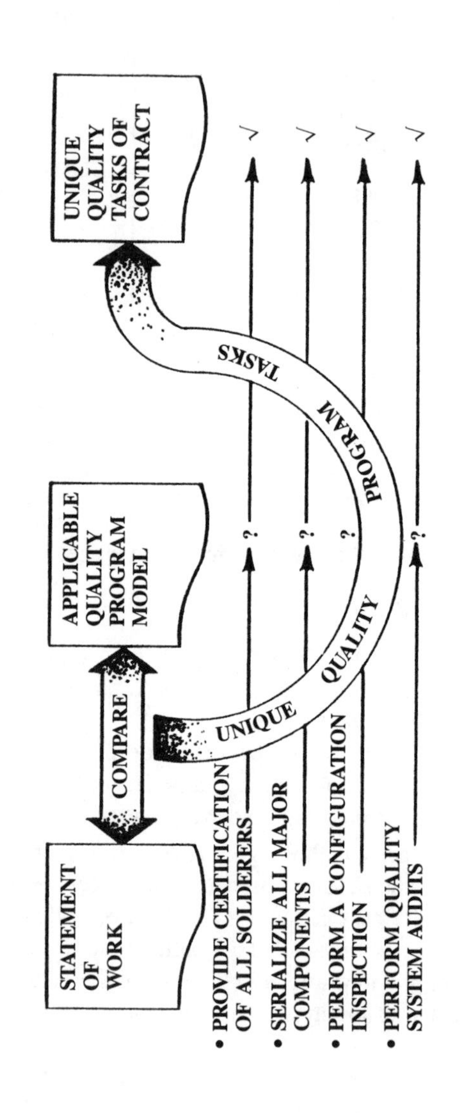

FIGURE 6. Idenfify Unique Requirements

As part of the review of the contract, the appropriate model phase was determined, such as advance development, engineering development, etc. On the surface this would appear to have been a rather simple task. Frequently, it was not. On numerous occasions the procuring government agency used a program phase term that was different in title and meaning from those commonly utilized. Also, in recent years we have acquired contracts that were phase hybrids spanning two of the classical D.o.D. phases to save development time and money. The real test of phase definition was to thoroughly assess the objectives or mission of the contract and compare them to D.o.D. 5000.1 milestones. This enabled identification of the correct phase.

Now for the tailoring task.

After the correct phase model was selected and all the contract quality requirements were identified, a sorting process was performed. Contract Quality requirements that were in addition to MIL-Q-9858A requirements were segregated. These requirements were isolated into individual subjects and were identified as "unique" Quality requirements. Each unique requirement was addressed with the same kind of implementing procedure as the MIL-Q-9858A subjects in the model's Quality Assurance Program Instruction. (See Figure 7.)

APPLY THE SAME RATIONALE TO EACH TASK BEYOND MIL-Q-9858A AS YOU DID TO ALL THE OTHERS WITHIN MIL-Q-9858A SCOPE:

1. **DESCRIBE EACH UNIQUE TASK IN WORK STATEMENT FASHION**
2. **DEVELOP THE METHOD/PROCEDURE YOU PLAN TO EMPLOY**
3. **STAY IN CONTEXT WITH THE APPROPRIATE LIFE CYCLE PHASE**
4. **USE THE SAME PROCEDURE FORMAT AS CONTAINED IN THE SELECTED MODEL**

F233683 7119

FIGURE 7. Address Unique Contract Specifics

Quality requirements that modified MIL-Q-9858A were addressed on a document known as a Quality Assurance Notice (QAN). The QAN clearly identified where and how in the model a MIL-Q-9858A subject was modified. The next step was then to change that MIL-Q-9858A Quality Program Instruction accordingly.

Once this task was completed, we revisited the Quality Requirements Matrix in the front of the model and tailored in the unique Quality requirements and their corresponding Quality Program Instruction numbers. A similar tailoring task was performed on the Quality Program Milestone Charts.

The next task involved the addition of the program name and product identity at appropriate places throughout the developing plan, including text, unique program requirements and the QANs. This last bit of tailoring was done in a manner

Result: A Quality Assurance Program Plan Designed to Phase, Tailored to Program, and Has Management Utility for the Life of the Contract

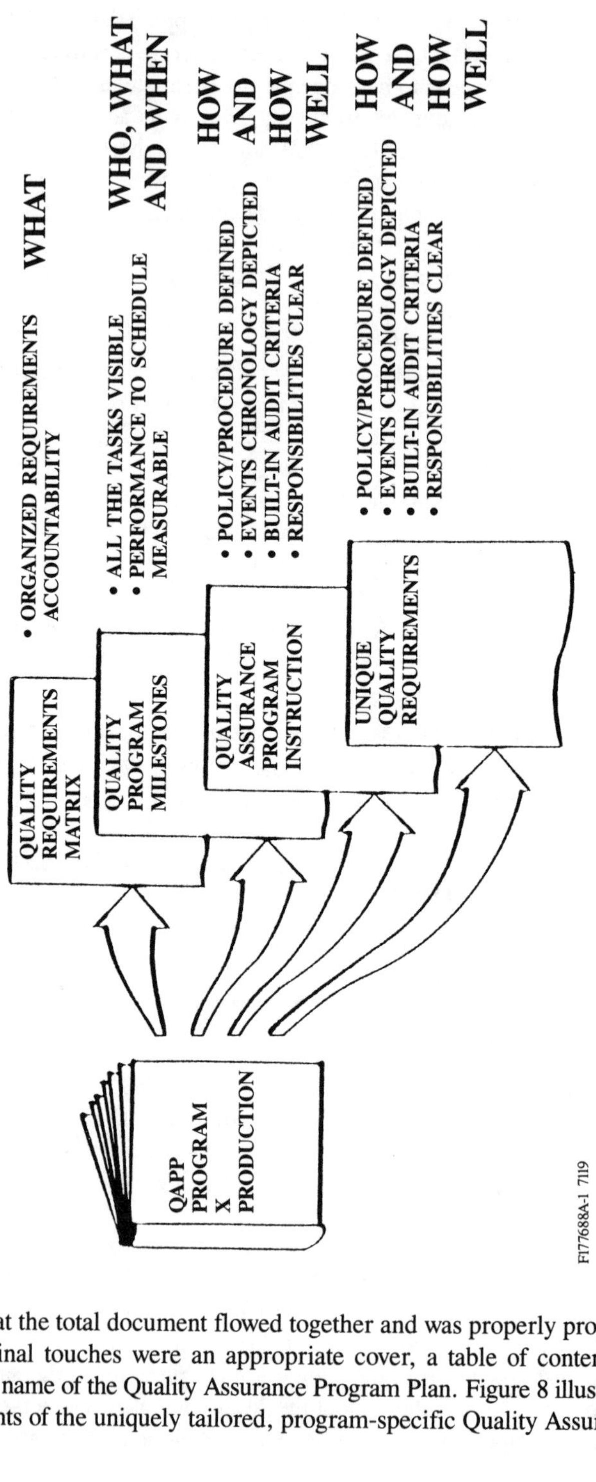

QAPP
PROGRAM
X
PRODUCTION

QUALITY REQUIREMENTS MATRIX
- ORGANIZED REQUIREMENTS ACCOUNTABILITY

WHAT

QUALITY PROGRAM MILESTONES
- ALL THE TASKS VISIBLE
- PERFORMANCE TO SCHEDULE MEASURABLE

WHO, WHAT AND WHEN

QUALITY ASSURANCE PROGRAM INSTRUCTION
- POLICY/PROCEDURE DEFINED
- EVENTS CHRONOLOGY DEPICTED
- BUILT-IN AUDIT CRITERIA
- RESPONSIBILITIES CLEAR

HOW AND HOW WELL

UNIQUE QUALITY REQUIREMENTS
- POLICY/PROCEDURE DEFINED
- EVENTS CHRONOLOGY DEPICTED
- BUILT-IN AUDIT CRITERIA
- RESPONSIBILITIES CLEAR

HOW AND HOW WELL

FI77688A-1 7119

FIGURE 8. The Completed Quality Assurance Program Plan (QAPP)

which assured that the total document flowed together and was properly program identified. The final touches were an appropriate cover, a table of contents, a foreword, and the name of the Quality Assurance Program Plan. Figure 8 illustrates the final ingredients of the uniquely tailored, program-specific Quality Assurance Program Plan.

Once the Quality Assurance Program Plan (QAPP) has been developed, it is coordinated with top management and both the procuring and regulating agencies of the customer to assure that all the requirements have been adequately addressed in a cost effective and well controlled manner. The appropriate signatures were acquired and the QAPPs were released as the top Quality program controlling management tool. Figure 9 reemphasizes the major ingredients of each QAPP.

- **HANDOUT**
- **BRIEFLY REVIEW**
- **COMMENTS**

- **DESIGNED TO D.O.D. PHASE**
- **TAILORED TO STINGER**
- **PROCEDURE TO REQUIREMENT CORRELATION**
- **MILESTONES FOR PROGRESS VISIBILITY**
- **AUDIT CRITERIA FOR COMPLIANCE ASSURANCE**
- **A VIABLE MANAGEMENT WORKING DOCUMENT**

F209387 729

FIGURE 9. Completed Quality Plan

Figure 10 further illustrates the primary role of the tailored QAPP in the Quality procedural hierarchy at General Dynamics Pomona Division.

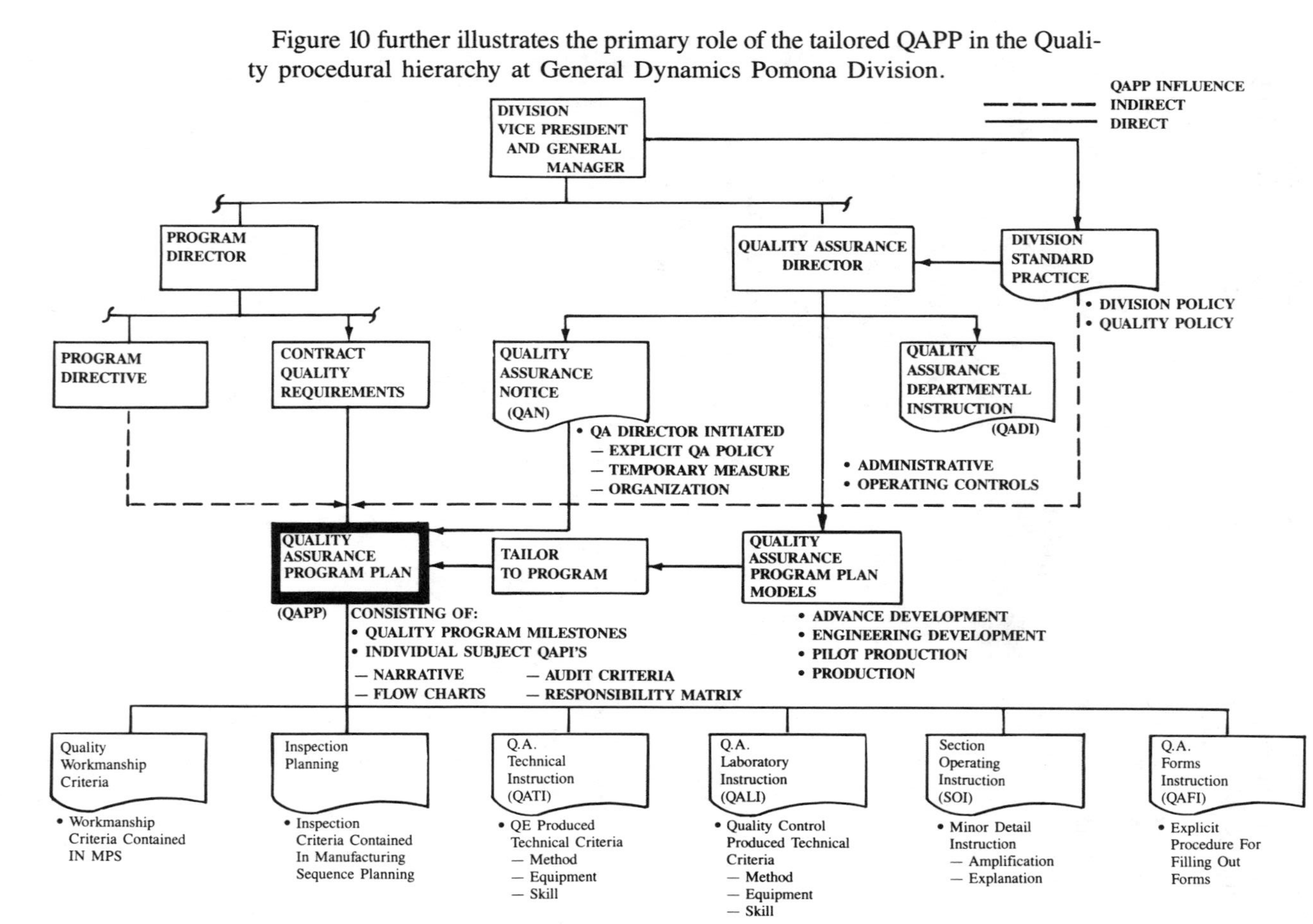

FIGURE 10. Quality Assurance Procedural Hierarchy; Organization and Documentation

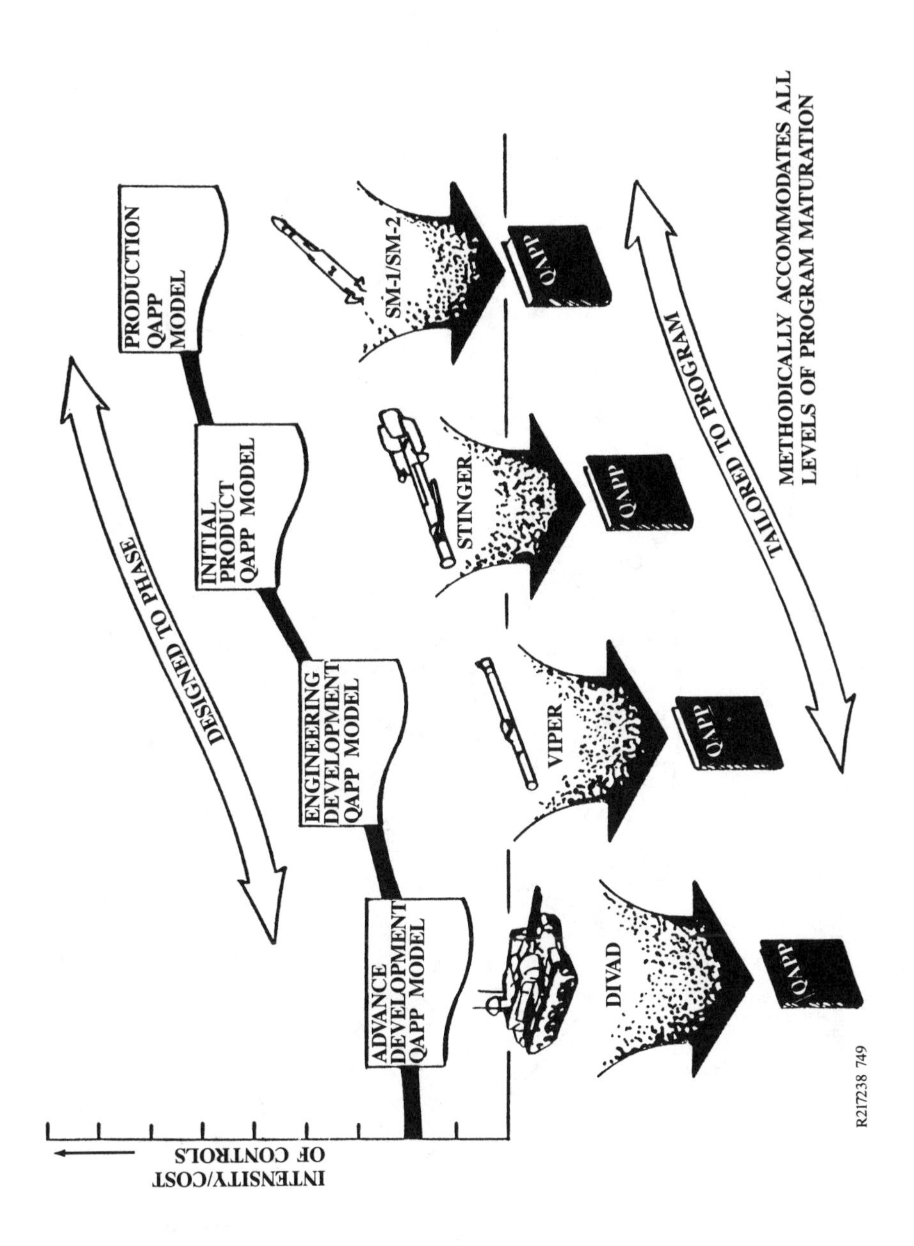

FIGURE 11. QAPP Fits the Program Needs

General Dynamics has been using this modeling and tailoring concept for two years now and in that time has realized a Quality program planning cost avoidance in excess of five hundred thousand dollars. Additionally, each product line has been provided with a Quality Assurance Program Plan that meets the specific needs of the program and its current phase of maturity.

Our Quality Program Modeling and Tailoring Technique Has Resulted In Quality Assurance Program Plans For Each Of Our Programs That Have Proven To Be Cost Effective Working Tools Of Quality Management

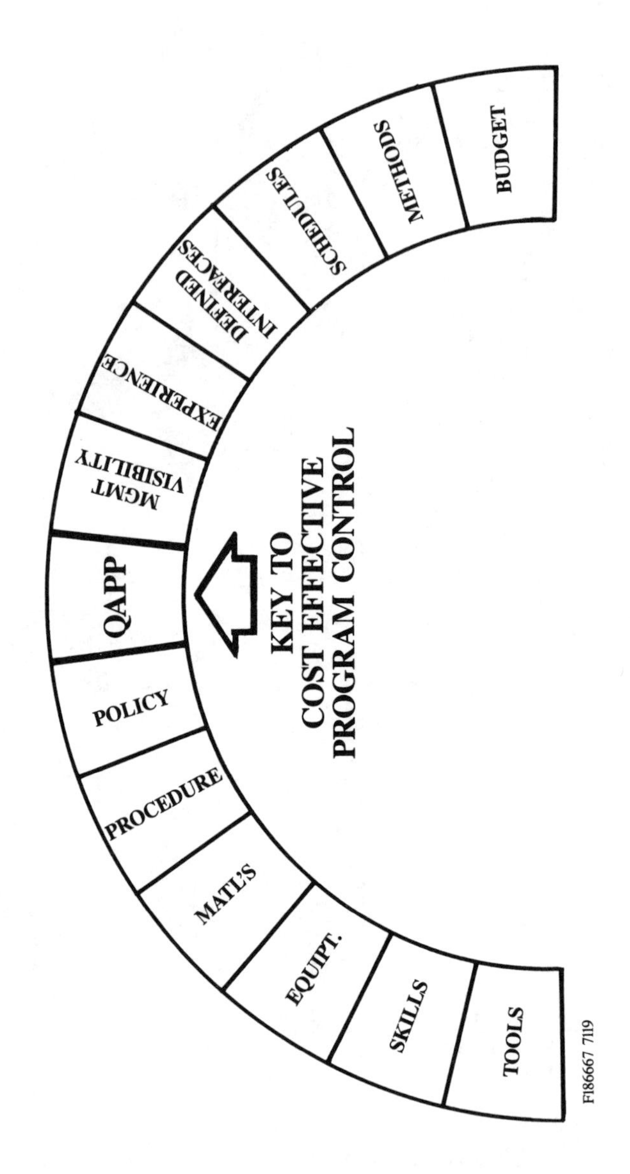

FIGURE 12. The Working QAPP is the Keystone

Figure 11 is a general overview of the total concept of phase-designed Quality program models tailored to each program's needs.

In conclusion, General Dynamics/Pomona has learned that the Quality program modeling technique is a very cost effective approach to tailoring, is methodical, satisfies the customer, and provides program management with a viable managerial working document. (See Figure 12.)

QUALITY/RELIABILITY CHALLENGES FOR THE 1980'S

T. Gurunatha, P. Eng.
Xerox Corporation

ABSTRACT

With the fast growth of industry, the gap between the customer and the designer is widening rapidly. Cost of quality is also growing proportionately. Reducing cost will be a challenge in the 1980's. This paper discusses the application of *"Ideal Design of Effective and Logical (IDEAL) Product Assurance Systems"* to this task. The present concept of Quality Assurance in many companies as a "Quality Reporting Department" does not leave much room for identifying the exact cause of problems. Responsibility for corrective action is typically spread among several departments within Design Engineering and Manufacturing. Customer satisfaction is often forgotten at this stage of the game. Quality Assurance Departments must be provided with technical expertise to overcome this problem.

The cost of quality (scrap, rework, inspection, loss of sales etc.) is running up to 15% of sales in some companies. Increasing profits and/or improving business stability in the 1980's will greatly depend on reducing this cost. In most companies there is little attention paid to controlling this cost before it happens nor is it reported correctly after it happens. Activities which contribute to such costs are spread among several departments. Since the Quality Assurance Department is the only department which communicates technically or otherwise with all levels of most departments in an organization, the challenge of monitoring reduction of this cost will fall in its jurisdiction. This activity demands technical expertise in the Q.A. Department to measure effectiveness of approaches to solving problems by other departments.

The IDEALS (Ideal Design of Effective and Logical Systems) concept (1), which is a systematic investigation of contemplated and present product assurance systems, is the easiest and most effective approach for achieving the goal of reducing quality cost. The basic rules in this approach are:

- Define the problem (base of the triangle shown in Figure 1)
- Find the theoretically ideal solution (tip of the triangle)
- Start working from the top down in the triangle instead of the traditional bottom up approach.

This concept can be applied to all problems, immaterial of their size or the type of business.

PICTORIAL PRESENTATION OF RELATIVE QUALITY COST

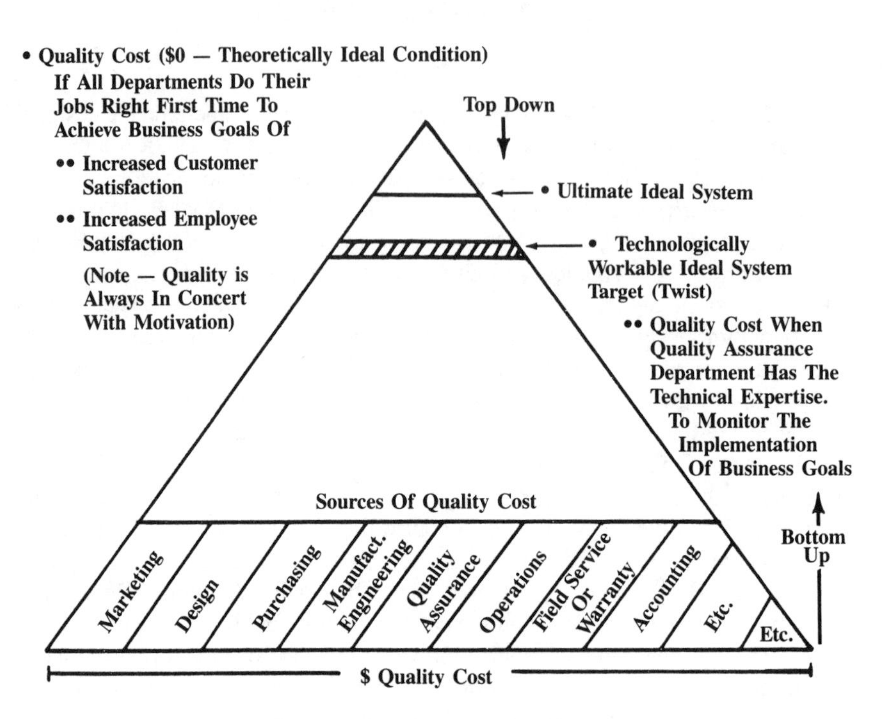

FIGURE 1

APPLICATIONS TO VARIOUS RELIABILITY/QUALITY ASSURANCE ACTIVITIES

Key places where modern technology can be adopted to reduce quality cost are:

1. Design, Development, & Testing
2. Manufacturing
   a. Receiving Inspection
   b. Component Fabrication and Assembly
   c. Vendor Surveys
3. Organization Structure and Quality Cost Reporting

1. *Design, Development, & Testing*

A new product suffers in reliability (or increase in life cycle cost) at manufacturing and in field due to lack of:

- Proper communication to the working level people through drawings (indicating level of criticality on every dimension). Such drawings are the language of engineers.
- Predicting (scientifically) manufacturability of components and/or assemblies at the Design stage.
- Probabilistic approach to tolerancing interface and/or interacting assemblies.

The above problems can be overcome by conducting design reviews by Product Assurance Groups (encompassing Reliability & Quality Assurance) to make sure everybody understands and works toward the goal of the business.

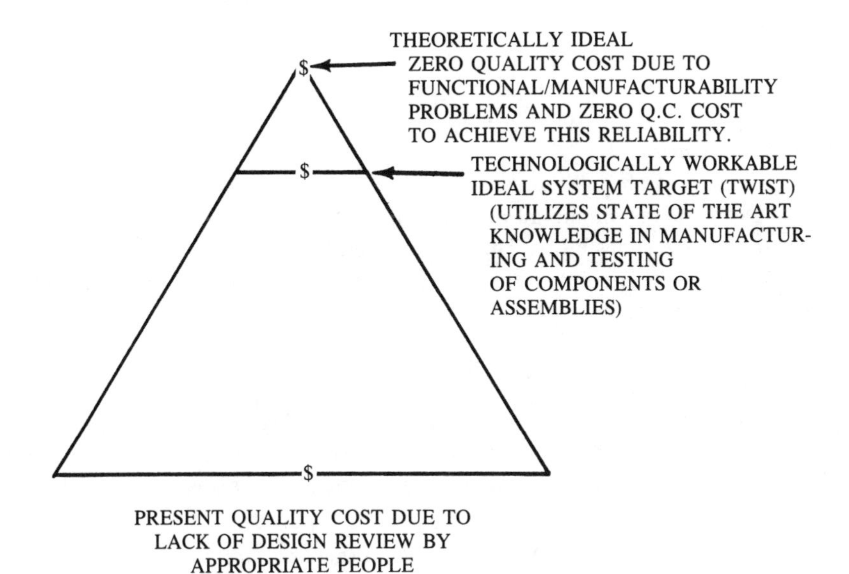

FIGURE 2

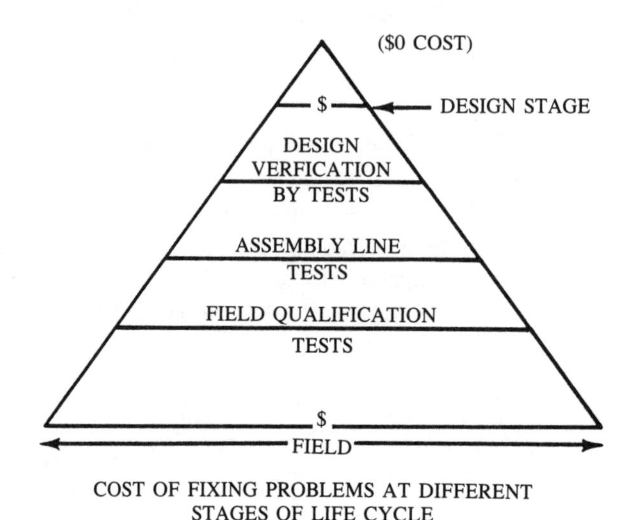

COST OF FIXING PROBLEMS AT DIFFERENT
STAGES OF LIFE CYCLE

FIGURE 3

2. *Manufacturing*

This segment of quality cost is rising rapidly in many companies due to the increased thinking among all departments that quality is the Quality Assurance Department's responsibility. This has resulted in more and more inspection and rework. It is like trimming a tree (Quality/Reliability problem) instead of pulling the tree out by the roots (exact cause of problem).

a. *Receiving Inspection*

Adaptation of single sampling plans and quoting acceptable quality levels (A.Q.L.) has resulted in training purchasing departments and suppliers to do things wrong a certain percentage of the time. This is one of the biggest contributors to increasing quality cost.

We should be looking at the Receiving Inspection function as a support function to evaluate process capability of new suppliers or items and where necessary, adopt sequential sampling and other statistical methods. (2).

b. *Component Fabrication and Assembly*

Traditional methods of measuring the Manufacturing Department's performance as meeting targets do not allow room for improvements. Measurement should be made of the *percentage below the historically lowest in-house*

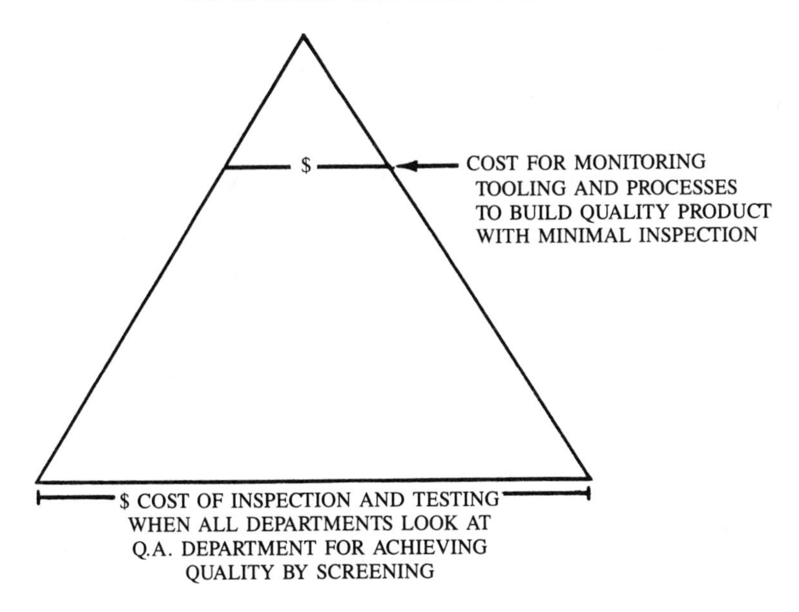

FIGURE 4

or field record (target). This helps the business goal of customer acceptability to flow right down to the working level people, and initiates utilization of modern technology.

The following measures should be taken to reduce this segment of quality cost:

- Monitor the tooling and processes in production of components or assemblies to repeatedly produce high quality products.
- Provide inspection services for training new operators, for establishing process or tooling capabilities, and for monitoring quality, instead of taking 100% inspection for granted.
- Automate end item testing to monitor all aspects of design intent.
- Apply statistics for determining the direction in which the process has shifted from established standards. This can be done much more economically now by using computers.

c. *Vendor Surveys*

Have you wondered why the price quotations for the same component differs by considerable amounts (assuming we obtain same quality product from both)? You will be surprised to note that the supplier with low quotations will be working from bottom up.

Vendor surveys should take into consideration the technical expertise available and its utilization in the past for promoting customer satisfaction. Enforced quality record keeping does not necessarily result in a quality product. The statement "Operation is successful but the patient died" should be remembered. We could have all kinds of records indicating conformance to specification but the product may not function, as the inspection or test procedure may not have been written with the required technical expertise.

3. **Organization Structure and Reporting of Quality Cost**

Traditionally, quality assurance organizations are growing bottom up (from workers to big bosses — see Figure 5). In some organizations it stops at little bosses. The lower down that the flow of accurate quality information stops, the less the recognition given to the cost of quality. In such organizations quality control is looked at as a hindrance to production. When things go wrong they will point fingers at Quality Control Department. When everything goes fine, the rest of the departments take credit.

One of the best ways to overcome this problem is to document dollars wasted due to "people not doing things right first time" and flagging the departments which are contributing most towards damaging the stability of the business. This will awaken the big bosses and the president.

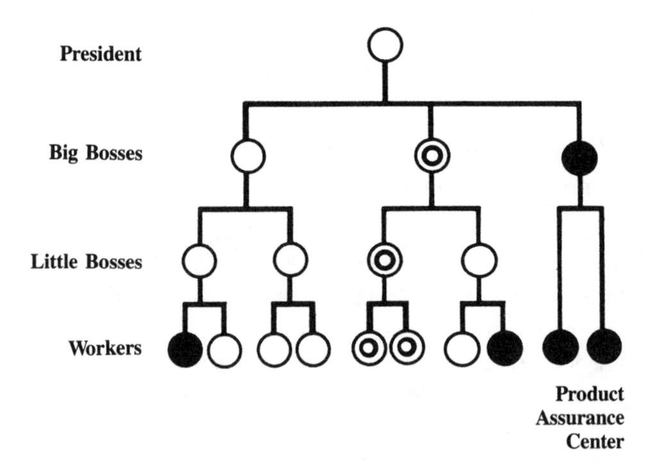

FIGURE 5

Quality cost is misinterpreted in many companies as the cost of running the Quality Assurance Department.

Quality cost is the cost of not doing things right the first time by all personnel involved in the life cycle of a product.

The purpose of reporting quality cost is:

- To measure effectiveness of management policies in reducing cost of quality to a minimum.
- To identify departments which are most effective in "doing their job right the first time" and to identify those which are not.
- To assist management in future planning of quality and reliability assurance activities.

CONCLUSION

The challenge for Quality Assurance in the 1980's will be to look at the product life cycle as a whole instead of looking at segmented activities as in the past. Responsibilities will shift from control to prevention, and the best way to get recognition of prevention activities is to document the savings in dollars. Product assurance activities will be dynamic, requiring state of the art technical expertise and accounting know-how. Just as a balance sheet gives management a picture of how well or how bad management performed after the fact, Quality Assurance can review other department performance in doing things right the first time, and enforce corrective action. This will help businesses to face the 1980's competitive market.

REFERENCES

1. Nadler, Gerald, *Work Systems Design: The Ideals Concept,* Richard D. Irwin, Inc., Homewood, Illinois, pp.22.
2. Duncan, Acheson J., *Quality Control and Industrial Statistics,* Richard D. Irwin, Inc., Homewood, Illinois, pp. 272.

BIBLIOGRAPHY

1. Anthony, Robert and Hekimian, James, *Operations Cost Control,* Richard D. Irwin, Inc., Homewood, Illinois, 1967.
2. Crosby, Philip B., *The Art of Getting Your Own Sweet Way,* McGraw-Hill Book Company, 1972.
3. Crosby, Philip B., *Quality is Free,* McGraw-Hill Book Company, 1979.
4. Wai-Chung Chow, William, *Cost Reduction in Product Design,* Van Nostrand Reinhold Company, 1978.
5. American Society for Quality Control, *Guide for Reducing Quality Costs,* Milwaukee.
6. Kapur, K. C. & Lamberson, L. R., *Reliability in Engineering Design,* John Wiley & Sons, 1977.

COMPUTER ISOLATION OF SIGNIFICANT QUALITY COSTS

J. E. Mayben, Manager, Product Assurance
General Dynamics Fort Worth Division
Fort Worth, Texas

ABSTRACT

This paper presents practical examples of how and why the computer should be utilized for identification and isolation of significant quality costs. The examples illustrate why it is good management to apply the Pareto principle for priorization of significant cost impacts and reallocation of resources to effect positive reduction of these costs. The fundamental elements of the examples are used to develop characteristics of a computerized management system which will provide rapid isolation of adverse quality cost trends before they exceed control limits. Methods and techniques for rapid resolution of potential cost problems thus identified are set forth and discussed in detail.

INTRODUCTION

Quality costs are those specifically associated with the achievement of product quality wherein product quality is defined as conformance to all product requirements specified by product specifications and drawings. Historically, these quality costs have been categorized into prevention costs, appraisal costs, internal failure costs, and external failure costs. Of prime importance is the fact that quality costs can be used as a measure of the effectiveness of the Quality Assurance (QA) system in each phase of product life.

The philosophy of quality cost management is to reduce failure costs by attacking the problems causing these costs, reducing appraisal costs in accordance with the results achieved, and investing in prevention activities to the extent necessary to achieve maximum overall cost effectiveness. This philosophy is based on the premise that for each failure there is an assignable cause; causes are discoverable and preventable; and prevention is more economical than correction.

Accomplishments are achieved through the principle of "cause and effect." As failures are revealed through appraisal actions, they can be examined for cause and eliminated through corrective action. The further into the life of a product a failure is discovered, the more expensive it is to correct. As failure costs are reduced, appraisal efforts can then also be reduced in statistically sound increments. The experience of this improvement can be applied through prevention activities to all new work.

By utilizing the concept of quality cost as a measure of QA effectiveness, QA principles must be applied throughout all phases of product life to minimize the cost impact due to poor product quality. The ultimate usefulness of quality costs will be achieved when they routinely become an integral part of the strategic planning of a corporation. Progress can then be seen as a measure of quality costs, over time, against a base of sales or total project costs as incurred.

TECHNIQUE FOR COMPUTER UTILIZATION

The application of the computer for acquisition and analysis of quality and product data has provided an excellent means to accurately and rapidly assess quality costs. By pre-programming of threshold and statistically sound control limits, QA management is able to isolate significant quality costs. This provides the basis for reallocation of resources for these priority areas while the computer maintains a constant vigil for other quality costs which might exceed pre-set limits.

As opposed to manual techniques of selected process studies or "the squeaking wheel" concept, the computer is used to monitor product quality levels at selected points in manufacture, test, and field use. This provides a real time basis for early identification and resolution of problems on a cost effective basis over the entire product line of a company or a division. Simply stated, the computer is used to apply the Pareto principle for isolation of significant quality cost areas in order to provide rapid corrective/preventive action in these areas. Correctly applied and utilized, this technique prevents "surprises" at top management levels and places QA in the "partner" role as opposed to the classical "policeman" role.

Three examples wherein the computer is used to isolate significant quality costs, and the resultant actions, will be discussed in depth. Although overlaps in the various categories of quality costs will exist in each example, generally the first example is directed at prevention costs, the second at appraisal costs, and the third at failure costs.

EXAMPLE 1. PRODUCTION START UP ON A MAJOR WEAPON SYSTEM

Classically, the production acquisition cost of a major weapon system by Department of Defense exceeds the desired or targeted costs allocated for the system. Analysis of the actual cost history shows that the difference between the targeted versus the actual costs is due, in large part, to the results of unsolved, repeat problems. The key, then, to reduction of production start up costs with subsequent savings on all production of the weapon system, is to rapidly identify and solve the repeating problems.

The discussion in this example will be restricted to the assembly task of one major structural component of a weapon system. The program was aimed at isolating significant problems and getting these problems resolved as soon as possible. These problem areas were restricted to those within QA responsibility for correction and further, to those areas which could be readily quantified using the QA data base. Figure 1 shows the learning curve for a weapon system which was used as the "baseline" in the comparison study of this program. The learning curve shows the direct labor man-hours broken down by industrial engineering into specific contributing parts.

The four areas which could be readily quantified from the QA data base for computer analysis by Quality Engineering were insufficient or inferior parts, inspection pickups (inspection findings), completing work of prior shops, and improved design of assembly and methods. These four areas accounted for 38% of the first ship cost on the baseline weapon system structural component while they represented 43.5% on the computer analyzed weapon system structural component. Further, 41.5% of the 43.5% was documented on material review action data forms which will be referred to as "MRB Rework Hours" in the remainder of this example.

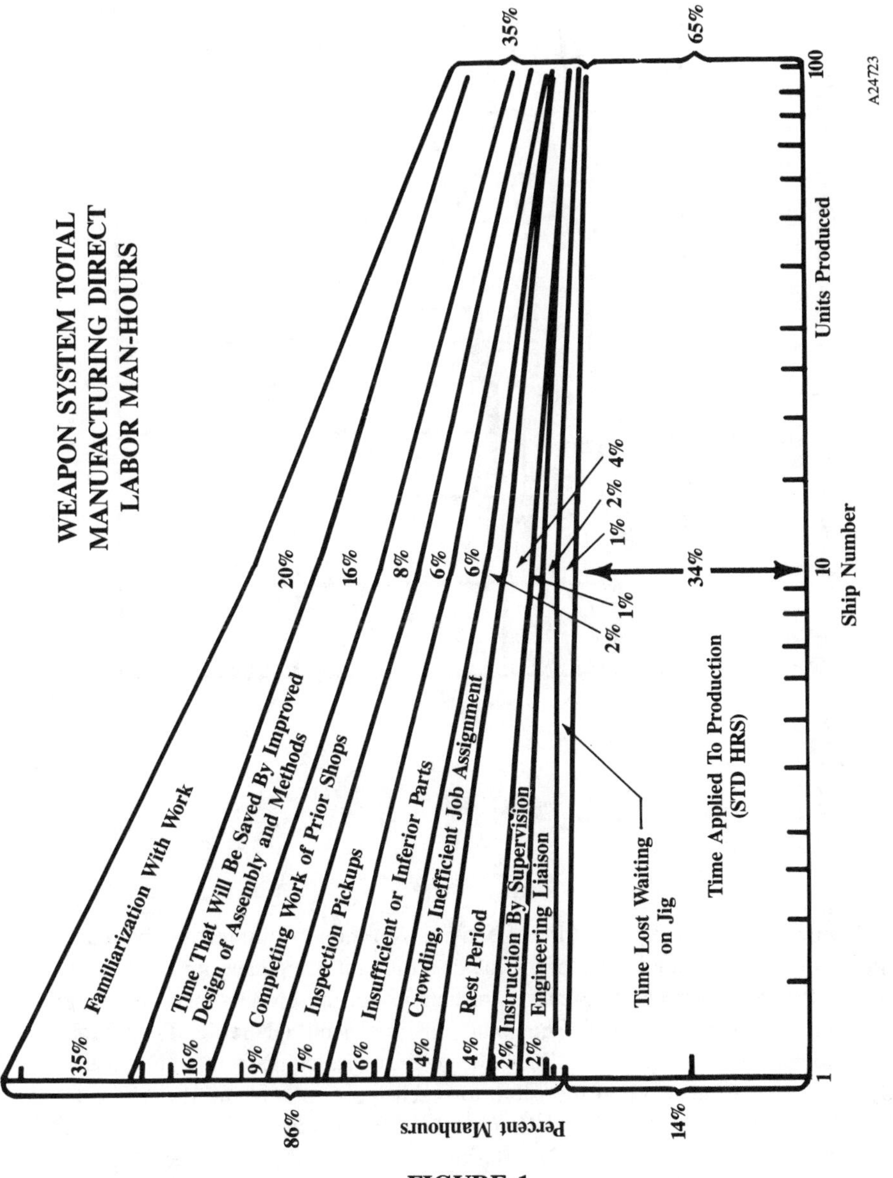

FIGURE 1

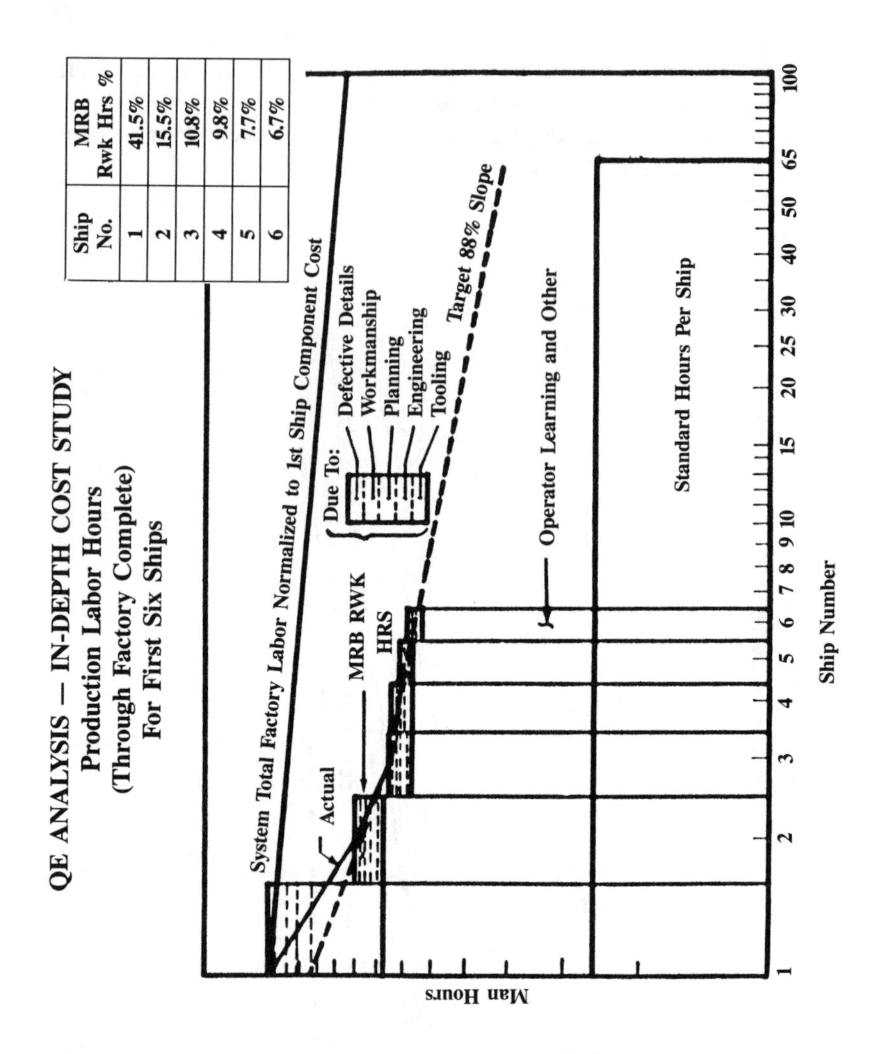

FIGURE 2

Figure 2 shows the results of the effectiveness of rapidly eliminating the repeat problems in the assembly task of the major structural component for the first six ships. It is a typical learning curve presentation of the production labor hours through factory completion. The computer analysis identified the significant problems, determined their contribution to the cost for each component, and, basically, which department was responsible for each proportion of the rework costs. The inset table shows that the MRB rework hours decreased from 41.5 percent on No. 1 Ship to 6.7 percent on No. 6 Ship. Also, when the component learning curve slope is compared to the slope of the baseline system learning curve, normalized to first ship cost, it is readily apparent that the cost reduction is due to the effectiveness of the QA system. Had the problems been allowed to repeat, the baseline system learning curve would have been approximately duplicated while building the new component.

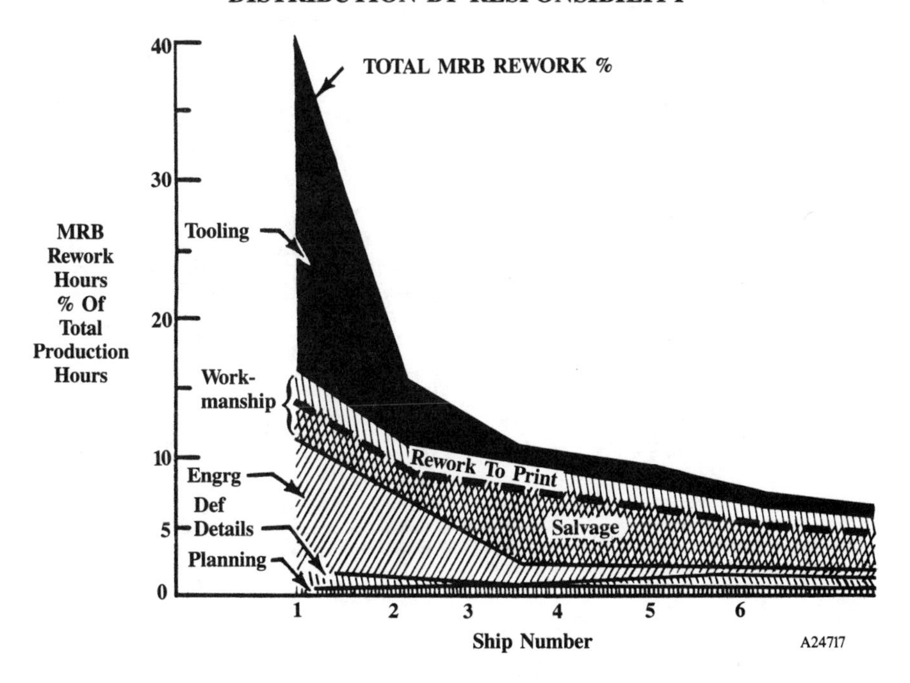

FIGURE 3

Additional analyses were performed by Quality Engineering on the MRB rework hours in terms of percent distribution by responsible department. As shown by Figure 3, Tooling was the greatest contributor, followed by Engineering, Shop (workmanship), Defective Details, and Planning. These problems were rapidly corrected by Ship No. 6 so that the only significant area was that of workmanship.

These rapid results were achieved by Quality Engineering utilizing the system shown in Figure 4.

SIGNIFICANT AND REPETITIVE PROBLEM ANALYSIS

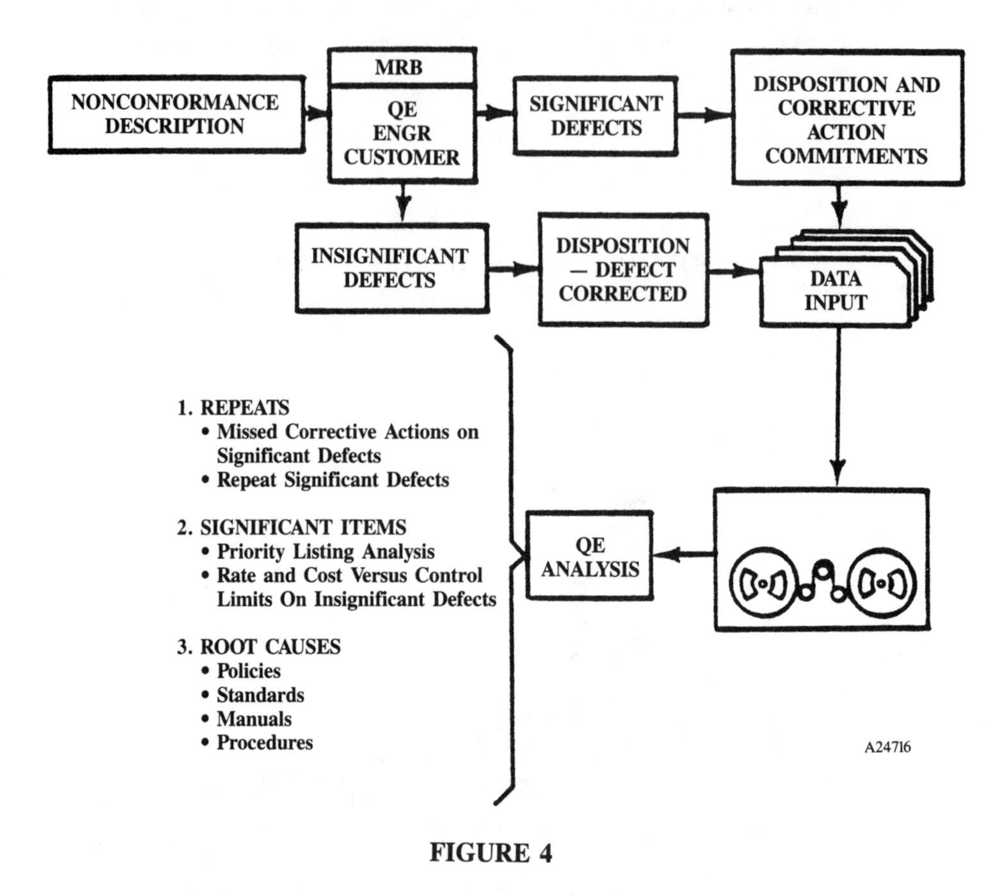

FIGURE 4

SIGNIFICANT AND REPETITIVE PROBLEM ANALYSIS

Nonconformances were presented to the MRB where the significant defects were segregated from the insignificant. In both cases, the hardware was corrected and the history of the nonconformance entered into the computer. Utilizing selected computer programs based on discrete coding of input QA data, the Quality Engineer used the computer for various analyses. The data was reviewed to find if any corrective actions were missed on ineffective or significant defects and whether significant defects repeated. Also the "significant" tabulation was reviewed along with an analysis of the priority listing of problems. In addition, the rate and cost quantities were compared against control limits on insignificant defects. As the final step in the analysis, a determination was made into what caused the defects or nonconformances; whether policies, standards, manuals, or procedures were inadequate; and what changes were required to prevent future recurrences of the problem on the current and next program.

One of the computer analysis routines available to the Quality Engineer was a search mode by significant quality costs by defect code. This revealed that the primary MRB problem was holes which were mislocated/omitted, which accounted for almost 33% of the rework costs and defects on the first ship. Further computer analysis showed that tooling was responsible for 61% of the rework hours due to holes mislocated/omitted, and that 86% of these hours were caused by mislocated pilot holes in detail parts.

The cause of the problem in the detail parts was that the tolerance buildup at assembly mislocated pre-drilled pilot holes in some instances and both mating parts had pilot holes in other instances. In the Quality Engineering analysis, the initial corrective action was to fix the various detail part tools to correctly position pilot holes and to correct planning to eliminate duplicate pilot holes in mating parts. A review of tooling and planning standards and manuals revealed that no published guidelines existed for pilot hole utilization and callout. Final preventive action consisted of guidelines for tool planners and tool project engineers regarding pilot hole use.

Of prime significance is the fact that classical part number analysis in all probability would not have disclosed this "family" or "system" type problem. The computer program enabled summary data analysis at multi-selected layers of various matrices of part numbers, defect codes, rework hours, and departments.

The primary Engineering problem was "wrong fastener called out," which was the fourth-ranked MRB problem in terms of rework costs. The causes along with the initial corrective action and final corrective action are shown in Figure 5. Without the computer, manual data searches to isolate these significant quality cost problems and to effect root cause corrrective/preventive action probably would have been cost prohibitive.

EXAMPLE 2. APPRAISAL COSTS OPTIMIZATION FOR ELECTRONIC EQUIPMENT FABRICATION

The quality costs associated with defective electronic parts and components are well-known throughout industry and the using military branches. Many good things have been done in the last several years to improve parts quality and end item reliability through better production techniques and widely applied environmental screening. However, latent manufacturing defects, poorly screened or non-screened parts, and part substitutions continue to expose electronics manufacturers to recurring problems. These problems are usually spread over a large quantity of part numbers and unless end item screening is effectively performed, occur after delivery during field use. This, of course, is the most expensive place for failure to occur. Various studies have indicated that the cost due to failures during the life cycle can be bracketed as shown in Table I.

End item manufacturers and acquisition agencies are often shortsighted by procuring non-screened or commercial parts which result in high failure costs. Often, only a portion of this money is needed to offset the higher procurement costs to effect significant savings during manufacture and use of the end item. Discussion in this example cites such a case as experienced during the initial production of various Line Replaceable Units (LRUs) of the Stores Management Set of a major weapon system.

NO. 4 MRB PROBLEM — WRONG FASTENER CALLED OUT
Present And Future Actions

CAUSE:
(1) Design and Draftperson Errors
(2) Unfamiliar With Design
(3) Insufficient Time To Check Out Thoroughly

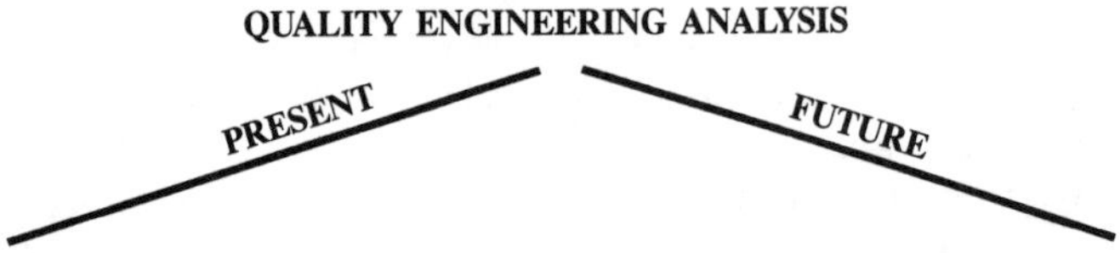

QUALITY ENGINEERING ANALYSIS

PRESENT

CORRECTIVE ACTION
1. Corrected Drawings

FUTURE

CORRECTIVE ACTION
1. Develop checklist for drafters and drawing checkers. Train job shoppers in design reqts.
2. Evaluate adequacy of existing procedures on design reviews as relates to production and quality participation.
3. Trade study to see if first ship is best place to catch these types errors.
4. Verify actions and input "memory" of computer for next program.

FIGURE 5

TABLE I

FAILURE COSTS DURING ELECTRONICS LIFE CYCLE

Failure Point	Estimated Failure Cost Range
Parts Screening on Receipt	\$ 1.00 to \$ 25.00
PCB Level	\$ 50.00 to \$ 100.00
LRU End Item, Burn-In & ATP	\$100.00 to \$ 200.00
Field Use	\$250.00 to \$2500.00

Even though parts reportedly screened per MIL-M-38510 were procured and used, component failure rates (principally ICs) were being experienced at the 8-15% levels causing 100% rework of PCBs which failed during acceptance testing. (Part population was 100 to 200 ICs per PCB.) Although some IC failures were chronic by part number and board location created by design considerations, analysis indicated that the parts generally were not being screened, or were being improperly screened, by the parts manufacturers. Failure losses during end item manufacture were appraised at about $1,000,000 per year.

Using computer analysis routines based on the in-process QA data base, studies were conducted which revealed that re-screening the parts upon receipt coupled with Destructive Physical Analysis (DPA) by lot and Failure Analysis (FA) of removed parts, the in-process failure rate could be reduced to one percent or less. In order to accomplish these tasks, both facilities and manpower had to be acquired and allocated, which was reflected in increased appraisal costs. The cost savings analyses are shown in Table II, and reveal that payback would occur within a two-year period and effect an annual savings of over $400,000 subsequent to payback. Thus by spending approximately 75¢ per device in appraisal costs, not only was the liability of failure costs converted into capital assets, monies saved beyond this were also reflected in higher profits.

TABLE II
PAYBACK ANALYSIS FOR PARTS SCREENING & FAILURE ANALYSIS

Item	Present Method	Proposed Method	Savings
IC Tester	$ 660,000	$367,000	$293,000
DPA & FA Labs	$ 523,000	$370,000	$153,000
Total, First Year	$1,183,000	$737,000	$446,000
Second Year	–	–	$480,000
Third Year	–	–	$410,000

It was stated earlier that one element of the philosophy of quality cost management is to reduce appraisal costs in accordance with results achieved. In the example under discussion, it was not sound economics to continue to pay for double screening of the parts, and obviously it was not sound economics to implicitly trust the part suppliers to perform MIL-SPEC screening. What was needed, and subsequently developed, was a continuous computerized system for monitoring part failures from the in-process QA data to isolate significant quality costs to effect specific corrective action and to adjust receiving and/or source controls as required.

A Failure Analysis Control System (FACS) was developed specifically for electronic parts removed during the PCB and LRU manufacture and test. The system provides for monitoring and trending all parts which are removed and replaced to correct a test failure during processing and test. FACS determines the parts which must be analyzed and provides a closed loop system to assure timely corrective action to prevent recurrence. The system provides a priority listing by part number of removed parts with the part having the highest quality cost impact listed first, as shown in Table III.

TABLE III

PIECE PART REMOVAL RATES SUMMARY FOR ELECTRONICS FAB DEPARTMENT (FOR PERIOD MAR 80 THROUGH AUG 80, UNBURDENED COSTS)

Rank	Estimated Removal Costs	Removal Rank (%)	Part No. & Nomenclature
1	$34,806	4.5	M6106/21-001 Relay
2	33,857	9.3	JANTX2N3868, MIL-S-19500/350 Transistor
3	21,127	2.5	54LS251/88313 (Sub. for M38510 IC)
4	19,366	7.9	C4915-1, Microcircuit
5	19,096	2.5	M39016/15-005L Relay
6	17,064	1.7	C8728-1 Microcircuit
7	15,845	3.6	M6106/24-001 Relay
8	12,053	15.1	C8899-1 D/A Converter
9	11,647	1.9	C8857-1 E Prom
10	11,241	1.6	M39016/15-004L Relay
~	~	~	~
44	406	0.8	C10725 Microcircuit
45	271	0.3	M38510/30109BEB Microcircuit

FACS personnel require failure analysis to be conducted on each part on the priority listing to effect positive corrective action. Significant failures at specific board locations often point up design deficiencies and the need for additional derating or selection of a higher reliability part. Trend charts, as shown in Figure 6, are also provided by the computer to monitor the priority parts on the "hit parade" to determine effectiveness of actions taken and to identify the need for any additional actions.

In addition to the FACS, the computer is utilized to monitor each PCB performance level as well as burn-in and final test results of LRUs. Computer programs based on discrete coding are used to provide priority problem areas for in-depth analysis by Quality Engineering. Corrective action is then taken based on economic trade studies which take into account the repetitive nature of each problem. Even though corrective action may be deferred, the hardware is always reworked to drawing and specifications.

TYPICAL PIECE PART REMOVAL TREND CHART

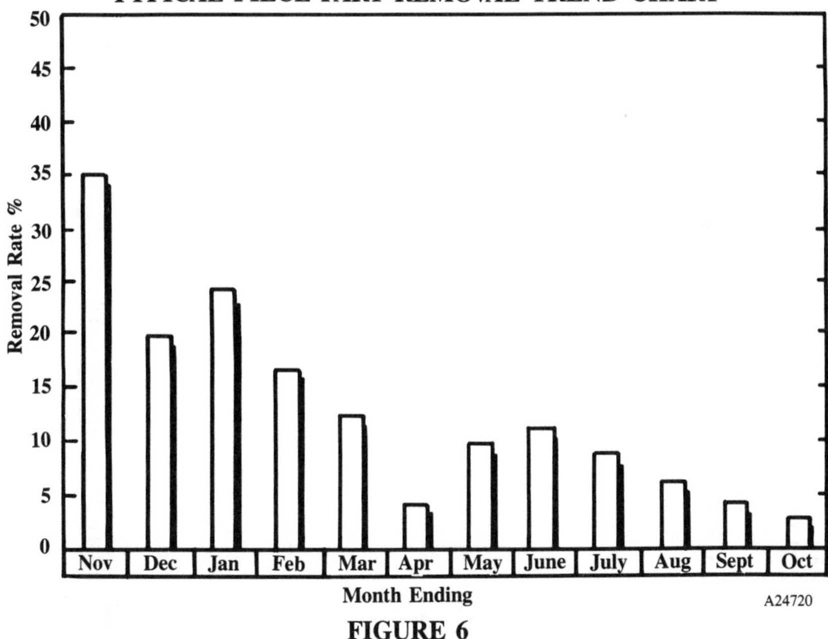

FIGURE 6

EXAMPLE 3. ISOLATION OF SIGNIFICANT QUALITY COSTS DURING WEAPON SYSTEM CHECKOUT

The discussion in this example will be limited to the failures within the functional systems and components which occur during the installations, operations, and delivery of a major weapon system, specifically a jet fighter aircraft.

The components in and of themselves are complex and incorporate the latest state-of-the-art technology, which complicates the task of isolating significant problem areas. This task is further complicated by the fact that many systems "talk" to each other and are heavily dependent upon imbedded software. In addition, most of the avionic systems contain built-in test (BIT) and self-test capability. Since each component undergoes ground operations as well as flight operations by both company and customer pilots, system "funnies" can become elusive and Could Not Duplicates (CNDs) can account for a large share of component removals.

Compounding the total picture is the fact that equipment removals are directly affected by maintenance and troubleshooting policy, adequacy of test instructions, thoroughness of the factory Automatic Test Equipment (ATE), and the proficiency levels of the test operators. Aircraft systems operations and checkout begins progressively as early as 75 days before delivery with the last 30 days dedicated to the totally integrated weapon system which can mask developing problems. Often the pressures of meeting scheduled delivery dates and expedited equipment installations due to shortages also contribute to equipment removals and system malfunctions.

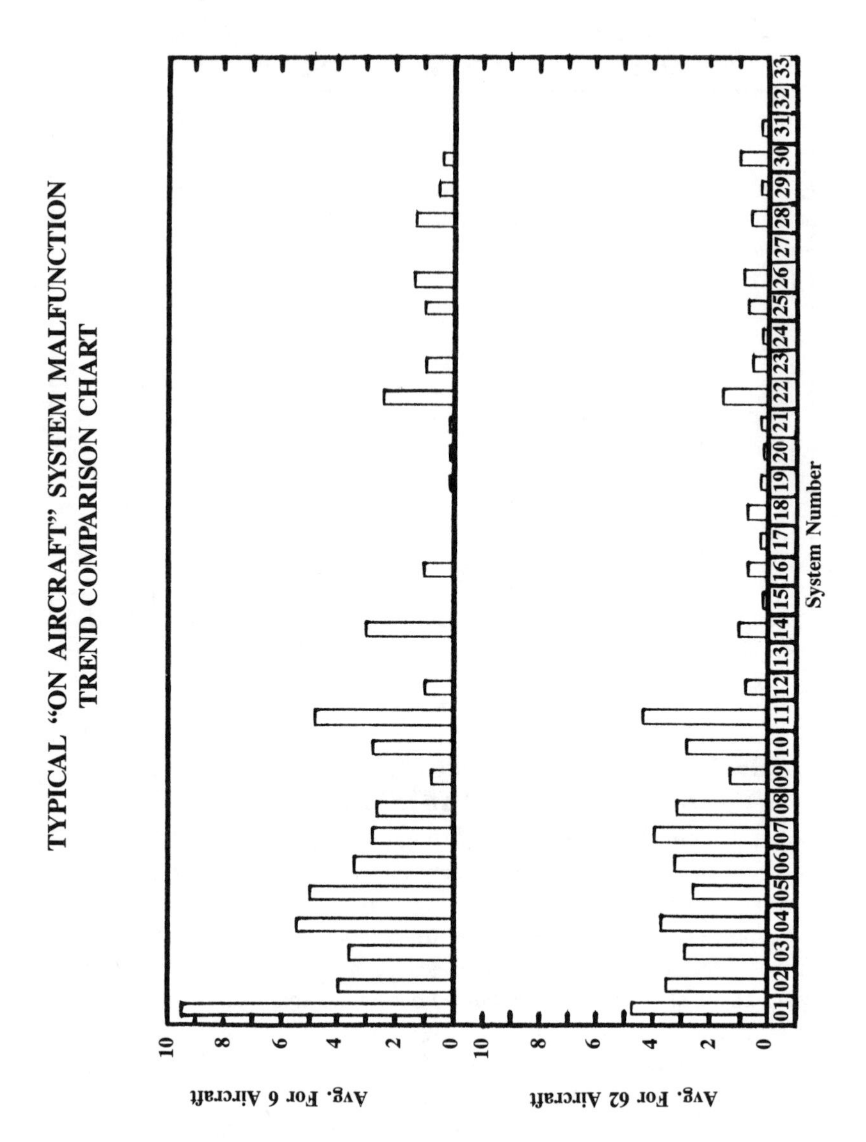

FIGURE 7

Based on the above discussion, it can be seen that a cadre of well-trained Systems Quality Engineers is needed to isolate significant problems and to determine and correct root causes of malfunctions. Resolution efforts of many problems extend over a long period of time and often require a combination of various actions from suppliers, design, and production. Systems technical knowledge along with effective interpersonal skills on the part of the Systems Quality Engineer are necessary to achieve problem resolution as rapidly as possible. Utilization of specialized computer programs enables quality engineering to identify equipment which has exceeded pre-set control limits for investigation and corrective action as necessary.

The total task consisted of utilizing the computer to monitor the performance levels of 600 functional components per aircraft from initial installation and operation through delivery to the customer. At a delivery rate of 20 aircraft per month, the computer had to track and report on 12,000 specific equipments each month. The significant quality cost drivers selected consisted of (1) on aircraft adjustments, and (2) equipment removals. The computer program was structured to report at the Work Unit Code (WUC) system level and by part number within the WUC system.

Reporting of the on aircraft adjustments was established on a bi-weekly basis, wherein the trends for the last six aircraft processed were compared to the baseline of the preceding 62 aircraft for each WUC system. A typical system trend chart is shown in Figure 7.

As can be seen, priority systems are readily identified in terms of highest quantity of malfunctions on specific systems and, also, in terms of those systems which are degrading in terms of increasing quantities compared to the baseline. Systems Quality Engineering resources are allocated to the priority systems for determination of needed actions. Additional computer programs based on discrete coding are available to Quality Engineering to single out and isolate the problems within the WUC system. Specific actions can then be defined, implemented, and tracked via the trend charts for effectiveness.

"Equipment removals" was identified as the primary quality cost driver at the weapon system level. The removal of a piece of equipment at the weapon system level not only dictates replacement, retest, and in a number of cases reflight, at the aircraft level, it also requires that the removed item be retested, repaired, failure analyzed, etc., to the extent that it can be returned to a serviceable condition. Removals also demand a ready bank of spare assets so that the weapon system may continue to completion. This further drives up the total costs. It is obvious that the earlier a problem is recognized and solved, the smaller the total cost impact of the problem will be.

Equipment removals are reported monthly based on removals per aircraft. The removal rate for each part numbered equipment is reported for the current month, the past month, and also over the average of the last 66 aircraft. The computer prioritizes significant equipment removal rates based on the current month's data as shown in Table IV, with the highest rate being listed first. The computer only lists equipment with removal rates in excess of 10%. Data analysis has shown that this listing of equipment is about 15% of the total equipments being monitored and they account for 80% of the total removals and rework costs associated with the removals.

TABLE IV
F-16 Equipment Removal Rate Summary
GD Fort Worth Production Operations

REMOVAL RATE*				
66	THIS	LAST	WUC	Equipment Nomenclature
.57	.35	.30	42GAA	A/C NICD Storage Battery
.48	.35	.30	75DCO	Central Interface Unit
.46	.17	.32	74EAO	Indicator Unit
.45	.25	.10	74BAO	Pilot Display Unit
.45	.20	.15	14AAO	Flight Control Computer
.36	.40	.30	14FBO	Electronic Component Assy
.35	.19	.15	42JBA	A/C Inverter Storage Battery
∫	∫	∫	∫	∫
.13	.10	0.00	23OOO	Turbofan Power Plant
.13	.10	.10	14ADO	Flight Control Panel Assy
.12	0.00	.10	41AAA	Bleed Air Reg Valve (13 Stage)
.11	.05	.13	44AAH	Navigation Inlet Light
.11	.06	.11	14BBO	HT/Flap Servoactuator Assy
.10	.10	.15	76DAO	Control Panel-Chaff/Flare
.10	.05	0.00	74ACO	Radar Transmitter

*"66" — last 66 aircraft; "THIS" — this month;
"LAST" — last month.

This listing is addressed by Systems Quality Engineering on a 100% basis for needed actions to reduce the removal rate as quickly as possible. *Early* production removal rates were in excess of 2.00 per aircraft for some equipments. As shown in Table IV, the *current* highest removal rate is .57. In addition to the computerized listing shown in Table IV, the computer also provides removal rate trend charts on each listed equipment which provides Systems Quality Engineering and QA Management a method to determine effectiveness of actions taken as well as any adverse changes that need to be addressed. A typical example is shown in Figure 8.

With the computerized program monitoring and reporting equipment removals and notifying Systems Quality Engineering of the need for early action, the Quality Engineer can request additional data, failure analyses, and monitor production line operations as necessary to assure that the problem is correctly and accurately defined prior to a crisis. The cause can then be isolated along with a determination of the actions necessary to correct the problem and prevent recurrence. A monthly summary report for the top ten to fifteen equipments is provided to upper management to assure priority support of the corrective actions being taken.

The net annual quality costs savings due to reduced equipment removal rates during 1980 were conservatively estimated at $580,000 for the domestic production line of the aircraft. This represents removal/replacement cost at the aircraft level only and does not include any component retest/rework/analysis costs. It is believed that this significant accomplishment could not have been achieved without the use of the computer.

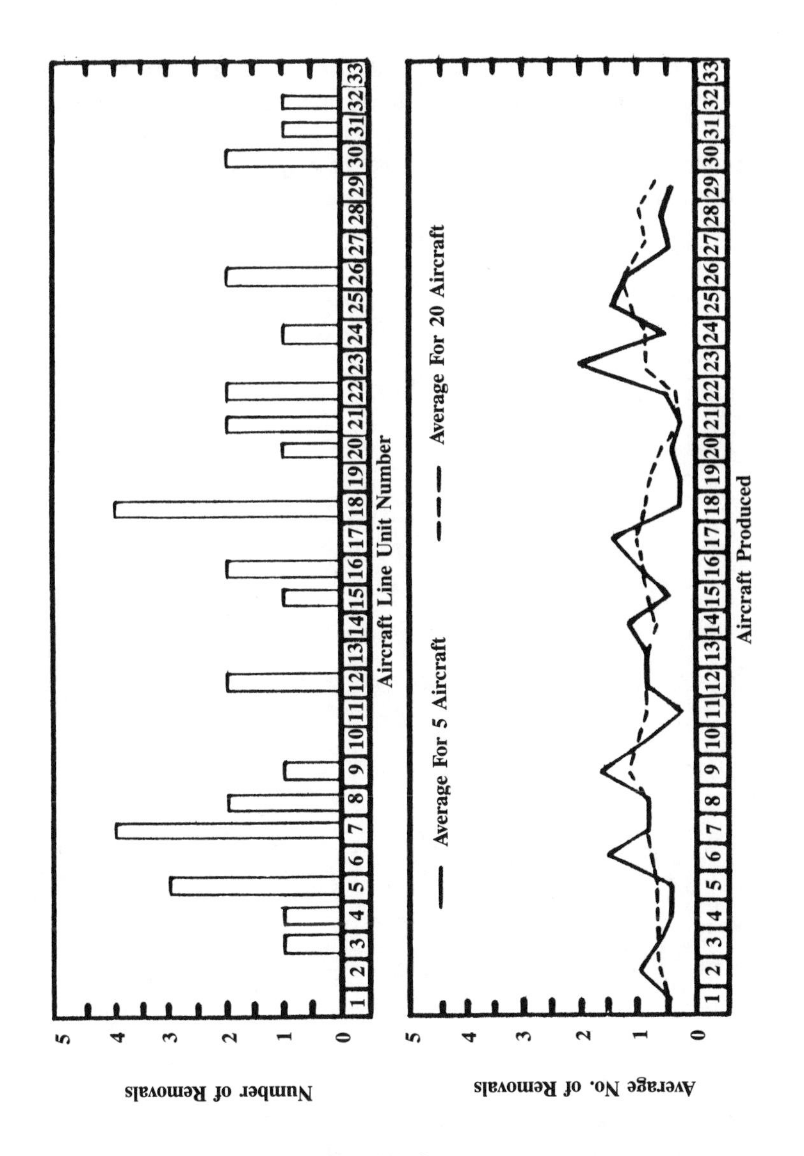

FIGURE 8

CONCLUSION

Although significant advancements have been made in the utilization of computerized techniques for the isolation of significant quality costs, the use of the computer for routine in-depth analysis of technical production problems is still limited in many high-technology, low-rate production systems, such as in the aerospace industry. However, voice data input methods, coupled with better coding techniques, may greatly enhance the development of specific computer programs needed for workmanship and equipment problem analysis. This will essentially free the Quality Engineer from manually analyzing and classifying defects and malfunction data allowing him to allocate more time to priority problems. The major breakthrough for QA management in the use of the computer is that, once programmed, significant problems cannot be suppressed, though they may be temporarily ignored.

HOW TO EFFECTIVELY IMPLEMENT A QUALITY COST SYSTEM

Dr. Alvin O. Gunneson
Vice President & Director — Quality, Worldwide
Revlon, Inc., Route 27, Edison, New Jersey

ABSTRACT

Most quality professionals are thoroughly familiar with the theory and practice of quality costs, yet many have not been able to implement a truly effective cost of quality program. This paper will deal with the day-to-day trials and tribulations of putting in a cost of quality system. The paper will discuss the research which is necessary before a program is conceived, the information and presentation required to sell the concept, and the introductions which must be conducted with subordinates, peers and superiors. The paper will discuss the reactions which can be expected as the presentations are made, the questions which are likely to be asked, and the most appropriate answers to provide.

COST OF QUALITY CONCEPTS

Philosophy

The most practical measurement of all is plain old-fashioned money. Defect levels and percent defects fail to get anyone excited or stagger anyone's imagination. Substantial dollar figures identified and available to be converted to profit will get the attention of the most complacent manager. It has long been said that if you do not measure something, you do not control it, much less reduce it. Few managers say they are not interested in reducing error costs, yet most companies in the world do not have a cost of quality measurement program. Most companies do not have cost of quality programs for several reasons. Many believe that a cost of quality program will involve extensive accounting systems and additional people on the payroll. Other people believe that their present cost accounting system is sufficient to identify the areas that need improvement and others are simply too busy fighting problems of the day to adequately consider a program that could be used as a tool to prevent some of their problems.

It should be understood that a cost of quality program is an excellent tool in the management of a business. A cost of quality program can provide a measure of the overall management health of a business in terms of quality. It could provide, in single number, a measure of the error-related activity of all departments within the corporation. A cost of quality program should, therefore, be the very first part of any quality improvement activity. When the total cost of quality number is presented, it can readily be seen in dollar terms how much improvement is possible, and it further provides management with an indicator of where they are, so they may receive recognition for the improvements that they make.

Besides being an overall management indicator of the quality effectiveness of a company, cost of quality numbers provide priorities for mustering corrective action in needed areas. They provide justification in dollar terms for expenditures which may be necessary for equipment, additional people or training programs. Perhaps most importantly, they provide recognition to line management people for their contribution to profit. They have previously been able to report reductions in defect levels. Now they can report dollar savings which is of great concern to those who run the businesses of the world.

An incorrect philosophy is for companies to delay the implementation of a cost of quality program until quality improvements have been made in order to report low cost of quality numbers. This generally serves only to delay the program because the required improvements are seldom made and if they are made, inadequate recognition is received. The correct philosophy should be to post the numbers first. Make them as high as possible to stimulate maximum improvement activity. A high cost of quality will generate maximum participation from the executives which will result in maximum quality reductions and maximum recognition.

The cost of quality program in the company should always be presented as a positive one. The program can easily be construed as a negative force exposing the high degree of waste, error, and expenditures which would be unnecessary in a company well-managed for quality. For this reason, care should be taken to inform management that this is a tool for improving operations and gaining recognition for them. No one should ever be chastised for the initial high cost of quality. Improvement should be a requirement. Only when the cost of quality numbers fail to improve should there be any management pressure brought to bear on any individual or department. Of course when the cost of quality improves, recognition should be commensurate.

Do not let anyone tell you about the economics of quality. Let them tell you how much money it costs to do their jobs now. Then show them how much less it could cost to do things correctly. The economics of quality will then be very clear. It is always cheaper to do the job correctly the first time. If we do not know how much it is costing us for the portion of our work that we are doing incorrectly, it is very unlikely that we will have much initiative to improve.

Perhaps the most important philosophy of all is that a cost of quality program should be simple and it should be practical. It should be modified to suit the existing accounting procedures in a company as much as possible. When initially launching a cost of quality program, one must take the utmost care in not becoming bogged down in the details and in the trivia. A new cost of quality program should not attempt to identify all costs beyond the theoretically perfect company. Rather, it should concentrate on the readily available quality cost contributors. Sufficient detail should be used so that a measurement base will be valid for two or three years. After that point, the program can be reevaluated and additional, more complex and more detailed measurements can be brought in to readjust the base. Most companies will have plenty to do in reducing the costs that will be exhibited by running a basic operations quality cost system.

Quality Costs

There are many descriptions offered for the cost of quality. The following descriptions are presented to facilitate understanding the wide scope of activities which may be chargeable to quality costs.

The quality cost of a company is the difference between the actual cost of making and selling products and the reduced cost, if there was no possibility of failure of the products or errors of the people in the manufacture, sale, or use of these products.

These costs may be segregated into three categories as follows:

- Appraisal Costs — The costs of inspecting, testing, and analyzing products to determine their conformance to requirements.
- Failure Costs — The costs resulting from the failure of products to conform to the requirements during their manufacture or use.
- Prevention Costs — The costs incurred in an effort to reduce failure costs and, consequently, appraisal costs.

The preceding analysis and definition restrict quality costs directly associated with the products of a company.

However, in principle, restricting quality costs in this manner could be considered parochial. Quality costs should be construed as any cost relating to mistakes, defects, and failures, made by anyone in the company, which hamper the operation of that company. Now we have expanded accountability for error costs from the production areas into the white collar offices where the more far-reaching and costly errors are committed.

A wider definition should be, "The quality cost of a company is the difference between the actual operating cost of a company and the operating cost if there were no failures in its products and systems, no mistakes by its staff, and no possibility of failure or mistake."

For most companies, this wider definition would result in quality costs much higher than the more restricted definition, and much higher values of such ratios as quality cost to sales. A wider definition is important to indicate the much larger possibilities for the improvement available to a company by application of quality control principles than would appear from the restricted definition. It is expected that cost of quality measurement will begin in the manufacturing areas; but it can be seen that its use as a management tool is virtually unlimited.

In summary, because of product defects and employee error, there is the need for product appraisal with resultant failure and appraisal costs. An effort to reduce failure and appraisal costs may require expenditures for prevention. The result is that the company's actual costs in manufacturing and selling are higher than they would be if there were no errors, and if products did not fail inspection or tests during manufacture or use by the customer. The actual cost beyond the theoretical cost of the perfect company is the cost of quality, reported in prevention appraisal and failure cost categories.

Responsibilities

If a cost of quality program is to be successful, the responsibilities described herein must be followed to the letter. Any other approach is very likely to produce a mediocre program at best.

Instruction to begin a cost of quality program in a plant must be given by the chief executive officer of the facility or the corporation. The responsibility for the program and the requirements of the program must come from the controller. Quality Department and other people may serve as technical advisors, motivators and general consultants to the program. Quality Department people may also do a considerable amount of the training and development of cost-gathering procedures, but this activity should all be in ghostwriter form as the cost of quality program must be perceived as a normal accounting function of the corporation. The Quality Department people will serve to keep the reporting honest. They will provide technical advice and will spearhead the corrective action required to reduce the cost of quality. Each plant controller, therefore, will be responsible for the preparation and the submission of the cost of quality report. The accounting departments will provide the cost data and participate in the tabulation as required. The controller and the quality manager should jointly prepare an internal procedure to define the exact department inputs and the responsibilities for reporting costs which are not accumulated through existing accounting systems. Where detailed costs are impractical to report, estimating and proportioning are acceptable if the estimating guidelines are established jointly by the controller and the quality manager.

Cost Accounting

Care should be taken when beginning a cost of quality program to assure that all of the readily available error categories are included in the report. Sufficient time should be taken to establish a reporting baseline which need not be changed for two or three years. Conversely, care should also be taken to assure that the implementation of the program does not get bogged down with an enormous number of minor costs. Minor costs measured precisely are difficult and expensive to identify and will result in little additional savings because at this point everyone will be busy reducing the high-cost items.

A method of determining the measurement base could be for the quality manager and the controller to first study the standard categories and elements of quality costs. They should identify those costs which apply to their business and discard those which do not. They should also add any cost elements peculiar to their business. This preliminary list of quality costs could then be the subject of a meeting among the department heads to establish what the final measurement base will be.

As mentioned earlier, the method of gathering the costs may be determined by the individual plants as a part, or extension, of existing cost identification and reporting systems. In areas where detailed costing would be impractical for any plant, estimating or averaging on a consistent basis is acceptable. It should be noted on the form that the amount is an estimate. While estimating is acceptable, it must not be excessive and the operation must work toward reducing the amount of estimation by establishing standard systems and procedures for measuring and reporting the cost of quality.

In some instances, the existing cost accumulation procedures already provide the specific data required. Where existing, this data can be supplied by the controller's department. In other cases, percentages of existing department labor totals can be apportioned to quality costs and certain categories of quality cost can be segregated from existing cost accumulation categories that include nonquality as well as quality costs. The controller, quality manager and department heads involved must jointly develop the apportionment, segregation, and estimating methods appropriate for each situation.

The standard reporting categories of prevention, appraisal, and failure costs should be considered fixed for reporting uniformity. The line accounts within these categories may be modified as applicable to individual plants. Some line accounts may be combined and reported in a consolidated number. Some plants may have cost areas not listed in a standard cost of quality procedure which represent error costs for them. These costs should be inserted in the proper cost categories and reported. When developing the accounting procedure, remember, the intent of the cost of quality measurement and reporting system is to identify error costs so that they may be controlled and reduced. If a given task is performed because someone or something did not perform correctly, that task is a cost of quality.

For multiplant operations, the cost of quality reports may be consolidated on a regional or group level before being forwarded to the corporate controller's office. Individual plant reports should be attached to the consolidated report.

Launching the Program

The two most difficult aspects of launching a cost of quality program is to obtain enthusiastic participation from plant management in implementing the program and to determine which operating costs will be measured and reported.

The latter simply takes some skull sessions wading through operating costs. The former takes a presentation to management which shows them how they may convert error costs to profit in a practical manner. Practical generally means without addition to staff and without bogging existing staff down with complex additional tasks. If there must be additions to staff or a substantial increase in accounting systems, the payback must be established clearly for the first year.

What follows is not a schematic to success which must be followed to the letter. It is, rather, a series of steps and thoughts which have proven successful if implemented in accordance with the culture and business style of a given corporation.

It is suggested that these concepts be used as a basis to tailor one's own ideas and techniques compatible with one's own personality, management style, and culture of the business.

Determining the Need

The first step in launching a cost of quality program is to determine the need for the program. This must be presented to senior management in a manner which will justify the effort and which will interest them in participating in the program. If the need is not presented well, the program will be perceived as another make work program and it will not be successful. If enthusiasm cannot be generated for the program, it should be postponed and presented at a later date with another strategy.

Sufficient information for effectively presenting the need for a cost of quality program can usually be gathered in any business within a day or two. While gathering the cost data for the presentation, one need not, and indeed should not, detail all the quality costs in the business. The complexity of detailed data will complicate selling the program. One should simply gather the major costs, being careful to be obviously conservative so that no one can question the integrity of the numbers. This means having mental notes of quality costs not included in the numbers being presented in the event a discussion ensues. This eliminates any possibility of being accused of overstating the numbers. It also serves to further excite management about additional cost reduction possibilities beyond those substantial opportunities being presented.

The research for the management presentation should begin by gathering all of the obvious failure costs. Included here should be the cost of scrap, rework, and nonstandard labor. The cost of purchased material rejects, warranty payments, handling customer complaints and the cost of storing and handling defective, excess, and obsolete material. These costs are usually sufficient to justify a cost of quality program without having to include other detailed and apportioned costs. For appraisal costs, take the cost of the quality department and any production inspection and tests costs. Prevention costs will be more difficult to locate because very likely there is little prevention activity being performed. Once this data is compiled, you have the basis for making a presentation to the senior staff.

The quality manager should spend time with accounting, the controller, and other cost gathering and reporting groups within the corporation to learn what financial data is presently available and what reporting is being done.

One may be surprised to learn that most of the costs are available requiring perhaps only some change in format. Remember, it is important to not request changes in accounting procedures to facilitate cost of quality reporting unless it is absolutely necessary.

From this information, a brief outline of how the costs may be gathered and reported should be developed. The purpose of doing this is to enable the quality manager to be conversant with how this program can be implemented practically and simply without creating enormous amounts of work for other people. The greatest deterent to implementing a cost of quality program is the fear that it will create additional work for the presently overworked and understaffed departments.

Conditioning Middle Management

Now that quality cost information has been gathered for presentation to the senior staff, it is suggested that some work be done with middle management before the presentation. This is advised to prevent operation's resistance to the program. This is particularly important if the senior staff asks them their opinion. It is always dangerous to leave this base uncovered because others asking for middle management's opinion probably will not present the merits of a cost of quality program well.

To win middle and peer manager's desire for the program, some concepts will need to be developed to inform them of the benefits of participating in the program. It is particularly important to show how the program will help make their work easier and help them gain recognition for their efforts. It should be recognized that consciously or subconsciously, management has a fear of anything that will measure waste in their departments. This includes a great many general managers and division presidents as well. Management may also be concerned about reporting a high cost of quality from the point of hurting individual pride all the way to fear of reprisals from corporate, the board of directors or government and consumer groups. They may be concerned about these numbers from legal viewpoint also. The person seeking to implement a cost of quality program must be aware that these fears exist and should address them to prevent any anxieties from making implementation more difficult.

Capitalize on the desire of management to convert expenses to profit with improved product quality as a bonus. Effective cost saving and operations improvement programs are fundamental to good management. Managers using this program effectively will look good at salary and operations review time. Managers are all working for that much coveted promotion and salary increase. Arrange your thoughts and comments to show how implementing bold, aggressive, prominent programs are a mark of a good manager on the way up to bigger and better things. Discuss how cost of quality can show, in dollar terms, that managers are, indeed, successfully converting waste to profit with the resultant improvement in product quality and sales. It is natural to be concerned about a high error or scrap rate being posted for all to see. The positive aspects of the program and the potential for recognition will overcome this fear if the presentation is correct.

To generate added interest, department managers could be told that for the first time they will know, in dollar terms, how much of their cost is caused by other departments. Operation managers will be able to determine now how much the engineering design changes are affecting their operation; how much the purchasing department is impacting operations with substandard purchased material; and how much maintenance is impacting them by equipment which breaks down through improper maintenance. Conversely, department heads supporting operations could be told that they will know, in dollar terms, how much their activities are or are not impacting the operations. Traditionally, people get blamed for more of the operating problems than they really cause. Further, having a measure of this sort will provide information to have employees improve and facilitate reporting this improvement at year-end. It will be a vehicle to clearly show contribution to the profit of the corporation.

In summary, the cost of quality numbers will give management three major operating tools. First, it will give them priorities, in dollar terms, on where to direct their efforts for corrective action. We all know that a standard product with a high percent defective, may not be nearly as costly as an exotic product with a low percent defective. Second, the cost of quality will provide the justification in dollar terms for additional expenditure, additional equipment, additional training, and so on. Third, and perhaps most importantly to the individual, cost of quality will enable the line people to show their contribution to quality improvement in dollar terms. Earlier, management was able to say, defect levels or scrap levels were reduced by 1%, 2%, or 3%. Now they will be able to say $50,000, or whatever, was contributed to profit this month. That statement will provide considerably more impact at merit increase negotiation time.

These concepts are not only for the initial introduction to management, but are to be used continually, over and over again, like a broken record, to everyone who will listen. Eventually, even the greatest skeptic will believe in the value of the cost of quality program and begin to advocate it like it was his or her idea. At that point, you know you have arrived.

Developing the Presentation

Having brought these points to the attention of management within the plant, we must deal with the senior management, the general manager, the controller, the group executives, corporate executives, and so on. It is essential that when talking with these people, you must be able to visualize the business from as close to their perspective as possible. Learn to talk their language if possible. Take care to make statements to which the only reasonable answer can be a yes. If there is a possibility of them saying no or disagreeing, the question could be rephrased. Your presentation statements and questions should build a very solid foundation in terms of building within them a proper mental attitude for the acceptance of the cost of quality program proposal. One should always be careful to understand or learn the personal preferences, needs, fears, and prejudices of the key people one must deal with. One should always try to understand the business from the other person's point of view. Most managers come to meetings already burdened with heavy tasks, hardships, apprehensions, and desires.

Before finalizing the presentation, find a quiet place where you can think for a while and put yourself in their position, individually. What does each individual want to hear, what are their greatest needs, what will most help them enhance the things they are striving for? Put their needs first and yours secondary. Understand they really don't care what you want. What they want will get their attention.

Executives must show continued growth in their operation, continued growth in sales and continued growth in profit. They must show they are improving operations. They will not be interested in things which require spending money. If expenditures are required, it would be best to save those until they have had a chance to experience some success with the cost of quality. It would be wise to indicate that initial savings could be realized with the existing people on the staff. Show that there is no need at this time, to buy sophisticated, fancy equipment or to increase the staff or to do anything else which is expensive. It should be said that this is a program where we can convert an enormous amount of waste to profit simply by getting people who are already on the payroll to do what they are being paid to do — their jobs correctly the first time. Once profits are realized from this effort, management can, at their discretion, reinvest a portion for additional improvement.

Presentation statements to the senior staff could include the following general philosophies: We are all tasked with improving costs and reducing waste. No one could disagree with that. In order to control waste and to reduce waste, we must know how much it is and be able to measure it. At this point, there is no overall measurement system which tells us what our total cost of error is. No one can argue these statements. A prudent manager must consider it embarrassing that he doesn't know what these costs are. Now you really have their attention because no one wants to be embarrassed. If we only measure quality concepts and present them in various expense categories, they will improve, simply because we've given the costs visibility. Managers are competitive. They like to compete with themselves and with others. They like to see improvement. Now, if you go further and question waste and insist on improvement, your people will strive to comply. The conditions will improve while cost of quality becomes less. All it takes for the first year or two is a few minutes of the general manager's time each day or week asking about the progress, demanding improvement, and recognizing achievement.

Management Presentation

Once you have these concepts and the cost accounting system straight in your mind and are very conversant with them, you are ready to begin your sale to management. The big question is where do you begin.

The first step is to go to the general manager and tell him that you have been thinking of ways to prevent defects, ways to convert expenses to profit or if more appropriate, ways to reduce scrap costs, rework costs, delay costs, and general errors. State that you have taken time to learn and understand the cost of quality system that the ASQC and other advanced organizations have been successful with. State that after carefully studying this, you believe that it would be a very beneficial tool for your corporation if it is implemented into the structure of the operation without disruption and with maximum results of minimum effort. The general manager can readily be expected to say, tell me more. At this point, ask if a half hour is available now; if not, reschedule the discussion. This is the most important presentation, and it must have the general manager's full attention. When you receive his undivided attention, the knowledge you gathered earlier becomes essential. Explain that the cost of quality includes the total costs of error in the operation, the cost of inspecting and evaluating and appraising a product to determine how much error there is, and the cost of the actions required to prevent error. Explain that most plants without a cost of quality program exhibit relatively high error costs when they start. Because they have this error, they need a substantial number of inspectors and appraisers to prevent the errors from getting into the product and the field. Explain that companies normally show very little expenditure in the prevention of defects area. The purpose of this program is to reduce failure costs and subsequently appraisal costs by taking preventive measures. As failure costs diminish, there will be less need for inspection, then appraisal costs can also be reduced. There need not be an initial expenditure for prevention. The initial failure cost reductions can be achieved simply through getting management to do what they are being paid to do; train the workers; provide procedures; calibrate their equipment; give proper instructions and enforce the rules. Savings from this alone will be enormous. A fraction of those savings can be put towards defect prevention for more sophisticated equipment, more sophisticated training, and better systems. Failure costs will continue to tumble at a rate substantially higher than the investment in prevention.

You could continue that you have taken the liberty of tabulating some of the basic costs of quality in the plant. Be sure to point out that your numbers are very conservative and that there are many categories that haven't been considered. Then you should list the costs you have tabulated. The total of those costs should be divided by total manufacturing costs. This will usually show that quality costs are 20-25% of manufacturing costs. This will first produce shock, then enormous interest. The interest will increase further when you state that companies can, and do, operate with a 2%-5% cost of quality as the result of running an effective cost of quality system.

The presentation should close with assurances that you have mentioned this program to the controller and the department heads and they agree the program is available. It should also be stated that many of the costs to be monitored are available from the standard accounting system and what is not available is easy to gather and makes sense to measure as good business practice.

Implementing the Program

If the groundwork was done properly and the presentation done well, the chief executive will direct that a cost of quality program be launched in the facility. He will charge the controller with the responsibility to head the effort and to do the reporting. All others will be directed to participate in this worthwhile effort.

The next step will be for the controller and the quality manager to review the outline of the quality cost system developed for the presentation. They should review the costs which are available under the present accounting system and decide where additions to the existing system will be needed to enable reporting all quality costs on a regular basis. It might be wise to spend more time with the controller discussing the costs that should be reported to be sure that everything which is practical to measure has been considered.

The next step should be to call a cost of quality meeting with all the department heads in the facility. This meeting will require at least two hours and should be devoted to training and brainstorming. The first part of the meeting should be run jointly by the quality manager and the controller. This should be dedicated to informing them about the cost of quality requirements and the value of the program to them and to the company. This will not be an unfamiliar topic because the quality manager will have briefed each person individually during the gathering of the data and while preparing the management presentation. The audience should be receptive at this point. Relating the senior staff's enthusiasm about the program should only serve to evaluate their desire to participate.

The next part of the meeting should be dedicated to teaching the operating heads more about the subject of quality costs. The presentation given to senior management could be conducted again for this group with more detail inserted to assist them in understanding the role that they will play in gathering the data and using it to improve operations, costs and quality.

The last section of the meeting should be a review of the quality cost reporting procedure outline to make it most practical for everyone involved. It will not be unusual to drop some costs and add others as a result of the group decision. The most important part of the meeting will be to have unanimous agreement on which costs will be reported, an understanding that this is an important management tool, and an enthusiastic desire to produce results with the program.

The brainstorming session should be conducted at the same meeting that quality costs are explained and discussed. Feelings toward the program at this first meeting should be running high and it will be a productive time to solicit information.

After everyone has agreed on the costs to be reported and the method of reporting, the cost of quality procedure can be finalized.

The final step in implementing the program should involve some training sessions for management on the elements of quality costs. There should be a case study developed where trainees will be required to complete a cost of quality form from the data presented in the case. This will verify that they understood the lecture material regarding the segregation of costs into the prevention, appraisal and failure categories. The exercise will also reinforce the learning.

After the procedure has been finalized and the people trained, the program can be launched. Reporting could be unofficial, with limited distribution for two or three months to work out any difficulties with the system.

"OUR ONLY OUTPUT IS INFORMATION"

R. W. Stalcup
Chief — Quality Planning and Analysis
Otis Engineering Corporation

ABSTRACT

The principal function of any manager, at any level, of any organization, is to make decisions — logical, rational decisions based on information. Quality Data Systems and Quality Cost Systems collect a lot of information, which, in itself, is worse than useless to a manager. Our systems must translate such information into a communication to the manager which will motivate him into an action decision. In other words, the Quality Cost Reporting System must develop that output, to each level of manager, which is vital information to his decision making.

INTRODUCTION

Basic to any business enterprise is the concept that accounting records must be kept and analyzed so that the owner will know the financial condition of the enterprise. As the business grows in size and complexity, the need for financial analysis to aid in the managerial process increases rapidly. One of the primary areas where this need exists, and unfortunately, is often overlooked, is in the management of the quality system in the business. Most quality professionals will agree that a Quality Cost System is a major control for the management of any company. Problems arise when top management fails to recognize the need; when the existing accounting system does not track the quality costs; or when we attempt to implement this important control. This paper will attempt to present a solution to the last problem in particular and give you ideas about the first two.

QUALITY COST SYSTEM AS A CONTROL

One reason for the problems encountered in the development and implementation of Quality Cost Systems stems from the fact that we do not approach the task as the development of a system control, but as something else, such as a cost reduction generator, a hammer to use on production management, or as a departmental status builder. While those things will probably result from a properly designed system, they are a fringe benefit and not a primary consideration in system design.

We design for control so that concerned managers can make the necessary decisions about quality operations and the costs of quality. In essence, this means that control is exercised through decisions based on the information received. This, of course, translates to mean that the output of a control is information which requires a decision by a manager. Hence, the title of this paper, "Our Only Output Is Information."

In viewing the output of the Quality Cost System as a control, it becomes apparent that the input of data and the way we process and present it will influence the decision making process. The message to be output, or the communication which requires a decision, i.e. action, on the part of a manager thus becomes the guiding force in the system design. The decisions that need to be made to control the organization thus need to be identified and the system then designed to produce the proper information for each required decision unit or control area. This infers that we need a listing of decision questions. Some samples are listed later as an aid in formulating these questions.

DECISION QUESTION PROCEDURE

If we are to achieve our purpose of designing a system control, then we need to examine this area very closely. There are several approaches that may be taken. The following is one that seems to be both effective and comprehensive.

Make a chart for each level of management in the company. Some find it more effective to start at the top and work down, others start at lowest level of inquiry and work up through the levels. Try whatever seems best for your situation. On the left side start a listing of the kinds of decisions relative to quality that are a normal function at that level, i.e. what trends, exceptions, variations, activities, etc., are of interest at this level. To the right of each listing, list the information, i.e. the specific cost and other relevant data, that is necessary for that decision. To the right of that show how the information should be combined to provide the decision for action. Continue the process until complete. Now go through the process of formulation of the decision questions for each management level. For management reporting consider the vital, not the insignificant. When the decision questions are complete, take the chart and the questions to the manager(s) concerned and go over it with them. Also show them the charts for the levels above and below the one that concerns them. Seek further inputs by asking; (1) if he had all the shown information regularly and readily available, would it be beneficial, (2) what other information would be needed for each decision, and (3) what other decisions does he make. If there are too many managers in the company, then select one or more from each level of each division for this review.

SAMPLE DECISION QUESTIONS

1. Does the trend or size of our external failure costs indicate that a problem exists in outgoing quality performance?
    a. What kind of problem i.e. design, conformance or utilization?
    b. What product/activity line and/or customer/field area are principal contributors?
    c. What are the related appraisal costs?
    d. Is there a similar trend in internal failure costs for the same product/activity?
    e. What are related prevention costs?
2. Is the trend in overall internal failure costs meeting expectations?
    a. Is it maintaining its proper relation to production costs?
    b. What production area(s) have highest cost trends when related to productive labor?

c. Which units have highest failure costs?
d. Are there any significant failure cost variations in any area?

3. What are the relationships of quality cost to various bases such as total sales, productive labor, unit costs, etc. for the company, product line or production unit?

A careful analysis of the decision questions for a particular company, its product lines, its productive units, and organizational elements will indicate how the cost system should be designed.

PACKAGE REVIEW

The task of review for completemess of the decision question package can now be combined with that of determining the specific data required and its source. Let me recommend that you use the Quality Operating Cost charts in the ASQC publication "Quality Control — What & How" as a guide for this. (Cross reference the elements of each cost category, i.e. Prevention, Appraisal, Internal Failure and External Failure, with the information listing of the decision question charts (perhaps these should be consolidated) and make sure that vital elements and probable sources have been considered.

DATA SOURCES

The next task of course, is to determine the data, or information, required as inputs to produce the desired outputs. Once these have been identified, a search for existing data available, its method of collection and source should be made. In our search for existing data, it is important that Accounting, Production Control, Procurement and the Central Processing or EDP activities be checked out thoroughly. Although the data needed may not presently be generated or identified as such, it is frequently contained in the transaction records of these activities and all that need be done is to develop an extraction method. It is also well to note that in addition to specific cost data, we will need considerable other information which is also usually in these records. (Information related to population, such as number of orders, number of items ordered, initiated, completed, rejected, repaired, scrapped, entered into inventory, shipped, handled, inspected, etc., when properly mixed with the costs associated in the transactions, can result in output information which quickly communicates to the concerned manager a need for action).

This then leaves us with a listing of the balance of data needed as input to the system. From this we now identify the sources of the needed data and start the development of a reporting/collecting activity or system which will provide the input. As we accomplish this task, we may be faced with the necessity to make some sort of trade-offs in sources and types of data. We may even have to trade off a decision question response because the cost to acquire the input may be excessive in relation to its benefit in the decision process.

SYSTEM ANALYSIS AND DESIGN

Once we have identified all the data needed, what is available in existing records, and what new sources and collection system we need to provide all the necessary inputs, we go back to our decision questions and begin to match inputs to required outputs. This tells the System Analyst how to design the system to obtain the necessary outputs. The system design must be such that it develops the necessary information for each level of management which will use the output. It is important that multiple sort and compilation capability, including Pareto Analysis, of the data be designed into the system so that we can research back to a cause to pinpoint a specific spot for corrective action. This, of course, provides us with immediate cost reduction data, thereby permitting periodic compilation and reporting of savings generated by the system needs to be built into the system since it provides the necessary feedback to the manager involved, i.e. that the control established by the Quality Cost System is, in fact, accomplishing the desired control.

CONCLUSION

It is recognized that within any company there may be limitations, such as cost availability of systems analysts, programmers, or even computer capability which may prevent the design and development of a complete product Quality Cost System, let alone a system for all organizational elements. However, we recommend that for whatever part of the system you undertake, follow these concepts and you will get the desired results. In fact, if you now have a system, or a partial system, examining it in the way described may likely point out some ways to improve its capability.

Remember, a quality cost system is a management control. It provides to the decision maker the specific communication he needs for action decisions. Therefore, "Our only output is information."

COST IMPROVEMENT THROUGH QUALITY IMPROVEMENT

Frank Scanlon
Secretary — Director Quality
The Hartford Insurance Group

ABSTRACT

This paper is a continuation of the 1980 paper "Cost Reduction Through Quality Management." The Hartford over the past five years has taken many of the principles developed by ITT for manufacturing operations and translated these ideas into a workable quality improvement program for the insurance industry. The program itself is basic enough to be adapted to administrative functions in both service and manufacturing companies. This presentation will include:

- A Film "Why Me?"
- The development of a practical quality improvement program including the six basic ingredients.
- Quality communications.
- Quality recognition.

INTRODUCTION

Building a practical quality improvement program for service companies and the administrative functions within manufacturing companies that involve people can become a part of your overall quality system. This is the new frontier for quality professionals and is the logical area for the quality profession to expand and grow in the future. As part of the management of your company, it is expected that you will take advantage of this cost effective program.

My paper presented at the 34th Technical Conference in Atlanta identified the methodology necessary to develop quality costs. This paper is geared to the development of a quality improvement program which will measure the percentage of improvement or slippage from an agreed upon quality base. The base was developed using the information in the previous paper.

The need.

Several years ago, ITT realized that approximately 50% of its income was derived from non-manufacturing companies and that more than 50% of the employees of a manufacturing company were employed in non-assembly line functions thereby making it necessary to address quality within these functions. The Hartford, which is a subsidiary of ITT, started to build programs and systems addressing the issue of administrative quality. This paper is an overview only of the various quality projects and programs within The Hartford.

To explain each segment of our program in detail would require two to three days. The Hartford's Quality Program was developed by modifying those items that worked in the manufacturing process over the years and translating them into language which insurance company management and employees could understand. This paper will emphasize the similarities of the manufacturing and administrative quality programs with little emphasis on the differences. Most of what I present to you, you will recognize, but it will be in a form you probably have not heard before.

Your administrative quality program can:

- Be measured.
- Guarantee quality improvement.
- Be cost effective.

Just like manufacturing programs, administrative quality programs can give your company the cost and service advantage over competitors.

Quality Improvement vs. Quality Control.

I am identifying quality control as a type of quality program that will "ensure" conformance on each product or service being handled today. Quality improvement on the other hand means analyzing today's results so that preventative action can take place to ensure that tomorrow's product or service is better.

The major impact of an administrative quality improvement program is threefold:

- First, identify for managers and supervisors their biggest problem in order that;
- Two, they can correct the *cause* of the error for the purpose of;
- Three, preventing these errors from recurring as frequently in the future.

The consistent application of these three steps will ensure quality improvement.

RESPONSIBILITY

It is important for you and your senior management to understand that a quality professional cannot be held responsible for the quality of managers, supervisors and technicians. The quality professional can only *build* the program which managers can utilize for the purpose of error detection, and then, after data analysis, use to develop practical corrective action plans.

DEFINITION OF QUALITY IN ADMINISTRATION

Quality in the administrative area means exactly the same thing as it does in manufacturing, that is, "quality is conformance to a standard."

BUILDING AN ADMINISTRATIVE QUALITY PROGRAM

Over the years, quality programs have become so sophisticated that they now have become difficult to understand by the non-quality professional. My suggestion is that we go back to basics and build our quality program with six ingredients that can be easily understood and utilized by all managers and supervisors. They are as follows:

1. *Commitment/Leadership.* Although this basic term sounds like "motherhood," it is absolutely impossible to build a practical quality improvement program without the commitment and leadership of senior management. Commitment is not sufficient! The commitment must be complemented by the *visible* leadership of senior management. This is the easiest ingredient for a quality professional to sell.
2. *Sampling.* Once you have determined the credibility and tolerance levels that will be acceptable, and you know the existing error rate, published sampling tables can be utilized. Sampling in lieu of 100% checking is recommended, as:
   — Sampling will fulfill the requirements of the program.
   — 100% checking is costly in terms of dollars and timeliness.
   — 100% checking is no more than 85% effective.
   — There is a tendency for individuals to commit more errors when they know 100% checking program is in effect.
3. *Checking.* This simply means those items which have been selected as part of the sampling process are verified for conformance to the written instructions. They are either accurate, that is, they conform to the standards, or in error, and do not conform.
4. *Recording.* The purpose of recording data is to provide the information necessary to perform an analysis and develop the error trending. Proper recording will permit you to determine who is making errors and whether or not a specific unit is improving or deteriorating. This process includes identifying errors through: sampling (processing), rejects by other units within the company (internal), and rejects by outside sources such as customers and vendors (external).
5. *Analyze.* Analyzing the recorded data will enable a manager/supervisor to categorize the errors by type and individual. This is the only way to identify a unit's major problem.
6. *Corrective Action.* This is the payoff of an effective quality program. The only way in which this can be accomplished is by integrating people into the corrective action process. The only person who knows why the error is made is the person who created the error; therefore, the only person who can identify the real cause of error is the person who made it. Effective corrective action addresses the *causes* of errors, not the errors themselves.

HOW DOES THE SYSTEM WORK

Now that we have defined the program, what makes it work? Communications. Without communications, written or oral, it is impossible for the individual staff member to know the standard. We, like most companies, have effective communication downward; however, for real communication to take place, it must be upward as well. And to make communication most effective, we must also generate lateral communication. This can be easily accomplished if each employee knows the answer to the following questions:

a. What is my job?
b. Where do I fit in general terms in the organizational structure?

c. Who gives me my work? What do they do? How do they do it? How do their errors impact me?
d. When I finish my work, who receives my end product? What do they do? How do they do it? How do my errors impact them?

THE HEART OF A QUALITY PROGRAM

The heart of a quality program is recognition — recognition of deserving employees. In 1976, we surveyed 10% of our employees, and came up with five suggestions for improving our quality recognition program. The first was to add a monetary value to the awards. The third was to establish "unit recognition" as well as individual recognition. The fourth was to increase local control of the program and fifth was to step up publicity. Surprisingly enough, the second item, right behind the addition of monetary value, was to be recognized by first-line managers for doing the job right. You have an opportunity to build quality programs for positive reasons. As we analyze errors, we not only find out what our biggest problem is, but also what our greatest asset is. First-line managers and supervisors must be sensitive to the quality work being performed by individuals and recognize them accordingly. This visible recognition of deserving individuals will have a favorable impact on all staff members within a given unit and as a result, it will have a positive impact on the entire quality program.

SUMMARY

This paper obviously is not the answer to all non-conformance problems within the administrative areas. We have not even started to discuss areas of:

- Improved error detection devices within the process; for example, data processing editing.
- The cost of administrative errors which are never caught; for example, the billing of a customer for 20% less than the value of the product may never be detected and it could be the total profit of a given sale.
- The subject of building in quality "in the administrative area."
- Better quality costing.

These are all necessary and the basis for additional papers.

I believe the reasons effective administrative quality programs have not been built in the past are:

- Terminology differences between manufacturing and administration;
- Nonfatal consequences of administrative errors;
- Accepting favorable trends as a measure of successful programs;
- Quality professionals are not sure how to approach this task.

Hopefully, this presentation will assist you in overcoming some of these obstacles and enable you to build a viable administrative quality improvement program for your company.

Quality is a very serious business! Fifteen to 25 percent of your paperwork is being redone because someone has made an error and that is costly! In addition, it adds to the frustration of your employees. Although quality is a very serious business, it cannot be advanced in an uptight environment.

QUALITY COST — A KEY TO PRODUCTIVITY

H. J. Harrington
International Business Machines Corporation
General Products Division
San Jose, California

ABSTRACT

The marriage of quality, productivity, and customer satisfaction is an essential part of every business. Unfortunately, quality professionals have the narrow concept that quality means merely conformance to specification. This notion has allowed the United States to slip far behind their competition in many product lines, such as autos, TVs, and semiconductors. To overcome this adverse position, the quality professional must place equal emphasis on direct and indirect quality costs.

INTRODUCTION

Today, American business faces its biggest challenge since Pearl Harbor.

At that time, the problem was how to increase our manufacturing capabilities to meet the military needs of an all-out global conflict. Today, we are involved in another conflict, one whose end results can be just as, or more, disastrous than World War II. This time it is a fight to hold our world leadership role as an industrial power. This battle does not use bullets, bombs, or tanks. It uses items like TVs, autos, and semi-conductors.

The deciding factor as to whether we win or lose is the price and quality of our products. America today is being drained dry with a negative balance of trade. Within the U.S., we are shifting away from our own products to products produced in other countries because we believe they produce a higher quality product.

Our challenge is to increase American productivity and simultaneously improve the quality of the end product. It has been a rude awakening for American businessmen who have only a financial and marketing background to suddenly realize that product quality is a strategic weapon, often more critical than price in today's worldwide battle for the world market.

Japan turned the corner, improving quality and increasing its rate of production. In fact, Japan's production is increasing at a rate of more than four times the U.S. rate. More than 20 years ago, a trademark saying "Made in Japan" represented a cheap imitation of a quality American product. But today, Japan is setting the quality standard for many products like TVs and cameras.

Japan came to us for help, and it listened and learned from us. Now we must take a long, hard look at Japan to see, as they say in the Pontiac Commercial, "Do they know something we don't?" In today's environment GOOD ENOUGH IS NOT GOOD ENOUGH, ONLY EXCELLENCE in everything we do will allow us to get our share of the market place.

PRODUCTIVITY

Productivity Growth Rates

Very simply, productivity improvements occur when we get more out for what we put in, or produce more with the same resources. It doesn't mean working harder, although sometimes working harder may result in short-term productivity improvements. Long-term productivity advancements are accomplished by working smarter, not harder, by finding or creating better, more imaginative ways of using our available resources. Productivity measures the efficiency of an operation, and applies not only to manufacturing, but also to white collar jobs. It is normally measured as the ratio of output (goods, materials, or services) to input (material, capital, labor, resources). Productivity growth rate is measured by the change in output per employee hour or the rates of output over input.

Figure 1 shows the productivity growth rate for seven leading countries throughout the world between 1972 and 1977.

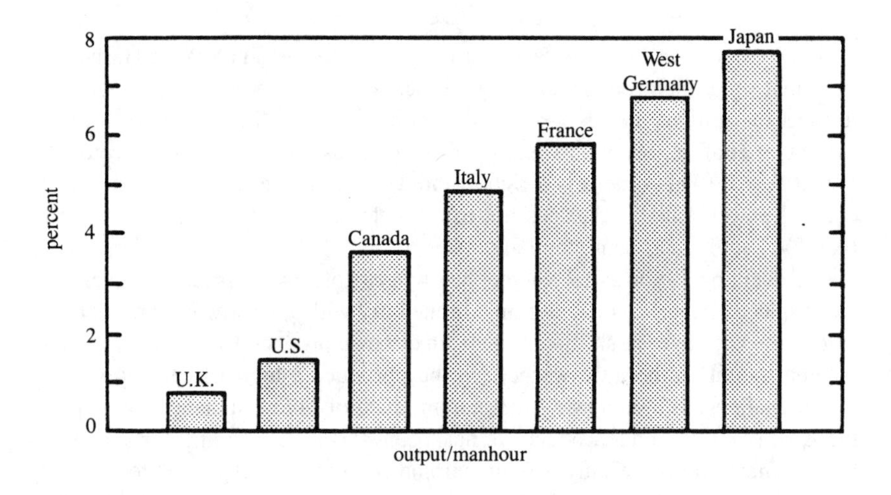

FIGURE 1. Average annual productivity growth 1972-1977 (source: Bureau of Labor Statistics)

It is no surprise that Japan is number one, and that West Germany is number two, but it is shocking to see that France, Italy, and Canada all are ahead of the United States. And if we look at some other data, we find that Belgium, Norway, Switzerland, and Denmark are also ahead of the United States. Since World War II, labor productivity in the private, nonfarming sector rose at about 2.3 percent per year. In this post-war era, it has risen every year, with the exception of 1969, when it declined a small amount; in 1974, when it fell 3 percent; and in 1979, when it plummeted to a minus 2 percent, a drop of 2½ percent. Figure 2 graphically portrays this trend.

A close analysis of Figure 2 indicates that the output per employee hour increased from 108 to 117 over the past 10 years (1969 to 1979), less than 1 percent per year. Figure 3 highlights this problem by plotting the percent growth rate for the total private sector since 1948. Note that the data points between 1948 and 1977 are grouped in yearly averages, while the 1978 and 1979 data points are actual yearly values. In the third quarter of 1980, manufacturing productivity decreased 2.7 percent at an annual rate. At the same time, there was a 10.7 percent reduction in output and an 8.3 percent reduction in the number of hours worked. This was the fourth consecutive quarter of declining productivity in the manufacturing sectors — the longest period of declining productivity since 1955/56.

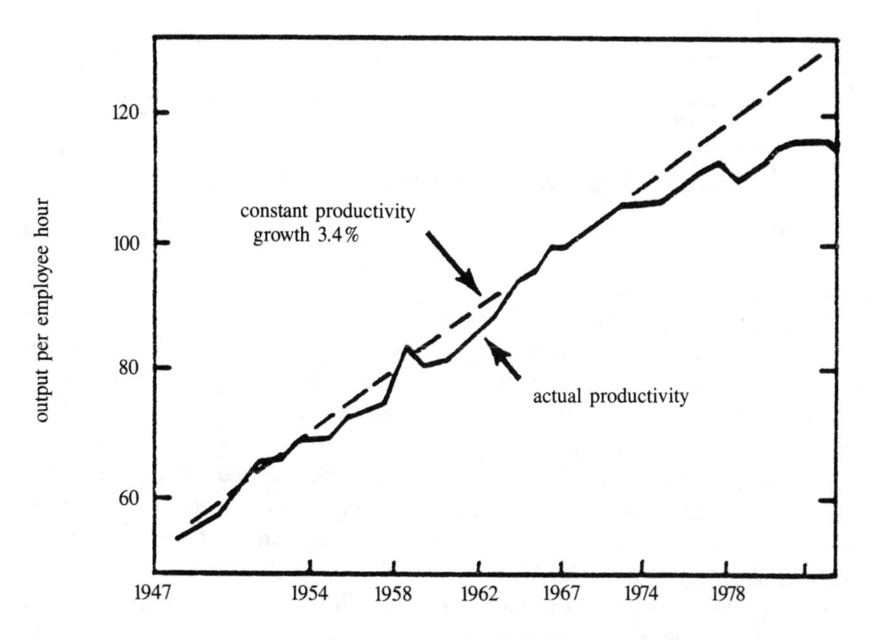

FIGURE 2. United States productivity growth

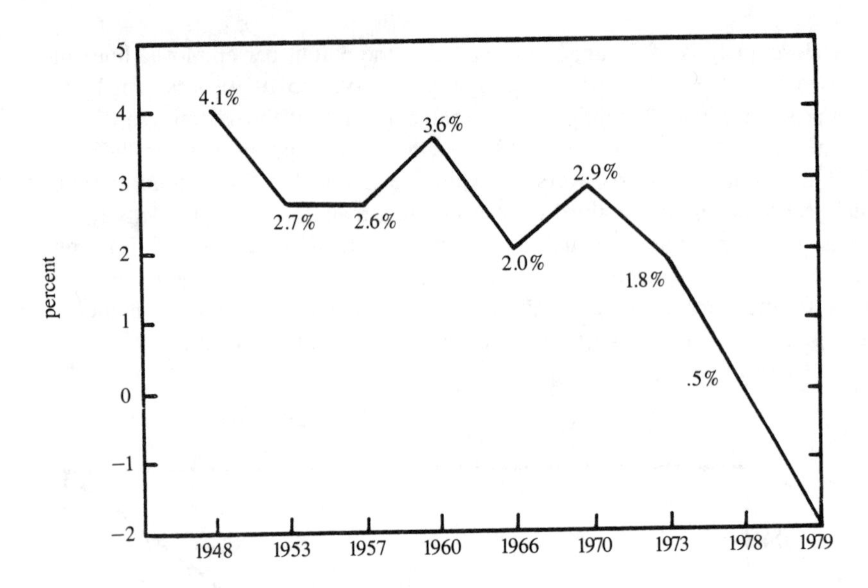

FIGURE 3. United States productivity growth rate

The Productivity Race

In a recent survey, 74 percent of the managers and 71 percent of the nonmanagers surveyed felt that we can improve our productivity. To offset some of the doom and gloom, let me state that the U.S. is still the most productive nation in the world. Our output per employee hour surpasses all other countries in the world. But if we don't do something about it now — right now — within ten years the U.S. will lag behind many other countries. An estimate made by the American Productivity Center, Inc., indicates that, if something does not change, many countries will be more productive per person than the U.S. in the next ten years. France will be ahead of us by 1986, Germany by 1987, Japan by 1988, and Canada by 1990. By 1991, France, Japan, and Germany will be out-producing the U.S. by over eight percent per year per person.

Productivity by Work Sectors

The productivity data presented in this report includes both manufacturing and nonmanufacturing industries. It excludes both agricultural and government employees. Although not including these two major sections of the work force probably slightly distorts the picture, they tend to offset each other. Since World War II, farm production gains have averaged 5.5 percent, but this has been offset because the government segment expanded extensively in the post-war era, while the agricultural employment decreased. Figure 4 shows the productivity annual rate of change, divided into individual components, and as a total.

A very interesting fact comes into view by closely analyzing these graphs. The nonmanufacturing portion of the productivity change has had an increasingly strong effect on the total productivity change. This is brought about by the shift of employment ratio from manufacturing to the service industries. In the late 1940s and early 1950s, one third of all non-agricultural employment was in the manufacturing sector, and now it accounts for less than one fourth of the total work force.

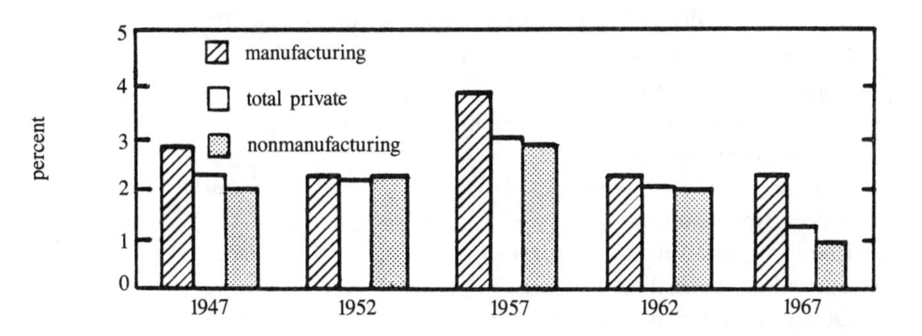

FIGURE 4. Comparison of annual productivity

Effect of Productivity on your Pocketbooks

Why do we care about increased productivity? What does it do for us? Well, it can cure many of the personal and govenment problems that face us today. Increased productivity can:

1. Improve our present standard of living
2. Increase the value of the dollar
3. Reduce inflation
4. Reduce unemployment
5. Strengthen our position in the domestic market
6. Further job security
7. Make us more competitive on the international market
8. Improve returns on investments
9. Increase our buying power.

Now let us take a close look at how the lack of productivity decreases our buying power. It is only logical that, if we are able to produce five percent more products this year than we did last year, our value to the company has increased by at least five percent, without increasing the product's cost. But if we receive an eight percent salary increase, the additional three percent cost must be passed on to the consumer, contributing to increasing the Consumer's Price Index. Table I shows how this very simple concept is reflected at the national level. When the percent productivity change is subtracted from the percent compensation change, the productivity delta is obtained. Comparing the productivity delta to the Consumer Price Index change for the same year proves that there is a direct relationship between the productivity delta and changes in the Consumer Price Index.

This table clearly shows the impact on our buying power when productivity increases do not keep pace with compensation increases.

HOW TO IMPROVE PRODUCTIVITY

Now that we understand the problem and have some idea of the gain that can be derived from solving it, let's try to look at what can be done to increase productivity. The following is a list of some things that can improve productivity:

1. Lessening of government regulation
2. Capital equipment investment
3. Research and development expenditures
4. Management awareness
5. Creative problem solving
6. Automation/Robotics
7. Employee involvement (teamwork)
8. Expanding international markets
9. Doing the job right the first time.

TABLE 1

	% Compensation Increase	% Productivity Change	Compensation Productivity Delta	Change In Consumer Price Index
1968	7.6	3.3	4.3	4.7
1969	6.8	0.2	6.6	6.1
1970	7.1	0.7	6.4	5.5
1971	6.7	3.4	3.3	3.4
1972	6.3	3.4	2.9	3.4
1973	8.2	1.9	6.3	8.8
1974	9.1	−3.0	12.1	12.2
1975	9.9	2.1	7.8	7.0
1976	8.8	3.5	5.3	4.8
1977	8.0	1.6	6.4	6.8
1978	8.5	0.5	8.0	9.0
1979	9.3	−2.0	11.3	13.3
TOTAL	96.3	15.6	80.7	85.0

PRODUCTIVITY AND THE QUALITY PROFESSIONAL

Aren't quality and increased production in conflict with each other? If we are going to do the job right, don't we have to slow down, reducing your productivity? The answer to all of these questions is, "NO!" Almost everything the quality professional does complements productivity. At least three of our major activities are designed to improve productivity, reduce cost, and decrease manufacturing cycle time. They are:

- Statistical process control
- Defect/error prevention
- Statistical sampling plans and techniques.

Val Feigenbaum estimates that the "hidden plan" in the U.S. firms may account for from 15 to as much as 40 percent of the manufacturing plant's production capabilities. He defined the "hidden plan" as the personnel and equipment required to rework, restart, rebuild, or replace defective products. Consider what the impact on productivity would be if the "hidden plan" costs were reduced from 25 percent to 15 percent. It is easy to see that an immediate 10 percent productivity increase would be obtained, but in addition, the people and equipment previously part of the "hidden plan" can be redirected to produce new error-free products, further increasing our productivity.

Defect/Error Prevention

A major portion of the quality engineer's time and effort is directed toward corrective action and defect prevention. Basically, one of the major reasons for the Quality Assurance function is that an unacceptable error rate would exist without an independent evaluation of the engineering and manufacturing cycle.

What are some of the causes of errors?

- People
- Equipment Failures (Equipment designed, built and maintained by people)
- Materials Failures (Normally caused by failure of the process that was designed by people or poor selection of input material by people)
- Process Failures (Less than 100 percent yield from the process designed by people)
- Design Failures (Engineering designs that don't take into consideration all the stress, usage application, or tolerance extremes).

We can go on with these examples of errors, but by now we should see the point I'm trying to make. That is, people either directly cause errors, or can implement controls to eliminate them. This means that the only errors we live with are the ones that we choose to live with. I've often heard Phil Crosby say, "It's always cheaper to do it right the first time." But it's even cheaper not to do it at all. In our own zest to achieve "zero defects," we can have a tendency to over control. I do not disagree with Phil that zero defects should be our objective; but there is a need to temper this objective by looking at both sides of the coin, for sometimes the cost of doing it right the first time is more than the repair costs.

Quality Cost Analysis

An approach often used to reduce quality cost is cutting the quality function's budget which, in turn, often increases the total quality-related costs. Before we can embark on a program to reduce quality-related cost, we need to know what we are expending. This is where a quality-cost analysis system can most effectively be used. Figures 5 and 6 portray two standard cost curves for the same process. The internal and external defect cost levels for both graphs are the same at zero quality costs, but from that point on, the curves take a radically different shape. In Figure 5 (Case 1), the controllable quality costs are largely applied to appraisal quality costs, and in Figure 6 (Case 2), a large portion of the total controllable QA costs are directed to the prevention activity. (Controllable quality costs are defined as prevention quality costs plus appraisal quality costs.)

Prevention costs include all the costs designated to help the employee do the job right the first time. It is obvious that spending quality dollars in doing the job right the first time decreases the internal and external failure rate, and at the same time, appraisal costs can be reduced because:

- Reduced inspection levels can be implemented faster
- Reduced QA inspection time in reinspecting rejected lots.

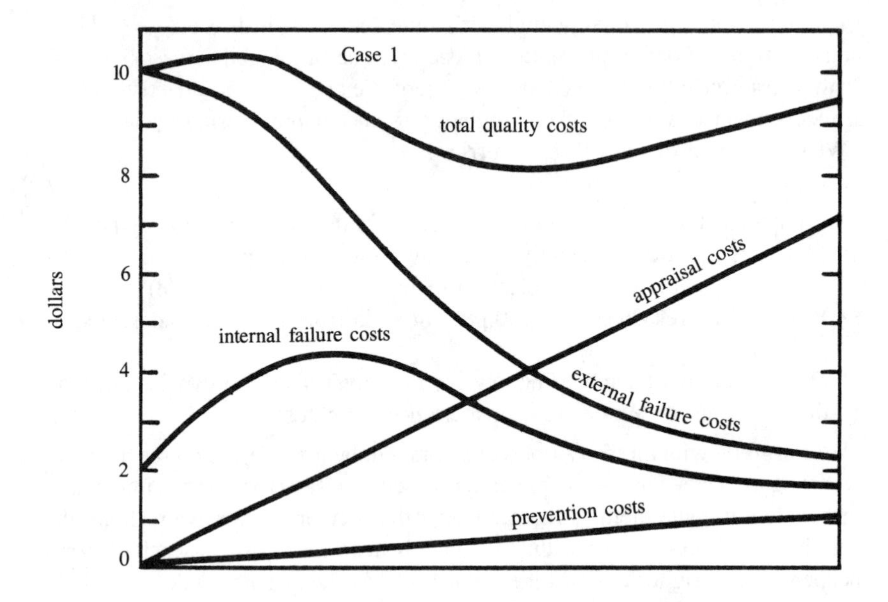

FIGURE 5. Case 1 with expenditures in preventive activities

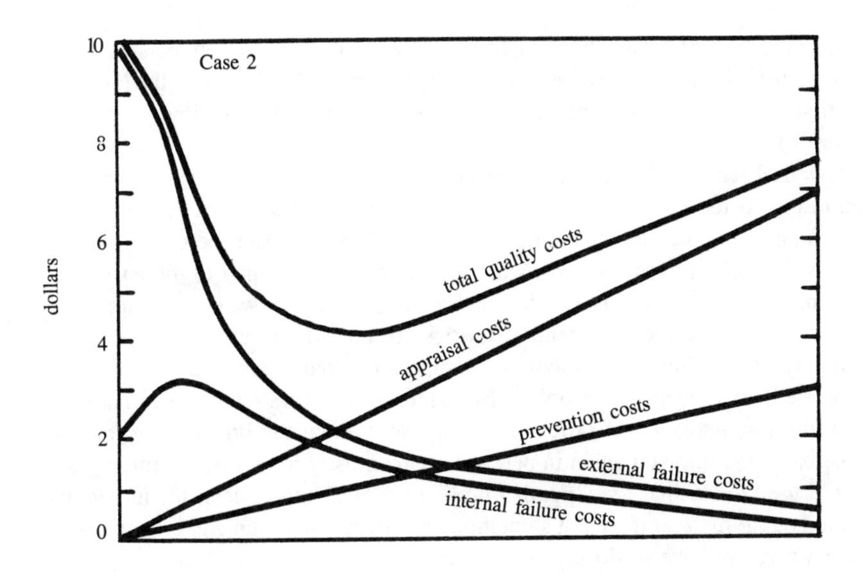

FIGURE 6. Case 2 with increased expenditures in prevention activities

Figure 7 shows two bar graphs that represent the process quality costs for Case 1 (Figure 5) and Case 2 (Figure 6), at the optimum operating point. These bar graphs clearly show how the quality costs for a typical process were cut from 8 million dollars to 4.25 million dollars by stressing "doing the job right the first time," and controlling — not over controlling the process.

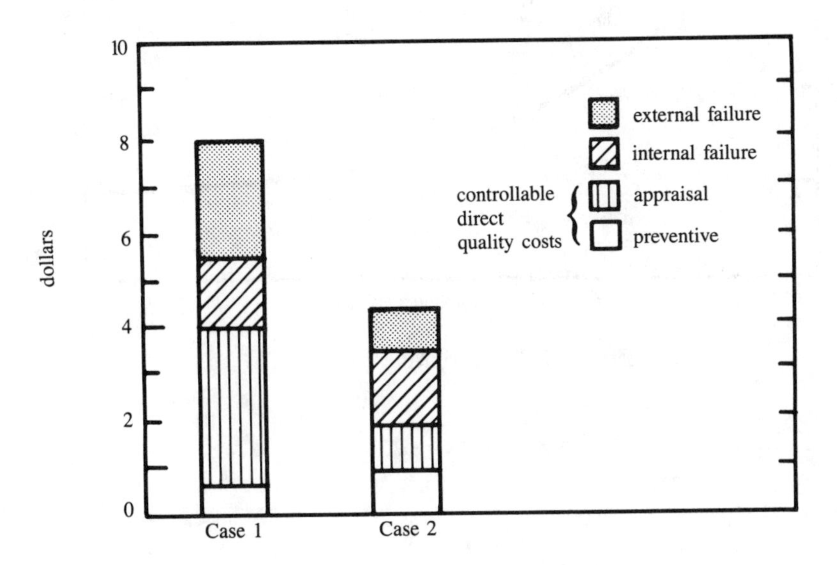

FIGURE 7. Optimum quality costs operation point

To really simplify this whole quality costs system, it can simply be stated that prevention activities are activities that affect the person's ability to do the job right the first time. Or to say it another way, activity that affects first time yields (see Figure 8).

Appraisal activities are activities that prevent defective products from being shipped to the customer or to higher level assemblies (see Figure 9). Appraisal activities do not reduce the defect rate; they only detect a higher percentage of the defects that are present in the product before they are shipped to the customer.

Using this as a basis, it is easy to see that the only reason we need an appraisal activity is because the prevention activities are not wholly effective. When the appraisal system finds a problem, it is imperative that corrective action is implemented, to prevent the problem from recurring. How many times have we worked on and solved the problem that has shut down the line, only to have it come back and bite you again in perhaps six months, a year, or even three years later? I used to say that I've "put that problem to bed" until one of my inspectors pointed out to me that if we put something to bed, it can get up and bite us again. What we really have to do is bury the problem so it will never come back.

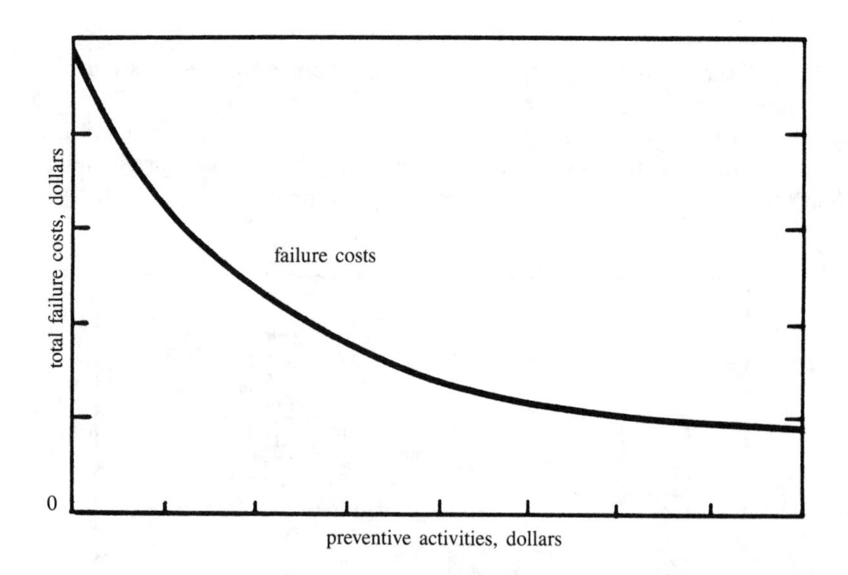

FIGURE 8. Preventive Costs

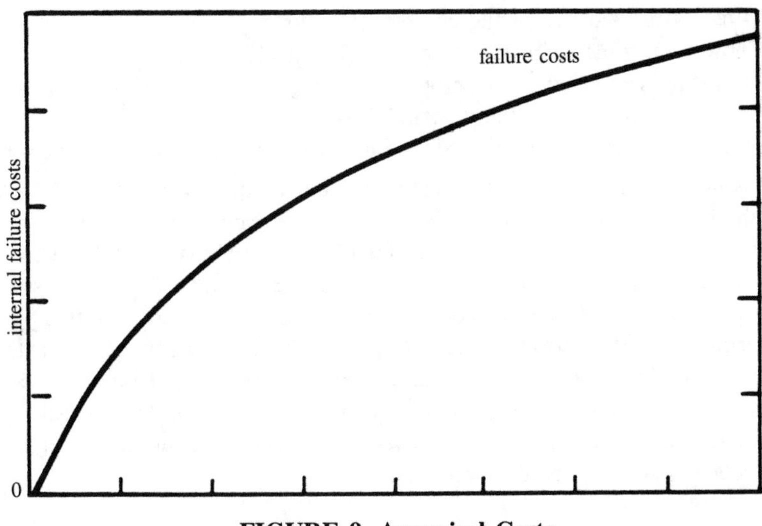

FIGURE 9. Appraisal Costs

Although we have classified a wide variety of items as prevention quality costs, one quality cost still needs to be added. This is the one of the manufacturing equipment cost devoted to doing the job right the first time. In the past, we readily added manufacturing test equipment cost into the quality cost curve, but shouldn't some parts of the equipment cost that prevents defects be added to the quality prevention costs? For example, Japan has invested many yen in automated equipment for two reasons:

- Increased productivity
- Increased quality.

Much of this equipment could not have been justified by considering either one of the two factors alone. Now, where does the cost of automated production equipment (not measurement equipment) purchased to improve process yield (quality) fit into the quality cost model? Well, it logically follows that at least a portion of this cost should be included in the preventive cost, as it is dollars invested to do it right the first time.

Indirect Quality Costs

Indirect quality costs are defined as those costs not directly measurable in the company ledger, but are part of the product life cycle cost. They consist of three elements:

- Customer-incurred quality cost
- Customer-dissatisfaction quality cost
- Loss of reputation cost.

Customer-incurred quality cost is the cost that the customer incurs when the product fails to function. Typical customer-incurred quality costs are:

- Loss of productivity while equipment is down
- Travel and time spent to return defective merchandise
- Overtime expended to make up production because equipment was down
- Repair costs after warranty period is over
- Backup equipment maintained so that it can be used when equipment fails.

It is not uncommon to see the customer-incurred quality cost exceed the total purchase price of the item during the product's life cycle. A good example of this is the repair cost during the life cycle of a television set. Figure 10, Case 1A, and Case 2A, shows how the optimum cost operating points increase when customer-incurred quality costs are considered. In this case, a very conservative picture is portrayed, as the customer-incurred quality cost is only equal to the external failure cost. This example vividly shows how the effects of reducing field failure rates drastically improve total product quality cost. Most of the external failure cost and customer-incurred cost involved manpower that could have and should have been utilized to increase productivity.

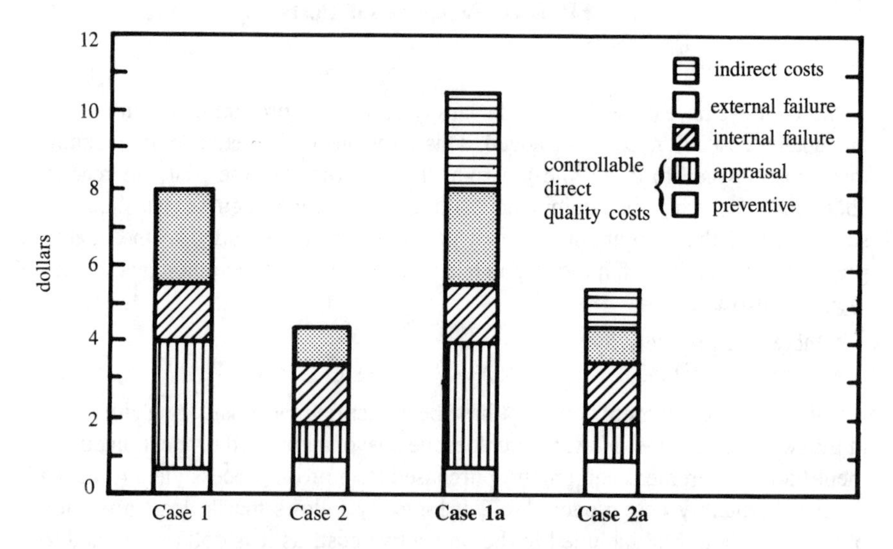

FIGURE 10. Optimum quality costs operating point

In my 1972 paper entitled, "The Shadow Over The Quality Cost Curve," I projected that the indirect quality cost factors would become a major factor in product advertisement and purchasing decision. This projection has come to pass as the auto industry can verify. In just over ten years, an almost nonexistent Japanese automotive industry has become the largest in the world, based on giving the customer a high quality, dependable car that performs to customer needs. In 1979, over 45 percent of the cars sold in California were Japanese made. It was projected that it would top 50 percent in 1980.

In 1980, Japan's automakers planned on producing 10.6 million vehicles, 200,000 more than American companies. In 1979, 4.5 million vehicles were exported, bringing into Japan $15 billion, an increase of 11 percent. Toyota Motor Company alone has captured 4.8 percent of the U.S. market.

Figure 11 shows the last three years' sales trend for U.S.-made cars and trucks. Figure 12 divides this trend into three major classifications. Note that small car sales held relatively constant over the three-year period, but trucks and large and intermediate cars dropped drastically. This drastic drop in U.S. sales was caused by two factors:

- Quality of foreign-made cars
- Detroit not meeting the needs of the buying public.

Today, Detroit is making cars that meet or exceed the Japanese gas mileage, with more room inside, at less cost; but the public is buying from Japan — why? The difference is the perceived quality of the product being delivered. Now, we may be able to argue and present data that this is not true, but that isn't important to anyone but the quality professional who is trying to save his job. *IN TRUTH, I CONTEND THAT OUR CUSTOMER'S PERCEPTION OF OUR PRODUCT IS MORE IMPORTANT THAN FACT.* The only fact that is important is our customer's perception of our product.

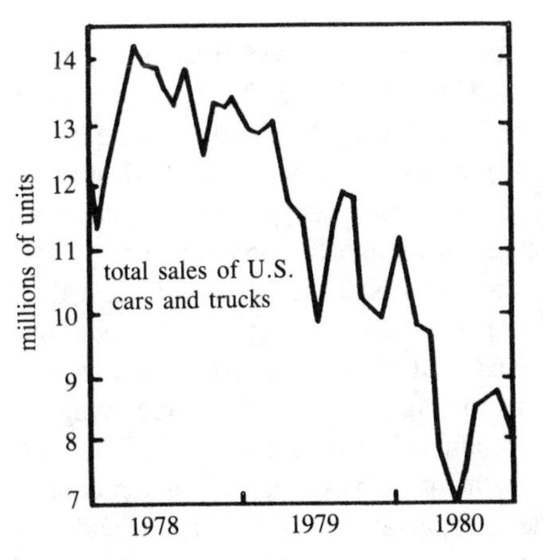

FIGURE 11. Dropping car and truck sales of U.S.-made vehicles

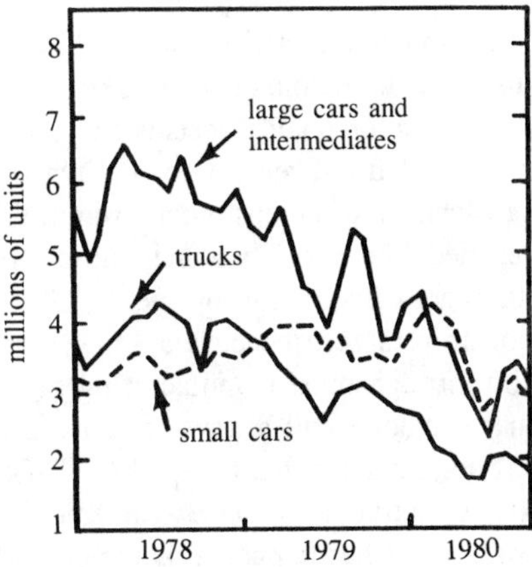

FIGURE 12. U.S.-made vehicles by major classification

To help us understand what is happening in this world market, let us look at the customer-dissatisfaction cost curve, Figure 13.

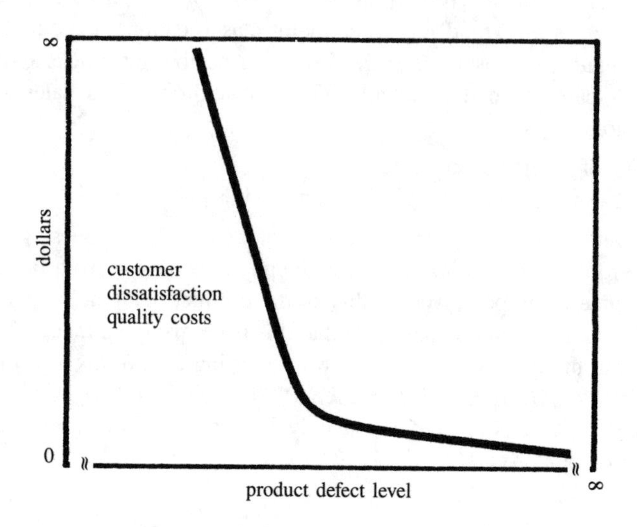

FIGURE 13. Customer dissatisfaction costs

Customer dissatisfaciton is a binary thing. Customers are either satisfied or dissatisfied. Seldom do we find one that is in between. Figure 13 portrays customer-dissatisfaction quality costs versus product defect level. The sharp decrease in cost per small improvement in product quality reflects the binary classification of satisfaction in the customer's mind. Once a customer acceptance level has been reached, the curve becomes almost flat, even though the product quality level undergoes major improvements. The customer-dissatisfaction quality cost does not drop to zero because a small percentage of maverick articles (lemons) can slip through most manufacturing processes, and because human nature assures that someone will always expect more from a product than he is receiving.

I don't believe that the quality of U.S. products has suddenly dropped. In fact, this is not true, as we have weathered the apathy of the hippie generation remarkably well. What has happened is the customer's expectations have changed, and they now require a much better product to satisfy their demands and expectations. I am sure all of us have been part of a conversation that went something like this: "When I used to buy American cars, I had a list as long as my arm of things that had to be corrected at my first checkup. But with the Japanese car, I never had to take it back to the dealer." What has happened? The customer's dissatisfaction curve has moved to the right, but our quality level has remained constant (see Figure 14).

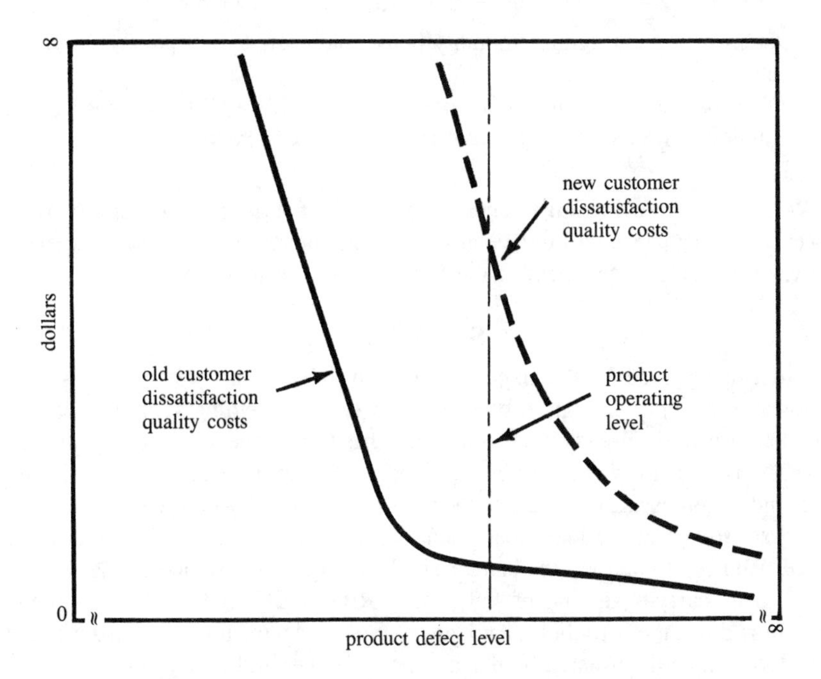

FIGURE 14. Customer dissatisfaction costs

As the customer expectation level increased without an accompanying increase in product quality, our American companies began to lose customers. When we have a satisfied customer, it's easy to keep him, as long as our product remains price compatible; and, even if it is a little more expensive, the American public tends to stay with a product that has a proven track record. But, once we have lost the confidence of that customer because he is not satisfied with our product performance, it is next to impossible to get him back in the fold. In addition, this infection has a tendency to spread to his family members.

Action/Effect Analysis

As we develop plans to alter the present quality system, we must complete an "ACTION/EFFECT ANALYSIS" of each expenditure, to ensure that the end results benefit the corporation, not just the Quality Assurance function. All too often, we decrease Quality Assurance's expenditures, in hopes of decreasing quality costs; but the net effect is an increase in quality costs, as yields drop, or as customer returns increase.

A very simple example of an Action/Effect Analysis follows:

- *Action* — We can reduce monthly household costs by not paying the mortgage payment.
- *Effect* — The bank will repossess our home, and we will lose much of the equity that we have in the house, making it necessary for us to rent another home.

We are not improving our financial condition by not paying our monthly mortgage payments; as we will reduce our equity and increase our monthly expenses if we rent a home equivalent to the home we are presently occupying.

SUMMARY

Unfortunately, the American public seems to sit back until near-disaster strikes before it gets angry enough to band together to do something about a situation, but that disaster is already here. Something has to be done and I believe that the mandate is clear for the quality professional to rise up in force, to change its image, and regain its rightful place in the world manufacturing community; produce the best product at the least cost, using its own highly-skilled American labor force. We need to pick up the banner and beat the drum. We need to get behind Uncle Sam, carrying the flag of "DOING IT RIGHT THE FIRST TIME," jointly marching down the battlefield that leads to top management and the manufacturing floor. Make them listen by playing the tune over and over again, until they are in step, and our battle cry can be heard on their lips, "Do It Right The First Time."

QUALITY COST ANALYSIS: A PRODUCTIVITY MEASURE

L. James Esterby, Director, Quality Assurance
Victor Equipment Company
Denton, Texas

ABSTRACT

Productivity is the measure of how well resources are brought together in organizations and utilized for accomplishing a set of results. It is a combination of effectiveness; i.e., how well a set of results is achieved without regard to cost and efficiency; i.e., how well resources are utilized or minimized. This can be related mathematically as:

$$\text{Productivity Index} = \frac{\text{output obtained}}{\text{input expended}} = \frac{\text{performance achieved}}{\text{resources consumed}} = \frac{\text{effectiveness}}{\text{efficiency}}$$

Quality cost analysis measures the efficiency of resource utilization and, thus, when combined with a measure of output, provides a measure of productivity. This paper examines this technique.

INTRODUCTION

The quality function has been in a state of transition over time. During and after the Second World War a move from an inspection orientation to one of statistical quality control evolved. Along with the national concern for cost containment came the recognition among the quality practitioners of the cost effects of quality and thus the development of quality costs analysis. Now, in response to the depletion of resources and the need for improved productivity, it is incumbent upon the quality profession to assist in this effort, its measurement and control.

BASIC QUALITY COST CONCEPT

The basic concept of quality costs is recognition and organization of certain quality-related costs to gain knowledge of their major contributing segments and of the direction of their trends.[1] This definition implies two conditions. First that product quality carries with it a cost association that needs to be recognized and controlled. Secondly, that such costs are time variant and therefore need to be considered and tracked over time. Costs, so identified, are extremely valuable to management in measurement, analysis, and budgeting.

Quality costs have been categorized into the four elements of prevention appraisal, internal failure, and external failure. These elements have been defined as follows:

1. Prevention: Costs associated with personnel engaged in designing, implementing and maintaining the quality system.

2. Appraisal: Costs associated with measuring, evaluating or auditing products, components and purchased materials to assure conformance with quality standards and performance requirements.

3. Internal Failure: Costs associated with defective products, components and materials that fail to meet quality requirements and cause manufacturing losses. Note that these costs occur prior to shipment of the product to customers. Historically, this element has been the largest component of quality costs.

4. External Failure: Costs generated by defective products being shipped to customers. This element has historically been the most elusive costs to identify and record.

Throughout this paper the term failure costs shall mean the summation of internal and external failure costs unless specifically stated otherwise. Aggregated failure costs, with rare exception, comprise the vast majority of total quality costs.

The four elements of quality costs have been found to have the relationship depicted in Figure 1.

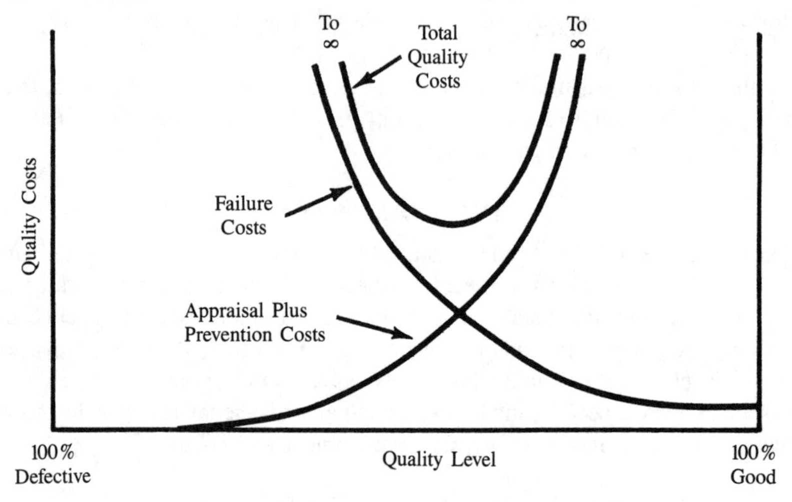

FIGURE 1. QUALITY COST RELATIONSHIP

As product quality tends toward the 100% defective level, failure costs tend toward infinity while appraisal plus prevention costs tend toward zero. As product quality tends toward the 100% good level, failure costs tend toward zero while appraisal plus prevention costs tend toward infinity. Total quality costs are the algebraic sum of appraisal, prevention and failure costs at each point along the axis. The total quality cost curve is seen to have a minimum point; the point at which the other two curves intersect. It is desirable to have the organization operate at this point of minimum total quality costs.

BASIC PRODUCTIVITY CONCEPT

Productivity is the measure of how well resources are brought together in organizations and utilized for accomplishing a set of results.[2] Productivity has, to a great extent, been a tool of the macroeconomists. The resources utilized have generally been gross man hours or total capital stocks and the results have been the gross national product of a country or the gross product of an industry.

There are two elements to this definition. Effectiveness relates to how well an objective is obtained. For example, one might wish to dig a foundation for a building. This could be accomplished by either a man with a shovel or a man with an earth mover. In this case both are equally effective, as effectiveness is concerned only with the achievement of the desired result without regard (at least serious regard) for the costs involved. Both can produce a hole in the ground.

The second element in productivity is efficiency; i.e., how well resources are utilized in achieving the stated result. Here we are concerned with the total costs of all inputs used. In the case cited above, should the foundation be tiny, the shovel might be appropriate and the earth mover totally unwieldy, whereas for a large foundation the shovel might require such a long time to complete that the total costs would far exceed the total costs of the earth mover.

Productivity is a combination of effectiveness and efficiency. This combination takes the form of a ratio and is called the productivity index:[3]

$$\text{Productivity Index} = \frac{\text{output obtained}}{\text{input expended}} = \frac{\text{performance achieved}}{\text{resources consumed}} = \frac{\text{effectiveness}}{\text{efficiency}}$$

In its most general form a productivity index (PI) is a function of output (Q) and the inputs of labor (L), capital stocks (K), and intermediate products (X); e.g., purchased components, in the form:

$$PI = \frac{Q}{L + K + X} \quad (1)$$

where each of these factors is an aggregate of various components. In particular, K includes both created capital of structures, equipment, and inventories and the capital of land and natural resources. Labor and capital are considered primary inputs while intermediate products are considered a secondary input since it is a function of its primary inputs L_x and K_x.

In its most used form, a productivity index is based on a single input factor (thus the reported national productivity of dollars of gross national product per labor hour). This form may not be the most complete but is the most manageable form from a data collection and computation standpoint.

QUALITY PRODUCTIVITY CONCEPT

We may now combine these two basic concepts, quality costs analysis and productivity, into a quality productivity relationship. Since our interest is in determining the effect of product/service quality on productivity, those factors not affected directly by quality will be assumed to remain constant.

Labor is an aggregate of many components; most commonly direct and indirect labor. Quality affects direct labor through failure costs and indirect labor through appraisal and prevention cost (given the accounting systems most often used in this country). But, more appropriately for this discussion, labor can be considered an aggregate of effective labor (L_e), labor that produce totally conforming or satisfactory products or services, failure labor (L_f), the labor component of failure costs, appraisal labor (L_a), and prevention labor (L_p). Thus:

$$L = L_e + L_f + L_p + L_a \tag{2}$$

Capital is also an aggregate of many components, most commonly current assets and long term assets, again given current accounting methods. Quality has its primary effects on inventories through scrap and accounts receivable through withheld payments for substandard products or services. Here again, for purposes of this discussion, capital can be considered an aggregate of effective capital (K_e), capital that produce totally conforming or satisfactory products or services, failure capital (K_f), e.g., scrapped inventories or withheld accounts receivable, and appraisal capital (K_a) inspection and test equipment. Thus:

$$K = (K_e + K_f + K_a) \tag{3}$$

Intermediate products (X) could be evaluated through the primary factors of labor and capital. However, due to the significance of procured material (intermediate products) on quality, they will be considered separately. Intermediate products must then be considered the aggregate of effective intermediate products (X_e) and failed intermediated products (X_f). Thus:

$$X = X_e + X_f \tag{4}$$

Output (Q) is also an aggregate in the real world comprised of a variety of products and services. An examination of this aggregation has been the subject of numerous papers in itself and does not play a part in the examination of the effects of quality on productivity. Therefore, for the purpose at hand, output shall be considered a singular component. This would, in fact, be true on the firm or departmental level where only one good or service is produced.

If we now substitute the results of equations (2), (3), and (4) into (1) we have:

$$PI = \frac{Q}{L_e + L_f + L_p + L_a + K_e + K_f + K_a + X_e + X_f} \tag{5}$$

We now define:

Effective Input: $I_e = L_e + K_e + X_e$ (6)

Prevention Costs: $P = L_p$ (7)

Appraisal Costs: $A = L_a + K_a$ (8)

Failure Costs: $F = L_f + K_f + X_f$ (9)

Substituting these relationships into equation (5) we now have a quality productivity index (PI_q) that allows us to relate quality and productivity. Thus:

$$PI_q = \frac{Q}{I_e + P + A + F} \tag{10}$$

At this point it is appropriate to examine the use of failure costs as an input to the production process. Failure cost may more commonly be viewed as a negative output; e.g., scrap. However, from the standpoint of analyzing productivity we should consider only those outputs that effectively reach the consumer. In this regard, failure costs become an inefficient use of resources and are thus an input to productivity.

What then is the effect on productivity as the components of quality costs are varied? Figure 2 presents the quality cost relationship again but with three levels of quality identified.

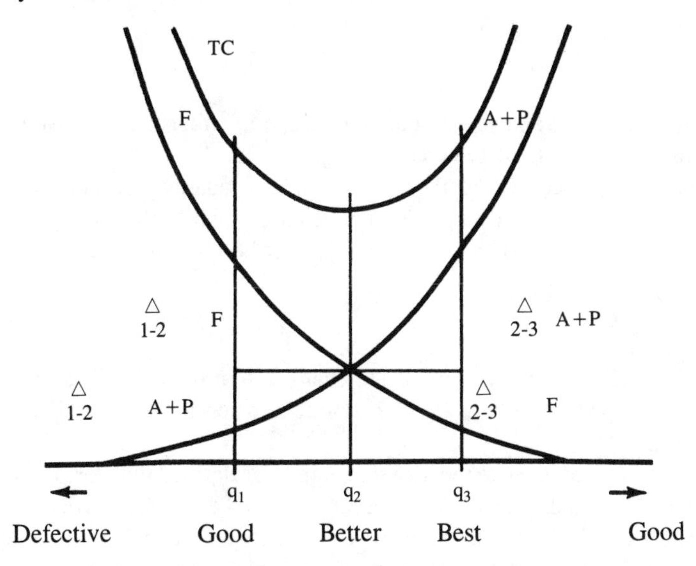

FIGURE 2. Quality Costs Relationship

These three levels have the relationship that $q_1 < q_2 < q_3$, or in the merchandizer's terminology; good (q_1), better (q_2), and best (q_3).

Starting at level q_1 and as the level of quality improves (toward q_2) the change in failure costs ($\triangle F_{1\text{-}2}$) is greater than the change in appraisal plus prevention ($\triangle A+P_{1\text{-}2}$). It is noted also that the change in failure costs is a negative quantity while the change in prevention plus appraisal is a positive quantity. Putting these changes together we have an improvement in productivity in going from level q_1 to q_2:

$$PI_{q1} = \frac{Q}{I_e + P_1 + A_1 + F_1} < PI_{q2} =$$

$$\frac{Q}{I_e + (P_1 + A_1 + \triangle A_{1\text{-}2} + \triangle P_{1\text{-}2}) + (F_1 - \triangle F_{1\text{-}2})} \quad (11)$$

In moving further toward q_3 we find that the change in appraisal plus prevention $\triangle A+P_{2\text{-}3}$ is greater than the change in failure costs $\triangle F_{2\text{-}3}$. Here again, the change in failure costs is a negative quantity while the change in prevention plus appraisal is positive. Thus in going from q_2 to q_3, we have experienced a decrease in productivity:

$$PI_{q2} \frac{Q}{I_e + P_2 + A_2 + F_2} > PI_{q3} =$$

$$\frac{Q}{I_e + (P_2 + A_2 + \triangle A+P_{2\text{-}3}) + (F_2 - \triangle F_{2\text{-}3})} \quad (12)$$

Both equations (11) and (12) assume no change in either the effective inputs (I_e) or the output (Q) due to the change in quality level.

This result is in accord with the objective of basic quality costs concept of reducing total quality costs. If output is unaffected by quality the productivity index will improve so long as the algebraic sum of the changes is less than zero:

$$PI_{q1} > PI_{q2} \text{ if } (\triangle F_{1\text{-}2} + \triangle P_{1\text{-}2} + \triangle A_{1\text{-}2}) < 0 \quad (13)$$

The index will remain the same if the changes sum to zero. Thus, with output constant, the productivity index is maximized when the total quality costs are minimized.

We need now investigate the idea that output, as defined thus far, is unchanged by a change in quality level. With failure costs taken as an inefficient input of labor and capital, the numbers of goods or amounts of services produced will not change. However, let us consider the merchandizers' terminology of good, better and best. Most often this is applied to three models of a given goods offered at a particular time having three levels of price such that:

$$\text{if: } q_3 > q_2 > q_1 \text{ , then: } p_3 > p_2 > p_1 \quad (14)$$

This relation holds for services also; e.g., the extents to which a lawyer will prepare a will or contract or the detail made available by a programmer in a computer program. The price inherently varies with the extent.

With this information we modify our definition of output from the single factor quantity (Q) to total revenue (TR) that is comprised of a constant level of output (0) and a variable price (p):

$$TR = p0 \quad (15)$$

The productivity index now becomes:

$$PI_q = \frac{TR}{I_e + F + P + A} = \frac{p0}{I_e + F + P + A} \qquad (16)$$

It can be seen that an improvement in the productivity index can occur if a change in quality allows or induces an associated change in price even if the denominator of the productivity index remains constant. Thus:

$$\text{if: } q_2 > q_1 \text{ , then: } p_2 > p_1 \text{ , and then: } PI_{q2} > PI_{q1} \qquad (17)$$

At this point one might argue that the denominator cannot remain constant if quality changes. That, in fact, a change in the factors of the denominator is a prerequisite for a change in quality. Generally, this is true. However, in this context, quality is the customer-perceived quality. Perceived quality is both qualitative and quantitative as opposed to a technically quantitative level of quality. It is affected by many things, some even outside the control of the firm such as money supply, competition, time, and perhaps most of all, an ever-changing expectation by the consumer himself. Perceived quality can change without a prerequisite change in technical or real quality. That which is "better" at this place and time may be "good" or "best" as conditions change. Such a change in perceived quality can, at times, be achieved merely by an arbitrary change in price by the seller; i.e., if the price is higher the product must be better.

Considering the effect of quality on both the numerator and denominator produces an interesting result: the increase in the productivity index may continue on beyond the point of minimum total quality costs as quality (perceived quality) is improved. Consider quality levels q_1 and q_3 in Figure 2. In going from q_1 to q_3 we have certainly passed the point of minimum total quality costs (q_2). The total quality costs at these two points are equal and thus the denominators of the productivity indexes are equal. However, we have an increase in real quality (and an associated increase in perceived quality). At this point a higher price can be charged for q_3 than q_1 thus resulting in a higher total revenue and higher productivity index. This indicates a maximum productivity index somewhere between q_2 and q_3. The location of that point becomes empirical due to the many factors affecting price, not the least of which is the firms' own propensity to raise prices given an increase in real or perceived quality.

To this point we have investigated a productivity index that is appropriate to the firm but is too broad for most applications to a given plant, department, or product. Measured over time, this index can provide an indication of the general direction of the firms' quality productivity. It has demonstrated that quality has a decided effect on productivity.

While this index may be unwieldly to apply below the firm level, the components are not. Figure 3 presents a hypothetical firm of three departments and three products.

DEPARTMENTS

PRODUCTS	1	2	3
1	Q_{11} I_{11} F_{11} A_{11} P_{11}	Q_{12} I_{12} F_{12} A_{12} P_{12}	Q_{13} I_{13} F_{13} A_{13} P_{13}
2	Q_{21} I_{21} F_{21} A_{21} P_{21}	Q_{22} I_{22} F_{22} A_{22} P_{22}	Q_{23} I_{23} F_{23} A_{23} P_{23}
3	Q_{31} I_{31} F_{31} A_{31} P_{31}	Q_{32} I_{32} F_{32} A_{32} P_{32}	Q_{33} I_{33} F_{33} A_{33} P_{33}

Productivity Components Matrix

FIGURE 3.

Here departments and products are used in their most generic manner. Departments may be the drilling department in a manufacturing firm, reservations in an airline, admitting in a hospital, or underwriting in an insurance firm. Products may be a welding torch for a manufacturing firm, the Dallas to Denver route for an airline, an appendectomy for a hospital, or a whole life policy for the insurance firm. For each product-department combination there is an associated output (Q_{ij}), input (I_{ij}), failure cost (F_{ij}), prevention cost (P_{ij}), and appraisal cost (A_{ij}) generated. Whether these five components can be identified and measured for each product-department combination is a function of the sophistication of the accounting system. In most cases not all of this data will be available in this form. However, certain of these totals are to be found. For example, total production of product one at finished goods is recorded. Failure cost totals are usually found in the form of scrap and rework by department as is warranty by-product. Appraisal costs (inspection and test) are usually associated with a department or group of departments, say, manufacturing, assembly and shipping. Prevention (usually engineering) is often associated with a product or product family, say, all welding torches or counter top ranges.

Measurements, and therefore controls, can be accomplished using this data. The measure of productivity to be used here is:

$$\text{Productivity Index (PI)} = \frac{\text{Performance Achieved}}{\text{Resources Consumed}}$$

For the departments these become:

a) Departmental Productivity: $P.I._d = \frac{\text{Department output total}}{\text{Department input total}}$

b) Departmental Failure Rate: $P.I._{df} = \frac{\text{Department failure total}}{\text{Department output total}}$

c) Appraisal Efficiency: $P.I._{ae} = \frac{\text{Department output total}}{\text{Department appraisal total}}$

d) Appraisal Effectiveness: $P.I._{af} = \frac{\text{Department failure total}}{\text{Department appraisal total}}$

These last two indexes may require groupings of the production department depending upon the organization of the appraisal function as to how it logically services the production departments. In the use of this index it must be recognized that appraisal without associated prevention or correction will not likely result in improvement and is thus highly dependent on other input factors. It is also noted that the sum of the departmental appraisal totals used here may not equal total appraisal, as some appraisal costs may be appropriate only to the firm as a whole.

For the product these become:

a) Product Productivity: $P.I._p = \dfrac{\text{Product output total}}{\text{Product input total}}$

b) Product Failure Rate: $P.I._{pf} = \dfrac{\text{Product failure total}}{\text{Product output total}}$

c) Prevention Efficiency: $P.I._{pe} = \dfrac{\text{Product output total}}{\text{Product prevention total}}$

d) Prevention Effectiveness: $P.I._{pf} = \dfrac{\text{Product failure total}}{\text{Product prevention total}}$

These last two indexes may require groups of product totals into product line totals depending upon the organization of the prevention function and how it logically services the products. As stated above, it must be remembered that prevention without an appraisal effort will not likely result in improvement and is thus highly dependent on other input factors. It need also be noted that the sum of the product prevention totals used here may not equal total prevention, as some prevention costs may be appropriate only to the firm as a whole.

For the entire firm these become:

a) Firm Productivity: $P.I._f = \dfrac{\text{Output total}}{\text{Input total}}$

b) Firm Failure Rate: $P.I._{ff} = \dfrac{\text{Failure total}}{\text{Output total}}$

c) Quality Costs Effectiveness: $P.I._{QC} =$

$$\frac{\text{Failure total}}{\text{Appraisal totals + prevention totals}}$$

Wherever failure total is used in the foregoing it may be desired to use the components of failure; i.e., scrap, rework, warranty, etc. to provide a more detailed measure.

We have now just calculated a number of productivity indexes; ratios. What do we do with them? What is the standard, the goal, the acceptable level? The primary use is the comparison of a given index against itself over time; time series. Much can be learned from "eyeballing" these data plotted over time. Better still are some of the well documented statistical techniques including control charts.

The departmental and product indexes may be compared one to another and to the indexes for the entire firm either for a given time or over a time series. Care is required here in that certain departments or products may be inherently more or less productive than others. For example, a drill press department is inherently less productive, on a labor basis, than is a department of numeric controlled automatic machines. The same condition exists if the products of concern are, say, doorstops and computers. Before applying these comparisons one should reasonably expect them to be comparable.

How well should we do? This is basically a question of capital (including labor) budgeting. Capital budgeting is, in essence, an application of the classical proposition from the economic theory of the firm; namely, a firm should operate at the point where its marginal revenue is just equal to its marginal cost.[4] That is to say that we should continue to invest in productivity improvements so long as we receive one or more dollars in either cost savings or increased revenues for each dollar invested. Any level stated as "acceptable" may well prevent the full utilization of resources and cause the firm to operate at a productivity level less than maximized for the current state of technology.

We have considered several measures of quality productivity on a firm, departmental and product basis. Is this a cook book? No! Productivity measurement is difficult. There are several reasons:

1. Except for the case of the single product, single department firm that remains in that state over time (a very unlikely case), all components of the productivity measurement are aggregates. The aggregation of, say, a toothbrush, lawn mower, and computer, is difficult in any given period of time. It is even more so over time as the product or department mix varies.
2. Work process are complex and unwieldly and thus any measure tends to simplify the actual case. Cost elements are so intertwined that they may not be measurable on an independent basis. One may need several measures or ratios to achieve a clear picture of what is occurring. Here the problem becomes one of economics: how much measurement can I afford to provide the yardstick with which I wish to control?
3. The work is measured after it is in progress. The process is designed to accomplish the output objective usually without regard for the measurement of its components. The measurement process then needs to be designed to fit the work process. Often the measurement process suffers. To provide the measurements needed for productivity indexes the measurement and work process both may require revision. The measurement process might also conflict with the accounting practices required by law and convention. A dual accounting system, one for financial reporting and one for management, can be costly.

CONCLUSION

Quality cost analysis is a productivity measurement. It does, and has, addressed the efficient utilization of resources. The emphasis at the level of the firm, industry and nation has turned from cost control to total resource control. Quality costs analysis can aid in this new direction. Productivity relates the effectiveness in achieving objectives with the efficiency in the use of resources. Quality plays a positive role in both effectiveness and efficiency.

REFERENCES

[1]Cost Effectiveness Committee, ASQC, *Quality Costs — What and How,* American Society for Quality Control, Milwaukee, Wisconsin, Second Edition, pp. 5, 1971.

[2]Mali, Paul, *Improving Total Productivity,* John Wiley and Sons, New York, pp. 6, 1978.

[3]Ibid, pp. 7.

[4]Weston, J. F. and Brighan, E. F., *Managerial Finance,* The Dryden Press, Hinsdale, Illinois, Sixth Edition, pp. 286, 1978.

BIBLIOGRAPHY

Brighan, E. F. and Pappas, J. L., *Managerial Economics,* The Dryden Press, Hinsdale, Illinois, Second Edition, 1976.

Committee on National Statistics, Assembly of Behavioral and Social Sciences, National Research Council, *Measurement and Interpretation of Productivity,* National Academy of Sciences, Washington, D.C., 1979.

Juran, J. M., *Quality Control Handbook,* McGraw-Hill, New York, Third Edition, 1974.

Kendrick, J. W. and Vaccara, B. N., *New Developments in Productivity Measurement and Analysis,* The University of Chicago Press, Chicago, Illinois, 1980.

SELECTION OF MIL-STD-105D PLANS BASED ON COSTS

Dr. Burton S. Liebesman
Bell Telephone Laboratories
Holmdel, New Jersey

ABSTRACT

The selection of an AQL value is an important step in the establishment of a source inspection program. A cost model is developed for selecting the AQL value for MIL-STD-105D plans based on sampling costs, field failure costs, and price differentials. The last quantity refers to a change in price as a function of contractual quality requirements. Simple decision procedures are established for two cases: (1) price differentials known, and (2) price differentials unknown.

Tables of decision parameters are available to facilitate the decision process. The decision maker must supply the costs and price differentials when known.

INTRODUCTION

The AQL is an index for the sampling plans of Military Standard 105D (MIL-STD-105D)[5]. Only a limited number of AQL values exist to choose from. Hence, methods are needed to help the decision maker select the best sampling plan. An investigation of some recent examples indicate that a variety of methods exist. In one case, the criteria used were based on the statistical properties of the plans and resulted in the selection of the AQL value just smaller than the quality standard that had been set previously. In a second case, the AQL selected was the one closest to the standard. Finally, in a third situation the result was the selection of the next smaller AQL value. The procedures used in these examples did not consider the various costs resulting from the use of the sampling plan. In this paper, a model is developed for the selection of sampling plans. The point of view is that of the purchaser and hence sampling costs are not included in the price of the product.

The Literature of Sampling Plan Cost Models

Numerous authors have developed models for the selection of sampling plans. J. C. McKechnie[11] describes a quick method of calculating an AQL value using the cost of inspection and the cost of a defective. His method suffers from three major deficiencies: (1) It fails to take into account the characteristics of the sampling plan (i.e., MIL-STD-105D); (2) It does not include certain pertinent costs such as the cost of processing rejected lots and price differentials for different quality levels; and (3) It does not try to minimize total cost, but rather finds the AQL value that gives the same cost as 100 percent inspection.

J. W. Enell[6], on the other hand, selects the AQL value which equalizes the sampling costs and the expected cost of defectives. His method suffers from first two deficiencies of McKechnie's.

Other authors[1, 4, 8, 9, 10] consider a variety of costs and develop models based on the minimization of the sum of these costs. However, they fail to restrict their plans to those of MIL-STD-105D.

Finally, Brown and Rutemiller[2, 3] do consider the three deficiencies seen in McKechnie's paper, but their approach is based on simulation and does not allow quality changes by the supplier in response to the dynamics of MIL-STD-105D. In addition, simulation is costly, especially when one considers the large number of plans in MIL-STD-105D. And simulation may not give results which are sufficiently accurate.

SUMMARY

My purposes in developing the cost model described in this paper were to extend the work of the authors cited above, eliminate the deficiencies, and obtain procedures that would be simple and inexpensive to use.

The full cost equation is developed first and then simplified by a set of assumptions.* This results in the following equation for the long run average cost per unit shipped:

$$AC[Q,AQL] = \frac{k_N}{N} IC[Q,AQL] + \frac{k_F}{N} DC[Q,AQL] + PQ[Q]$$

where†,

k_N is the lot inspection cost; k_F is the expected field cost of all defectives in the lot; IC and DC are tabulated factors; Q is either the standard or an AQL value; N is the average lot size; AQL is the AQL value of a contending plan; and PQ is the price of the product when produced at quality Q.

The long run average cost equation is used in two ways. First, when $PQ[AQL_i]$ is known we compute $AC[AQL_i, AQL_i]$ for all i where $PQ[AQL_i]$ are known and select the value of i that minimizes $AC[AQL_i, AQL_i]$. The equation for AC is converted to a form that uses two quantities which have been tabulated to simplify the calculations. In the second case, $PQ[AQL_i]$ is unknown and it is assumed that Q equals the standard (S) which is bracketed by two values AQL_j and AQL_{j+1}. In this situation, p_j, the ratio of the expected field cost (K_F) to sampling cost (K_N), is calculated. This is compared to a tabulated quantity p_j^* and if $p_j \geq p_j^*$ the plan with AQL_j is chosen; otherwise the plan with AQL_{j+1} is chosen.

THE FULL COST MODEL

The full cost model includes all terms that contribute to the costs associated with the selection of an AQL sampling plan from MIL-STD-105D. The total cost to the customer for a given quality, intensity‡ and lot size is:

$$TC[q_I,I] = SC[q_I,I] + FC[q_I,I] + N \cdot PQ[q_I] + GC[q_I,I]$$

where, $SC[q_I,I]$ is the sampling cost at product quality q_I and plan intensity I; $FC[q_I,I]$ is the field cost of defectives at product quality q_I and plan intensity I; $PQ[q_I]$ is the price of the product when it is produced at quality q_I (price differentials); $GC[q_I,I]$ is the general staff overhead cost; and N is the average lot size.

I will state a set of 10 assumptions which will result in a simplified cost equation. These assumptions are based on practical experience with source inspection.

Assumption 1 is that the overhead cost of the general staff is a constant with respect to the AQL value. GC is set to 0 because a constant will not affect the selection of the optimum AQL value.

Assumption 2 is that the contractual price of quality is a function of either the stated AQL value or the standard (S), but not the sampling intensity. Then, $PQ[q_I] = PQ[q]$ where, q = either the AQL value or the standard (S). The basic equation for the total cost becomes:

$$TC[q_I,I] = SC[q_I,I] + FC[q_I,I] + N \cdot PQ[q] \quad (1)$$

SAMPLING COSTS

Let us look at the expected value of the sampling cost.

$$E(SC[q_I,I]) = k_1[n_I] + k_2[N,n_I](1-P_A[q_I,I]) + k_3[q_I,I] \quad (2)$$

*The results are very robust to these assumptions. Hence, the procedures may be usable even when some assumptions are not met exactly.

† Appendix I contains a definition of all terms used in this paper.

‡ For MIL-STD-105D there are three intensities, normal, reduced and tightened. A fourth intensity, discontinue, implies no inspection or acceptance of the product.

where, $k_1[n_I]$ = cost to the customer of sampling using sample size n_I; $k_2[N,n_I]$ = cost to the customer of processing a rejected lot; $P_A[q_I,I]$ = probability of acceptance at product quality q_I and intensity I; and $k_3[q_I,I]$ = cost to the customer of replacing found defectives when quality is q_I.

Assumption 3 is that the inspection cost is a step function of the sample size. This is because the cost to the customer of source inspection is based on the time spent by the inspector, which is generally in increments of one-quarter or one-half day. Let k_N be the cost of inspecting a lot during normal intensity. It is assumed that reduced intensity results in the inspector spending half as much time as under normal inspection. Then, the cost of reduced inspection is $0.5k_N$. It is also assumed that when the tightened sample size exceeds the normal sample size, the inspector spends 25 percent more time. Thus,

$$\begin{aligned} k_1[n_I] &= k_N && \text{(normal inspection)} \\ &= \frac{k_N}{2} && \text{(reduced inspection)} \\ &= (COT)k_N && \text{(tightened inspection)} \end{aligned} \quad (3)$$

where, COT = 1.00 when $n_t = n_N$, and COT = 1.25 when $n_T > n_N$.

Assumption 4 is that all rejected lots are not screened but are resubmitted at an improved quality. The cost of reinspecting these lots equals the cost of the original inspection.

$$k_2[N,n_I] = k_1(n_I) \quad (4)$$

Assumption 5 is that the cost to the customer of replacing a found defective is zero. Thus,

$$k_3[n_I, q_I] = 0 \tag{5}$$

Substituting (3), (4) and (5) into (2) we get a simplified equation for the expected value of the sampling costs at intensity I:

$$E(SC[q_I, I]) = k_1[n_I](2 - P_A[q_I, I]) \tag{6}$$

FIELD FAILURE COSTS

The expected cost of replacing defectives that reach the field is:

$$E(FC[q_I, I]) = k_4[q_I, I]P_A[q_I, I] + k_4[q_F, I](1 - P_A[q_I, I])P_A[q_F, I] \tag{7}$$

where, $k_4[q,I] = C_F(N - n_I)q$; C_F = cost of a field failure; $P_A[q,I]$ = probability of acceptance when quality is q_I or q_F and intensity is I; q_I = quality of a lot when intensity is I; and q_F = quality of reinspected lots.

Assumption 6 is that the quality during reduced inspection equals the quality during normal inspection, while the quality during tightened inspection (q_T) is 0.5 AQL:

$$q_I = q_N \text{ (normal and reduced inspection)}$$
$$= 0.5 \text{ AQL (tightened inspection)}$$

The assumption that $q_T = 0.5$ AQL is based on the fact that under this quality the probability of acceptance is very high and the supplier has a small chance of going to the discontinue phase. The chance of going to the discontinue phase increases rapidly when q_T gets worse than 0.5 AQL.

Assumption 7 is that the quality of a resubmitted lot (q_F) equals 0.5 AQL and that all resubmitted lots are accepted because the probability of acceptance for quality equal to 0.5 AQL is very close to 1.

Using assumptions 6 and 7, the equation for expected field costs becomes:

$$E(FC[q_I, I]) = C_F(N - n_I)\{P_A[q_I, I]q_I + (1 - P_A[q_I, I])(0.5AQL)\} \tag{9}$$

TOTAL COST

The next set of assumptions will enable us to combine all costs in a meaningful way.

Assumption 8. A cycle of operations consists of a sequence of periods designated N, R, T and D for normal, reduced, tightened and discontinued phases. Shown below is a typical cycle:

$$N \rightarrow R \rightarrow N \rightarrow R \ldots \rightarrow R \rightarrow N \rightarrow T \rightarrow (D \text{ or } N)$$

All cycles end when the system switches to the tightened phase followed by a switch either to discontinue or back to the normal phase. The expected number of lots inspected during the normal, reduced and tightened portions of the cycle are E_N, E_R and E_T respectively. The expected total cost during a cycle is just the total cost per inspection during each portion multiplied by the expected number of lots inspected during that portion. The calculations will be done in terms of the expected total cost per unit shipped to the field. During a cycle, the number of units shipped is $N(E_N+E_R+E_T)$. Then, the expected total cost per unit shipped is:

$$TCU[q_N,AQL] = \frac{E_N TC[q_N,N]+E_R TC[q_N,R]+E_T TC[0.5AQL,T]}{N(E_R+E_R+E_T)} \qquad (10)$$

Assumption 9 is that quality provided by the suppliers is at one of two levels and that quality remains at a given level for an entire cycle. Each level has a distribution of quality for lots being inspected, with the mean values being Q (good level) and 3Q (poor level) respectively. In addition, I assume that the good level provides 75 percent of all units shipped and the poor level provides 25 percent of all units shipped.* This gives an average quality entering the inspection process of 1.5Q. Thus, q in equation (10) will have values Q and 3Q. Assumption 9 is equivalent to calculating the costs at the two levels of quality and weighting the results by 75 and .25.

Assumption 10 is that when the quality is 3Q, the reduced phase is not entered. This is because the probability of going to reduced inspection is very small and this assumption simplified the analysis considerably.

The equation for the long run average cost per unit shipped becomes:

$$AC[Q,AQL] = 0.75TCU[Q,AQL] + 0.25TCU[3Q,AQL]$$

or

$$AC[Q,AQL] = \frac{0.75}{N}(SCI[Q,AQL] + FCI[Q,AQL]) + \frac{0.25}{N}(SC3[3Q,AQL] + FC3[3Q,AQL]) + PQ[Q] \qquad (11)$$

where, SCl[Q,AQL] = sampling cost when quality is Q and the plan used is one with an AQL value equal to AQL; FCl[Q,AQL] = field cost when quality is Q and the plan used is one with an AQL value equal to AQL; SC3[3Q,AQL] = sampling cost when quality is 3Q and the plan used is one with an AQL value equal to AQL; and FC3[3Q,AQL] = field cost when quality is 3Q and the plan used is one with an AQL value equal to AQL.

*Each level consists of many states which are part of a Markov chain. Only those states of one level which end a cycle communicate with states of the other level. I assume that 75 percent of the units are produced when the Markov chain is in one of the states of the good quality level and 25 percent are produced in one of the states of the bad quality level. No assumption is made concerning the frequency of occurrence of cycles of each type.

The equation for SCl[Q,AQL] is obtained by substituting equations (3) and (6) into equation (1) for each intensity; substituting the three values of $TC[q_I,I]$ with $q_N = Q$, $q_R = Q$ and $q_T = 0.5$ AQL into equations (10) to obtain TCU[Q,AQL]; and separating the sampling cost from the other costs.

$$SC1[Q,AQL] = \frac{\{E_N k_N[2-P_{AN}] + \frac{E_R k_N}{2}[2-P_{AR}] + E_T(COT)k_N[2-P_{AT}]\}}{E_N+E_R+E_T} \tag{12}$$

where, P_{AN}, P_{AR} and P_{AT} are the probabilities of acceptance when $q_N = q_R = Q$, $q_T = 0.5$ AQL; and E_N, E_R and E_T are the expected number of lots inspected under the three intensities.

The equation for SC3[3Q,AQL] is obtained by substituting equations (3) and (6) into equation (1) for normal and tightened intensity; substituting the two values of $TC[q_I,I]$ with $q_N = 3Q$ and $q_T = 0.5$ AQL into equation (10) to obtain TCU[Q,AQL]; and separating the sampling cost from the other costs.

$$SC3[3Q,AQL] = \frac{E_{N3}k_N(2-P_{AN3})+E_T(COT)k_N[2-P_{AT}]}{E_{N3}+E_T} \tag{13}$$

where, P_{AN3} and E_{N3} are the probability of acceptance and the expected number of lots inspected during a cycle when quality is 3Q.*

The equation for FC1[Q,AQL] is obtained by substituting equation (9) into equation (1) for each intensity; substituting the three values of $TC[q_I,I]$ with $q_N = Q$, $q_R = Q$ and $q_T = 0.5$ AQL into equation (10) to obtain TCU[Q,AQL]; and separating the field costs from the other costs.

$$FC1[Q,AQL] = \frac{C_F\{E_N(N-n_N)\,[P_{AN}Q + 0.5AQL(1-P_{AN})] + E_R(N-n_R)\,[P_{AR}Q + 0.5AQL(1-P_{AR})] + E_T(N-n_T)\,(0.5AQL)\}}{E_N+E_R+E_T} \tag{14}$$

The equation for FC3[3Q,AQL] is obtained by substituting equations (9) into equation (1) for normal and tightened intensity; substituting the two values of $TC[q_I,I]$ with $q_N = 3Q$ and $q_T = 0.5$ AQL into equation (10) to obtain TCU[3Q,AQL]; and separating the field costs from the other costs.

$$FC3[3Q,AQL] = \frac{C_F\{E_{N3}(N-n_N)\,[3QP_{AN3} + 0.5AQL(1-P_{AN3})] + E_T(N-n_T\,(0.5AQL)\}}{E_{N3}+E_T} \tag{15}$$

The per unit inspection and field failure costs can be factored out of equation (11) so that we obtain the simplified cost equation:

$$AC[Q,AQL] = \frac{k_N}{N}\, IC[Q,AQL] + \frac{k_F}{N}\, DC[Q,AQL] + PQ[Q] \tag{16}$$

where, $k_F = C_F Q(N-n_N)$;

$$IC[Q,AQL] = \frac{0.75}{k_N} SC1[Q,AQL] + \frac{0.25}{k_N} SC3[3Q,AQL]; \quad (17)$$

$$DC[Q,AQL] = \frac{0.75}{k_F} FC1[Q,AQL] + \frac{0.25}{k_F} FC3[3Q,AQL]. \quad (18)$$

*Note that E_T is used here because the quality during the tightened phase is assumed to be the same under Q and 3Q.

The terms IC and DC do not depend upon the product being inspected, but instead are functions of the operating characteristic curves of the sampling plan under consideration.† We can caluculate IC and DC using equations (12) through (15), (17) and (18), tabulate their values and provide these tables for the decision maker. He can then use equation (16) to calculate the long run average cost per unit shipped for quality Q and AQL value AQL.

Case 1: Price of Quality Known

In case 1 the price of quality is known for some values of AQL. The decision maker calculates $AC[AQL_i\ AQL_i]$ for all i where $PQ[AQL_i]$ is known. The value of AQL_i that minimizes AC is selected as the optimum AQL value. Equations (16), (17) and (18) are used directly in the calculations.

Case 2: Price of Quality Unknown

In case 2 where $PQ[AQL_i]$ is unknown, it is assumed that the supplier operates at the standard (Q=S). We find the quantities AQL_j and AQL_{j+1} such that

$$AQL_j \leq S < AQL_{j+1} \quad (19)$$

Then we compare $AC[S,AQL_j]$ with $AC[S,AQL_{j+1}]$. Since PQ[S] is unknown, but will appear in each equation we can assume that PQ[S] = 0. Then,

$$AC[S,AQL_j] = \frac{k_{Nj}}{N} IC_j + \frac{k_{Fj}}{N} DC_j \quad (20)$$

and

$$AC[S,AQL_{j+1}] = \frac{k_{N,j+1}}{N} IC_{j+1} + \frac{k_{F,j+1}}{N} DC_{j+1} \quad (21)$$

where, $IC_j = IC[S,AQL_j]$; $IC_{j+1} = IC[S,AQL_{j+1}]$; $DC_j = DC[S,AQL_j]$; and $DC_{j+1} = DC[S,AQL_{j+1}]$.

Equation (20) can be rewritten as follows:

$$AC[S,AQL_j] = \frac{k_{Nj}}{N} (IC_j + \varrho_j\ DC_j) \quad (22)$$

$$\text{where, } \varrho_j = \frac{k_{Fj}}{k_{Nj}} = \frac{C_F S(N-n_{Nj})}{k_{Nj}} \quad (23)$$

In addition, equation (21) can be rewritten as follows:

$$AC[S,AQL_{j+1}] = \frac{k_{N,j+1}}{N} IC_{j+1} + \frac{k_{Nj}}{N}\left[\frac{N-n_{N,j+1}}{N-n_{Nj}}\right] p_jDC_{j+1} \quad (24)$$

Equating $AC[S,AQL_j]$ to $AC[S,AQL_{j+1}]$ we obtain a value of p_j where the long run average costs are equal. This quantity is called p_j^*.

$$\frac{k_{Nj}}{n}\left[IC_j + p_j^*DC_j\right] = \frac{k_{N,j+1}}{N} IC_{j+1} + \frac{k_{Nj}}{N}\left[\frac{N-n_{N,j+1}}{N-n_{Nj}}\right] p_j^*DC_{j+1}$$

Solving for p_j^*:

†Note that by dividing by k_N and k_F in equations (17) and (18) these product dependent terms were removed from IC and DC respectively. This leaves quantities which are functions of the quality, the AQL value and the probability of acceptance.

$$p_j^* = \frac{\frac{k_{N,j+1}}{k_{Nj}} IC_{j+1} - IC_j}{DC_j - \left[\frac{N-n_{N,j+1}}{N-n_{Nj}}\right] DC_{j+1}} \quad (25)$$

Equation (25) can be used to calculate the decision variable p_j^* which then can be used to select the AQL value:

If $p_j \geq p_j^*$ select AQL_j, and

if $p_j < p_j^*$ select AQL_{j+1}.

Note that p_j^* is a function of S and AQL because IC and DC are functions of these quantities.

EXAMPLES

In example 1, the price of quality is known. Equation (16) is used to calculate $AC[AQL_i,AQL_i]$ for all values of i where $PQ[AQL_i]$ is known. The value of AQL_i that minimizes AC is selected as the optimum value.

The parameters for example 1 are N = 850, k_N = \$240.00, C_F = \$10.00 and n_N = 80. Table I gives the price of quality (PQ), the cost factors (IC and DC) and the AQL values. The minimum value of AC occurs for an AQL = 1.50. Note that if C_F = \$150.00, the optimum AQL value is 1.00 percent.

In example 2, the price of quality is unknown. Equations (19), (23), (25) and (26) apply. The parameters of this example are N = 500, k_N = \$126.00, C_F =\$150.00, S = 0.65 percent and n_N = 50. The value of ϱ_j from equation (23) is 3.25. The value of ϱ_j^* is obtained from Table II. Enter this table with sample size code letter H(N = 500) and find ϱ^* under column S/AQL = 1 (since S = AQL = 0.65). The value of $\varrho_j^* = 0.90$. Since $\varrho_j > \varrho_j^*$ we select AQL = 0.65 percent. If C_F were $10.00 instead of $150.00 the choice would be AQL = 1.00 percent.

TABULATED DECISION PARAMETERS

The parameters IC, DC and ϱ_j^* have been tabulated in Table II for the plans of MIL-STD-105D with acceptance number = 1. The tables for the other acceptance numbers have been completed and will be published in the near future.

FUTURE WORK

Reference[5] states that "unless otherwise specified, Inspection Level II will be used." As an extension of the current work, the inspection level could be considered variable. Since sample size is strongly a factor of the inspection level costs could be saved by using a level requiring a smaller sample size.

ACKNOWLEDGEMENTS

I would like to thank H. Cautin, G. G. Brush, and M. S. Phadke of Bell Telephone Laboratories for their help in developing this cost model.

REFERENCES

1. Breakwell, J. V., "Economically Optimum Acceptance Tests," *Journal of American Statistical Association*, June 1956, pp. 243-256.
2. Brown, G. G., and Rutemiller, H. C., "A Cost Analysis of Sampling Inspection Under Military Standard 105D," *Naval Research Logistics Quarterly,* 20, 1970, pp. 181-199.
3. Brown, G. G., and Rutemiller, H. C., "Tables for Determining Expected Cost Per Unit Under MIL-STD-105D Single Sampling Schemes," *AIIE Transactions,* 1974, pp. 135-142.
4. Dawes, E. W., "Optimizing Attribute Sampling Costs — A Case Study," *1972 ASQC Technical Conference Transactions,* pp. 181-187.
5. Department of Defense, "Sampling Procedures and Tables for Inspection by Attributes," MIL-STD-105D, April 1963, Washington, DC.
6. Enell, J. W., "Which Sampling Plan Should I Choose?," *Industrial Quality Control,* May 1954, pp. 99-100.
7. Flehinger, B. J., and Miller, J. M., "Incentive Contracts and Price Differentials," in *OR and Reliability* (Ed D. Grouchko), Gordon and Breach, Inc., New York, 1971, pp. 121-143.
8. Guenther, W. C., "On the Determination of Single Sampling Attribute Plans Based Upon a Linear Model and Prior Distribution," *Technometrics,* 13, 1971, pp. 483-498.
9. Hald, A., "The Compound Hypergeometric Distribution and a System of Single Sampling Plans Based on Prior Distributions and Costs," *Technometrics,* 2, 1960, pp. 275-340.

10. Hald, A., "Bayesian Single Sampling Attribute Plans for Continuous Prior Distributions," *Technometrics,* 14, 1968, pp. 667-683.
11. McKechnie, J. C., "Minimum Cost Inspection Levels," *Quality,* May 1977, p. 38.

Table 1

Parameters of Example 1

AQL	PQ	IC	DC	AC
0.65	2.50	1.087	0.971	2.864
1.00	1.50	0.831	1.053	1.830
1.50	1.00	0.906	1.001	1.392
2.50	1.00	0.813	1.016	1.460
4.00	1.00	0.804	0.957	1.574

Table II

Tabulated Decision Parameters (ac = 1)

Sample Size Code Letter	IC(AQL,AQL)	DC(AQL,AQL)	S/AQL 1.0	1.2	1.4	1.6
			p*			
B	1.076	0.993	0.5	0.5	0.6	0.6
C	1.076	0.992	0.6	0.6	0.6	0.7
D	1.093	0.973	0.7	0.7	0.7	0.8
E	1.083	0.983	0.8	0.8	0.9	0.9
F	1.094	0.971	0.8	0.8	0.9	0.9
G	1.085	0.980	0.9	0.9	0.9	1.0
H	1.087	0.977	0.9	0.9	1.0	1.0
J	1.085	0.976	1.0	1.1	1.1	1.1
K	1.085	0.975	1.1	1.2	1.2	1.2
L	1.088	0.969	1.3	1.3	1.3	1.3
M	1.085	0.971	1.3	1.3	1.3	1.3
N	1.087	0.969	1.4	1.4	1.4	1.3
P	1.085	0.971	1.4	1.4	1.4	1.4
Q	1.085	0.971	1.4	1.4	1.4	1.4

APPENDIX I
Definition of Symbols

ac = Acceptance number.

AC[Q,AQL] = Long run average cost per unit shipped.

AQL = Acceptable quality level; AQL value of a contending plan.

C_F = Cost of a field failure.

$$COT = \begin{Bmatrix} 1 & \text{when } n_T = n_N \\ 1.25 & \text{when } n_T > n_N \end{Bmatrix}$$

$DC[Q,AQL]$ = Defective cost factor when product quality is on the average Q and a plan with an AQL value equal to AQL is used.

DC_j = $DC[S,AQL_j]$; $DC_{j+1} = DC[S,AQL_{j+1}]$

E_N = Expected number of lots inspected during the normal part of the cycle when $q = Q$.

E_{N3} = Expected number of lots inspected during the normal part of the cycle when $q = 3Q$.

E_R = Expected number of lots inspected during the reduced part of the cycle.

E_T = Expected number of lots inspected during the tightened part of the cycle.

$FC[q_I,I]$ = Field cost of defectives at quality q_I and intensity I.

$FC1[Q,AQL]$ = Field cost when quality is Q and the plan used is one with an AQL value equal to AQL.

$GC[q_I,I]$ = General staff overhead cost.

$IC[Q,AQL]$ = Inspection cost factor when quality submitted for inspection is on the average Q and a plan with an AQL value equal to AQL is used.

IC_j = $IC[S,AQL_j]$; $IC_{j+1} = IC[S,AQL_{j+1}]$

$k_1[n_I]$ = Cost to the customer of sampling using sample size n_I.

$k_2[N,n_I]$ = Cost to the customer of processing a rejected lot.

$k_3[q_I,I]$ = Cost to the customer of replacing found defectives when quality is q_I.

$k_4[q_I]$ = Field cost of all defectives in a lot when quality is q_I.

k_F = Expected field cost of all defectives in a lot when Quality = Q.

k_N = Cost of inspecting a lot during normal intensity.

N = Average lot size.

n_I = Sample size when intensity is I.

n_j,n_{j+1} = Sample sizes under normal intensity for the case of price of quality unknown and $AQL_j \leq S > AQL_{j+1}$.

n_N = Sample size under normal intensity.

n_R = Sample size under reduced intensity.

n_T = Sample size under tightened intensity.

$P_A[q,I]$ = Probability of acceptance at product quality q_I or q_F and intensity I.

$PQ[q]$ = The price of a product when it is produced at quality q; often called price differentials [7].

P_{AN} = Probability of acceptance when quality is Q and the intensity is normal.

P_{AR} = Probability of acceptance when quality is Q and the intensity is reduced.
P_{AT} = Probability of acceptance when quality is Q and the intensity is tightened.
P_{AN3} = Probability of acceptance when quality is 3Q and the intensity is normal.
Q = Mean of the distribution of good quality; either the standard or an AQL value.
q = Quality.
q_F = Quality of reinspected lots.
q_I = Quality when intensity is I.
q_N = Quality when intensity is normal.
q_T = Quality when intensity is tightened.
ϱ_j = Ratio of expected field cost to sampling cost.
ϱ_j^* = Decision variable when PQ is unknown.
S = Standard.
$SC[q_I,I]$ = Sampling cost at product quality q_I and plan intensity I.
SC1[Q,AQL] = Sampling cost when quantity is Q and the plan used is one with an AQL value equal to AQL.
SC3[3Q,AQL] = Sampling cost when quality is 3Q and the plan used is one with an AQL value equal to AQL.
$TC[q_I,I]$ = Total cost at product quality q_I and plan intensity is I.
TCU[q,AQL] = Expected total cost per unit shipped when the quality during the normal phase is q and the plan has an AQL value equal to AQL.

THE COST OF SOFTWARE QUALITY ASSURANCE

W. D. Goeller, Quality Manager-Prime Programs
Vought Corporation
Dallas, Texas

ABSTRACT

Software Quality Cost will be developed by first identifying the primary cost drivers and then discussing the methods that can be applied in measuring these cost drivers.

In order to have a common ground of understanding for the Software Quality Assurance System that will be used in developing Software Quality Cost, MIL-S-52779 will be the base system document. Since this is a new field, with a limited sampling of actual experience, this paper will contain a mixture of actual experience coupled with theoretical development of cost needs. Experience will be the first choice in developing the Software Quality Cost measurement system, when available.

INTRODUCTION

For any of us to understand each other, there must be a common ground of reference. In this paper, the common ground will be the military Software Quality Assurance Specification, MIL-S-52779. The specification provides a control system that is applicable through all phases of software development and use. The control system guidelines that the military specification provides have sufficient flexibility to use in most applications.

Of course, the question of software quality costs may seem premature since most companies are only now struggling to conceptualize a quality assurance effort for software. Our experience with quality costs for hardware have led us to a circular argument; i.e., if we could measure the cost, we could estimate the savings that would accrue, thus helping justify the cost of changes that enhance the control systems needed to measure the costs. We are attempting to build into the software development process the basis for future measures which will circumvent this same problem for software quality costs.

There are distinctions between hardware and software that must be recognized. In software, your purpose is to build a single high quality item, not a production of many items. The normal quality measuring systems that show percent defective do not apply to this one-of-a-kind product. Unfortunately, a single unit of "bad" software gets high visibility since it is usually responsible for producing or controlling many items. To complicate the problem further, general management has little, if any, hands-on experience with this relatively new product.

SOFTWARE COST CHARACTERISTICS

In order to help identify where the cost drivers occur, a literature search was conducted to determine if there were times in the life cycle of the software where the cost of change varied significantly. D. J. Riefer of Software Management Consultants has accumulated data that is shown in Table I.

Table I.

Phase	Cost per Change	Factor
Design	$45-50	X
Test	$375	7.5X
Maintenance	$1500	30.0X

As the table shows, the manintenance phase is obviously the first area to prospect for gold. Dr. E. Burton Swanson of UCLA[1] conducted a survey scoping the magnitude of this cost. He received 487 responses that reveal 49% of their software applications effort is spent on maintenance of operable software with 43% spent on new development. The remainder was on incidental costs. The effort called maintenance was subdivided into the categories of perfective, adaptive, and corrective with corrective being defined (as you might expect) as a response to a failure of any kind. The definition of perfective maintenance was given as refining for efficiency or other user enhancements; adaptive maintenance was defined as in anticipation of potential failure.

If this discussion of maintenance types seems out of place in a Quality Assurance paper, it is because we are used to working with hardware. In 1975, the Rome Air Development Center (RADC) contracted for the design of a center that would acquire, analyze and disseminate information on software technology. In 1978, that center was established and named the Data & Analysis Center for Software (DACS). This center has published the DACS Software Engineering Glossary. The DACS Software Engineering Glossary does not define "corrective action;" instead, it defines "corrective maintenance" as specifically intended to eliminate a fault. Whether we call them faults, failures, potential failures, corrective maintenance or corrective action, the basic, underlying situation is that operable software requires much change.

The 30 to 1 ratio for cost of change shown in Table I was established in a commercial software environment. To be thorough, we looked for a comparable ratio for military software; i.e., command, control, communications software. An unpublished Department of Defense study of a large scale military procurement cited a 37 to 1 ratio for cost of software changes in the maintenance phase[2]. The ratio is about what we would expect; a slightly higher ratio for military applications since the military has a very formal approach embodied in Directive 5000.29 to manage computer resources as elements of major importance. Clearly, an early quality assurance effort with the objective of preventing/reducing changes can have a high payoff.

The study summarized in Table I analyzed changes to find assignable causes. Their conclusion was that 80% of the changes were traceable to either design deficiencies or requirements deficiencies. A work by Alberts[3] looking at known errors indicates that design errors occur most frequently. Design error frequencies ranged from 46% to 65% of the total errors. This finding is further substantiated by Boehm[4] in his work on Software Engineering. To help frame the concept of a design error for software, it is a change that requires a corresponding change to the specification used by the programmer. It is often characterized by a lack of understanding (communication) of an algorithm which results in the wrong problem being solved.

The testing of software is not to be confused with acceptance testing of hardware. Hardware functions can be thoroughly exercised. A paper by Dunn & Ullman[5] states that software is functionally non-testable due to the large number of paths through the structure. Figure 1 shows a simple structure that contains approximately 10^{20} paths that would have to be exercised to totally assure its correctness.

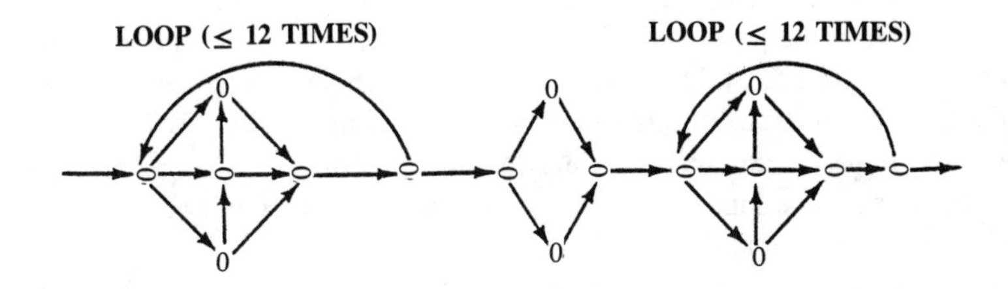

FIGURE 1.

At the rate of one path exercised per nanosecond, testing would be 62% through today, if it had begun at the birth of Christ. Simply said, we are unable to test all possible paths.

This inability to thoroughly test software does not mean that testing is not beneficial. Over half of the errors (54%)[3] are not uncovered during the design phase. The testing is the only chance to detect these errors before operating the system in the customer's shop.

A view of latent software errors is given by IBM Vice President Bob Evans who shows the progress of software integrity from 1967 to 1977[6]. The 1967 era programmer produced 500 "good" lines of software per year with an estimated 7.5 errors remaining after delivery. Based on the same criteria, in 1977 performance improved to 1,000 "good" lines of software per year, with an estimated 1.5 remaining errors. In other words, we have dropped from 1.5% error to 0.15%, a factor of 10.

To summarize:

1. Errors found later in the software life cycle are the most costly to repair.
2. Software test is in its infancy and is very difficult to optimize.
3. Delivered software will have latent errors.

With this background, we can now look at the software life cycle to determine the cost drivers and develop or identify a measure that can be applied to gauge improvement.

COST DRIVERS

Dividing the software life cycle into two distinct activities of development (predelivery) and maintenance (post-delivery), Alberts[3] quotes maintenance to be just over 50% of total costs, while Boehm[4] quotes two separate studies at General Motors and General Telephone & Electronics. These studies show maintenance to be approximately 75% at General Motors and 60% at General Telephone & Electronics. Two Air Force system studies averaged 69% for maintenance. Perhaps even more graphic is one on-board aircraft computer development where costs were $75 per instruction versus $4000 per instruction for maintenance. These averaged together would show an expected maintenance cost of 63.5% of the total cost. Clearly, maintenance costs are the number one driver of software life cycle costs. Maintenance, whether corrective, perfective or adaptive, requires a thorough understanding of the existing software which dictates thorough documentation. This documentation must be kept current with the software so some kind of configuration management activity can be used to identify and control changes.

Figure 2 indicates the division of cost between maintenance and development. The division of cost is based upon the rationale developed above.

In order to determine the amount of cost that test absorbs from the development phase, Alberts[3] study was used. His study indicated that 48% of the development phase costs are due to testing. Figure 3 gives this relationship.

Combining the information developed here with Alberts[3] study, the relationship of the software drivers is derived. Figure 4 provides the graphic illustration of this relationship.

A third cost driver was identified by Bell & Thayer[2] as the requirements definition portion of the design phase. They simply state that, "requirements are incorrect, ambiguous, inconsistent or simply missing." To combat this, software quality assurance must have frequent reviews of requirements and review of the design as it progresses to implementation. Peer reviews are a normal activity here and would seem to be very similar to the productive Quality Circles used so effectively in Japan.

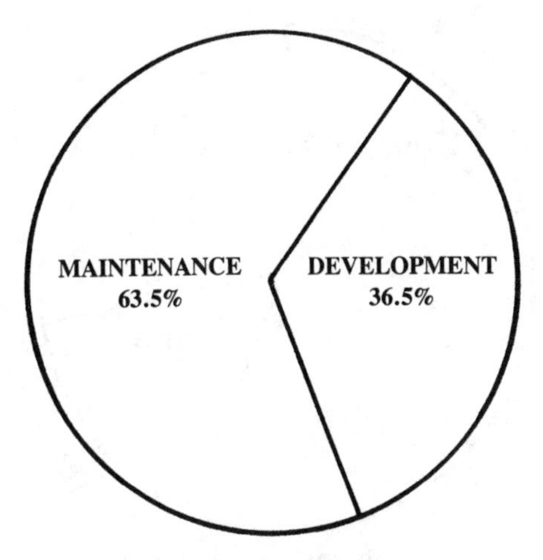

MAINTENANCE TO DEVELOPMENT
COST RELATIONSHIP

FIGURE 2.

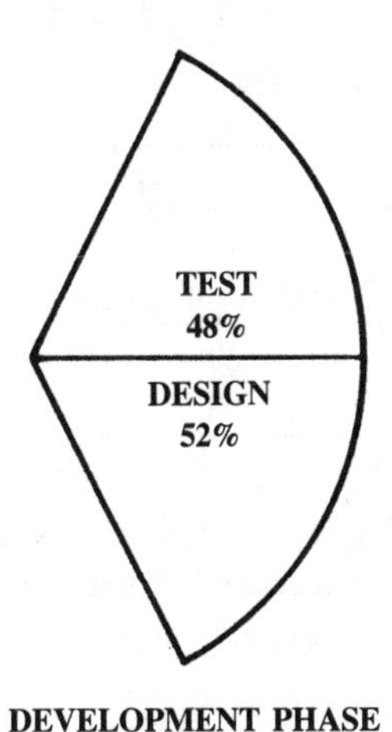

DEVELOPMENT PHASE

FIGURE 3.

In a classic experiment performed by G.M. Weinberg[7], two different programmers were given a 20 page specification from which to produce software. The specifications were identical, except for one sentence. In one specification, that sentence stated the software must be operable in the shortest time; the other specification sentence stated the desire for efficient execution. Table II shows the result of the two programmer's efforts.

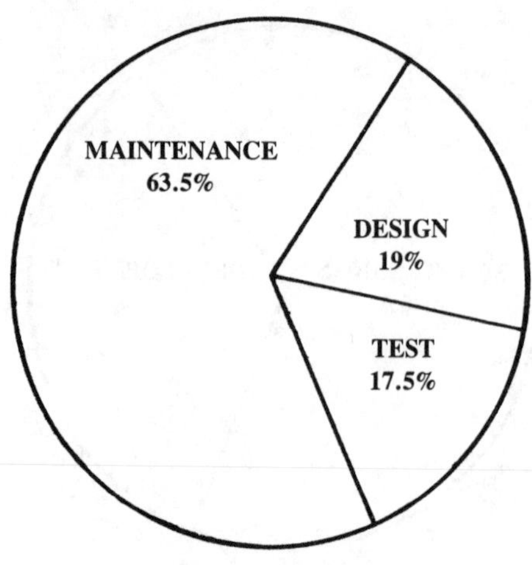

COST DRIVERS OF SOFTWARE

FIGURE 4.

Table II.

	Results	
Objective	**Mean Number of Runs**	**Mean Execution Time**
Prompt (P)	29	6
Efficient (E)	69	1

This experiment was followed up by a second involving five programming teams, again given the same basic specification, but with five different stated objectives. As before, each team ranked on top for its objective (Table III).

Table III.

	Ranking				
Group Objective	**Core**	**Output Clarity**	**Program Clarity**	**State-ments**	**Hours**
Minimum Core	1	4	4	2	5
Output Clarity	5	1	1-2	5	2-3
Program Clarity	3	2	1-2	3	4
Minimum Statements	2	5	3	1	2-3
Minimum Hours	4	3	5	4	1

This work allows us to state (oversimplified) that you get what you ask for.

MEASUREMENT

One of the objectives stated in the Abstract is to design a system to measure the Quality Cost of Software. There are two significant cost areas. The first is the cost of the Quality Assurance departmental personnel and facilities. From Guide Systems Quality Assurance Project (MA220)[8], an industry-wide survey showed that the average number of Quality Assurance Staff to programmers is 1 to 15. This survey indicated a wide variation in the ratios and suggested an effective ratio depends on other factors, such as:

- ratio of new development to maintenance
- number of functions other than Quality Assurance performed by the staff
- budget constraints
- corporate demand for Quality.

The list of respondents of this survey covers many of the nations's top 100 corporations and appears to be an adequate beginning point when attempting to size a Quality Assurance Software Organization.

The other area of Software Quality Assurance cost that will be investigated is the cost of errors. In order to understand the effect of errors on cost, we must return to the previously developed cost drivers of maintenance, test and design requirements. The matrix developed below in Table IV gives a relation of these cost drivers to a set of basic objectives usually associated with each phase. A measurement identification follows the objectives.

Table IV.

COST DRIVERS VS MEASUREMENT

Cost Drivers	Objectives	Measurement
Design Requirements	• Clear Unambigious Requirements	• Errors from Peer/QA Design Review
	• Design Reflects Requirements	• Errors from Peer/QA Design Review
	• Program Consistency	• Errors from Peer/QA Design Review
Test	• Test Planned to Demonstrate Requirements	• Errors from Test Plan Review
	• Stress Software to Extremes of Capability	• Errors from Test Plan Review
Maintenance	• Minimum Corrective Maintenance	• Post-Delivery Errors
	• Easily Adaptive	• N/A
	• Easily Perfective	• N/A

The table develops three basic areas of measurement. These are errors found during peer/QA design reviews, errors from test plan reviews, and post-delivery errors.

In order to associate a cost for each of these error determination sources, the accounting department should establish special charge numbers for redoing tasks that these errors cause. Each area can then plot the error on an associated cost/error chart similar to that shown in Figure 5.

Upon gathering the information displayed in Figure 5, the need exists to use it to optimize our resource allocation. An objective should certainly be the correction of errors before delivery. This has a double advantage. The company saves, first money by reducing errors during the costly mainenance phase, and second, its reputation with its customer.

To understand how to use the cost information that is obtained, let us return to the definition of the three cost drivers and recall there were basically two divisions. These were development and maintenance. Development was a combination of test and design as noted in Figure 2. If we were able to determine the optimum development cost that was coupled with the optimum maintenance cost, we should be at or near the lowest possible program cost. Figure 6 is developed for that purpose.

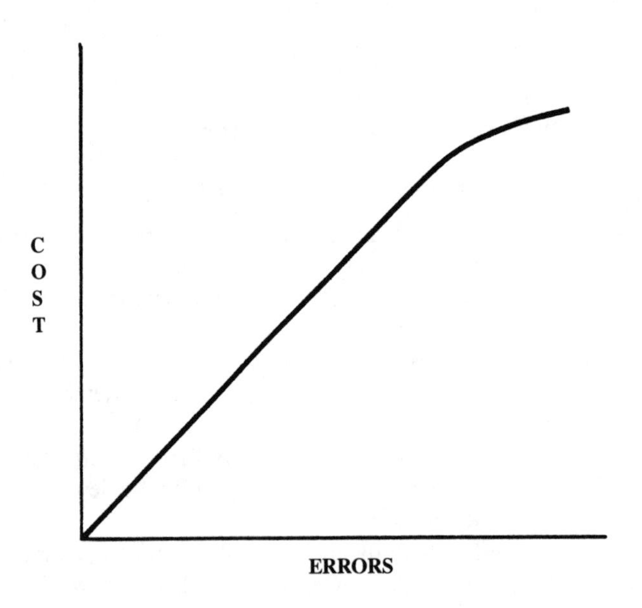

COST-ERROR CHART

FIGURE 5.

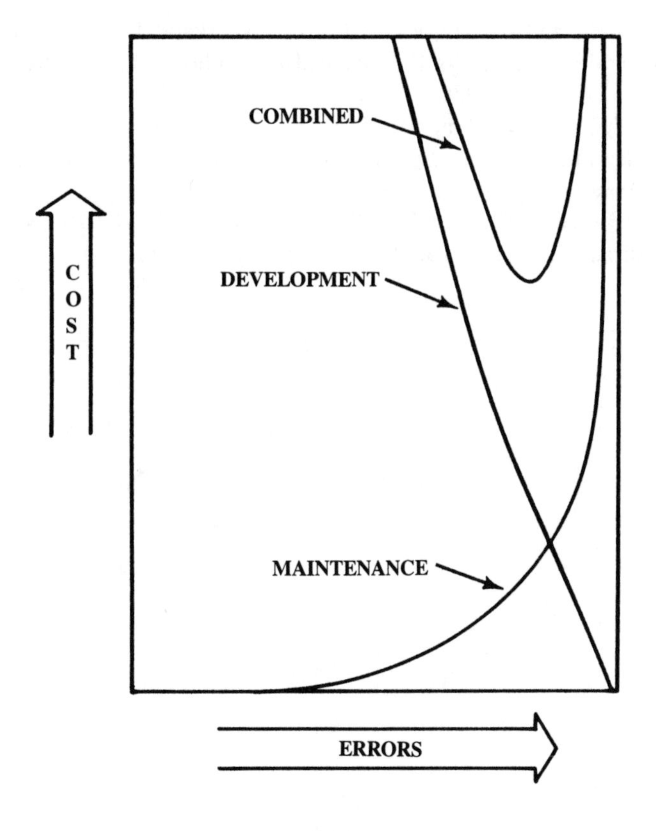

FIGURE 6.

If extreme effort were concentrated in the development phase (design and test), it is expected that very few errors would get to the user and conversely, if the development costs were held to a minimum, we would expect a large number of errors. This is certainly what has been developed earlier in this paper. Also, we noted that a high number of latent errors caused a severe increase in the cost of maintenance. Thus, we can develop a maintenance curve with an asymptotic condition at the right hand side of our Figure 6. Many of you will recognize this as a standard economic curve that applies to many characteristics in business. The trick is to ride in the saddle. The key to development is to track both your costs and errors generated by the three cost drivers.

(1) *Computerworld,* May 26, 1980
(2) D. J. Riefer, Software Management Consultants, Torrance, California
(3) Alberts, David S., "The Economics of Software Quality Assurance", National Computer Conference, pp. 433-442, 1976
(4) Boehm, B. W., "Software Engineering", TRW, Redondo Beach, California, pp. 19-59, 1976

(5) Dunn, Robert H., and Ullman, Richard S., ''A Workable Software Quality Reliability Plan'', Proceedings 1978 Annual Reliability and Maintainability Syposium, pp. 210-217, 1978
(6) *Information System News,* May 5, 1980
(7) Weinberg, Gerald M., "The Psychology of Improved Programming Performance", *Datamation,* pp. 82-85, November 1972
(8) "A Guide to Implementing Systems Quality Assurance", GUIDE Systems Quality Assurance Project (MA220), pp. 41-42, December 1979

GUIDE FOR MANAGING VENDOR QUALITY COSTS

W. O. Winchell, Manager
Customer Satisfaction Activities
Consumer Relations & Service Staff
3044 W. Grand Blvd., Room 2-236
Detroit, Michigan

ABSTRACT

In this tutorial, key portions of the new ASQC publication "Guide For Managing Vendor Quality Costs" will be reviewed. Covered first will be the fundamentals regarding quality costs found in vendor relationships. Following this will be suggestions as how to effectively manage these quality costs. Through starting joint programs with key vendors, significant quality problems can be solved quickly and permanently. Also if conditions permit, encouraging vendors to start their own quality cost program can ultimately result in valuable improvements. Most important of all is that no quality cost program is effective without prompt corrective action.

DISCUSSION

Our talk today will discuss key portions of the new ASQC publication — "Guide For Managing Vendor Quality Costs." Vendor quality cost is an important subject to all of us, because in most cases, we not only buy from vendors but we are a vendor to someone else. Key to first understanding and then managing vendor quality costs is recognizing that much of these costs are hidden. By hidden, we mean that these costs are not really easily identified or perhaps not even seen as vendor related quality costs.

For the purpose of this talk, we will look at these hidden costs from just one perspective — that of the buyer. From the buyer's viewpoint, hidden vendor quality costs occur in basic ways.

The first type of cost is quality cost inside the vendor's facility. For the most part it is much the same as the buyer's quality cost. However, it is included as part of the piece price the buyer pays for the part. The vendor, perhaps rightly so, protects the identification of this investment for competitive reasons. The buyer is furnished the piece price alone and, if this price is competitive, he's usually satisfied with this knowledge.

The second type of hidden vendor quality cost is that which buyer spends, but does not usually segregate, to insure the quality of a product at the vendor's facility. This may be in the form of sending quality engineers to the vendor's plant to help him solve a crisis.

The third type of hidden vendor quality costs is that which the buyer spends in-house, again not usually segregated, specifically for vendors. An example of this may be process engineering required to repair or correct a purchased part not quite up-to-specification. Many of us call this fire-fighting to keep the line going.

Of course, there are also visible costs that the buyer can and sometimes does segregate to specific vendors. An example of this type of cost is receiving inspection effort.

For a typical manufacturing firm, these vendor related costs are a huge chunk of money — perhaps in the range of 10-20% of the selling price to your customer. Through proper control of these hidden and visible costs that are vendor related, we should achieve several results:

- The first is to optimize our profits.
- The second is to insure a product that is "fit-for-use" by our customer. "Fitness-for-use" of our products will insure our success in the marketplace. If we don't meet our customers' requirements, our competitors will, and we will not survive in the long run.

The basic purpose of my talk is to suggest to you methods by which you can control both the hidden and visible quality costs incurred in your vendor relationships.

Very simply, two ways will be emphasized. The first way is to talk your vendors into adopting a quality cost program. However, discretion must be used since all vendors cannot support a quality cost program. They may be too small or special circumstances may prevent successfully accomplishing it.

The second way is to use your quality cost program to identify high magnitude vendor problems for resolution. Before we discuss these methods in detail, let us first review the basic principles of quality costs and how they apply to our vendors.

The American Society for Quality Control publishes a document titled "Quality Costs — What & How," authored by the Quality Cost Committee. Included in this publication are the essential elements that make up a quality cost system, recommended definitions and methods of installing and gaining acceptance.

In this publication the ASQC recognizes four basic categories of cost. They are identified as:

- *Prevention* — which primarily are the costs resulting from planning to prevent defects in products.
- *Appraisal* — costs relative to the conformance to quality standards.
- *Failure* — costs are the results of poor planning and include the failure of products to meet the established or required quality standards.
  - failure costs are either *internal,* which are mainly manufacturing losses, or *external,* which are due to defective products shipped to the customer.

  Each category of prevention, appraisal, internal and external failure contains many elements. Let's look at a few examples of elements in each category.

Prevention includes:
- Quality control administration
- Quality systems planning and measurement
- Vendor quality surveys
- Quality data analysis and feedback

Appraisal includes:
— All inspection and test costs
— Laboratory acceptance testing costs
— Vendor quality audits and surveillance

Internal Failure includes:
— Rework
— Scrap
— Retest and reinspection

External Failure includes:
— Returned material processing and handling
— Warranty replacement

The first part of the quality cost program, *that of measuring,* is the most difficult. Most businesses do not enjoy the sophisticated accounting system that permits interrogating for the cost of quality elements.

A system for collection of this information has to be designed to fit each specific company.

Once we have collected the information and categorized it, we can now *analyze.* The most obvious is to examine those quality elements for which the greatest costs have occurred. It is here that the most effective and dramatic cost savings can be realized. But we must also note the quality elements having little or no costs, since insufficient effort may be occurring. For example, little action in vendor evaluations may result in high receiving inspection costs.

Action, of course, is the result of analysis. All the good cost reporting in the world is wasted unless someone takes corrective action. The logical next step is to take action in areas needing improvements. Specific persons must be assigned specific improvements against a time scale to assure effective results.

Finally, we *control* the improvements in a continuing effort to do even better over the long term.

Now that we have reviewed quality costs in general, let us now discuss how they apply in the buyer-vendor relationship. Previously we discussed how the vendor related costs can be classified as either hidden or visible.

Now I would like to discuss each of the categories in detail. Hidden quality costs, as you may recall, are in three parts:

— The first are those which are incurred by the vendor at his plant.
— The second are incurred by the buyer in fixing problems at the vendor plant.
— The third are those costs — which are not usually allocated to vendors — incurred by the buyer as a result of potential or actual vendor problems.

Those quality costs incurred by the vendor at his plant are, for the most part, unknown to the buyer and, therefore, hidden. However, even though the magnitude is hidden, the type of cost is not. They are the same type quality costs as the buyer incurs. For example, the vendor certainly has prevention effort. If he manufactures the product, he has expenses related to the quality engineering of the product. Even if the vendor is a small shop, this task must be done by someone and may very well be handled by the Production Supervisor if the plant lacks a Quality Engineering Staff. Certainly effort is expended in the appraisal area even by the smallest vendors. Someone must inspect the product prior to shipping to the buyer. In a one man shop this is done by the man who made the part. Unfortunately, we all have failure costs whether we are a large shop or a small one. When the vendor makes a mistake in manufacturing, he must either rework the part or scrap it causing an internal failure expense. If the vendor sends it to the buyer, it may be rejected with the resulting cost being considered as external failure.

The second type of hidden quality cost — that which is incurred by the buyer in fixing problems at the vendor's facility — is usually not specifically allocated to vendors. Except for an awareness of our troublesome vendors, there is usually no tabulation of the cost of the effort expended. Therefore, the actual expense is hidden. One example of this cost is the buyer sending a Quality Engineer to a vendor to resolve a crisis. Another example is the auditing a buyer may do at the vendor's plant to insure that he has the proper quality controls to build a product to the buyer's specifications.

The last type of hidden quality cost is that at the buyer's plant incurred on behalf of a vendor. Like the second type of hidden cost, we are certainly aware of it, but we probably do not keep track of it. The magnitude of this expense is not known for specific vendors. Again, the actual expense is hidden. This type of cost may include, in the prevention area, the following:

- Preparing the specifications for packaging the product that is shipped by the vendor to the buyer.
- Specification and design of gauges that must be used in the buyer's receiving inspection and perhaps as well by the vendor prior to shipping.
- Designation of appropriate specifications that the vendor must follow in the manufacture of the product.

For the hidden appraisal costs, we may incur expense for:

- Certain inspection operations and quality control effort in the buyer's production line related specifically to a vendor product.
- Review of test and inspection data on vendor parts to determine acceptability for process in the buyer's plant.
- Calibration and maintenance of equipment necessary in the control of the quality of vendor parts.

In the internal failure area we may do troubleshooting or fire-fighting in order to use a vendor part that is not quite up to specification.

Finally, regarding external failure, there may be field engineering required to analyze and correct a problem caused by a vendor.

It must be remembered that this discussion of the types of vendor related hidden costs is by no means exhaustive. There are many more. Some of the types of costs not pointed out may be very significant depending on your individual situation.

There are, as discussed initially in this talk, vendor related quality costs that are apparent and perhaps much easier to identify in magnitude and assign to various vendors by the buyer.

These are called visible costs. They are primarily in the appraisal and failure areas.

Included in the visible quality cost category are the following:

— Receiving inspection.
— Laboratory acceptance testing.
— Scrap and rework that is vendor responsibility.
— Warranty replacement on parts supplied by vendor.

The visible costs, if tracked, are perhaps most significant because they can be good indicators of problem areas.

Now that we have looked at the various types of hidden and visible quality costs, let us discuss the possible methods through which we may optimize these expenditures.

The first step for the buyer in optimizing his vendor related quality costs is to determine what costs are important to him. It is suggested that comparing the relative magnitude of quality costs by category could be a good start. To do this, your own quality cost program could be invaluable. But, if you don't have a quality cost program, a special study could be initiated to determine this information. The ASQC publication "Quality Costs — What & How" would be a valuable guide in accomplishing this.

Let us discuss a hypothetical situation in which vendor scrap and rework are the biggest problem areas for the buyer. If the buyer makes the assumption that through improvements in vendor-caused scrap and rework warranty will be lowered, then for this company vendor-responsible scrap and rework are the important items in the vendor relationship.

The next step is determining which vendors are causing the problems. Very likely you will find that 20% of your vendors are causing 80% of your problems in scrap and rework. Now the buyer can narrow his effort to the vital few of the vendors and take appropriate action.

What is appropriate action? First and foremost, I would suggest that you promote that the "vital few" vendors start a quality cost program if appropriate. Discretion must be exercised before insisting on this. Some companies may be too small to support a quality cost program or to accomplish special studies. Special circumstances may exist in other companies that would prohibit this action. However, if a vendor finds that launching such a program or study is feasible, the costs most visible to the buying company will most likely be reduced by doing so. If these are reduced, the hidden costs expended by both the buying company and the "vital few" vendors should also be lowered. The result will be that the vendor's product and the buying company's product are more "fit for use". This should increase profits for both which is very desirable. Also, improved profitability for the vendor may eventually result in lower prices for the buyer in a competitive market.

What other action can be taken? Keeping track of vendor-related costs — both hidden and visible — may be a gigantic undertaking for the buying company. However, some way must be found to insure that progress is being made. It is recommended that the buying company track the important visible costs for each of the "vital few" vendors. It is submitted to you that, in most cases, you will find that if progress in reducing the visible costs is demonstrated, hidden costs are also decreasing.

What other action can be taken if we know the magnitude of the visible costs? It is possible that this information can be incorporated into a buyer's vendor rating system. The Guide for Managing Vendor Quality Costs describes a rating system using the cost of a purchased part and readily identifiable vendor quality cost. Although not theoretically perfect, it is a simple concept. Better yet, it has resulted in significant improvement in vendor quality. It is well worth your consideration.

A vendor relationship depends upon more than the traditional measures of meeting deliveries and receiving inspection rejections. It also must recognize the "fitness for use" of the vendor's products for which visible quality cost can be a valuable indicator.

Visible quality costs can also be used in other ways. Some companies debit vendors for the scrap and rework occurring in the buyer's plant to put the responsibility for failures where it hurts most — in the pocketbook. However, in the long run, they may be counterproductive in that some vendors may ask for a price increase to cover this situation. On the other hand, many buyers find it very effective to reduce the amount of business given to the offender and reward the good performer with a greater share of the "order pie".

Perhaps a far more positive method is to use the visible costs to identify vendor quality improvements that are needed. The buying company would initiate projects jointly with vendors to resolve the problems that are the source of high quality costs. The problem may be solved through an action by the buying company — perhaps the specifications are not correct or the vendor really doesn't know of the application of his product in the total package.

On the other hand, it may be that the vendor's manufacturing process needs upgrading, through perhaps better tooling, and the vendor must take some action. Through joint projects, using visible quality costs as facts, we can solve those problems and get better products. The results will be lower costs to both the vendor and buyer. The Guide for Managing Vendor Quality Costs contains an illustration of a successful joint project and ideas for how to start them.

Let's now summarize what we have suggested. Basically, use quality costs in your vendor relationships. Through this tool, find out what costs and vendors are most important to focus upon. Suggest that these vendors, if appropriate, adopt a quality cost program in order to obtain improvements in "fitness for use" of their products. Use your visible vendor quality costs as a basis for starting joint quality/reliability improvement projects with your suppliers.

Most important of all is that any quality cost program is not complete without an effective corrective action program. The mere act of collecting quality costs alone will do nothing but add cost for you or your company. Only through pin-pointing the important problems and solving these can we make progress.

QUALITY OPTIMIZATION VIA TOTAL QUALITY COSTS

A. H. Žaludová, National Research Institute for Machine Design, Prague
and
F. H. Žalud, Automobile Research Institute, Prague, Czechoslovakia

ABSTRACT

The modern concept of quality includes features of technical function, dependability and safety, ergonomic, aesthetic and ecological features, economic features such as material and energy consumption and others. The creation and realization of product quality involves costs to manufacturer, to user and to the community at large (total quality costs). A simple measure of the cost-benefit relationship as a function of measures of the most important individual quality characteristics leads to the possibility of formulating various optimization criteria. Application of such a model to problems of technological innovation relating to optimal life and reliability measures, environmental protection and resource conservation will be demonstrated on examples.

INTRODUCTION

Requirements on product quality are continually changing with changing economic and social conditions and with advancing technical progress. A few decades ago, adequate function and appearance constituted the main features of product quality. Higher demands on increasingly complex equipment gradually led to new requirements on dependable performance in time and on ergonomic features. In more recent times, the consumer-user and the community at large have been formulating and implementing, often through legislation, further requirements on product safety, ecological integrity, material and energy economy, labour productivity and other features extending far beyond traditional quality characteristics.

The complexity of problems involved necessitates not only a systems approach to their solution, but also a rational basis for setting quantitative targets for all quality features and for decision-making in general in the field of quality control and assurance. The tools of value analysis, cost-benefit analysis and economic optimization have so far been applied mostly from the manufacturer's viewpoint. Under contemporary conditions of depletion of material and energy resources, deterioration of the environment and other negative aspects of social and economic life, it seems logical to extend the principles of economic optimization to include overall costs to the community and society as a whole.

INTEGRATED QUALITY CONCEPT AND ITS QUANTIFICATION

The modern concept of product quality as the totality of features and characteristics that bear on its ability to satisfy a given need is illustrated graphically in Fig. 1, where the individual features have been classified into several wider groups. Group Gl contains the technical characteristics, such as dimensions, strength, temperature, speed, conductivity, power, essential for the main product function. Time-behaviour features, such as reliability, longevity, maintainability, availability, are grouped under the heading of dependability in group G2. Similarly, groups G3, G4, G5, G6 contain other generally acknowledged sets of related characteristics (aesthetic, ergonomic, ecological-safety and economic) into which can be assigned such quality characteristics as appearance, ease of handling, noise, fuel-consumption, etc.

Features of classes Gl and G2 are the main determinants of product technical effectiveness, while features of classes G3 to G6 may be considered as essential constraints. For engineering products of capital investment or individual consumption nature, the totality of all features is associated with costs to both manufacturer and user-consumer (see Fig. 1). The sum of these two types of costs constitutes total life-cycle costs or total user-consumer costs and forms an important criterion for assessing the economic effectiveness of the product concept, design, manufacture and use.

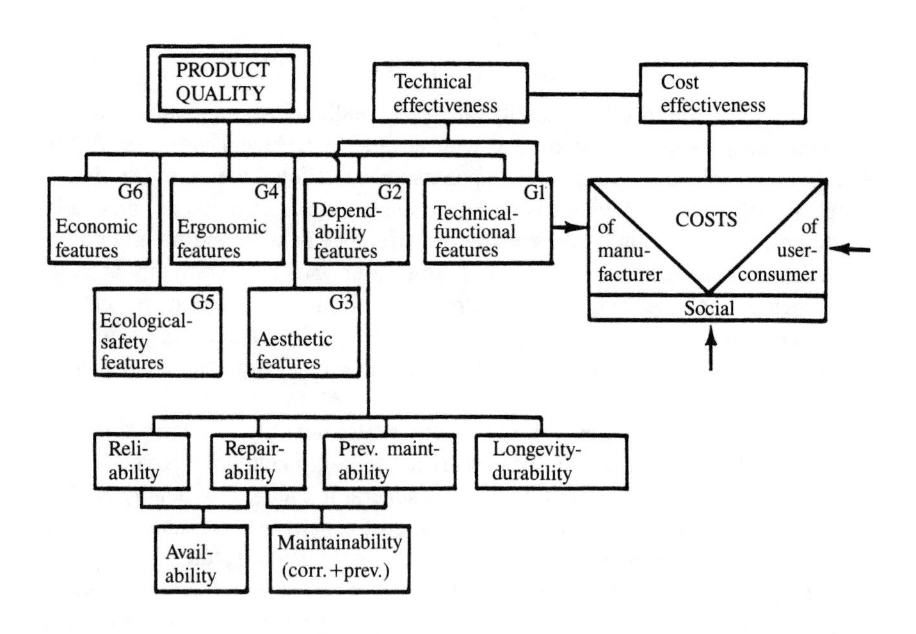

FIGURE 1

In conformance with current QC theory and practice, it is customary to quantify individual quality features by putting measures on them at the conceptual, design, manufacturing and usage stages of product life. At the design stage, these measures generally take the form of nominal values and tolerance limits for the individual features, at the manufacturing and usage stage, common measures are mean values, standard deviations, percentage nonconforming [1], [2].

Such measures may be denoted generally by K_j. where the second index may take on the meaning of required (r), specified in design documentation (d), achieved during manufacturing (m) or achieved in use (u). A measure of quality level is then obtained on the scale (0.1) by making a comparison of the measure K_j. at the design, manufacturing or use stage with some "standard" measure which may be that required by a national or international standard, by contractual specifications, etc. The most common measure of the level of manufactured quality and usage quality is the fraction conforming (ideally 1) or non-conforming (ideally zero).

TOTAL QUALITY COST-BENEFIT ANALYSIS

The traditional concept of quality costs as the expenditure incurred by prevention and appraisal activities and by losses due to failure to achieve specified quality arose in QC terminology essentially from the viewpoint of the manufacturer. In latter years, "failure" costs have been divided into internal and external failure costs, the former being those incurred within the manufacturing organization and the latter those arising outside (but affecting) the manufacturing organization, in particular, warranty costs.

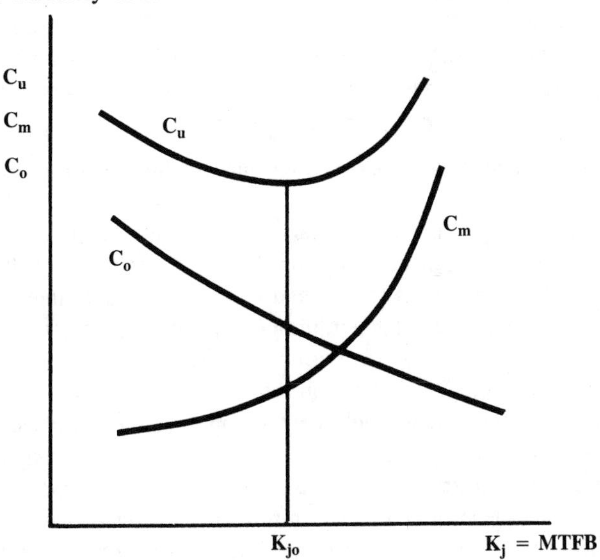

FIGURE 2
Minimizing C_u

The newer concept *total life-cycle costs* or *total user-consumer costs* developed in connection with the economic assessment of product dependability (reliability, maintainability, longevity, availability). These costs are generally assumed to comprise the purchasing price and user's operating and usage costs during the whole product life or some other specified period of time, including costs of preventive maintenance and repair, of damage and downtime due to failure. Total user's costs are obviously a function of the individual quality measures of the product. For example, for a given rated value of the output parameter, higher reliability measures, such as MTBF of a complex equipment will generally entail higher manufacturing costs and therefore a higher purchasing price (due to improved materials, addition of redundancies, etc.) and at the same time lower user's operating costs due to repair and downtime. Hence the possibility of optimizing quality measures (such as K_j = MTBF), so that total user's costs are minimized (see Fig. 2 and next Section) [5], [6].

Denoting the most important specified quality measures by $K_1, K_2, \ldots K_r$ (with deletion of the second index), the above can be expressed by the relation

$$C_u = C_m + C_0 \tag{1}$$

Where C_u = total user's costs during product useful life,
C_m = purchasing price (manufacturing costs),
C_o = user's operating and usage costs during product useful life

and where C_u, C_m, C_0 are all functions of $K_1, K_2, \ldots, K_r$.

In view of contemporary problems of deterioration of the environment and of depletion of material and energy resources, it seems logical to extend the above model in two directions. First, we augment total user's costs according to Equ. (1) by further component C_p denoting costs due to pollution and other negative effects on the environment, thus forming *total community (social) costs* in the form

$$C_s = C_m + C_o + C_p \,. \tag{2}$$

Secondly, we extend the number of important quality measures $K_1, K_2, \ldots,$ K_r
to include measures K_p of environmental deterioration (Fig. 1, group G5) and K_m, K_e of material and energy consumption (Fig. 1, group G6). Finally, we consider that all three cost elements C_m, C_o and C_p are functions of this extended set of quality measures and hence that a minimum may also be sought for total community costs according to Equ. (2) with respect to K_p, K_m and K_e for a given rated value of main technical output parameter.

In the above analysis, we have neglected the important circumstance that alterations in some of the quality measures $K_1, \ldots, K_r, K_m, K_e, K_p$ may also affect the total average useful effect E_s of the item during the period of time considered. This fact can be remedied by introducing a so-called integral quality measure k_s of a product or complex system, defined as a measure of item effectiveness, i.e. as the ratio of the total average useful effect (benefit) E_s of the item during a specified period of time and the total average community costs total quality costs) relating to the same period of time. Hence,

$$K_s = E_s / C_s \quad \text{or} \quad K_s = E_s^* / C_s^* , \tag{3}$$

where $E_s^* = \sum_t / E_{st} (1 + i)^{-t}$, $C_s^* = \sum_t C_{st}(1 + i)^{-t}$

denote the "present" values of the total useful effect and total quality costs, adjusted for the future year t to the same (initial) year and i denotes the average annual rate of interest or cost of capital.

The most important quality measures which influence the numerator E_s relate to product technical parameters from the group of features G1 (e.g. mean or rated productivity, mean or rated power, mean or rated output) and to product dependability from the group G2 (e.g. mean life, availability factor, etc.). The effect E_s can be measured in units of useful output (e.g. MWh, ton-kilometers, man-kilometers, cubic meters, etc.) or in financial units.

Each of the components C_m, C_o, C_p of total quality costs may be broken down further into the following most important cost elements:

$$C_m = \sum_k C_{mk} \tag{4}$$

Where C_{mk}, $k = 1, 2, \ldots$ denote, for example, average unit costs of R and D, of material, of energy, of labour, of pollution abatement actions, of profit and other costs;

$$C_o = \sum_k C_{ok} , \tag{5}$$

where C_{ok}, $k = 1, 2, \ldots$ denote, for example, average unit operating costs of material, of energy, of labour, unit costs of repair, of preventive maintenance, unit losses due to downtime and other costs;

$$C_p = \sum_k C_{pk} , \tag{6}$$

where C_{pk} , $k = 1, 2, \ldots$ denote the unit losses due to various pollution factors such as noise, gaseous, liquid, solid particle emissions, corrosion, etc.

If E_s is expressed in the same units as C_s (i.e. financial units), then the measure K_s is non-dimensional and expresses the *gross* economic effect (benefit) per unit costs to the user and the community. In this case, the measure K_s may be replaced by the measure

$$K_s' = K_s - 1 = (E_s - C_s)/C_s , \tag{7}$$

expressing the *net* economic effect (benefit) per unit costs.

Since both C_s and E_s are functions of the specified quality measures K_j , $j = 1, 2, \ldots, r$ (e.g. rated power, mean life, MTBF, availability factor, mean level of SO_2 emission, factors of material and energy consumption, etc.), so also is the coefficient of effectiveness K_s. The measures K_j are all in turn functions of measures of design and manufacturing factors, operating conditions, maintenance policy, globally denoted by x_i, i = 1, 2, . . ., . Hence

$$K_s = \frac{E_s(K_1, \ldots K_r; x_1 \ldots x_\ell)}{C_s(K_1, \ldots K_r; x , \ldots x_\ell)} \cdot \tag{8}$$

In concrete cases, it will generally be possible to limit the number of relevant measures of type K_j, x_i to a vital few.

VARIOUS CRITERIA OF OPTIMIZATION

I. Criterion K_s (or K_s') = maximum

The most general criterion for the choice of optimal values of quality measures K_{jo}, x_{io} is the maximization of the gross (or net) benefit per unit cost K_s (or K_s'). Such values can be found theoretically by the solution of the set of $r+\ell$ equations of the type

$$\partial K_s / \partial K_j = 0 \quad \text{or} \quad \partial K_s' / \partial K_j = 0 , \tag{9}$$

$$\partial K_s / \partial x_i = 0 \quad \text{or} \quad \partial K_s' / \partial x_i = 0 , \tag{10}$$

$j = 1, 2, \ldots, r, \quad i = 1, 2, \ldots, \ell$.

Alternatively, it is possible to express the solution in terms of minimizing the reciprocal integral measure $1/K_s = C_s/E_s$ (or $1/K_s' = C_s/(E_s - C_s)$), i.e. of minimizing the costs per unit gross (or net) benefits.

Equation (9) for maximizing K_s with respect to K_j leads to the condition

$$\frac{\partial E_s / \partial K_j}{\partial C_s / \partial K_j} = \frac{\dot{E}_s}{\dot{C}_s} = \frac{E_s}{C_s} = K_s . \tag{11}$$

This means that the optimal values of $K_j = K_{jo}$ are those which lead to the ratio of the marginal benefits to marginal costs being equal to the value of K_s (see point A, Fig. 3a and 3b). Of prime interest, therefore, are the marginal costs and benefits associated with changing the level of measure K_j by one unit.

This type of problem has been dealt with in some detail by Sittig[3] and with special reference to the setting of optimal dependability measures in the publication[4].

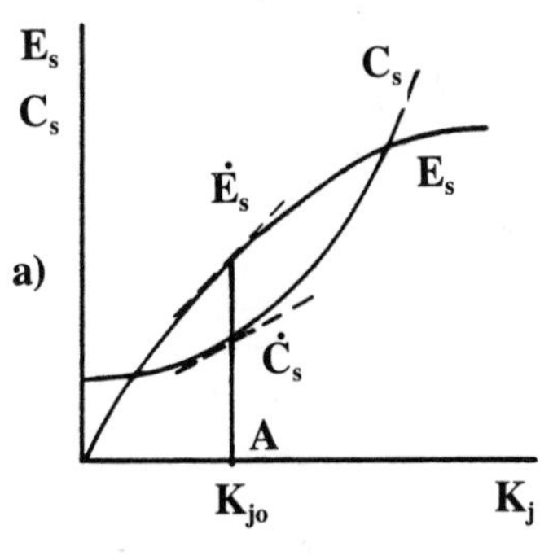

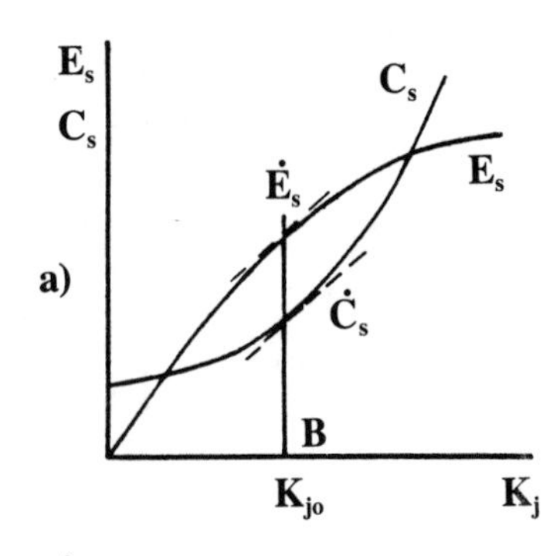

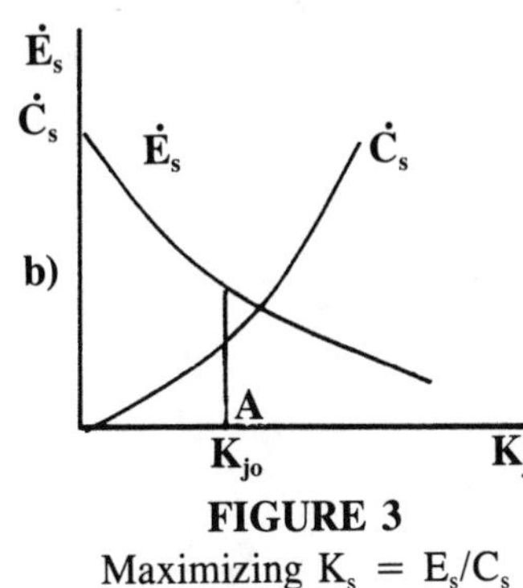

FIGURE 3
Maximizing $K_s = E_s/C_s$

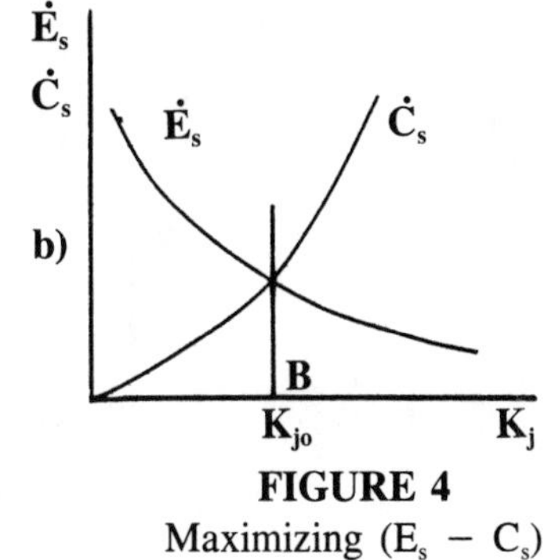

FIGURE 4
Maximizing $(E_s - C_s)$

II. Criterion $E_s - C_s$ = maximum

A further commonly used criterion for optimization of quality measures is the condition that the net economic benefit $E_s - C_s$ (numerator K_s' in Equ. (7)) should be a maximum, regardless of the total costs C_s. The difference between total benefits and total costs in a specified time period will be maximized when the condition

$$\partial E_s/\partial K_j = \partial C_s/\partial K_j \qquad \text{or} \qquad \dot{E}_s = \dot{C}_s \tag{12}$$

holds, i.e. for that value of $K_j = K_{jo}$ for which the tangents to the curves E_s and C_s are equal, or for which the marginal effect of altering K_j equals the marginal costs (see point B, Fig. 4a and 4b).

III. Criterion C_s = minimum

A third optimization criterion often used is the condition that total costs C_s should be minimized, regardless of E_s (or for constant E_s). This condition involves finding the optimal value $K_j = K_{jo}$ for which

$$\partial C_s/\partial K_j = 0 \qquad \text{or } \dot{C}_s = 0, \tag{13}$$

i.e. for which

$$\dot{C}_m = -(\dot{C}_O + \dot{C}_P) \ . \tag{14}$$

In this case, the marginal costs of an alteration (improvement) in K_j by one unit by the manufacturer equal the marginal reduction in costs to the user and/or the community. Figures 5a and 5b illustrate the case where $K_j = K_p$ denotes the

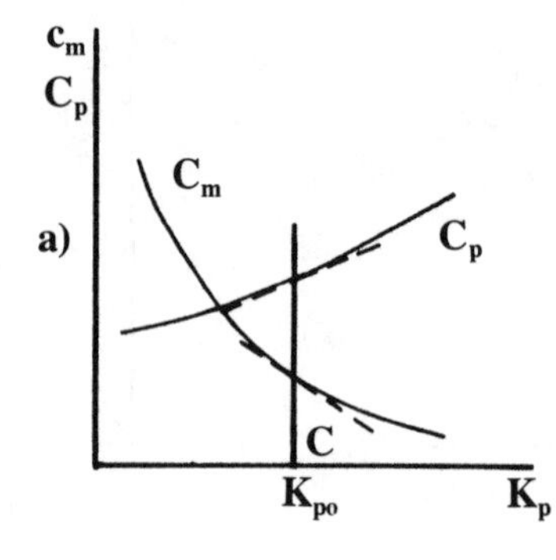

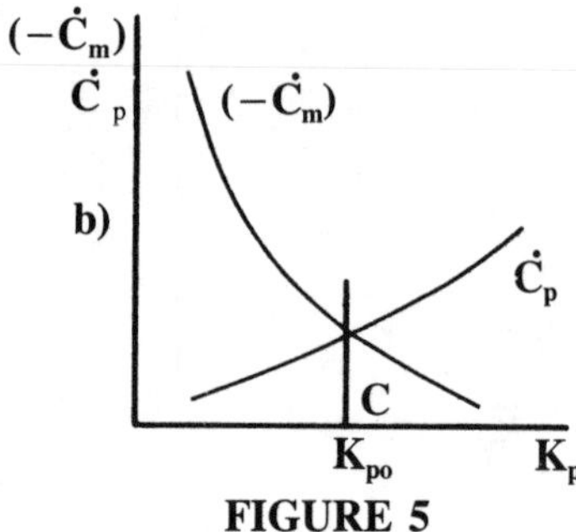

FIGURE 5

level of a particular pollutant affecting mainly manufacturer s costs C_m and damage losses C_p. Lower pollutant level $K_j = K_p$ increases C_m and decreases C_p. The optimal level is given by point C in Fig. 5a and 5b.

Criterion III is clearly a special case of criteria I and II, assuming E_s = constant with respect to the quality measure K_j under consideration. This will be approximately true for a particular grade and type of product with specified rated output parameter, provided changes in the measure K_j do not affect the useful effect E_s of the product. In this case, criteria I, II and III are equivalent. Such measures might be, for example, those relating to environment protection.

If on the contrary, K_j denotes a quality measure such as *mean life* T_a which directly affects the useful effect E_s, criterion III must be modified to the condition of minimizing the mean *specific* costs

$$c_s = C_s/T_a \tag{15}$$

per unit time interval with respect to the measure $K_j = T_a$. Considering that the total useful effect E_s may be expressed in terms of the mean specific effect $e_s = E_s/T_a$ = constant during life-time T_a, we see again that there is an equivalence of criterion III to criteria I and II, provided we use the modified condition of Equ. (15).

In the case of optimizing the measure of useful life T_a, modified criterion III leads to the condition

$$\dot{c}_s = \partial c_s/\partial T_a = \dot{c}_m + \dot{c}_o + \dot{c}_p = O \quad , \tag{16}$$

where $c_m = C_m/T_a$, $c_o = C_o/T_a$, $c_p = C_p/T_a$ are the mean specific costs of manufacturer, operation and pollution per unit time, and may or may not be functions of T_a. Some special cases are considered in the next section.

EXAMPLES AND COMMENTS

According to [4], the analytical expression for the integral measure of effectiveness K_s for utility road vehicles with non-continuous operation, taking discounting into consideration, is of the form

$$K_s = \frac{e_s T_a(1 - e^{-\varkappa T_c})/xT_c}{C_m + T_a\{c_{01} + c_{02}\lambda(T_a) + c_{03} + c_{04}\zeta(T_a)\}(1 - e^{-\varkappa T_c})/\varkappa T_c} \quad (17)$$

$$= \frac{e_s T_a(1 - e^{-\infty T_a})/\alpha T_a}{C_m + T_a\{c_{01} + c_{02}\lambda(T_a) + c_{03} + c_{04}\zeta(T_a)\}(1 - e^{-\infty T_a})/\infty T_a}, \quad (18)$$

where

T_a	= mean useful life / actual operating time /	[h]
T_c	= mean life in calendar time	[h]
$\lambda(T_a)$	= failure rate (renewal density)	[f/h]
$\zeta(T_a)$	= $\lambda(T_a) \cdot \theta_r(T_a)$ = downtime factor	[−]
$\theta(T_a)$	= mean downtime per failure	[h/f]
e_s	= mean specific useful effect per unit operating time	[$≠h]
C_m	= purchasing price	[$]
c_{01}	= mean constant operating costs per unit time	[$/h]
c_{02}	= mean repair costs per failure	[$/f]
c_{03}	= mean preventive maintenance costs per unit time	[S/h]
c_{04}	= mean costs (losses) per unit downtime	[$/h]
$\varkappa$	= $\ln(1 + i)/8760$ = index of investment rate of growth	[h^{-1}]
i	= annual rate of interest	[−]
∞	= $\varkappa T_c/T_a = \varkappa/\upsilon K_{tu}$	[h^{-1}]
υ	= operation intensity factor	[−]
K_{tu}	= coefficient of technical utilization.	[−]

This model contains four explicit dependability measures T_a, T_c, $\lambda(T_a)$, $\varrho(T_a)$, six economic parameters e_s, C_m, c_{01}, c_{02}, c_{03}, c_{04} and two constants $\varkappa$(or ∞) and i necessary to relate costs to their present value. In fact, C_m is also a function of the four dependability measures.

The model has been applied to the cost-benefit analysis of city buses. On the basis of field data analysis, it has been verified that the overall operating costs C_0 consist of two main components, the first corresponding to $(c_{01} + c_{03})T_a$ linearly dependent on time of bus operation and the second corresponding to

$\{c_{02}\lambda(T_a) + c_{o4}\,\varrho\,(T_a)\}\,T_a$ increasing non-linearly with operation time. Numerical results for a specific make of bus lead to the expression

$$K_s = \frac{150T_aR(T_a)}{300 + (59T_a + 2.6T_a^2)R(T_a)} = \frac{3T_aR(T_a)}{6 + (1.18T_a + 0.052T_a^2)R(T_a)},$$

where $R(T_a) = (1 - e^{-0.036T_a})\,/0.036T_a$ is the discounting factor, T_a is in units of 1000 hours and the costs factors are given in an arbitrary unit.

The simple solution without discounting using *Criterion I* is illustrated in Fig. 6 and yields the solution

$$(\partial E_s/\partial T_a)/(\partial C_s/\partial T_a) = \frac{3}{1.18 + 0.104T_a} = \frac{3T_a}{6 + 1.18T_a + 0.052T_a^2} = E_s/C_s\,,$$

i.e. $\quad 0.052T_{ao}^2 = 6, \qquad$ hence $T_{ao} = 10740$ hours.

Criterion III leads to the same solution

$$\partial c_s/\partial T_a = \partial(C_m/T_a + C_o/T_a)/\partial T_a = 0,$$

i.e. $\quad -6/T_{ao}^2 + 0.052 = 0, \qquad$ hence $\quad T_{ao} = 10740$ hours.

Criterion II leads to the solution

$$\partial E_s/\partial T_a = \partial C_s/\partial T_a,$$

i.e. $\qquad 3 = 1.18 + 0.104T_{ao}, \qquad$ hence $\quad T_{ao} = 17500$ hours.

The solution with discounting leads to practically the same value of T_{ao} for criteria I and III, buy to a lower value $T_{ao} = 15000$ hours for criterion II (see Fig. 6a and 6b).

For city buses operating with an average speed of say 30 km/hour, the optimal life given by points A and B in Fig. 6 correspond roughly to 320 000 km and 450000 km life. Two other interesting points to be considered in Fig. 6 are points D and F, representing limiting points of profitability.

In view of the complicated nature of the dependence of K_s on the measures K_j, the task of fixing optimal values K_{jo} is often solved numerically on a computer by calculating the value of K_s for different design modifications and deciding on the modification which gives the maximum value for K_s [7].

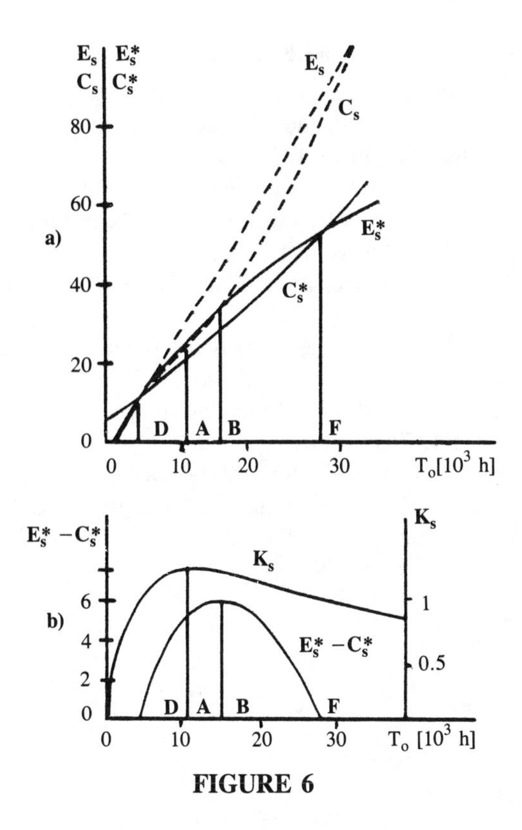

FIGURE 6

Criterion III is currently used to assess the social costs and benefits of regulatory actions relating to automobile emission controls and noise abatement[8], to safety precautions and health hazards in the nuclear energy field, to water and soil pollution from chemical plants and other industrial installations (see Fig. 5).

In connection with energy and material savings, it should be noted that improvements in life characteristics implicitly lead to reductions in energy and material in manufacturing, while improvements in reliability bring similar economies in the user field due to reduced repairs and spare parts consumption. Recent estimates of ecological damage assign about 50% of total losses to corrosive effects on property and materials. Hence the possibility of objectively evaluating the cost-benefit effect of anti-pollution modifications to "culprit" polluting equipment and the design measures aimed at improving the surface protection, and hence the life, of the "victims" of corrosion.

The integral measure K_s has many other applications. It is currently used, for example, to evaluate the economic effectiveness of realizing R and D projects[9]. Another useful adaptation in the automobile reliability field is the monitoring of specific repair costs $c'_{02} = C_{02}/E_s$ over consecutive time periods (e.g. every half-year) in order to judge the effectiveness of remedial actions. In this case, only one component of costs is the interest, and the measure of the useful effect E_s is taken simply to be the number of kilometers covered in the time period considered.

REFERENCES

1. ANSI/ASQC A3-1978 Quality Systems Terminology.
2. Žaludová, A. H. "Some reflections on quality terminology", EOQC Journal Quality (submitted September 1980).
3. Sittig, J. "Defining quality costs", EOQC Conf. Copenhagen, 1963.
4. "Methods of setting reliability specifications for technical equipment", GOST Guide, Moscow 1971 (in Russian).
5. Gryna, F. "User costs of poor product quality", Ph.D. Thesis, Univ. of Iowa, 1970.
6. Garcia-del-Valle, J. "Cost of usage versus cost of purchase — a change ahead", Proc. EOQC-IAQ Conf. Venice, 1975.
7. Žalud, F. H., Žaludová, A. H. "Changing trends in automobile reliability", Proc. FISITA Conf. Tokyo, 1976.
8. "RECAT Report — Cumulative regulatory effects on the cost of automotive transportation", Office of Science and Technology, USA, 1972.
9. "Methods for the evaluation of R and D projects", WG6, Eur. Ind. Res. and Management Assoc., Paris, 1970.

QUALITY COSTS CAN BE SOLD

John D. Breeze
Manager, Quality Assurance
Hermes Electronics Limited

ABSTRACT

The manufacturer of a medium volume military electronics product introduced a Quality Cost Program with cost categories defined in terms of "do it right the first time" concepts. A special definition was used for "external failure cost:" this definition may be helpful in other situations where product is submitted for acceptance by lot. After a careful assessment of overhead rates, it was decided to include only fringe benefits costs as a burden on labor. Initial estimates of Quality Cost were 12% of sales, a more detailed breakdown is presented, together with a discussion of methods used to gather data.

BACKGROUND

Hermes Electronics designs and manufactures electronics equipment and systems for communications and ocean engineering use. One of our major product lines is the Sonobuoy, an expendable data sensing and transmitting device used for various forms of ocean surveillance. Sonobuoys are manufactured in quantities of several hundreds per day and employ some hundreds of people in a very diverse manufacturing situation, including fabrication, printed wiring board assembly, welding, plating and coil winding. Sonobuoys are a "one-shot" device, for which warranty and service costs have no meaning. To all intents and purposes there are no salesmen, and units are not returned for replacement. The product is presented for acceptance in homogeneous lots of about 1000 units. A random sample is selected for testing under service conditions and the lot is accepted or rejected on the results of this test, after which the product is shipped and (almost) never heard of again. This means, of course, that there is no external failure cost, in the conventional sense.

GETTING STARTED

Some managers had been aware for some time of the potential benefits of operating a full-scale Quality Cost program. Some of the papers at the ASQC's Chicago Conference in 1978 added stimulus to our desire to get started. However, each time it was discussed, it always seemed to be necessary to wait until the latest series of changes were completed in the accounting system. Another hurdle to be overcome was the need to explain why the existing system of recording and reporting waste and rework cost was insufficient as a measure of Quality Cost. And, of course, there was always a rear guard action being fought to try to explain why "Quality Cost" was *not* the same thing as the "Cost of the Quality Assurance Department" (a topic of some considerable controversy, particularly at budget time!).

We finally moved off dead center during the plant summer vacation, when operations cease for a few weeks. Taking advantage of the fact that our fiscal year ends at that time, and with the co-operation of an interested controller, we had direct access to year-end financial data that was still fresh and warm. Furthermore, we were still surrounded by accountants and auditors who could be drawn into the discussions whenever they needed a mental change of pace. As a result of this work, we were able to put together a presentation to management that convinced everyone of the significance of our Quality Cost figures.

QUALITY COST CATEGORIES

The spadework involved in listing the relevant Quality Cost categories had been done soon after the 1978 conference, sparked off by L. Seder's presentation[1] at the meeting plus a careful study of the ASQC Handbook *Quality Costs — What and How*,[2] which is strongly recommended to anyone who is becoming interested in Quality Cost. Chapter 5 of the *Quality Control Handbook*[3] was also useful. The cost elements that we decided to work with are listed in Exhibit 1, which also shows where we obtained cost data for them.

In order to explain the Quality Cost categories to people in the company, we redefined them in terms of "Doing Things Right the First Time" (with grateful thanks, if appropriate, to the zero defects program). With the help of this slogan, the four main categories were explained as shown in Figure 1. These definitions were helpful and meaningful at virtually every level of the company and provided the key to explaining just how and why Quality Cost goes far beyond the cost of the quality assurance department.

PREVENTION COSTS	:	the costs of trying to ensure that we do it right the first time.
APPRAISAL COSTS	:	the costs of checking to see if we actually did do it right the first time.
INTERNAL FAILURE COSTS	:	the price we paid when we found we didn't do it right the first time.
EXTERNAL FAILURE COSTS	:	the price we paid for failing to discover we didn't do it right the first time.

FIGURE 1. Cost Definitions.

This system of definition gave us a way of identifying external failure cost that had previously not been recognized. It was mentioned earlier that this particular product is offered to the customer in homogeneous lots and is accepted or rejected on the basis of the performance of a random sample of units. Up to the moment of selection, the product remains under our control. As soon as the sample has been selected, no further operations take place until it is packaged and shipped. If the test results in a "reject" decision, a program of failure analysis, inspection and rework is carried out at our expense. We decided to separate all costs occurring after the selection of the test sample (except for packaging) and classify them as external failure costs. We also decided to include the cost of working capital tied up when a shipment was delayed as a result of failing the test.

SOURCES OF DATA

Locating sources of cost data turned out much less difficult than we had imagined. Many cost elements could be found as line items in the regular accounting reports, once they had been decoded from original accounting language. This was easily done, of course, during discussion with the controller. Other costs were obtained by identifying the actual people who did the work and extracting their payroll records. In some cases, groups of people involved in several Quality Cost related activities were asked to judge the proportion of their time spent in each category. Costs were then obtained by assigning their salaries and fringe benefits in the same proportion. This method was particularly appropriate for the design, manufacturing, and quality engineering groups that support the production operation and while it cannot be claimed to be accurate, it does provide an opportunity for another group of people to become involved in the Quality Cost program.

Some costs were allocated in proportion to the total effort. For example, in deciding the fraction of total inspection cost to be allocated to reinspection and retest, we assigned it in the same proportion that rework labor had to total direct labor. These labor costs were both available as line items in the accounting reports, whereas inspection had never been distinguished in this way.

The identification of the costs of "test" and "retest" resulted in an unexpected conclusion. Since our Production Test department is responsible for tuning and alignment of electronic circuits, plus troubleshooting, none of its costs were appropriate for these categories. It required a study of the production operation to identify the people who were actually doing "testing," in the sense of performing activities of judging product conformance by a testing operation.

OVERHEAD APPLICATION

Another point of heated discussion centered around the decision of whether or not to apply overhead burden to the labor and material cost estimates of the Quality Cost elements, particularly for scrap and waste. The ASQC Handbook recommends that overhead should be applied on labor content for both scrap and rework/repair costs. We found that this was inappropriate for our operation, because our burden rate includes the costs of many of the activities that are themselves Quality Cost elements. For example, our inspection and calibration expenses form a significant part of the labor burden and would be counted twice if we added the burden as proposed. In our case, we decided to add only the cost of the fringe benefit package to our labor costs when calculating Quality Costs.

Another reason for doing this is that when one makes the presentation of findings to management, one does not want to lose control by allowing a discussion to break out over whether or not overhead should be applied. By leaving it out, there can be little doubt that the final figures will be understated. And if our experience is typical, the final figures will be quite large enough to gain attention!

This may well be a conservative approach, but at least has the merit that the final figures retain credibility. The conclusion is that one should examine the way in which the burden rate is constructed before deciding to add it on, or not.

QUALITY COST ESTIMATES

Before the cost data was released for review, the rationale used in estimating and allocating costs was scrutinized by the company controller. Without this extra step, it was felt that opportunities might arise for the validity of the data to be criticized, at the expense of making the use for improvement. This done, the package was presented to the general manager together with some general recommendations for a program to improve our management of these costs.

The total amount for Quality Cost came to almost 14% of the cost of sales. This amount was not only much larger than the total profit for the year, but also represented an attention-getting dollar figure when expressed as a unit cost, being comparable with total direct labor. To give an idea of the results, Figure 2 and Exhibit 2 show the relative contributions of the Quality Cost categories and elements to the total amount.

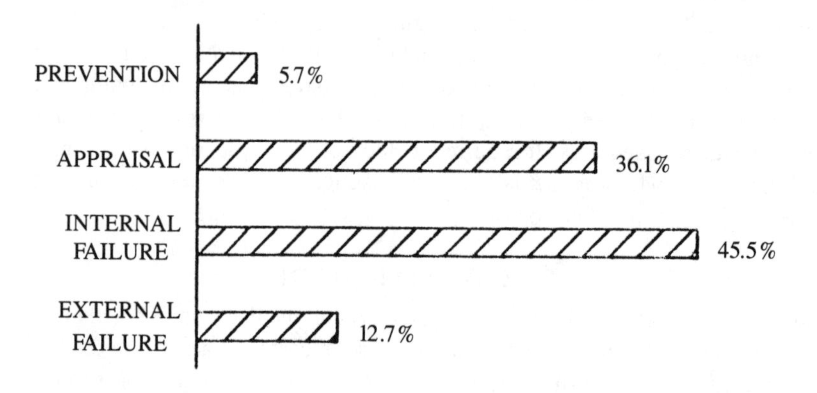

FIGURE 2. Quality Costs by Cost Category. (As % of Total Quality Cost)

Figure 2 shows the classic pattern of relatively high failure costs and relatively low prevention costs at the start of a Quality Cost program. When referring to the basic Quality Cost model in Juran's *Quality Control Handbook*,[4] there is no doubt that the situation must be ripe for opportunities for improvement.

Exhibit 2 gives one an urge to leap in and perform a Pareto Analysis on the high cost items. This urge should be resisted, because, of course, one should be equally interested in some of the very low cost elements. Furthermore, it is not inherently true that high cost elements are automatically out of line. It is very difficult to establish norms for Quality Cost performance, and one must be very cautious in drawing conclusions. The need for independence and objectivity is paramount, and yet here is one area where the quality assurance function cannot easily assume such a position, because it is involved in so much of the expenditures. The committee approach, involving representatives from engineering, manufacturing, industrial engineering and quality is the only reasonable approach. The problem is how to set it up and get it to work effectively.

LAUNCHING THE QUALITY COST PROGRAM

Our presentation to management was a formal exercise involving visible displays and handouts. Almost immediately, it aroused interest and even enthusiasm for an official Quality Cost program. The dollar figures were large enough to dwarf most of other individual cost elements of the business, and the relationship between prevention activities and total cost reduction was easy to explain. For example, most of the external failure costs, which had resulted from rework and retest of lots which had been rejected by the customer, could be attributed to changes in the product that had either been inadequately tested in the first place or which had gone unnoticed during manufacture. Then again, it was realized that although money was spent to collect in-plant data on defect rates and scrap, very little effort was devoted to analyzing the information and even less on goal-oriented action programs. We decided to launch our Quality Cost Program with a simple, 3-point program, for which we set some cost reduction targets:

(1) Pay much more attention to changes in design and process, both "planned" and "unplanned."
(2) Make better use of data already available from the existing defect and discrepancy reporting systems to identify high-loss areas.
(3) Identify causes and implement corrective action to reduce external failure costs.

These actions were intended to produce visible results from the existing cost reporting system. Having achieved that, it would be easier to move on to better ways of identifying and measuring the costs of categories that could only be estimated under the existing system of cost collection and reporting.

The program was announced through a series of seminars in various departments and by a special edition of the employees' newsletter. The slogan "Do it right the first time" was adopted; and the program was boosted at a general plant meeting, called for the purpose of informing all employees of the business activity for the coming year.

EARLY RESULTS

The program was accepted within the company and was generally supported. Naturally we found problems from the start, some of which are still being unravelled. For instance, a rigorous policy of problem solving and corrective action can make some unpleasant choices necessary when a sudden crisis arises. Do you drop everything and respond, or do you place it in the queue? What do you do when you find that your scrap reporting system isn't giving the true picture?

After two months, we had made no visible progress in reducing Quality Cost. Scrap costs were up almost 50% from a year ago and no problems of any significance had been fully solved. Our plan to implement effective remedial engineering and corrective action for all defects found at Final Inspection was floundering. While our external failure costs were very favorable, this could not be attributed to the Quality Cost Program, because the product had been manufactured before the program was announced. Furthermore, some extensive design changes were planned for introduction in the very near future, and if anything went wrong, the Quality Cost Program would be seriously undermined.

Well, as it happened, it was this situation that provided not only the opportunity to take advantage of the Quality Cost approach, but also to finish building the foundation under our Quality Cost Program and make it part of the operation. This is the way it happened.

One of the conclusions from the 3-point program was that all design changes had to be more closely scrutinized than had previously been the habit. It was decided, for example, that even running changes should be tried in parallel with prime production line before implementation. So, when a design review was held to evaluate the state of development of the latest series of design improvements, the increased level of attention to ferreting out possible problems resulted in a whole series of "what if's" and "how are you doing to's." A task force was set up to plan and implement the introduction of these changes; it met three times a week under the chairmanship of a product manager who had no line responsibility. (Later on it began to meet every day to deal with immediate questions and problems). It seems obvious now, but at the time it was a while before we became aware that we had created our Quality Cost action committee without even realizing it! Our program is now moving ahead well, and our scrap and rework costs are already down some 30% from last year.

In our next phase of activity, we plan to improve our system of Quality Cost reporting and to begin the detailed analysis of cost effectiveness of some of our regular activities.

SOME RECOMMENDATIONS

1. Get the support and involvement of the company controller. You need his help in generating credible Quality Cost estimates in the first place.
2. Be prepared to put a lot of effort into preparing a presentation on Quality Cost to your management group. Use visual aids to make the point.
3. Don't load your figures with overhead burden to make them look extra-large. If you haven't got a Quality Cost program already, you'll have no difficulty arriving at impressive numbers without risking your credibility.
4. Start out with a simple program of reform. You cannot expect to revolutionize the company overnight.

REFERENCES

1. Seder, L. "Avoiding Failures of Quality Cost Programs" (information received directly from the author).
2. A.S.Q.C. "Quality Costs — What and How" (American Society for Quality Control, 310 W. Wisconsin Ave., Milwaukee, Wisconsin, 2nd. edition. 1971).
3. Juran, J. M. et al (editors) *Quality Control Handbook* (McGraw-Hill, 3rd. edition, 1974).
4. Ibid, p. 5-12.

ACKNOWLEDGEMENTS

I wish to thank the management of Hermes Electronics for granting permission to publish this paper. My thanks, also, to those who helped develop the program and to Les Claydon and Peggy Dobson, who prepared the illustrations and manuscript.

QUALITY COSTS CAN BE SOLD: PART II

John D. Breeze
Manager, Quality Assurance
Hermes Electronics Limited

J.R. Farrell
General Manager
Hermes Electronics Limited

ABSTRACT

The introduction of a Quality Cost Program for a military electronics product was presented at the 1980 ASQC Technical Conference.[1] One year after introduction, the initial results and problems with the Program were studied and assessed. This paper presents these findings and points out some of the lessons that were learned.

BACKGROUND

The original start-up of our Quality Cost Program was described at the 1980 conference. That program was started by preparing an analysis of Quality Costs for 1979 based on Quality Cost elements that were derived from the ASQC Handbook "Quality Costs — What and How."[2] These Cost Elements can be seen here in Exhibit 2; a year later, we found no need to change them, or add to them, (but this may no longer hold true at the end of the present operating year). We found it appropriate to work with the definitions in Exhibit 1 as a good way of explaining the four basic Quality Cost categories of prevention, appraisal, internal failure and external failure. Again, these definitions survived a year of operation, and are not likely to change in the future.

The preliminary analysis of Quality Cost for 1979 was restricted to a specific product line, the Sonobuoy, an expendable sensing and transmitting device used for ocean surveillance. This analysis had shown that the observed total Quality Cost was at least 14% of the cost of sales and was a larger dollar amount than the total profit for the year.

With this background, a modest program of improvement was started, involving three planned activities:

1) Pay more attention to changes in design and process, both "planned" and "unplanned;"
2) Make better use of data already available from existing defect and discrepancy reporting systems;
3) Establish corrective action programs to reduce external failure costs.

Some details of the way in which we tackled these tasks are discussed below, together with the effects they had on Quality Cost for 1980. It will be seen that the activities and the results were not exactly what was hoped for! But first, we might look at the way in which the cost data was obtained for the second year of the program.

SOURCES OF QUALITY COST DATA

The methodology for gathering Quality Cost data for 1979 was discussed at the 1980 conference. Since the bulk of the costs was obtainable through accounting reports, the same scheme was followed, wherever possible, in 1980. Although we had hoped to collect a greater proportion of objective cost data through the use of time logs and job cost collection, only the activity of remedial engineering proved amenable to this approach. However, improvements had been made in the system of reporting scrap, and rework, so these items, which constituted some 28% of our Quality Cost, are believed to be more accurate this year.

In the 1980 presentation our reasons for not including overhead in labor figures were explained. At the end of 1980 we remained satisfied that it was appropriate only to include the fringe benefit portion of overhead as part of the Quality Cost.

When it came to comparing the results for the two years, 1979 and 1980, three specific points had to considered; the recognition of these points and the approach taken in dealing with them may be helpful to others.

1) The effect of inflation on the "value received" for each dollar expenditure was recognized by increasing the 1979 base data by 10% throughout. (It was coincidental that, for our operation, material costs and labor rates had escalated by essentially the same amounts). This made comparison of total dollar expenditures possible, both in relation to the cost per unit of product and in relation to the change in total product volume. Other enterprises would have to examine the escalations of material, labor, fringe benefit expense items and consider whether to adjust at mid year or year end.
2) The effect of the change in product volume could be suppressed, where necessary, in two ways; firstly, by considering cost per unit produced; secondly, by adjusting the previous year's data in direct proportion to the percentage change in volume. It was found that even the apparently variable costs did not follow a linear increase, and false conclusions could be drawn if this fact is not considered.
3) Changes in product design can affect the product standard cost, which is often used in the process of generating the accounting reports. This must also be considered as part of the basis for adjusting the prior year's data before making a comparison with current year information.

TARGETS FOR THE YEAR

The three major planned activities for the year were intended to involve certain additional expenditures and contribute specified cost reductions. These are illustrated in Figure 1, which also shows the actual results observed.

COST ELEMENT	PLANNED SAVING	ACTUAL SAVING
Scrap	20%	12%
Rework	25%	13%
Reinspection/Retest	25%	10%
External Failure	50%	48%
	PLANNED INCREASE	ACTUAL INCREASE
Data Acquisition/Analysis	+ 75%	NIL
Design Review/Testing	+400%	+100%
Process Control	+150%	+ 60%
Remedial Engineering	+ 30%	− 20%

FIGURE 1.
Cost Reductions and Additional Expenditures: Planned and Actual

The actual magnitude of the amounts involved can be inferred from the data in Exhibit 2, which shows the proportion of total Quality Cost contributed by each of these elements for the two years studied. Since the data does not show the kind of changes we were hoping for, it would seem that our good intentions were frustrated in some way, though not entirely, since we were able to achieve some fraction of the planned cost reductions, at a relatively modest level of increased expenditure. However, this data does not tell the whole story.

ACTUAL RESULTS ACHIEVED

Figure 2 shows the breakdown of Quality Costs by cost category for both years, 1979 and 1980. Prevention costs and internal failure costs remained almost the same; external failure costs fell by 4.6 percentage points and appraisal costs by 3.9 points. This seems to be an encouraging sign that is even more significant when we found that the total Quality Cost fell from 14% of the cost of sales to around 10.5% (after corrections were applied for inflation and change of standard cost). At the same time, the volume of product increased by around 30%, so the actual dollar amount of the difference between "planned" and "actual" Quality Cost was larger than expected, and in the desired direction of an overall reduction.

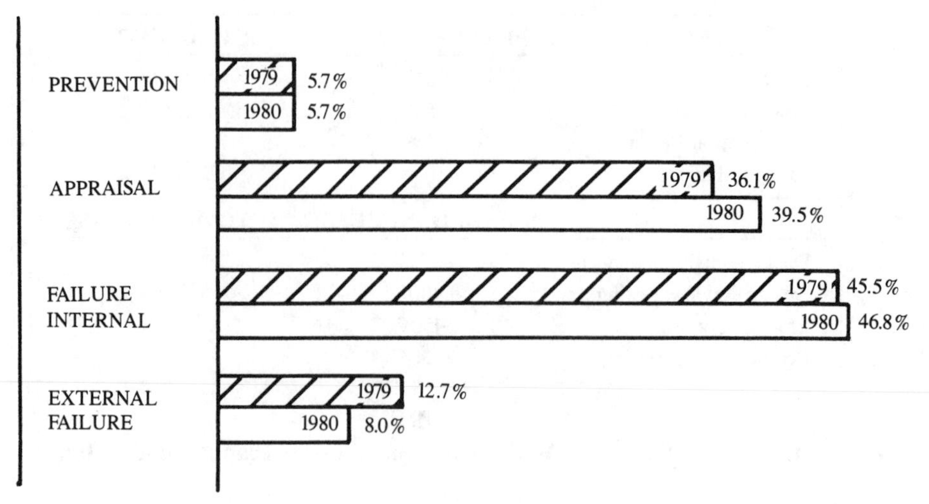

FIGURE 2.
Quality Costs by Cost Category
(As % of Total Quality Cost)

Exhibit 2 shows the changes that occured in each of the Quality Cost elements studied. It shows that noticeable cost reductions occurred in the areas of remedial engineering and almost all categories of external failure cost. At the same time, our estimates of quality planning and quality data analysis were also lower than for the previous year.

The explanation for these results seems to be as follows. Although our total product volume increased by 30%, the level of support activity remained almost the same as the previous year, since the staff employment changed by less than 5% in areas such as quality planning, remedial engineering and troubleshooting. At the same time, the Product line matured a further 12 months and, presumably, more of the cost-consuming problems were identified and eliminated. (This is still going on, actually, and there are many areas of significant defect rate still to be tackled).

As a tentative conclusion, it would appear that one should not make proportional extrapolations of Quality Cost as a function of product volume when the product is reasonably mature unless one increases the number of support staff. Indeed, our plans for 1981, which also anticipate an increase in product volume, recognize the fact that Quality Cost will tend to fall, even if we make no effort to control it, and our targets for the year are correspondingly more stringent.

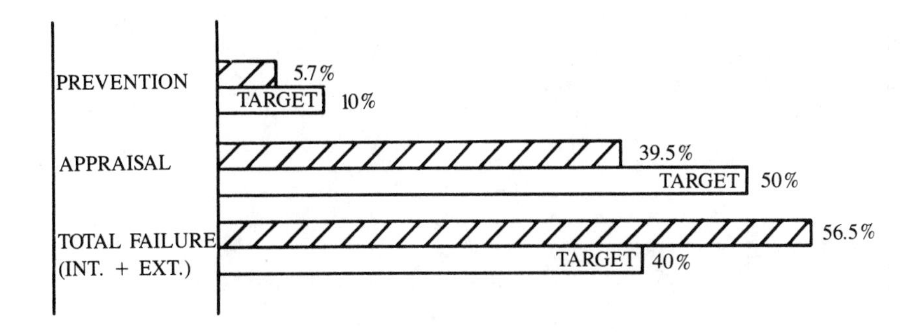

FIGURE 3. Quality Cost Targets for Future Progress

PLANS FOR THE COMING YEAR

Our plans last year focussed on achieving certain dollar changes in specified Quality Cost Categories. We found that such plans could be confounded by volume changes and ongoing product maturity. This year, our strategy will attempt to begin a radical shift in the overall distribution of Quality Cost among the categories, as illustrated in Figure 3. At the same time, we want to achieve an overall reduction of Quality Cost towards 7% of the cost of sales.

This approach has the support of all functional departments within the company. Furthermore, the recognition that the Quality Cost data for our second year of study had clear connections with the previous year, has attracted attention and confidence that Quality Cost is meaningful and can be controlled through management action. To this end, we will analyze Quality Cost data for each quarter of the coming year and establish specified areas of action for the supporting departments. More regular collection and analysis of cost data will likely result in some restructuring of the Quality Cost elements, such as the removal of those that make, relatively, very small cost contributions. Our progress along this path will be discussed at the ASQC Congress and in a subsequent presentation in this series.

NOTES

[1]Breeze, J. ''Quality Cost Can be Sold,'' ASQC Technical Conference Transactions, Atlanta (1980).

[2]''Quality Costs — What and How,'' 2nd. edition, American Society for Quality Control, 310 W. Wisconsin Avenue, Milwaukee, Wisconsin (1971).

PREVENTION	:	the costs of trying to ensure that we do it right the first time.
APPRAISAL COSTS	:	the costs of checking to see if we actually did do it right the first time.
INTERNAL FAILURE COSTS	:	the price we paid when we found we didn't do it right the first time.
EXTERNAL FAILURE COSTS	:	the price we paid for failing to discover we didn't do it right the first time.

EXHIBIT 1. Quality Cost Definitions

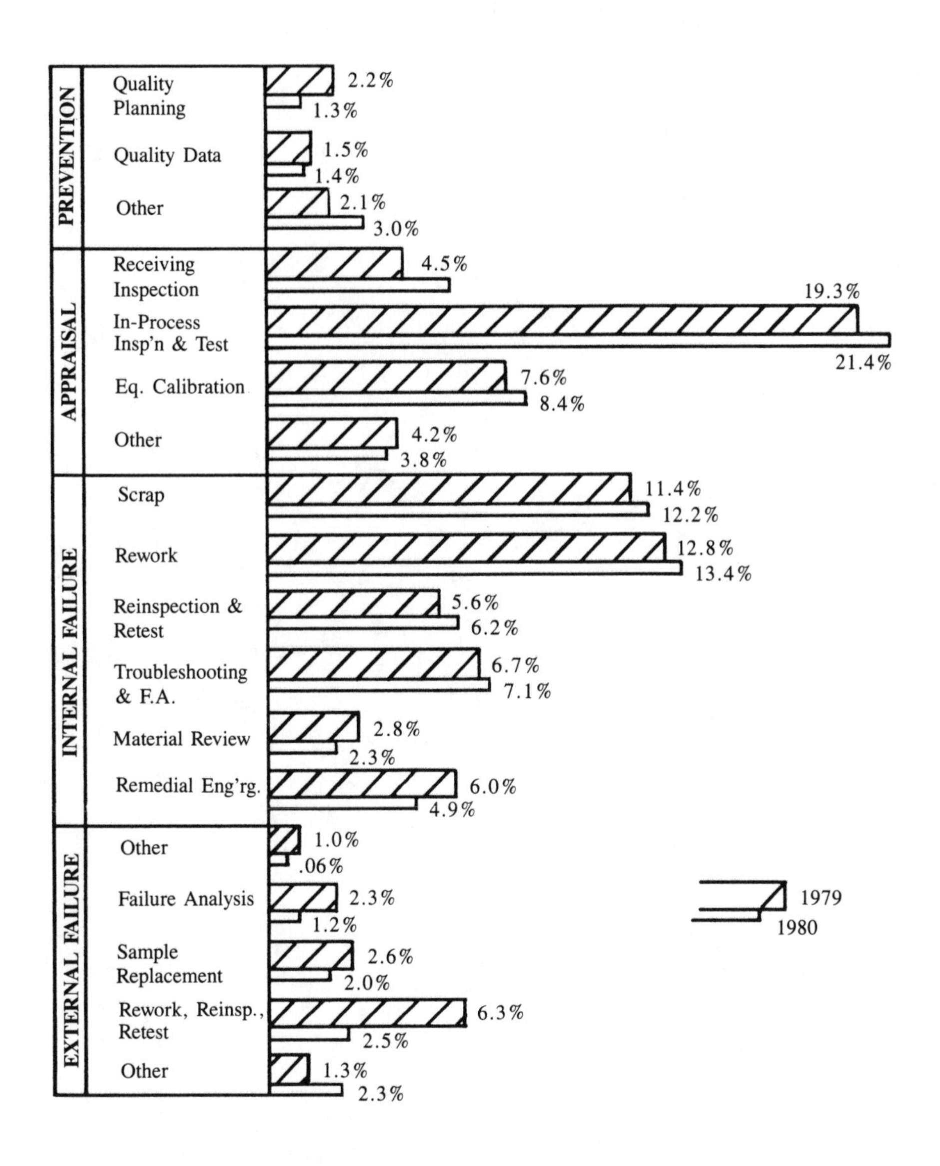

EXHIBIT 2. Quality Cost by Cost Element (As % of Total Quality Cost)

MINIMIZED COST SAMPLING TECHNIQUE

Dennis D. Lee, Senior Reliability Engineer
Sundstrand Hydro-Transmission
Ames, Iowa

ABSTRACT

Most sampling plans are based on some desired quality level and risk. If the major basis of a sampling plan is to minimize the total (inspection plus failure) costs to the company, a more methodical and precise approach can be made.

This can be accomplished by solving for the minimum point on the total cost curve (which is a summation of inspection cost and failure cost).

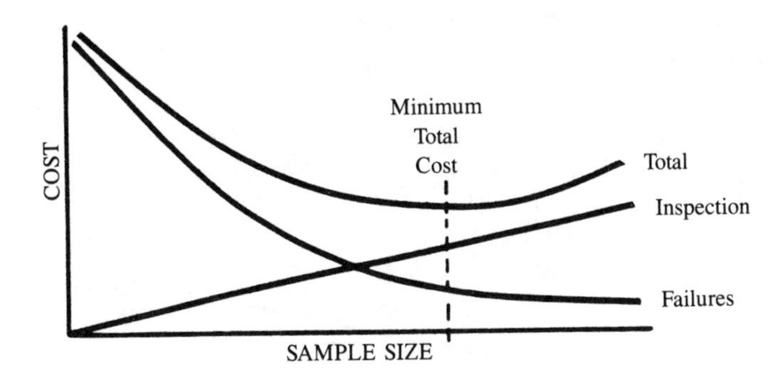

By estimating the cost of a failure to cost of inspection ratio and the approximate expected defect rate, one can determine the most economic sample size.

Some very interesting results come from this type of analysis. A series of charts have been generated to simplify sample size determination.

CONCEPT

It is quite common to establish inspection plans by determining some acceptable defect rate and risk, then, using some published plan such as MIL-STD 105, Dodge-Romig, etc., selecting an appropriate plan.

This approach has many advantages, is well documented, and familiar. However, how does one select the appropriate risk and defect rate? In many cases, inspection plans are picked very quickly with little specific data available and a minimum of consideration. Judgement, experience, past usage, and an estimate of criticality are all factored in, but without a methodical approach there is little chance that the most economical plan will be picked.

Many factors should enter into the selection of sample plans:

1) Functional criticality
2) Company image/quality goals
3) Customer relations
4) Safety
5) Manufacturing requirements
6) Cost

However, in many cases, cost is a major or the only consideration. In these cases, an attempt should be made to balance costs of inspection with costs associated with not inspecting such that a minimum total cost results.

Fortunately, a more methodical and precise approach can be used in these primarily cost balancing situations. By estimating a few parameters and applying the concept of minimum total cost (similar to the rationale behind economic order quantity formulas), an economic sample size, which yields lowest overall cost, can be determined quickly. It must be stressed that this type of analysis is recommended only for non-safety related parameters. In cases where other considerations are important, the most economical sampling plan may be determined only for reference and increased inspection levels added for the other considerations. The use of plans with less than the economic sample level would be hard to justify except on a short-term basis.

This concept is based on two premises:

1) As the number of pieces inspected increases, the cost of inspection rises.

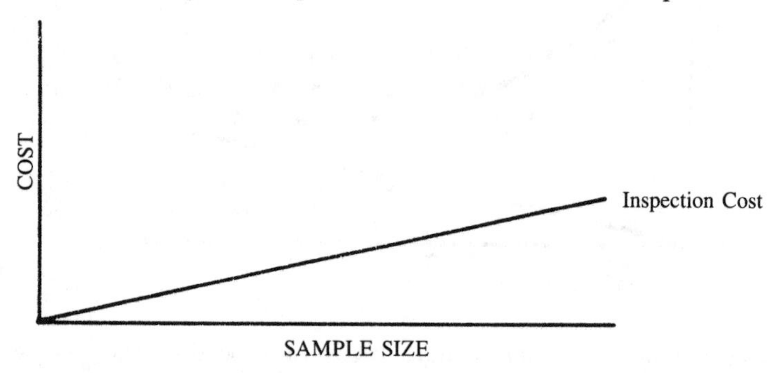

2) As the level of inspection rises, the outgoing quality improves and the total cost of failures decreases.

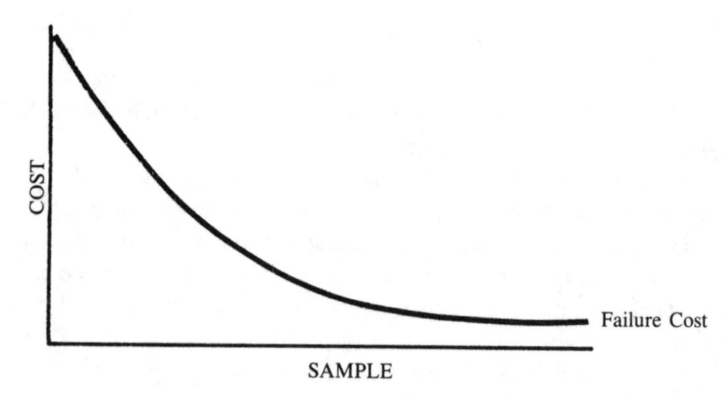

Combining these two items and adding a third curve, which is the total of the first two, yields:

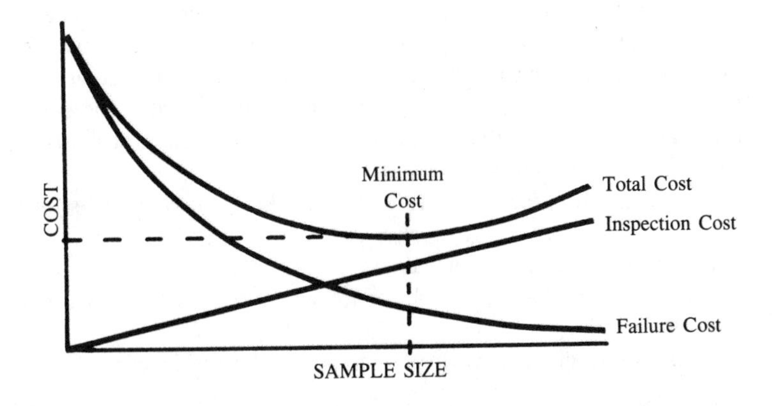

As can be seen, the total cost curve has a minimum point.

This means that for any given situation, there is a level of inspection which will yield minimum total cost.

To apply this concept, a model for each of the two cost situations (cost of inspection and cost of failures) must be determined. These models can be constructed with great complexity and exactitude or approximated with simple models. In most cases, the simple approach will yield results of sufficient accuracy, which can be used as an extremely useful tool in selection of inspection plans.

The use of simple programs in a computer will allow the creation of graphs or tables which can be easily used in actual day-to-day operations.

AN APPLICATION

This concept was applied at Sundstrand to guide the selection of inspection plans for a prototype program. Historically, the approach had been a limited inspection program due to manpower limitations. Use of economical sample sizes, in most cases, increased the level of inspection greatly but overall costs were lowered significantly.

One example:

Inspection cost per item	$ 10
Failure cost per failure	$ 150
Estimated defect rate	15%
Lot size	50
Total volume	300
Original plan — Total cost (1 sample/lot)	$6084
Economic sample — Total cost (18 samples/lot)	$1460
Savings	$4624

This illustrates that even in simple cases where the cost of failure is not high in relation to the cost of inspection, significant savings can result from this approach. Also, better utilization of manpower can result due to the prioritization of the workload.

The models selected were rather simple (see Appendix C), but adequate, and did not require a great deal of development effort. Basically, it was assumed that the cost of inspection was linearly related to the sample size and that the cost of failure was linearly related to the outgoing defect rate. The outgoing defect rate was dependent on the incoming defect rate and the probability of detection which varied with the sample size.

A computer program was written and used to generate graphs for various lot sizes.

Two graphs follow on Appendixes A and B. Appendix A is for lot size 20 and B is for lot size 500. The graph for lot size 500 is sufficiently accurate for lot sizes above 500.

To use these graphs, two parameters must be estimated:

1) Ratio of failure cost per failure to inspection cost per item
2) Incoming defect rate

As can be seen from the graphs, only reasonably accurate estimates of the parameters are needed. Except at very low defect levels, even plus or minus 30% has a small effect on the sample size due to the low slope of the curves.

To use these graphs, select proper lot size graph. Then estimate the incoming defect rate and the cost ratio. From this, the most economic sample is easily determined.

Interesting observations from these graphs:

1) In many cases, zero or low sample levels are most economic. This is true primarily at the lower cost ratios.
2) With the high cost ratio items (and lower cost ratios but higher defect levels), sample sizes increase as defect level improves — such as a lot rejected then returned after rework.

CONCLUSION

While this model is relatively simple, it reflects reality sufficiently to make it a valuable aid to sample plan selection. OC curves for these plans are identical to any single sample plan with no defects allowed.

It is easy to add whatever refinements or modifications are desired to suit a particular case. With the use of a computer to generate a data set and graphs or tables, this data can be easily used by most levels of inspection if a list of parts with cost ratios and defect rates is available.

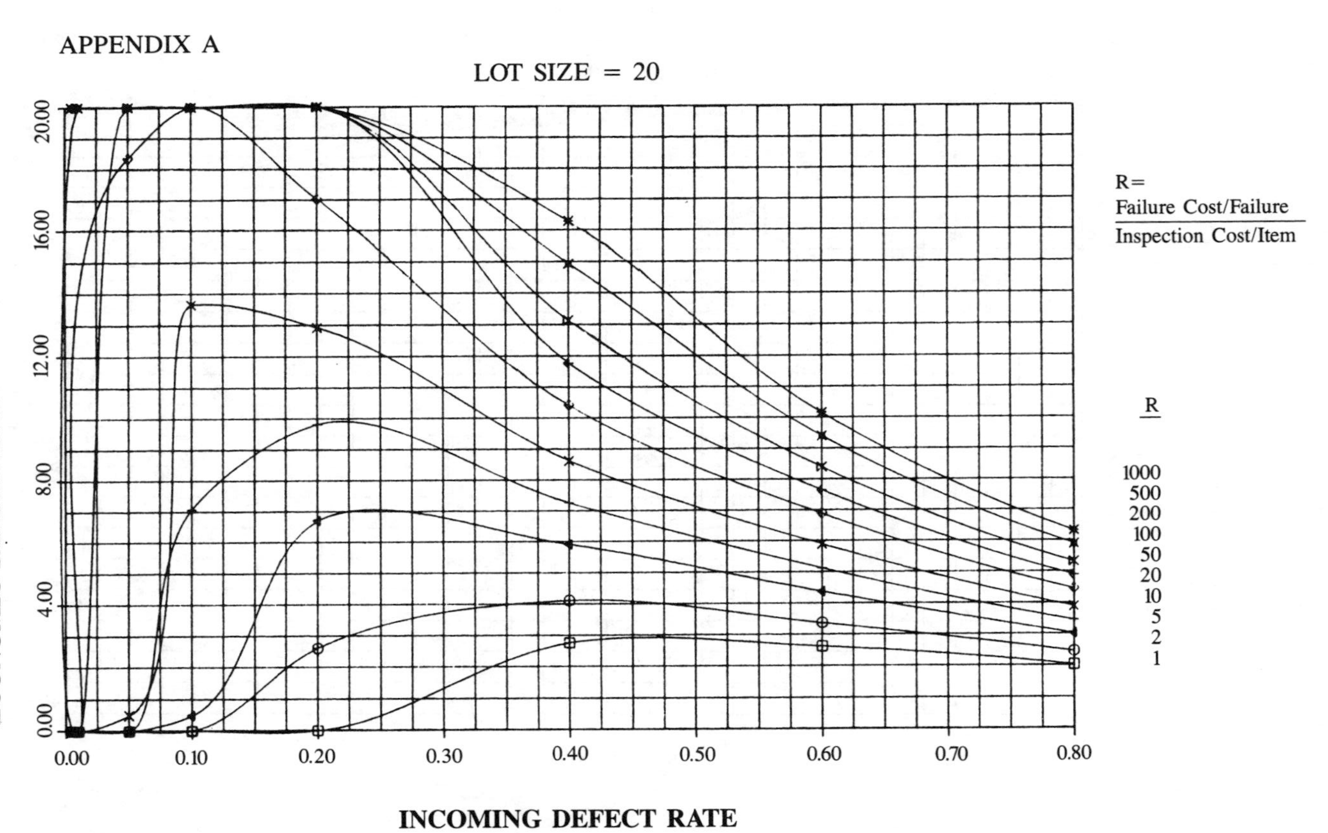
APPENDIX A
LOT SIZE = 20
R =
Failure Cost/Failure
Inspection Cost/Item
R
1000
500
200
100
50
20
10
5
2
1
ECONOMIC SAMPLE SIZE
0.00
4.00
8.00
12.00
16.00
20.00
INCOMING DEFECT RATE
0.00
0.10
0.20
0.30
0.40
0.50
0.60
0.70
0.80

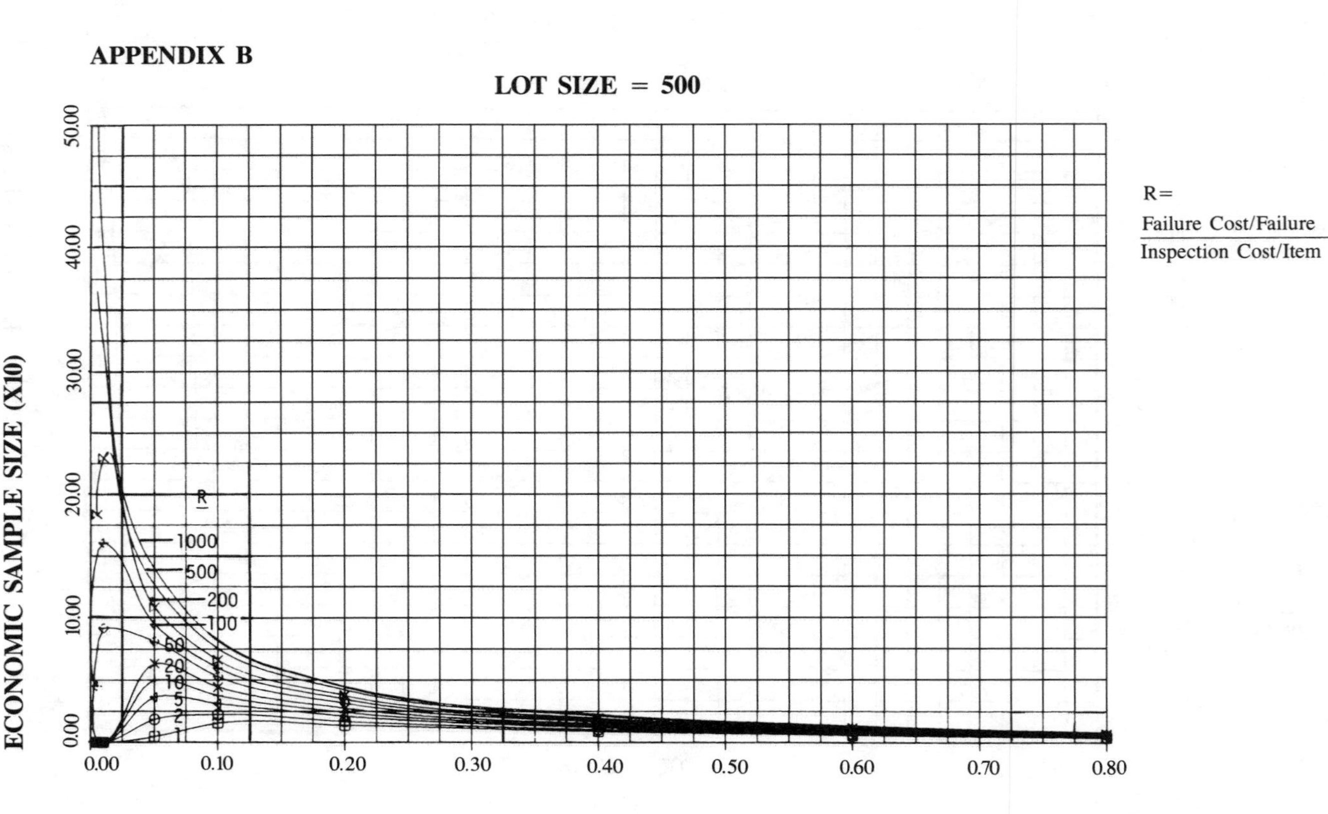

APPENDIX B
LOT SIZE = 500
ECONOMIC SAMPLE SIZE (X10)
50.00
40.00
30.00
20.00
10.00
0.00
0.00
0.10
0.20
0.30
0.40
0.50
0.60
0.70
0.80
INCOMING DEFECT RATE
R
1000
500
200
100
60
20
10
5
2
1
R =
Failure Cost/Failure
Inspection Cost/Item

APPENDIX C

1) To generate a model of total inspection cost, the following assumptions were made:

a) Entire sample is inspected regardless of when defects are found.

b) Cost of rejection of a lot is only the cost of inspection.

Defining: I = Estimated cost of inspection/item
N = Sample size
CI = Cost of inspection — total

The model selected is the cost of inspecting each item (I) times the number of items inspected (N) or:

$CI = NI$

An obvious enhancement here is to factor in probabilities of finding the defect early in the sample and terminating the inspection procedure at that point to reduce cost. An advantage to inspecting the entire sample is gaining a better estimate of the defect rate for use later in selecting the best plan.

2) Generate a model of total failure cost. Assumed:

a) Binominal distribution

b) Cost of failures is independent of rate of failure

c) Lots are rejected for one or more defects in sample

Defining: N = Sample size
P = Incoming defect rate (estimated)
L = Lot size
F = Cost/failure (can include in-house, field, warranty, etc.)
CF = Total failure cost

With the selected plan of only passing lots with zero defects (R=0) in the sample, the probability of accepting a lot with defect rate of P is:

$$= \binom{N}{R} P^R (1-P)^{N-R}$$

$$= (1-P)^N$$

Thus, the model selected is merely the total number of defects (L) (P) times the cost per each failure (F) times the probability of the defects not being caught $(I-P)^N$, or:

$$CF = (1-P)^N \; L \; P \; F$$

Combining the two models:

$$\text{Total cost} = NI + (1-P)^N \; L \; P \; F$$

Taking the first derivative, setting equal to 0, and solving for N yields the minimum cost sample size:

$N_{Min.\ Cost} = \text{Log}\ [\ -I/L\ /\ PF\ \text{Log}\ (1-P)]\ /\ \text{Log}\ 1-P$

As can be seen, the only inputs are:

Inspection cost/item	I
Failure cost/item	F
Lot size	L
Defect rate-incoming	P

Thanks to Dr. L. Genalo, Iowa State University, for his help.

GUIDE FOR REDUCING QUALITY COSTS

Ronald J. Williams
Leviton Manufacturing Company, Inc.
Little Neck, New York

ABSTRACT

Once the elements of quality cost have been defined and methods for analyzing and reporting them have been accomplished, there is a need for a technique for using those costs in programs to reduce costs and, thereby improve profits. This presentation will address the usefulness of Quality Costs, the Quality System, finding the problem area, the team approach, reducing failure costs, prevention of quality costs, reducing appraisal costs, and methods for measuring improvement.

GENERAL

Quality costs are incurred by all major functions in an organization so problem areas can exist anywhere. Careful analysis must be developed to attack them. When this need exists, a strategic program should be developed using input from all functions. The object of any quality cost program is to improve quality and reduce quality cost. Strategic planning is one method of doing this. This can be done by the following:

Review Past Performance And Present Position
A thorough review provides a realistic assessment of past performance, current conditions, and future potential. Typical indicators:

A. Customer complaints
B. Quality costs
C. Rejection rates
D. Scrap rates
E. Tests

One good tool is quality audits to determine weaknesses. You have to know where you have been and where you are now, to decide where you should be in the future.

Appraise The Environment
There are numerous environmental factors which may interact with quality program.

Set Objectives
From the knowledge and understanding achieved in the status review and the environmental information, the strengths and weaknesses of the quality program should be known. Specific objectives with completion dates should be established. Objectives should include maintaining the strength and improving the weak area. Identify *both short- and long-range* objectives.

Select A Strategy

Once clear objectives have been established, a definite strategy should be formulated and clearly stated.

- Consider all possible ways of accomplishing the objectives and develop a group of alternative strategies. Use:
  1. Pertinent records, reports, other information.
  2. Other company strategies.
  3. Experience and expertise in the field.
  4. Strategic planning of other departments.
- Evaluate all the alternatives that appear to have practical application. Consider:
  1. The strengths and weaknesses of the quality program determined from this review.
  2. The effect on other departments.
  3. The effect of possible customer or competitor changes.
  4. The reality of the situation.
- Select the final strategy and prepare the strategy statement. Review:
  1. The consistency with other company strategies.
  2. The communicable aspects of the strategy.
  3. The weakest element of the strategy.
  4. The most difficult part to implement.
  5. The most significant aspects of the strategy which are different from previous efforts.
  6. The economics of the strategy.

Thorough consideration of the steps for selecting a strategy will maximize the probability of attaining the goals.

Implement The Program

All prior efforts lead up to the implementation or action step. Key factors in effective programs:

- Assignment of action teams. Best utilization of manpower available.
- Awareness of team members of the importance of the project, their responsibility, commitment, and management support.
- Clear definition and description of the project (schedules, responsibilities).
- Review to see if project is still on schedule or if changes are needed.
- Flexibility in the project to maintain direction toward the established objectives.
- Reporting and feedback of the actions and results of the program.

Report And Evaluate The Plan

Analysis and measurement of the effectiveness of the program will provide guidance in managing the present program and direction in future strategic planning. The report should include three documented features:

A. Summary report of all strategies.
   - Lists of all strategies and due dates.
   - Descriptive statement.
   - Benefits.
   - Costs.
   - Approval signature.

B. Individual report of each strategy.
   - Name of strategy.
   - Statement of purpose and objectives.
   - Description.
   - Justification — need or lost improvement.
   - Total cost.
   - Schedule.
   - Approval signature.

C. Regular status report of all strategies.
   - Action to date on all projects.
   - Costs and benefits.
   - Adjustments to the plans.

The Quality system must involve more than inspection and test. Every department has a responsibility for assuring that the customer's quality requirements are met. The responsibilities must be clearly assigned. Quality is *everyone's* responsibility. These responsibilities usually are:

Marketing

- Liaison with customer.
- Determine the customer's quality requirements and acceptable quality levels.
- Assure that the product quality requirements and any quality system requirements are defined clearly and completely.
- Investigate the customer's opinion of the product quality and performance.
- Continually report this opinion to the department concerned.

Engineering

- Design quality and safety into the product.
- Design products that comply with the customer's quality requirements.

Manufacturing

- Challenge any drawings or specifications which would be too expensive or impractical to meet.
- Provide facilities capable of meeting the quality requirements.
- Manufacture and deliver products that comply with the drawings and specifications developed by engineering.

Quality Assurance

- Assure that the product meets the customer's quality requirements.
- Establish economical controls for preventing defective products.
- Assure that specific Quality Cost objectives are established and achieved.

Purchasing

- Select vendors who are capable of complying with our quality requirements.
- Keep vendors informed of current quality requirements.
- Work with vendors to correct quality problems.

Controller

- Keep management informed about the costs of quality.
- Provide cost breakdowns so that problem areas can be identified and corrective action justified.
- Provide accurate data on costs generated by in-plant operations and on those generated by returned material and product warranties.

Industrial Relations:

- Recognize the impact of the human factor on quality.
- Develop and implement recruiting, selection, placement, training, and upgrading procedures that will result in a work force capable of meeting customer quality requirements.
- Communicate regularly to employees on the need for, and the importance of, working to quality standards.

Finding the problem areas can easily be done, although it is not practical to establish any meaningful absolute standards for cost comparisons. A quality cost system should be tailored to a particular company's needs. There are many analytical techniques that can be used. Two of the common techniques are:

TREND — Comparing present cost levels to past levels.

PARETO — Involves listing the factors that contribute to the problem and rank ing them according to the magnitude of their contribution.

Once a problem has been identified, reported, and the involved personnel are committed for action, the job is far from complete. *Involvement must be planned, coordinated, scheduled, implemented, and followed up.* Two types of problems; those which one individual or department can correct with little or no outside help, and those requiring coordinated action from several activities in the organization.

Example:

First Type. Operator-controllable defects, design errors, and inspection errors.

Second Type. Product performance problems for which a cause is not known, defects are caused by a combination of factors not under the control of one department, and field failures of unknown cause. These are normally the most costly and are not easily solved. Causes may be numerous and unknown. Solution may require action from several sources. The investigation of the problem and planning of its solution must be coordinated and scheduled. Each project should be docketed and action scheduled. Meetings should be held regularly and minutes published.

What steps can be taken in any Quality Cost Program to reduce failure costs? Here are some examples:

1. Make all persons concerned aware of problem and its possible causes. Regular reports.
2. Create a desire in others to solve the mutual problem.
3. Plan and carry out a logical investigation of the problem with others involved (corrective action).
4. Follow up on action taken.

Up to this point, we have been discussing ways to find and solve problems which are already costing money. How much better it would be if those problems could have been prevented from happening! Prevention activities are of two basic types — those related to employee attitudes and resulting approach to their jobs, and those formal techniques which identify potential problems early in the product cycle and eliminate them before they become expensive.

Employee attitudes toward quality are determined in large part by their beliefs about what is really wanted by their boss and higher management of the business, and by the degree to which they are personally involved in the improvement program.

- Employees should be aware of all aspects of the program affecting them or their work. Planned verbal communications, posters, charts, reports, and displays have all been found to be effective.
- There should be a way for each employee to present his ideas for quality improvement. Each idea should be properly evaluated and the employees should be told of the results.
- Involving employees in goal setting sometimes results in higher goals than would have been set otherwise.
- Good ideas and achievement of performance in excess of established goals should be publicly recognized. There are many ways to recognize achievement, but remember, it must have been honestly earned, have meaning and value and be properly presented and publicized.

Prevention In Marketing

A. Customer requirements.
B. Prices reflect unusual requirements.
C. Warranty.
D. Product performance/customer satisfaction.

Prevention in Design

A. Safe for intended use.
B. Profitably manufactured.
C. Participating in problem investigations and promptly correcting design caused deficiencies.

Prevention in Quality Assurance

A. Should review contractual customer specification.

B. Quality requirements must be determined.

The costs of appraisal sometimes approach half of the total quality cost. Most quality cost programs concentrate on reducing failure cost first. Programs for appraisal cost improvement can also have significant impact. Several techniques for improving these costs inspection and test planning:

Operator Inspection Advantages

- The operator usually handles every piece coming off the line.
- He is thoroughly familiar with the item he is making.
- He is in a position to spot defects quickly.

100% In-Line Inspection Advantages

- It saves the cost of further processing of product that is likely to fail final inspection and test.
- It provides data on quality performance that can be used to take corrective action.

Disadvantages

- May become routine, with rejects being screened out, less emphasis will be placed on prevention of defects.
- It tends to duplicate inspection and increase inspection cost.

First-Piece Inspection Advantages

- Detects defects at beginning of run; therefore, corrections can be made before run is started.
- Cost savings on nondefective merchandise manufactured.

Patrol Inspection

Inspector patrols the operation at periodic intervals and inspects the item being produced. Faster response to problems, less defects can be made.

In-Process Acceptance Inspection

All items made at an operation in a given period of time are inspected as a lot before they can be released to the next operation.

Advantages

- It is possible to control the quality level at each stage of manufacturing.
- It provides data to help pinpoint problem areas.

Disadvantages

- It does not prevent defects since items have been completed.
- It delays the movement of parts.
- It is not easily applied to continuous processes because of the difficulty of forming lots without disrupting the continuity of production.

Improving Equipment and Methods
Many of the most profitable areas of saving of inspection and test cost lie in improved equipment and methods used to do the job. Inspection and tests are not usually measured and controlled to the extent production jobs are; they are not usually too effective. The first step in improving equipment and methods is to find high cost areas. Construct a Pareto distribution and then concentrate improvement efforts on high cost areas.

Statistical Quality Control
Powerful tools can be used to help achieve in-process control, capability studies, control charts, and sampling inspection.

Inspection Accuracy
There are failure costs associated with incorrect quality.

- Falsely rejecting acceptable material.
- Falsely accepting rejectable material.

Rating System = Submit a known number of good and bad units to an individual and rating his ability to separate the units correctly. The number of incorrect decisions is then multiplied by the cost of each wrong decision.

There are many sources of information on product performance which usually give sufficient data to determine whether the customer's needs are being met. It usually takes a combination of sources to get the needed data, and the optimum combination varies with the product and market. Some sources are:

- Change in quality costs.
- Field failure and repair reports.
- Installation phase reporting.
- Personal observation by company personnel.
- Life testing of company and competitor's finished product.
- Market research on customer opinion and user cost.
- Data from spare parts sales.
- Customer complaints.
- Outgoing product audits.

Reporting Product Performance
Reports of field problems should be prepared periodically. It will indicate how well the company is succeeding in satisfying the quality requirements of its customer, and how effective the quality procedures are.

CONCLUSION

When the highest cost areas are analyzed, many improvements projects become apparent. For example, high warranty costs are a trigger to rank customer failure problems for detail investigation, with the aim of looking into product design, process control, or inspection planning for the cures to the highest cost problems. A natural temptation due to high external failure costs might be to put more emphasis on appraisal efforts; but this approach may simply convert some external failure to internal failure (scrap, rework) and bear an increased inspection burden to do this. Regardless of what the category may be, by identifying it, it can be reduced. *The entire process of quality improvement and quality cost reduction necessarily is pursued on a problem-by-problem basis.* Any quality cost improvement depends on understanding cause and effect relationships, and the study of total quality costs is the most effective tool available to management.

IMPROVED PRODUCTIVITY THROUGH QUALITY MANAGEMENT

Frank Scanlon
The Hartford Insurance Group
Hartford, Conn.
and
John T. Hagan
ITT World Headquarters
New York, New York

INTRODUCTION

Do you know how many people in the United States are involved in the business of providing service? Estimates range from 60 to 75% of the working population. In addition to this large and growing number of workers, there is also a high percentage of personnel on the manufacturing side of industry that are involved totally in administrative tasks. This number exceeds 50% for many companies. Altogether, there is a large majority of working people in the United States employed for the sole purpose of providing service to others. And most of them are not familiar with the meaning of productivity or quality management.

We are here today to speak out to the management of American Industry. We have a message about quality management and it is for the managers of banks, insurance, financing, food service, lodging, real estate, publishing, broadcasting, retailing, leasing, repairs, utilities, hospitals, schools, transportation, communications, government and other service operations, as well as the managers of product manufacturing companies. In this era of strong foreign competition and growing customer demands for quality of service, the quality reputation of an organization is too precious an asset to leave to chance. Less than satisfactory service internal to an organization will result in higher operating costs; less than satisfactory service as delivered to the customer will have a negative impact on revenues.

Consistent service quality will not happen accidentally. An organized, scientific approach to Quality Management is required to maintain or develop a quality reputation. That usually means an investment in the study and application of quality control principles to all areas of service work. That is, the work of the people who deliver service to the customer (or do the manufacturing work for products) and the people who provide support service to these front-line operators. Unfortunately there are three obstacles to the straight-forward fulfillment of quality management programs in the service business. The first is that the managers of service businesses are, generally speaking, almost totally unfamiliar with the substance and business value of quality control principles. The second is that all too often, investment in the control programs are viewed as unnecessary expenses rather than programs with a pay back. They are seen as having a negative rather than a positive effect on productivity. The third obstacle is that customers are not genuinely listened to. Their complaints are seen as an irritant rather than an opportunity.

The authors hope to remove these obstacles by showing that a small investment in quality management can yield a large return in productivity and in the number of satisfied customers, the most important ingredient for future revenues. We will also show that the experience of quality management always contains added value in the form of improved communications and morale in every segment of the organization.

To understand the problem of unsatisfactory service quality and the solution of Quality Management, it is necessary to define the situation in terms that all parties can relate to. Unsatisfactory quality is described as, "undesirable results due to unwanted and unnecessary variations in performance." The degree of deficiency may be different but we've all experienced it in hotels, restaurants, rental cars, appliance repairs, home maintenance and many others. It's usually not an out-and-out failure — just a completely unsatisfactory situation as measured by our own personal standards.

The cause of the problem is almost always attributed to standards of performance being weak or non-existent. It's very noticeable when service personnel seem to know exactly what they are doing — such as at Disney World or McDonalds. At the other extreme, customers are forced to complain. The unwanted difference occurs when individuals, managers and supervisors, as well as workers, are free to set their own standards of performance. *Fortune Magazine,* in its article, "Towards Service Without a Snarl" (March 23, 1981) stated: "For all the improvements made possible by technology, the quality of service still often depends on the individual who delivers it. All too often he is underpaid, untrained, unmotivated, and half-educated." This applies equally as well to the individuals who provide back-up support for the deliverers.

The solution to this predicament is to establish a Quality Management system. Very simply stated, this means establishing performance standards, measuring performance against the standard and then developing a quality improvement program. This solution is based on the premise that no matter where you are today in terms of performance, you can always get better. The benefits of this solution are that the unsatisfactory service prevented through performance improvement will not cost anything in extra service or in loss of future customers.

To bridge the gap between this simple solution to quality-of-service problems and the real world of organizational pressures and priorities, our program will attempt to deal with all the realities of the business world as they are in fact. We will start with a sales pitch aimed at getting management interested. This will be followed by an explanation of the Quality Council and/or implementing team, establishment of performance standards and responsibilities, measuring performance, building "the improvement program" and implementing or managing the actions required for results. Then we will introduce the value of recognition, the use of quality costs, and finish with a summary of the results that can be expected.

MANAGEMENT COMMITMENT AND THE QUALITY COUNCIL

The job of the Quality Professional is to convince management that Quality is what they need and show them how to build a quality system that can be measured, is guaranteed to produce improvement, and is cost-effective. In order to do this, the Quality Professional must be enthusiastic, project a "can do" attitude, and be specific with regard to the benefits for management. He must also build a very practical Quality Program.

The benefits to be used in selling management are:

Improved Image — quality improvement improves your reputation with employees and customers.

Improved Productivity — quality improvement means less rework and reductions in your work process, thereby, improving productivity.

Reduced Expenses — less rework requires less labor and material costs.

Improved marketability of services which will increase market share/customer base — improved quality gives you the expense and service advantages over your competitors.

Management of quality and quality costs — a quality improvement program will enable you to act rather than react to problems, to manage rather than be the victim of.

Improved employee environment — quality improvement reduces employee frustrations leading to improvements in human relations, morale and self-worth of employees.

Improved profitability — all the above leads to this positive end result.

Although quality improvement is an economical means of achieving these benefits, quality is not free. The price the general manager must pay is:

— Visible ongoing quality leadership
— A quality policy that is meaningful and practical
— A budget to support up-front needs
— Appointment of a quality council to carry out the quality policy

Visible ongoing commitment is a non-transferable function and does not end with the appointment of a team to carry out the quality policy. It is important that employees recognize their chief executive officer as the leader in quality.

There is no one best Quality policy. The words should be those which are most easily understood by the management and employees of an organization. A good example is the Hartford's quality policy, which is: "To perform each job or service in exact accordance with existing job requirements or standards." In setting a Quality policy, it should be remembered that the policy itself and the definition of "quality" is applicable to all employees of an organization from the chief executive officer to the most recently hired individual. An area in which management often gets into trouble with a quality policy is to imply that the policy is only applicable to line type individuals in the manufacturing, administrative or service areas. It must be clearly understood that a quality policy is applicable to engineers, planners, all of management, marketing and general managers themselves.

Appointment of a quality council by the general manager is a transfer of the quality program work task and not a transfer of the responsibility for visible leadership. The Quality Council should be made up of the same senior executives that the chief executive trusts in leading and managing major company functions — the individuals whose decisions and ideas make the company what it is today. The duties of the quality council are:

Determine *what* should be done to implement the company quality policy.

Determine *who* will be responsible for carrying out the actions decided by the council.

Provide a communication link between the council, the general manager, employees and the Quality function.

Assist *all* Quality representatives (Corporate Office and Profit Centers) in identifying and resolving internal or interdepartmental implementation problems.

Assign Quality task teams to audit existing programs and to explore specific problems or opportunities.

The only limitations on the Quality Council's charter is that the program must:

— Adhere to the Quality policy.
— Provide quality improvement.
— Develop meaningful measurements.

The role of the Quality Professional is to help create a system, attitude and environment that can assist all managers in carrying out the quality policy. This responsibility is carried out in conjunction with the quality council.

SETTING QUALITY STANDARDS

The first task of the Quality Council is to define the service quality standards. This means to clearly define exactly what is intended to be delivered to the customer. A service quality standard should be very similar to a product description or specification which, together with price, provides a specific choice to prospective customers. The exact requirements of the quality standard, the consistency with which it is met and the price for which it is delivered will have the greatest bearing on the growth or diminution of your customer base. There are exceptions to every business rule such as, in this case, the initial attraction of something new or different, or the choice location of a particular hotel — but in the long haul, the quality of service and price will make the difference. It is well also to note that it is not possible to be a price leader in any industry without first having established a consistent and satisfactory quality reputation.

In the process of setting standards for quality the most important input is that which comes from the customer. It is always wrong to assume or guess about what the customer really wants or expects. The rolls of the bankrupt are replete with many examples of this mistake. Perhaps the best known example is the Edsel. On the other hand, understanding the customers' needs and expectations is no easy task. Customers are more sophisticated today. They are also better educated, more discriminating and more demanding than at any time in the past.

To truly understand the customer, an organized study of the marketplace must be undertaken from different points of view. Some of the most successful avenues of study include the following:

Market Research — done independently or as part of an industry association or agency.

Public Opinion Polls — conducted by industry associations or agencies representing the customer or consumer.

Review of Competitive Activities — an investment in comparing yourself with your competition.

Analysis of Complaints and Compliments — learning from your customers' direct communication with you.

The objective of this study and analysis should be to allow you to establish quality standards or customer expectations for your product or service that are equal to, better than, or different in an appealing way from your competition. It can also be used to develop a special niche in the marketplace that can best be served by you. If you should discover that what is normally expected is considerably more than you are willing or able to deliver, your problem is considerably beyond the solution being presented in this paper.

The next step in the definition of quality standards is to convert the agreed upon customer expectations into well-defined, specific requirements. This is needed for two reasons. The first is to serve as the finished criteria or target against which the entire service process must be designed. The other is that acceptance criteria or targets can be established for all operations leading up to the end product or service. As an example, Japan Air Lines Co. has established key quality requirements for flight operations (see Att. A).

Note that there is a target and control limit for each measurable performance standard established. The control limits are the acceptance standards for the operation and the targets are the (1980) goals for the improvement program.

To understand the service process as it affects the quality standard, consider the control limit of 3.00 baggage irregularities per 10,000 checked bags in the above example. In order for this standard to be met, every activity involved in the handling of baggage must be clearly defined. If it is not defined, merely left to the individuals involved to determine how to do the job, or to the whims of a supervisor on a given day, an infinite amount of variation will occur in the end results. This is in direct conflict with the need for consistency in operations in order to achieve predictable and consistent results. In manufacturing companies, detailed process sheets and operating procedures are used to assure the production of conforming end products. The same approach, detailed service process sheets, are needed for the service industry. These process specifications can then become the basis for employee position descriptions and overall training programs.

As detailed service process specifications are prepared back up the line from the delivered service to each activity required to achieve the desired end result, it will become clear that these specifications actually spell out the standards of performance for all personnel involved in the service process. Measurements of performance against these standards will indicate quite clearly whether or not the end result can be achieved. It is like examining all of the sub-assemblies that go into an end product. If any of them fail to meet their quality standards, it is not possible for the end product to meet its requirements. Even a succulent hamburger, to meet its quality standard, needs the right ingredients, cooked exactly to order and served in a timely manner in attractive surroundings.

In short, establishing service quality standards means first determining the quality image that you want your customers to experience. It should be an image that he will remember, return to and tell others about. Next, it means determining the exact standards of performance that each employee must fulfill in order that the desired image can be achieved and a quality reputation be established and maintained.

MEASURING PERFORMANCE

Once quality standards are established, the next task of the Quality Council is to measure performance against these standards. Measurements are needed for three very important reasons. The first is to determine where a business stands against its standards in order to identify and justify specific improvement needs. The second is to establish a base-line for measurement of improvement progress and the third is to provide input for the identification of specific problems for improvement action.

It should be noted here that, generally speaking, people do not like to be measured. Foremost among the reasons is management's misuse, in a destructive rather than constructive sense, of measured results. Also, people never before measured may be well aware of the fact that they are under-producing. On the other hand, without measurements, management will not be able to identify performance problems and explore their causes. They will also miss opportunities to reward superior performance, a genuine quality improvement asset. Unfortunately, lack of measurement always penalizes the good performer and rewards the bad.

Quality measurement is an activity where a clear-cut level of investment must be determined. It is usually not practical or economical to measure detailed performance against each standard. One practical approach is to examine flow charts of service processes as they integrate to achieve the final result and to then carefully determine key points within the overall process. Key measures are those which provide the most appropriate and direct appraisal of the achievement of incremental performance objectives. For example, sampling inspection of hotel rooms ready for occupancy can serve as a measure of overall housekeeping and maintenance activities. Analysis of credit memos is a good measure in the distribution business. A sample survey of restaurant customers, after they have been served, can be a measure of order taking, food preparation, and the timeliness and quality of delivered service. Where a system is involved, such as a communications network, sample messages tracked through the network can measure how well each segment is performing.

One of the most important performance measures is the direct input of the customer. Some investment in this area is mandatory. Only about one in fifty customers will say what he really thinks, so it is very important to listen. Some of the methods employed are customer questionnaires, customer comment log books, telephone surveys, and analysis of complaint/compliment letters and phone calls. It is vital to progress that all customer complaints be objectively analyzed and resolved fairly and in a timely manner. If the concern for customer satisfaction is genuine, no customer need ever be lost.

Once it has been decided what performance measures are to be taken, the next step is to arrange for the data to be collected and organized for effective use. This generally means arrangements that show percentages of satisfactory performance at key points in the process and Pareto arrays of the frequency and distribution of all (measured) nonconformances to standard. Analysis of these arrays will lead directly to the identification of those specific elements of the service process in need of improvement, the perfect input for development of the Improvement Program.

BUILDING A QUALITY MANAGEMENT SYSTEM

The prerequisites for building a quality management system already discussed include general management commitment and personal leadership, establishment of a quality policy and a quality council to carry it out, setting quality standards and measuring performance against these standards. Additional prerequisites include understanding of the following quality elements by both management and employees:

A. Definition of Quality

Quality can never be achieved when everyone has his own idea of what it means. In terms of measured performance, quality can only mean conformance to a standard. That is, quality can only be achieved when the end-product or service, as delivered to the customer, looks and performs exactly like the instructions say it should. Quality is a noun, the result of clearly defined efforts. It is not an adjective — the degree of definition or effort.

B. First Line Manager's Role in Achieving Quality

Individual job specifications must be clear and understood by all employees that are part of the process. The end product must meet these specifications. If the specification is not correct, management must change it.

It is impossible to achieve quality if the individual hired cannot perform the functions necessary to meet the specification. It is the responsibility of first line management to hire the proper individual. Even with the most comprehensive testing available, management is not always sure that they have hired the right person. Therefore, it is a quality program failure when management fails to take remedial action when it becomes clear that they did not hire the right person.

If the individual hired to meet the job specification is not trained properly, then quality cannot happen. It is management's responsibility to train people properly.

Without communications, quality cannot happen. It is the responsibility of the first line manager to make certain communications not only take place, but are also effective.

C. Responsibilities for Quality

a. Quality is the responsibility of the individual who was hired to do the job. This is a non-transferrable responsibility. Management interviews people to fill a particular job and it is expected the person will do the job correctly. On the second day of work, people should not be told what their acceptable error factor is. Management cannot do the job for their employees, nor can they accept as a matter of course that errors are inevitable.

b. Managing quality is the responsibility of the first line supervisor. The management of quality includes:
   - Knowing the current quality level within the unit through measurement.
   - Knowing what the major error cause is through analysis.
   - Have in place a practical corrective action system which addresses the *elimination* of this error cause.

c. Monitoring quality is the responsibility of middle management. That is, they are responsible to know what the level of quality is and what corrective actions are in place to improve quality.

d. Questioning quality and quality measurements is the responsibility of the general manager. It is necessary for the general manager to show the visible leadership and involvement in the quality system by asking questions where slippage occurs, by challenging results when they are in conflict to what department heads and customers are saying and to recognize truly outstanding quality improvement.

e. Auditing is the responsibility of an independent functional and/or internal operation which specifically has the responsibility to take a snapshot profile of the quality level of a given function. This is the objective view of the quality performance of a given unit/function.

D. Quality Goal

The objective of a quality improvement program is to improve quality . . . period. The purpose of a quality management system is to identify for the first line manager what his major problem is in order that he can investigate the *cause* of the problem for the purpose of developing a preventative type of corrective action plan to *eliminate* the cause.

E. Quality Control versus Quality Improvement

Generally, the purpose of a quality control program is to develop a method to "ensure" end products are going on correctly. A quality improvement program is a program based on historical fact, that is, analyzing what has been done for the purpose of investigating error causes to ensure the same type of errors will not recur as frequently in the future, thereby, gaining consistent quality improvement. Quality improvement can be summarized as "do it right the first time, next time and every time."

With all of the prerequisites complete, we can now *build the Quality System.* At best this is a "cookbook" approach to building a Quality System. In applying it to your company, take into consideration the specialness of your company's product, service, process and personality and translate the recommendations into a language that can best be understood by your management and employees. An effective quality improvement program is something that must be unique to your company. Quality improvement is essentially achieved through a program of ongoing quality measurements, analysis and corrective action. The elements of the Quality System are as follows:

1. Sampling (Quality Measurements)
   Once we have identified the current level of quality performance and have made the management decision as to the target levels and variances that are acceptable (or we want to pay for), a standard sampling approach can be utilized for ongoing measurements. Sampling is done in lieu of 100% checking for the following reasons:
   - Purpose of the quality improvement program is to identify error causes through trending. Sampling will do this for us.
   - 100% checking is only 85% effective.
   - The cost of 100% checking in dollars and time is prohibitive.
   - Studies have shown if people know they are being 100% checked, they tend to make more errors rather than less.
2. Checking
   Very simply stated, the items which have been sampled out of the total work population are verified back to the written standard to determine if they meet the job specification or quality standard for that portion of the service process. If the item meets this standard — it is quality; if not — it is not quality.
3. Recording
   All errors identified regardless of source should be recorded as to where they occurred and if possible who made them. This includes errors which are detected in the processing (sampling checking method), errors from work rejected internally as a part of normal operations, and errors contained in service rejected externally by customers.
4. Analyze results and identify major problems
   Measurement data to be of real value must be analyzed and interpreted — differentiating between cause and effect. Examine and interpret each measure. Then, the results can be used for determining necessary corrective action.

Timeliness is of prime importance when problems exist. The results of data analysis, with proper identification of the major problems, must be provided to responsible personnel as quickly as possible. Quite frequently the real causes of a problem are unknown — and the measurements reflect only the effects. If the analyses and feed back are in "real-time" then there is a better probability that the causes of the problem can be identified and corrected.

Graphical displays of data are very effective. First, because the graphical displays are usually "self analyzing" (i.e., trend charts/graphs) and second, management awareness, understanding and acceptance of the data is enhanced by the graphical display. In this way, the data is used as a measure of performance.

Periodically (weekly or monthly) a Quality Status Report should be prepared and distributed to the management of the business. This report should include a narrative interpretation of results and actions currently required, in additon to the typical summaries (table, trend charts, etc.) of measurement data.

5. Corrective Action

   In order to go from analysis to corrective action, to truly address the cause, it is necessary to integrate people as part of the quality cycle. The best person to identify why an error was made is the person who created it; therefore, a practical corrective action plan must include direct input from the error source. When the error source is not known, investigative problem solving techniques must be brought into play. Every undesirable situation has a cause but it is sometimes hidden in a complexity of interactive functions. Until sorted out for corrective action, these hidden causes can plague a company for long periods of time. In order to achieve practical corrective action, the following is necessary:

   - Organize investigative and corrective action teams — identify the real problems.
   - Develop priority action plans — need to spend dollars wisely.
   - Get action commitments and dates — assure that everyone understands the plan.
   - Document and monitor actions taken — if expected results do not occur, continue the investigation.

Corrective action is the pay-off step. Without corrective change, the best you can hope for is the status quo. But with a viable corrective action system, constantly fed by employee recommendations, analyses of performance measurements and customer problems or complaints, performance improvement is assured. And performance with fewer defects will always be less costly and more on schedule.

6. Verify action results
   - Conduct follow-up audits. Avoid the tendency to assume because some action was taken, the result was the correction of a problem. Action does not always mean accomplishment.
   - Verify effectiveness. Corrective action is not effective if it only fixes the symptoms. It must go to the real cause of error.
   - Close out effective results. If the problem is really fixed, close it out and, thereby, close the loop of having achieved performance improvement. You are now at an improved and more profitable level of performance.

IMPLEMENTATION AND MANAGEMENT OF THE IMPROVEMENT PROGRAM

Once agreement has been reached on the exact make-up of the Quality Improvement Program, the next step is to plan, schedule and promote the implementation phase. This phase should start with a communication to all employees of management's commitment to Quality and their intent to initiate an "improvement program."

The kick-off event for the program can be done quietly or with a lot of fanfare. For example, Management can communicate to the employees through a simple, written pronouncement circulated throughout the organization, or they can organize a series of promotional gatherings using audio-visual aids. It depends on the personality and style of the company. The important criteria is that every employee receive the message and that they become keenly aware that "how we collectively do our work" is going to be different.

When quality is viewed as conformance to a standard, it becomes readily apparent that the achievement of quality is dependent upon each and every employee, from manager to operator, performing up to the standards of their job. That is why Quality Improvement is a companywide program. If the chain of quality is broken in any one area, the effect will be felt in the end results. In this respect, quality is no different than cost or schedule. That is, variations in the performance of individuals will have an effect on cost, schedule and quality. The dilemma to be avoided here is that if cost or schedule objectives are achieved at the expense of quality, there will always be a penalty to be paid in future sales.

It is important at the outset of Quality Improvement to reaffirm basic responsibilities. Each person must be held accountable for the quality (conformance) of the work performed at his station or desk. It is also advisable to encourage personal commitments to quality to go along with management's commitment. It would be an ideal situation if each employee committed to 100% error free work.

Following the kick-off or initiating event, the next step is the education and training of all involved personnel. Whenever Quality Improvement has not been an integral part of the company operations, its introduction will require a series of education and training programs for each element of organization as they are to become involved in the improvement program designed by the Quality Council. Usually, the key to this phase of the program will be the education of middle managers and first-line supervision. The workers normally respond with a positive attitude.

Training management personnel to respond differently to "quality" is not a task to be taken lightly. Basic attitudes about work and mind-sets that have developed over many years will be involved. There has to be a strong emphasis on the value and practicality of the program. And it must build a strong desire to identify and eliminate the causes of errors and nonstandard performance. It must also get into the mechanics of measurement, analysis, problem solving techniques and corrective action. The best approaches involve the use of workshops where the managers and supervisors actually gain the experience of doing quality improvement.

The challenging aspects of management training for Quality should provide a good justification for having a professional quality manager. As previously noted, a Quality Professional would be a valuable asset in developing the quality improvement program. He would be even more valuable in getting the program successfully implemented. In fact, it is not feasible to seriously consider a quality improvement program for a major enterprise without also considering the need for a key person to mold and guide the program through the intricate peculiarities of the specific enterprise.

In considering the role of the Quality Professional, the following responsibilities should be included:

1. Assure that quality is a built-in ingredient of all service processes and support systems by developing system quality procedures as part of the mechanism of company operations.
2. Become the internal "voice of the customer" through responsibility for the coordination and control of company responses to and resolution of all customer complaints and problems.
3. Assure that customer experience is summarized and used as important input to the establishment and improvement of service quality standards.
4. Provide education and training programs as needed for each element of organization to assure that all personnel from executives to operators understand the objectives of quality improvement and their individual responsibilities for the quality of the company's operations and product or service.
5. Establish and conduct a monitoring program to assure adequate quality controls of supplies and internal performance.
6. Keep abreast of quality control technology developments and assure maximum utilization of new ideas in the company Quality Improvement Program.

While management of the Quality Improvement Program is fundamentally the responsibility of the Quality Council, use of a Quality Professional will make the entire effort run more smoothly. The quality manager can act as a catalyst and help to crystalize the meaning of quality in each management area. He can also supply the additional leadership and day to day guidance necessary to keep the program on track. It is difficult to conceive of the implementation and management of a Quality Improvement Program without a full time quality manager.

COMMUNICATIONS

Quality cannot happen without communications. Communications is the life blood of a quality system. In order for communication to be effective, it must be reciprocal, that is, there must be a sender, a receiver and feedback to the sender to make certain the original communications and/or standards were fully understood. To make communications/quality even more effective, we must try to generate communications horizontally as well as vertically. Vertical communications usually means that there is a network in place to push communications downward, but in the spirit of quality improvement, there must also be a guarantee there will be an upward flow of communications. The horizontal communications needed are the communications that take place between individuals of the organization at a peer level, especially when they are performing different functions. Increased horizontal communications will also help employees to understand their job better. Management consultants make many dollars in this area by recommending some of these activities under the title of job awareness, job enrichment, job enlargement, etc. Basically speaking, instant quality and productivity improvement could be achieved if each employee of the company knew the answers to the following questions:

What is my job? Where do I fit in general terms in the organizational structure? Who do I receive my work from? What do they do? Why do they do it? How do their errors impact me? After I finish my work task, who receives the work? What do they do? Why do they do it? How do my errors impact them?

Do not leave communications to chance. The Quality Professional can help identify communication gaps, but every manager and supervisor must participate in completing the communications cycle. Further, upward communications can be stimulated through short-term, highly visible programs, such as Buck a Day, Quality Improvement 15 and ZD/30 and then through a more permanent type of upward communications, such as Quality Circles and other participative management schemes.

RECOGNITION

Recognition is the heart of a quality program. Recognition that is honest, deserving and sincere gives management the tool to truly integrate people as part of their quality program.

In an attempt to improve quality recognition at The Hartford, a questionnaire was sent to 10% of our employees countrywide for their input. The survey pointed out five areas in which our employees recommended improvement. They are as follows:

In order of priority —

1. To add monetary award.
3. To have unit awards as well as individual awards to stimulate the team approach.
4. More local control.
5. More publicity.

These four recommendations have been incorporated into Hartford's formal quality recognition program. Of more significance in this survey was the *second* most frequently mentioned item right behind monetary award — "All I want is to be recognized by my boss for doing the job right." Spontaneous recognition between first line managers and the employees is the most effective and inexpensive recognition available, and it should be done in front of as many peers as possible.

THE COST OF QUALITY

Cost of Quality techniques have not been universally applied to the Service Industry. This does not mean, however, that Quality Professionals have not been active in this area. In fact, the many efforts at application of Quality Costs to service and administrative activities could provide enough material for another paper. There is a need for development of Quality Cost Programs to support and justify the quality improvement programs for the service area. For this paper, however, we can merely share some of the conceptual thoughts that would probably underlie any serious development efforts.

One of the principle differences between a service business and a manufacturing business is the make-up of the customer base. Most manufacturing businesses operate from a large backlog of orders heavily dependent upon a few key customers, whereas service companies operate from a large customer base with many repeat customers. In the service business, more customers almost always mean more profit. For example, to a hotel more customers means a higher occupancy rate; to a telephone company, more calls per network; to a loan company, more contracts per employee. Because of this direct relationship, maintenance of the customer base is crucial to the survival and growth of a service business.

As amply supported in the main body of this paper, many service businesses now realize that "service quality" is a major factor in maintaining customer base and repeat customers. And they are coming to realize that a comprehensive quality improvement program is the best way to assure service quality achievement. To enhance the use of quality improvement programs, to lend credence to their business value and to provide cost justifications to the actions they propose, a supportive quality cost system should be developed.

The cost of Quality was formulated many years ago in the manufacturing sector of American Industry to assist the professional quality manager in getting the quality system understood and accepted. It was based on the premise that "there are no economics of quality" and it is intended to help sell the quality system on the basis of its impact on profit and loss. By identifying those costs directly affected, in a positive or negative way, by the quality program, the way was paved for better communications between the Quality function and other functions, especially general management. The exact same concept can be applied to the service business.

In applying the concept of quality costs to the service industry, it is not nearly as important to the implementation of quality improvement as it is to the knowledge and support of top management. When the chief executive can visualize a program to improve customer satisfaction as being equally beneficial to profitability, it is easier to get his full support and participation. It can also lend support to the acquisition of a Quality Professional.

Service quality costs, in concept, are considered to exist in two distinct parts:

1. That portion of operating costs caused by "inadequate conformance to performance standards" — costs resulting from customer rejections or complaints, as well as costs incurred due to internal errors or substandard performance, requiring some redoing of work.
2. Lost revenues due to unsatisfactory service quality — sales not achieved because of unhappy or lost customers.

Neither part of service quality costs, as defined above, can be measured directly. Therefore, the quality cost system will need to provide for an indirect means of identification. One way would be to estimate the values. Another would be to measure the difference between optimum or perfect performance and actual performance in both areas. These differences or estimates could then be identified as "quality costs" and targeted for improvement efforts. The real value of this measurement that actually exists and the fact that actual results will be a matter of record.

In the internal operations area, quality cost measurements can be based on examining each element of operation budget as being made up of two parts. The first part is what the budget would be if the work could be performed perfectly. This is called the "ideal budget." The second part is the difference between the ideal budget and the actual budget (based on recent past history and current expectations). This difference can be called the "quality variance."

An alternative measure of quality costs in the operations area would be to estimate the costs related to quality. This estimate should include the known costs of errors and substandard work-failure costs; the cost of quality inspections or monitoring — appraisal costs; and the costs associated with the implementation of corrective action — prevention costs. The total of these three cost estimates are then designated as operations quality costs and targeted for improvement. This method is a close parallel to the concept of quality costs for manufacturing companies.

In the area of lost revenues, there is no precedent in the manufacturing sector quality cost system. Maximum achievable revenues can be estimated on the basis of market research, competitor evaluations, actual customer experiences, capacity limitations and other factors. Revenues are also affected by pricing and there is a direct relationship between quality leadership and price leadership which should also be taken into consideration.

For service quality costs, it is suggested that maximum possible revenues be carefully estimated and identified as "ideal revenues." Then a quality variance for revenues can be identified as the difference between budgeted (expected) revenues and ideal revenues. This variation, like the operations cost quality variance, can then be viewed as another opportunity for business improvement.

Establishing quality costs as a measure of the difference between an ideal budget and actual costs or revenues does not imply that the business can be run perfectly. It merely emphasizes the potential for improvement that actually exists. It can then be used to focus attention on and justify the quality improvement program.

SUMMARY

The purpose of this paper was to provide an overview from the inception of the idea through implementation of a practical Quality Improvement Program that will guarantee improved productivity through quality management. It took us from the selling to management leading to the commitment and policy to the establishment of the quality organization, the setting of quality standards, the measuring of implementation and managing of the quality program through communications and recognition. It also introduced some concepts associated with service/administrative quality costs.

Some key points to remember are:

Make quality improvement an ongoing part of your operations.
An effective program is not something you turn on and off.
Make it part of the woodwork.

— Customize it to your unique organizational needs. Service businesses differ greatly in structure and organization. There is no standard approach.
— Get all personnel to participate. The people deep inside a system are the ones that really know the weakness and soft spots. They are an untapped reservoir of knowledge and value. It is urgent they be involved.
— Keep people fully involved and informed. Communications is the life-blood of the quality improvement program.
— Recognize and applaud outstanding performance. Recognition truly is the heart of a quality program and gives management the opportunity of building quality for positive reasons.
— We have just introduced concepts regarding quality costs in the service and administrative areas. Much more work is required in this area.

Once again, it warrants repeating that this presentation represents only the skeleton of a quality improvement program and it must be translated into the terminology which is most applicable to your operation and to the understanding of your management and employees. You must develop a unique Quality Improvement program for your company. You can achieve improved productivity through quality management.

References:

"Quality Management For Service Industries," J. T. Hagan and E. W. Karlin, ASQC Technical Conference, 1979

"Quality Costs For Service Industries," J. T. Hagen, E. W. Karlin and L. H. Arrington, ASQC Technical Conference, 1979

"Cost Reduction Through Quality Management," Frank Scanlon, ASQC Technical Conference, 1980

ATTACHMENT A

TABLE I EXAMPLES OF COMPANY LEVEL TARGETS OF QUALITY (1980)

Quality	Quality Indices		Classification		Target	Control Unit	Definition
Service (OPN)	Rate of Flt. Opn. Irregularities		A		0.74	0.95	Nbr. of Flt. Opn. Irreg. Per 1,000 Hrs. Flown
			B		0.21	0.28	Nbr. of Flt. Opn. Irreg. Per 100 Departures
	Rate of Schedule Performance		International		99.0	98.6	Actual Nbr. of Departures / Scheduled Nbr. of Dept. x 100%
			Domestic		99.0	98.0	
	Rate of On-Time Dept.	Over-All	International		82.0	79.5	Nbr. of On-Time Departures / Nbr. of Actual Departures x 100%
			Domestic		90.0	87.2	
		Speci-fied	Stn. Delay	Intern'l	6.30	7.52	Nbr. of Delayed Dept. (Exclude Code "D" Delay) / Nbr. of Actual Departures x 100%
				Domestic	5.00	6.75	
			Maint. Delay	Intern'l	1.47	1.74	Nbr. of Delayed Dept. Due To Technical & Maint. / Nbr. of Actual Departures x 100%
				Domestic	0.90	1.05	
	Rate of On-Time Arrival	Over-All	International		77.0	74.0	Nbr. of Delayed Arrivals / Nbr. of Actual Arrivals x 100%
			Domestic		79.0	73.5	
		Rate of Long Delayed Arr.		Intern'l	0.70	1.44	Ttl. Nbr. of Long Delayed Arrivals / Ttl. Nbr. of Actual Arr. x 100%
		Avr. of Delayed Duration		Intern'l	1.32	1.64	Ttl. Duration of Long Delayed Arrivals / Ttl. Nbr. of Long Delayed Arrivals
Service (PAX)	Complaint Passenger Rate		Intern'l	To	0.40	0.54	Nbr. of Complaints Per 10,000 Pax
				APO	0.70	0.84	
				Flt	0.70	0.86	
			Domestic	To APO Flt	Pending		Ttl. Nbr. of Pax Negative Reply / Ttl. Nbr. of Pax Replyed
	Baggage Irregularity Rate		International		15.7	18.7	Nbr. of Baggage Irreg. Per 10,000 Passenger
			Domestic		2.50	3.00	Nbr. of Baggage Irreg. Per 10,000 Checked Baggages

Excerpted from: "Service Quality Improvement by Circle Activities" by Hideo Takakuwa — 1981 Quality Congress.

STUDY COSTS & IMPROVE PRODUCTIVITY

William J. Ortwein
Pratt & Whitney Aircraft
East Hartford, Connecticut

ABSTRACT

Productivity changes are difficult to measure when standard values are not available. Variable inputs to generate changing levels of output complicate the problem. Where output is the product and input is resources, standard labor and material are known and measurable. Other costs which have a definite impact on profits, however, are more difficult to measure. A quality cost study is a tool to measure a portion of those difficult to measure costs.

This paper will explain a traditional cost pricing structure and identify areas of concentration for reducing nonstandard costs by indicating where the application of resources could be most rewarding.

GENERAL

It has been generally assumed that competition of all kinds is increasing. A general objective for one who actively directs affairs is to manage prudently and economically. Evidence of financial failure as a consequence of increasing competition may be found by examining companies, products, governments and their programs, and individuals. Although evidence of all types of failure is clear, the factors causing it are not. Relief for the profit squeeze on companies may very well rest on the ability of management to improve productivity.

Managers complain they cannot satisfy production schedules within the time frames, employee counts, and budgets imposed on them. Financial plans are generally based on sales which are difficult at best to project with any degree of accuracy in a great many industries. Some variability must be incorporated into all plans to allow for changing conditions. The application of variable budgets based on output, fixed budgets or profit plans with quarterly estimates, or some other similar approach must be employed to accommodate constantly changing market conditions. Assuming most organizations employ reasonable approaches to profit planning and the financial plans developed to achieve goals are generally realistic, managers must employ techniques which will assist in accomplishing the goals of satisfying production schedules within the time frames, employee counts, and budgets imposed.

We must learn to accomplish the same or increased output with the application of fewer resources. We must become more productive.

Attempts to measure productivity with the intent of establishing a course of action which will lead to the optimum utilization of available resources have resulted in a variety of studies and programs which have contributed to the profitability of business concerns.

In order to explore this approach of productivity improvement through a quality cost study, I would like to begin with a brief review of a simple financial cost and price structure.

TRADITIONAL COST AND PRICE STRUCTURE
FIGURE 1.

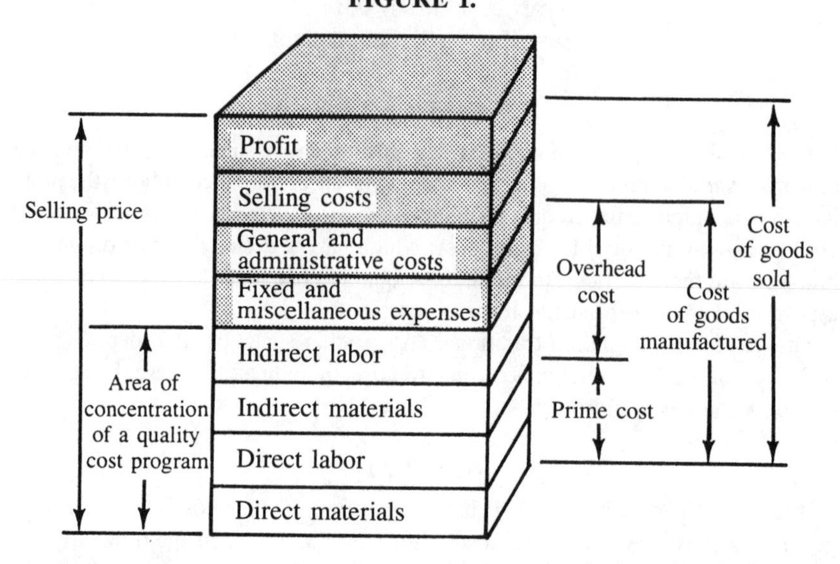

Figure 1 identifies, as is indicated on the left hand side, the compostion of the total selling price. There is no intent to imply because of the size of the individual blocks that each area encompasses the same amount of costs. The blocks merely lend themselves to pictorial view of some elusive concepts. The right side of the figure identifies some terminology referred to frequently in manufacturing companies.

The first grouping of costs is called prime costs. Prime costs consist of the total of direct labor and direct material and are referred to by that name because they are the basic input and an absolute necessity in order to produce an inventory or saleable product.

Direct materials consist of raw materials, semi-finished and finished parts which actually enter into and become part of the finished product. Confusion sometimes exists between direct materials and supplies. Supplies differ from direct materials in that supplies are used or consumed in the operation of the business but not directly in the product itself.

The second element of prime costs is direct labor, which is labor applied to convert direct materials into the finished product. Direct labor costs are those which can be specifically identified with a product or which vary so closely with the number of products that a direct relationship is presumed to be present. The wages and related costs of workers who assemble parts into a finished product, or who operate machines in the process of production, or who work on the product with tools would be considered as direct labor costs.

The second grouping of costs is referred to as overhead costs. They include all costs other than raw materials and direct labor which are associated with the manufacture of the product.

Indirect materials consist of supplies consumed in the manufacture of the product but not directly a part of the product itself as was mentioned before. Included in this category of overhead costs are items such as coolants, cutting oils, protective boxes for material handling, shipping supplies, perishable tools, heat, light, power, and the like.

Indirect labor represents wages and salaries earned by employees who do not work directly on the product itself but whose services are related to the process of production, such as foremen, process planners, truckmen, and janitors.

Fixed and miscellaneous expenses include depreciation which is the most well known fixed expense; however, taxes, rentals, and insurance on the assets used in the manufacturing process are also included in this category of expenses.

General and administrative costs (commonly referred to as G & A costs) is a catchall classification. Costs from operations of departments such as Financial, Personnel, Information Systems and Public Relations can be found in this category. These costs do not contribute directly to the product but are necessary for the successful conduct of a business concern.

Selling costs are those costs incurred in an effort to make sales and in transferring the completed product to the customer. Costs incurred "beyond the factory door" include warehouse costs, billing costs, and transportation costs.

Profit is the increase in owner's equity at the time of sale.

The traditional financial mechanism of a manufacturing concern organizes money activities similar to the structure identified in Figure 2.

The direct labor block of this structure is the section which is employed as the basis for most analysis when a campaign to increase productivity is initiated.

Productivity is generally measured through an Industrial Engineering Department on the basis of performance. Comparing actual hours to a predetermined standard for that operation establishes performance. Detailed studies are undertaken to establish a standard, and the results of such studies are extremely accurate. This method of identifying performance is an acceptable and proven approach. It also allows for variable productive output, standard labor (which could be less than half of all direct labor) as a base for direct labor variances. The next series of slides utilizes the same structure but concentrates on measurement parameters traditionally employed.

Direct labor variances consist of those direct labor operations which are difficult to measure. They include heat or surface treat operations, inspection, rework, etc., and performance which is hours expended in excess of standard hours. Standard direct labor plus direct labor variances added together equal total direct labor.

Total direct labor is then employed as a base for calculating overhead costs which, when added together, equal cost of goods manufactured.

Selling expenses and G & A are then calculated as a percent of the cost of goods manufactured to determine the cost of goods sold. The entire financial picture of a company relates directly back to the machinists out in the shop cutting chips.

FINANCIAL MEASUREMENT OF PERFORMANCE
FIGURE 2.

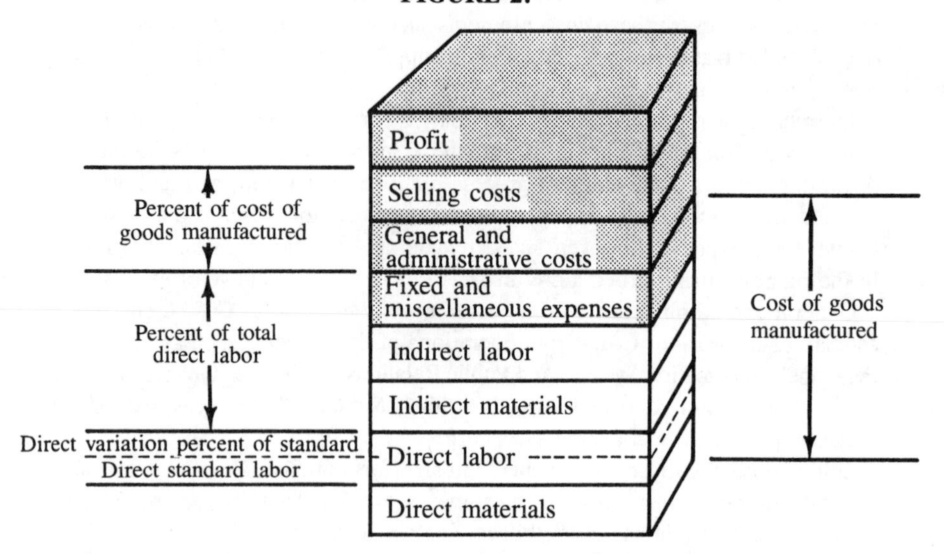

In general, we measure our entire effectiveness through the burdening of a small portion of our total labor force. If the workers' performance is good, as measured by Industrial Engineering, our productivity (generally assumed to be that of the entire operation) is considered to be good.

Employing the use of standards where possible to determine productivity is excellent in its application.

However, by what method can we accurately determine standards for inspectors involved in sampling plans, process control, surveillance, and the like? When is rework consuming an inordinate amount of available resources? What can be employed as a tool to measure effectiveness? How do we identify to management areas which appear to be eroding profits? *Consider a quality cost program.*

A Quality Cost program concentrates attention in areas which traditionally have consumed resources. The program highlights those areas where costs may be reduced or avoided without decreasing the value of the product to the consumer.

A program which identifies relationships that exist between categories and elements within the boundaries of a Quality Cost program, could result in increased productivity and earnings.

Expense categories included in the program segregate quality costs out of cost of goods manufactured. When compared to previous studies employing the same techniques, high cost areas are identified where investigative effort will present the most promising areas for future return.

The method of accumulating costs in a going concern requires some basic segregation within the structure identified in the cost/price chart. The company's chart of accounts defines in greater detail the costs incurred in the total operation.

The number of accounts in the chart of accounts as well as the account descriptions, varies from company to company. All charts of accounts are developed to suit the needs of that particular concern. A significant portion of the costs in a Quality Cost program are already identified as a result of previous requirements for internal control, federal regulations, contractual requirements, and the like. In addition to the detail found in the chart of accounts, job orders, work orders or some other similar system is usually available for further cost definition.

A Quality Cost program simply identifies a portion of the financial structure in a slightly different manner than have traditional financial methods. This program concentrates in areas where expenses are not incurred when the organization operates at 100% efficiency.

When implementing a quality cost program use the same nomenclature as present financial documents. Supervision and management will understand the program more readily if terms are familiar to them. Acceptance of the concept is enhanced when those who will use the program feel comfortable with the terminology.

Employ account descriptions from the chart of accounts, unit or department names, product line nomenclature and any other source of terminology which will lend itself well to analysis of your operation.

QUALITY COST CURVE

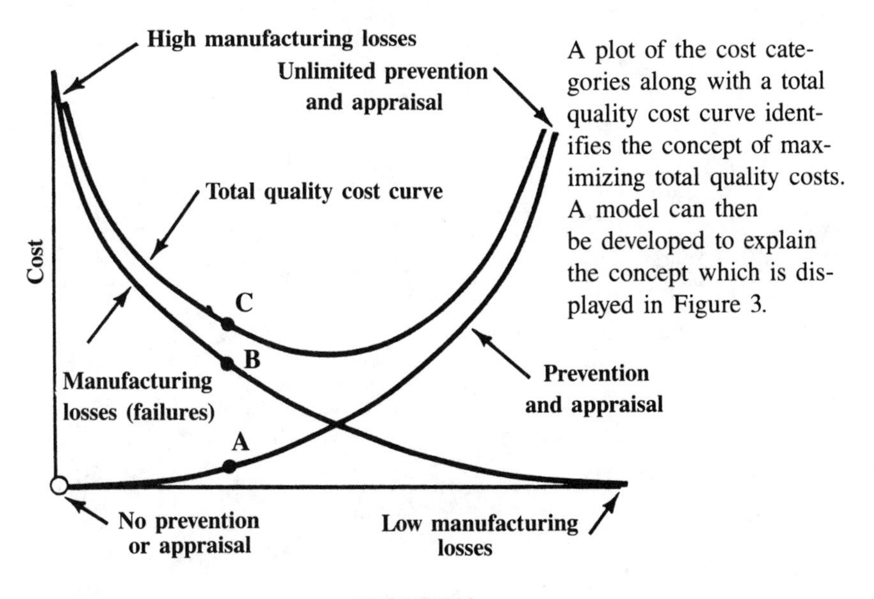

A plot of the cost categories along with a total quality cost curve identifies the concept of maximizing total quality costs. A model can then be developed to explain the concept which is displayed in Figure 3.

-FIGURE 3-

The application of this quality cost model identifies those resources which are consumed to assure fitness for use of the product.

Manufacturing losses are expected to be extremely high when little or no effort is devoted to prevention and appraisal activities. As ever increasing resources are expended on prevention and appraisal activities, manufacturing losses begin to diminish. They eventually reach a point where unlimited application of prevention and appraisal resources yield a minimal reduction in the cost of losses. Adding the two cost curves together at points A & B all along the curves generates a total quality cost curve, from a series of points at C.

The central portion of the curve represents the optimum application of quality resources. The time at which a company's program matures, when the optimum or indifference zone is reached, the primary function of a quality cost program is control. The program must continue in order to identify elements in the model which appear to display an unfavorable variation from that which is considered to be acceptable.

Studies concentrating on high dollar elements will identify areas where efforts are most likely to result in increased productivity. When improvements are implemented, the program is in a control mode.

The actual quality cost summary must relate to the manner in which a company conducts its business. In the interest of simplicity, the major cost categories in Figure 4 are identified by alpha characters *A* for prevention, *B* for appraisal, and *C* for manufacturing losses. I prefer to use the term manufacturing losses as opposed to quality losses because the term manufacturing losses more accurately identifies the source of the majority of unnecessary costs incurred in a manufacturing operation.

TOTAL QUALITY COSTS

		Dollars Expended (Thousands of Dollars)	Percent of Total Quality Costs
Prevention	A 1	$ 1.5	.4%
	A 2	18.4	4.9
	A 3	10.5	2.8
	A 4	4.1	1.1
	A 5	12.0	3.2
	A 6	5.6	1.5
	A 7	3.0	.8
	A 8	4.9	1.3
	Total	60.0	16.0
Appraisal	B 1	10.5	2.8
	B 2	23.6	6.3
	B 3	22.1	5.9
	B 4	4.9	1.3
	B 5	107.7	28.7
	B 6	32.6	8.7
	B 7	8.6	2.3
	Total	210.0	56.0
Losses	C 1	22.9	6.1
	C 2	6.8	1.8
	C 3	35.6	9.5
	C 4	19.9	5.3
	C 5	12.0	3.2
	C 6	7.8	2.1
	Total	105.0	28.0
Total Quality Costs:		**$375.0**	**100.0%**

FIGURE 4.

Total dollars are accumulated in each category through adding all costs in each element. Total categories add to equal total quality costs. The actual dollar amount is of some significance; however, its relationship to the whole is of greater consequence. The percent of total quality costs column readily identifies those areas which could present a high yield for effort expended. To expand on the concept, Figure 5 displays total quality costs as a percent of sales included in an ongoing program. This provides a comparison over time of the relative position which identifies total performance as well as trends generated by element and category.

QUALITY ELEMENTS AS A % OF QUALITY COSTS

		Period 1	Period 2	Period 3
Prevention	A 1	.4%	.6%	.5%
	A 2	4.9	4.9	4.7
	A 3	2.8	2.7	3.0
	A 4	1.1	1.3	1.5
	A 5	3.2	3.8	4.1
	A 6	1.5	2.8	2.9
	A 7	.8	1.0	.8
	A 8	1.3	.9	.8
	Total	16.0	18.0	18.3
Appraisal	B 1	2.8	2.7	2.8
	B 2	6.3	6.3	6.2
	B 3	5.9	5.7	5.8
	B 4	1.3	1.5	1.4
	B 5	28.7	29.0	29.1
	B 6	8.7	8.8	8.9
	B 7	2.3	3.0	2.8
	Total	56.0	57.0	57.0
Losses	C 1	6.1	5.5	5.8
	C 2	1.8	1.9	1.9
	C 3	9.5	8.6	8.0
	C 4	5.3	5.1	5.1
	C 5	3.2	2.9	2.9
	C 6	2.1	1.0	1.0
	Total	28.0	25.0	24.7
Total Quality Costs:		**100.0%**	**100.0%**	**100.0%**
Total Quality Costs As A % Of Sales		**6.78%**	**6.65%**	**6.58%**

FIGURE 5.

Figure 5 identifies continuously decreasing quality costs as a percent of sales. The trend is favorable and efforts to improve productivity appear to be effective. Concentration on high cost areas, such as B5 and B6, should continue to produce favorable results. Over a period of time, assume the area of indifference is reached. Remember the area of indifference is the central portion of the quality cost curve where the optimum application of quality resources is reached. Then, this chart will continue to identify total productivity of the quality function (including the manufacturing and quality departments among others) for control purposes. It will also continue to expose high dollar expenditure areas. These areas are ripe for improved productivity through improved technology, systems improvement, automation, and other programs.

To expand on the approach of reducing total costs by applying the "cost relevancy" concept, I will begin by defining the term cost.

Cost means economic sacrifice measured in terms of the standard monetary unit incurred as a consequence of business decisions.

Differing costs or cost concepts for various situations could be considered as the concept of "cost relevancy."

Cost relevancy in this application is identifying those costs generated through the expenditure of labor and materials which through reallocation can reduce total costs without reducing quality.

To explain this concept, consider a production operation which continuously generates characteristics that tend to hover around the high side of the tolerance band as in Figure 6. The reason for this is to allow sufficient stock on the item produced for further machining if blue print tolerances are not met. Scrap in this instance would most likely be held to a minimum; however, rework costs could be considered high.

In Figure 6, parts are represented by an "X." All parts above the (+) line (8 parts) are reworked at a cost of $15 a part. The part below the (−) line is scrap which at this point of the manufacturing process is worth $40.

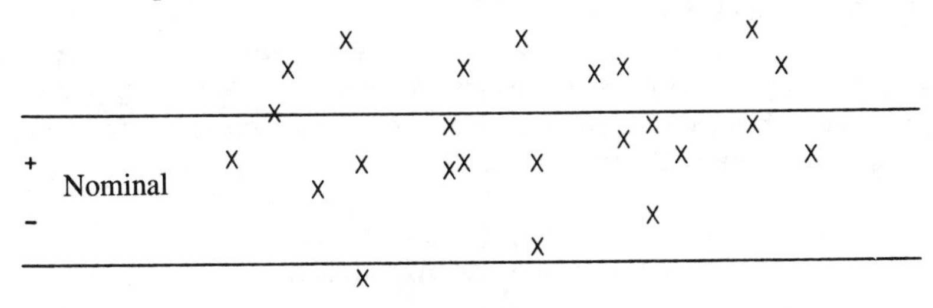

FIGURE 6.

Figure 7 identifies a shift of the process to fit more fully in the tolerance band which could eliminate almost all the rework (8 parts at $15 for a total of $120). Scrap costs are tripled by contributing an additional two parts, or $80 to scrap costs.

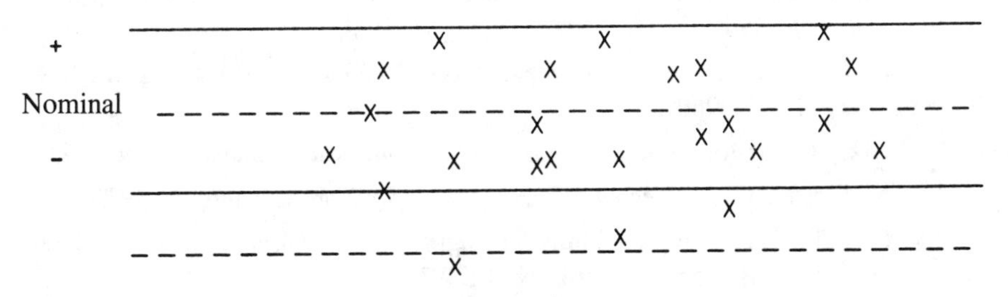

FIGURE 7.

The reallocation of resources which resulted in reduced rework and increased scrap arriving at a total cost lower than previously experienced demonstrates the "cost relevancy" concept. This concept is the basis for the traditional quality cost model referred to earlier. The term "cost" defined as economic sacrifice is measured in dollars. They are accumulated in a manufacturing concern's quality cost program and quantity resources (employees' salaries, materials, supplies, expenses, and the like) that are consumed in an effort to establish fitness for use. The maximum utilization of quality resources or the lowest point on the quality cost curve identifies the most productive quality program.

SUMMARY

The interrelationship of quality cost categories is evident as programs mature and data is analyzed and compared to other time periods. While analyzing data, recognize other manufacturing costs and their interrelation with quality costs. A program isolated on reducing quality cost may accomplish that task. However, if it increases nonquality costs by a greater amount than it reduces quality costs, improved productivity is not realized.

Do not permit the total financial picture to be overshadowed by any one program. Recognize all tools available to managers which may assist in improving productivity. A quality cost program is one method of measuring an intangible area. Avenues of approach are exposed where measurement techniques have been elusive. Trend analysis and Pareto Distribution are basic tools employed in the analysis of quality costs.

A sense of positive purpose and a resolve to alter course combined with this cost study could very well result in increased productivity. Relief from the profit squeeze, and a method to establish measurement parameters in an area which is difficult to measure is a definite possibility.

ACKNOWLEDGEMENT

For Permission To Publish:

Mr. A. E. Wegner, President

Mr. J. F. Dudley, Quality Assurance, Vice President

United Technologies Pratt & Whitney Aircraft Group, Manufacturing Division

Bibliography

1. Anthony, R. N., *Management Accounting Text and Cases,* Richard D. Darwin, Inc., 1964
2. Dickey, R. I. *Accountant's Cost Handbook,* The Roland Press Company, New York, N. Y., 1960
3. Gray, J. and Johnson, K. S., *Accounting and Management Action,* 1973
4. Nickerson, C. B., *Accounting Handbook for Non-Accountants,* 1975
5. Ostwald, P. F., *Cost Estimating for Engineering and Management,* Prentice-Hall, Inc., Englewood Cliffs, N. J., 1974

A PARTICIPATORY APPROACH TO QUALITY

Dr. Alvin O. Gunneson
Vice President & Director — Quality, Worldwide
Revlon, Inc.
Route 27
Edison, New Jersey

ABSTRACT

The quality of the western world's goods and services has slipped seriously in world stature in recent years. Quality has slipped in spite of unprecedented expenditures for quality and in spite of management building large, autonomous, technically competent quality departments.

The problem we face with quality today has been caused by the senior staff of western corporations and it must be corrected by them. They have become so occupied with growing the business, competing in the marketplace, and impressing the financial world that they have not taken the time to manage the business. The slipping quality of the western world's products and services is a manifestation of senior management's lack of attention to the nuts and bolts of management.

All of these are compelling reasons for quality improvement. This presentation shows how management must participate in quality and why formal and systematic quality improvement must be a regular part of management's business life.

TEXT

I am going to talk about quality today, but I am not going to talk about it in the traditional terms of the products. I am going to talk about it in terms of management. It is customary today to say that we can no longer accept inspecting quality into the product. We must build quality into the product. I would like to suggest today that building quality into the products is no longer sufficient. It is not sufficient because it has not worked well in the past. I believe our engineers and scientists and production people want to build quality into each and every product they produce, but only too frequently they cannot because management gets in the way. There are always compromises that must be incorporated due to budget, time or other constraints. I would like to start a new slogan for the 1980's — quality must be managed in. By managing for quality, I am talking about making it possible for employees to do their jobs correctly, about improved professionalism, about improved productivity — but most importantly, I am talking about the significant management change which is required for quality.

It should be very clear that productivity and quality go hand in hand and are inseparable. If products and services are produced correctly the first time, they need not be reworked or done again. Repeating any activity is certainly counterproductive. The high cost of error, delays and customer dissatisfaction is taking an enormous roll on productivity and on the competitive advantage of America. As a nation, we are being challenged in one commodity line after another by a very progressive industrial machine called Japan, Inc. Japan, Inc. is only the tip of the iceberg. There are many other Japans that are developing to become additional competition in the 1980's.

Why is it that America is facing the greatest challenge of its history in the field of quality and productivity? Is it that executives have not recognized the need for quality improvement? I don't believe that is the case. I think American executives have long been alarmed at the growth of competition from abroad. Is it that executives have not acted to improve quality? I believe that is not the case either. Executives in America have spent unprecedented millions in developing large, autonomous quality departments. There are more vice presidents of quality than ever before. Quality control budgets are not approved with barely a question from full commitment, yet, quality is still failing to keep pace with the competition. What is the cause? What else is needed?

I believe the continually touted mandate that top mangagement must support quality if there is to be significant improvement, is not the answer. Rather, I believe top management support is the problem, not the solution. We need to understand what this means psychologically and pragmatically. Management has been giving quality departments their full support while they go off to run their business. They leave the quality departments to create quality which we all know is impossible. CEOs purchase the latest quality programs, books and materials for the quality department in their eagerness to help and to support. The essential missing ingredient, however, is the chief executive's participation. Support is not enough. The executives of America at all levels must ask their quality departments where they can also participate in quality. They should insist that every person establish his own goals for quality improvement and be held accountable to CEO and staff, not the quality manager, for achieving them. There should be active recognition for performance and nonperformance in terms of quality for everyone. Executives from the chairman down must go beyond support and commitment to active participation. This participation must include changing from concentrating only on business factors to the inclusion of quality in meetings at all levels in the corporation, especially at the CEO level where activities usually revolve around sales, production and financial matters. The change must be from mental to physical for effective quality improvement. Most executives will talk at length on the importance of quality improvement, but unfortunately, employees can see little from the executives' actions to make them believe that quality improvement is as important as they say it is. People believe in one's seriousness by what is done, not by what is said. Until executives in America become consistent, talking, walking examples of quality, they will not be taken very seriously on this matter. It is unfortunate, but employees have been conditioned from decades of emphasis being placed strictly on production and financial matters at the expense of quality. They also know how conveniently these words are forgotten at the end of the month and the end of the quarter.

The responsibility for quality must be made very, very clear. It is in vogue to say that everyone in an organization is responsible for quality, but in actuality it becomes the quality department's responsibility. This becomes painfully clear whenever a company experiences problems with quality. If the problems are serious enough, it is usually the quality manager that is fired. This is truly unfortunate because the quality manager did not purchase the defective material. The quality manager did not produce the products incorrectly, nor did the quality manager ship the wrong products to the wrong customers. Yet, it is often the quality manager that is taken to task for these situations.

If there is ever a question as to whether additional executive participation is required for quality, or whether the responsibilities for quality are understood, I suggest a random number of employees be asked to prioritize the most important factors of their business. Phrased in another way, ask them what they will be in trouble for first, if they fail to perform. The vast majority will state that failing to make schedule is the first sin in the company and not staying within cost perimeters or budgets is the second. Quality will be somewhere behind those two, often far behind. The position of quality on that priority list is a fairly accurate barometer of how much executive participation for quality is required in the business. We all know executives are continually questioning schedule and cost. Their participation in those areas is very evident. They must make their participation in quality such that it is perceived to be equal in importance to schedule and cost. This is just good business because if products do not need to be reworked or remade, it will be easier to make schedules and cost goals.

With those introductory comments, what should be the approach to systematically achieve quality improvement? I believe the approach should include two major activities, one of assuring a proper quality department in every operation and another of creating an environment of defect prevention as opposed to detection. A quality department should be developed which progressively works from whatever stage of inspection activity it is conducting, to becoming completely defect prevention and audit oriented. The other, and definitely the most important activity, is for quality professionals to work within the management ranks creating an environment for defect prevention and productivity improvement while continually raising the professional stature of the quality profession.

Quality professionals should concentrate on management, as opposed to the product, getting them to perform the activities and to supply the wherewithall to make it reasonably possible for the employees to produce defect free work at ever increasing rates of productivity. Presently, management makes it virtually impossible for employees to produce error-free work. Purchasing people are buying materials and components without proper drawings and specifications. Inspectors are inspecting this material without adequate standards and criteria. Workers are producing America's products without formal training or adequate procedures, on uncalibrated worn out equipment. In spite of handicaps such as these imposed by management, our workers do an admirable job. Imagine how well they would do if management simply did what they are being paid to do — properly manage the grass roots of the business. Apparently oblivious to the real problem, management is continually trying to motivate the workers towards better performance. That could be considered insulting. Quality professionals must strive to have management realize that they must correct these fundamentally undesirable situations and provide the employees with the fundamental prerequisites for doing a proper job.

Quality executives must also concentrate on increasing their own professionalism and their abilities. Chief executives are continually searching for quality executives with proven management skills. The single, most prevalent complaint of CEOs is that quality managers must improve their management skills. Far too often, they consider the quality departments more poorly managed than the other departments in the business. If quality is to grow as a profession, it is incumbent on the managers to accelerate their management skills and improve the way they are perceived among the senior executive ranks. Technical talent runs high among quality professionals. The decade of the 80s requires having concentration on preparing for senior executive status and six figure salaries. From such a base, one can more readily influence the quality activities of America.

Quality improvement of the magnitude required in America will not happen without clear high-level authority. This authority should come in the form of a quality policy from the CEO, both written and spoken. Most companies have quality policies, but they generally do no more than satisfy customer requirements that such a policy exists. They are usually some good words on excellence and trying to be all things to all people. This may be acceptable for an introductory sentence, but I believe a quality policy should be a working document. It should set a performance standard for every person in the corporation. The policy should legislate that there be an effective quality assurance department in every operation facility, that formal quality improvement be an ongoing activity at every location and that cost of quality or cost of error reporting be a controller requirement.

I believe every company should have a quality orientation program documented in booklet form so that it can be given to all employees. This orientation booklet should state the principles, philosophies and disciplines for quality in the company and introduce the value and importance of aspects of the quality improvement program in the facilities. All new employees should be given the orientation and a copy of the booklet. This will serve to develop a culture of managers who will manage for quality with the same degree of fervor with which they manage for production and costs.

In order for quality improvement to be a reality, there should be a quality improvement program guide for people to follow. This guide should be flexible to allow each operation to adjust their quality improvement efforts to suit their own needs. The important part is that a formal quality improvement program be operational in each facility and that it not be on a "when I get time" basis.

If quality and productivity is to improve, there must be a very high performance standard that everyone must work towards. The standard I suggest is 100% performance. One hundred percent performance in terms of meeting quality requirements and in terms of performing all tasks correctly, all the time. Now, that sounds like a very impractical standard because we know that humans make mistakes and mistakes are normal. I would like to argue that this is a very practical standard, indeed. In fact, each and every one of us has 100% performance as the standard for people which we deal with in our personal lives. During the business portion of our lives, however, we are inclined to revert to acceptable quality levels and the "humans make mistakes" syndrome. As an example, how many times would you allow your employer to make a mistake in your paycheck? What is the acceptable quality level of that performance? How many times would you allow your bank to make mistakes in your banking affairs? Assume you went to the very best hospital in the world and they made a mistake during an operation. Would you say, well that is to be expected, mistakes are normal? No, I don't believe you would. I believe you would get angry, raise cain and undoubtedly hire a lawyer. There seems to be a double standard here. People have a standard of excellence or perfection for people they deal with in their personal lives and when they come back to work, they accept mistakes and errors with barely a question. In fact, some will explain at length how certain levels of rework is normal for a given industry and they go away intellectually pleased if they are running within the margin of error allowed. As long as margins for error are accepted as normal, one cannot expect the quantum improvement in quality which is needed. My point here is to ask that everyone make his personal standard his business standard. Demand 100% performance and you are going to receive it more often than not. Understand there will be mistakes, but do not accept them as normal. Question mistakes and error, ask what you can do to prevent them from being made in the future. Strive to correct the system which allowed the mistake to occur.

We have discussed the need for formal quality improvement. This is best achieved through the appointment of a quality improvement team. The team should be composed of a member of management from each of the operation departments in the company. Their first task should be to decide what the formal quality improvement program will be and to implement the program. In parallel, and subsequently, they should serve to correct the systems problems which affect quality and which prevent 100% performance. Once the error has been reduced in given operations, productivity will have improved commensurately. These operations can then be studied for further productivity improvement in terms of increasing process rates and automation while maintaining the low levels of error achieved. Increased process rates should not be attempted until error causes have been purged from the existing process and they are performing defect free.

If the quality improvement team is to achieve and report performance improvement, there must be a measurement system which identifies existing conditions and which records improvement. Quality measurement is usually relegated to factories or laboratories. This measurement system should be expanded to include the white-collar offices as well. It should measure and identify the 20% of the research scientists who provide 80% of the changes. It should identify the 20% of the managers who cause 80% of the delays and subsequent problems in the operating areas, and so on. When an area for improvement is identified and quantified, the team should concentrate on correcting the system until 100% performance is achieved. Then, continual measurement can be stopped in favor of occasional auditing to assure that the system remains in place. It is suggested that measurements be placed on charts and displayed in the area where the improvement is to occur. This visual display of conditions will be helpful in providing motivation for the required improvement. If one displays a performance chart of a group, there will usually be improvement against it even if nothing else is done. Harness that facet of human nature.

Once the areas to be improved have been identified and quantified, the cost of error or the cost of the problems should also be identified and assigned to them. Perhaps the best method of establishing priorities for quality improvement is to launch a total cost of quality program. This will identify in dollar terms where the greatest profit drain and consequently opportunities for improvement exist. Cost of quality programs have not enjoyed the amount of success they should have in America. I believe the main reasons are because they are perceived as a negative program and one which is very difficult to implement. It is quite natural that a manager whose cost of quality is 20% manufacturing costs is not anxious to advertise this to the world, particularly to the boss. Also, the many categories of error, prevention and appraisal can make the cost gathering and reporting process a laborious one. It is suggested that the program be considered a positive one in that it identifies the opportunity for improvement as opposed to how good or bad an organization is operating. I always say that corporate knows how good or bad an operation is and they really don't need cost of quality numbers to tell them that. What they need the numbers for is to identify where the improvements can be made and to credit people for the improvements. It should also serve to give headquarters a comfortable feeling that quality situations are under control and are improving. In implementing the program, one should not get bogged down in details. The major costs are usually available individually or as a part of other reporting. It is generally not a major operation to develop a cost of quality system compatible with the existing accounting system. The system becomes cumbersome when a controller takes the cost of quality procedure developed by the ASQC and attempts to implement it as written. This is not recommended and it is not the intent of the ASQC publication. Each company should use it as a guide to develop their own system in a way which is best for them. Everyone should know their cost of quality because if they do not know what it is, they can't control it much less reduce it. I am often asked what the return on a cost of quality program should be. Experience shows that cost of quality is 15% to 20% of maufacturing or operating costs for most companies. One should expect a reduction of 4% to 5% in the first year. This is very possible. A reduction of 3% to 4% in the second year and 2% to 3% in the third year is also about normal for an active quality improvement, cost of quality reduction program. A final thought on this subject is that the program should be a controller's program and not a quality department program. Quality departments can do the work behind the scenes, but they must not report the numbers. If quality reports them, they are often challenged, and receive little attention.

Up until this point, we have been talking about management participation, exclusively. We must not rule out employee participation, as this is a very important part of quality improvement. Quality circles are becoming a very popular means of achieving employee participation. I believe quality circles in America are by and large premature. I believe it could be insulting to our employees to ask them to meet in circles to identify and solve operating problems when management is delinquent in providing them with the very basics for doing their job properly. I think it would be much wiser for management, first, to form their quality improvement teams exclusively among management to correct the many known deficiencies that exist. Once management has cleaned up their act, so to speak, they can begin phasing line employees onto the teams so that a genuine participatory effort could be achieved between employees and management. I believe quality circles at this point are just another cop-out for management. They will purchase a quality circles program and give it to the quality manager with instructions that it be implemented, while they go on and run their business. This is just another example of management supporting but not participating in quality.

While the quality improvement team is operating, it is desirable, however, to have some parallel employee participation. This can be done by a problem identification program or an error cause identification system where employees fill out a suggestion type form stating their problem or identifying an error causing situation. These forms are then reviewed by the quality improvement team and corrective action assigned. This program differs from the standard suggestion program, as the people do not need to have an answer to the condition. They need only identify an error cause or have a problem in performing defect free work. There are many more people who have problems than who have bright ideas. This type program, therefore, will usually produce more activity than a standard suggestion program.

No quality improvement effort would be complete without a formal corrective action system. There are many documented corrective action systems, so I don't need to describe any in detail. I would like to offer my now familiar word of caution with corrective action systems as well. They should not become complex and loaded with paperwork. I would suggest that initial corrective action requests be verbal and only after several verbal requests have been ignored, should a corrective action form be issued. The system should be structured so that it will not be an honor to receive a corrective action notice. The number of corrective action notices received should be considered at merit review time. Only chronic problems and those requiring long term resolution should be put on the standard corrective action log.

The last step of any formal quality improvement effort should be a recognition program. We are very quick to condem, but we are relatively slow to recognize. A formal recognition system which reaches all levels of the organization should be developed to reward those who are outstanding performers for quality. The type of recognition is not nearly as important as the recognition itself. The quality improvement team can assign someone to research recognition programs and a recommendation can be made to general management.

CONCLUSION

In closing, I would like to leave you with two thoughts — the first is if management truly wants quality improvement, they should practice what they preach. In the beginning of the month and in the beginning of the quarter, management is inclined to boisterously advocate absolute conformance to quality requirements. By the middle of the month and the middle of the quarter, their comments regarding quality is not nearly so loud. By the end of the month or the end of the quarter, all bets are off. Many managers will ship anything they can get their hands on to meet their numbers. As long as this is the practice, no amount of words will convince anyone that quality is a serious matter. Last, the amount of success any company has with quality improvement is directly correlated to the amount of participation of the executives in the facility. Support is not enough! Whatever one may find in laboratories or factories in terms of performance of people, or quality of products, one can be sure it is a mirror image of the executives involved. If executives do not show concern, the employees will not be concerned. If executives are concerned about quality and set high standards, the employees will try to live up to those standards, quality will improve and everyone will be more productive and much happier.

Mr. CEO, YOUR COMPANY'S QUALITY POSTURE IS SHOWING!

Richard K. Dobbins, Principal Systems Engineer
Honeywell Inc. — Process Management Systems Division
Fort Washington, Pennsylvania

ABSTRACT

Your company's quality posture is the image presented to the world through your products and services. In an older company, it includes the "quality reputation" which has been established. In a new company, it is still an unknown quantity, with only hints derived from Marketing Brochures and Engineering specifications. It can be either a distinct market share asset or liability, with resulting influence upon sales. Positive quality postures are delicate and usually take years to achieve, yet can be quickly ruined. Intelligent management of the quality function can significantly improve your quality posture, with enhancement to market share and profitability.

QUALITY POSTURE — WHAT IS IT?

A Company Image To World

Just as with human beings, each company projects an image in everything it does, which is visible to all people who come in contact with that company, its products or services, and its personnel. This image, or bearing, is not only shaped by *what* the company does, but the *way* in which it does it, the *attitude* of it representatives, and the *fairness* in its dealings with others. In a word, it is the "posture" of the company.

In today's times, the consumer public is very "quality conscious", and uses that term in a much broader sense than strict conformance to specifications for products sold. From a customer viewpoint, therefore, the image projected by a company might be called its "Quality Posture". And this quality posture projection causes a reaction to be generated within the viewing public. The reaction might be favorable or not, with unlimited range between the extremes of "complete respect" to "complete contempt". In between these pure "whites" lie all the various gray shades of "complete indifference". Since these human reactions do exist to the projected quality posture of companies, it is reasonable that they play a significant role in business decisions affecting *your* company's products and services.

Reflection Of Customer Assessment

So, Mr. CEO, your company's posture is showing! How does it compare with the postures of your competitors? Is it an asset or liability regarding potential new accounts? Does it help or hinder you in obtaining repeat sales from your existing customer base? How does it affect the morale, productivity and turnover of your employees? Do you have any idea how your quality posture currently is perceived by the general public? And, do you have any specific strategy on how to improve your company's quality posture?

SHAPING YOUR QUALITY POSTURE

Quality Reputation

Let's consider what affects your quality posture, before we attempt to measure or improve it. Your company's quality reputation is of primary importance, providing you have been in business for any significant period. If your past experience to the world has been less than desirable, you have an adverse reputation which must be overcome. If your field track record has been good, you are fortunate, and must guard this advantage from being tarnished in the future. With many companies the quality reputation is mixed, varying greatly from one product to another, or with the various services offered or applications served. With a relatively new company or product line, you have not yet earned any viable track record, so you start off under a slightly negative influence as an "unknown quantity" to the skeptical general public.

Other Factors

But your quality posture is much more than your past quality reputation. Other significant factors include:

- The product claims, promises and inferences in your marketing literature and technical descriptions.
- The practical suitability of your products to the applications they are intended to serve.
- The competitive positioning of your products and services offered, compared to others which are available.
- The warranty policy of your company, if specially stated or implied, and comparison with actual warranty performance.
- Availability to field service assistance and replacement parts. Average time required from known need to fulfillment of need.
- Accessibility to technical assistance for both product and applicational problems.
- Relative pricing of your products and services compared to that offered by your competitors.
- Objective measurements, either theoretical or achieved, for reliability (MTBF), availability (%), and serviceability (MTTR).
- Adequacy and completeness of documentation regarding such topics as technical specifications, theory, wiring and installation, operation, preventive maintenance, servicing and troubleshooting, parts lists and recommended spare parts.
- The professionalism, knowledgeabililty, attitude and integrity of all personnel who have contacts with your customers or clients.

Dynamic Relationship

From the above listing, it is obvious that your quality posture is not purely a matter of quality policy, or past quality reputation. Nor is it within the complete control of your Quality Department! It should also be noted that your quality posture is not a static thing, but is constantly vulnerable to change, as the various shaping factors change. This means that at any given time, there might be a significant difference in ratings given your company by two different customers. They will each perceive your company performance as they have individually experienced it, and will measure it against their own priority levels of expectation. Recognize that the dynamic movement of quality posture measurements will tend to improve gradually, through evolution, as customers sense your improved performance. These ratings are delicate, however, since good ratings can be seriously impaired through most recent negative events, if significant enough to be communicated throughout an industry or application.

WHAT IS YOUR CURRENT QUALITY POSTURE?

Customer Viewpoint Needed

Recognition that there is such a thing as "quality posture", and that it can have an appreciable influence upon the financial health of your company, poses the question of "what is your current quality posture?" It cannot be precisely measured, and is completely independent of what your company Operations Manager or Director of Quality may believe (or proclaim) it to be. Really, it is the consensus of what your customer base and the general public perceive it to be. This means that the only way to measure it is through surveys of these groups.

Survey Goals

Any survey undertaken should have specific goals. Among those goals should be the intent to objectively determine:

- How your present (and past) customers assess your products/services overall, compared to your major competitors, and relative to their own specific needs. Explicit, defined ratings should be used with a broad range, to avoid large interpretation errors among respondents. (On a scale of 1 to 10, define what value represents "excellent", "fair", "unsatisfactory", "poor", and "very poor".)
- Same assessment data for major specific partial factors, such as technical features, workmanship, reliability, serviceability, etc. and/or delivery, documentation, field service, replacement parts availability, applicational assistance, etc.
- *Why* your products and services are so rated by your customers. It is important to try and determine this in all instances, not just in those areas where low ratings are given. This will help you recognize your strengths as well as your weaknesses.
- Whether your present customers would buy additional products/services from your company again, if the need arises.

Who Conducts Survey?

So a customer survey is needed! *Question:* Ok, how do you have the survey made? *Answer: VERY CAREFULLY.* Things to consider include:

- For best objectivity of overall competitive ranking, an independent firm should be utilized to conduct a "blind" survey, which will mask your company's partisan interest.
  - (+) This method will produce the least amount of specific information on *why* your company scored as it did, or what your particular weaknesses are.
  - (+) This may also block any follow-up with particular discontented clients, if their identity is protected to promote their participation.
- If your company surveys your own customers, you have the opportunity to go into greater depth of their reaction to your products and services.
  - (+) Survey may be biased by some who:
    - (1) Rate you higher than they would in a blind survey, since that will end the survey sooner, and you'll be pleased.
    - (2) Berate you severely for some small minor irritations not yet fully resolved, or not resolved exactly as they had desired.
  - (+) You may not learn as much about your customers' feelings towards your competitors.
  - (+) You may be pleasantly surprised about the positive affect which may be generated, simply because your company cared enough to survey customers about their feelings, gripes and recommendations!

Design Of Survey

The survey design is extremely important. Critique it carefully against the following considerations:

- Avoid too detailed a survey, or the length and complexity will discourage cooperation and truthful answers.
- Too general a survey, such as only an "overall" rating, will not give you sufficient data to improve your quality posture.
- Consider breaking it down into specific product groupings, or types of services, if you provide a broad range of offerings.
- Obtain certain ancillary data (such as: period of time products/services have been used; industry; application; geographical region; etc.) to help analyze meaningful differences in response.
- Identify the type of person(s) for whom you are designing the survey. If you sell equipment used by a large industrial company, should the survey be pointed towards the operator, foreman, service supervisor, production engineer, buyer, general foreman, or operations manager? Keep it restricted to questions which that individual is competent to answer, and which will be pertinent to your improvement needs.

Survey Methods

The method of conducting the survey can be extremely critical to the value of the results obtained. Take into consideration the following items:

- Mailed, written surveys have extremely poor response, with most ignored and thrown away, or filled in by incorrect or uninformed parties. A large expense will bring little fruitful information.
- In-person interviews by field personnel are very expensive, and lead to shading of unpleasant field responsibility problems to stress home office responsibility factors. High cost of this method will limit coverage obtained.
- Telephone surveys are usually most effective and productive, *IF:*
  - (+) You speak to the proper, knowledgeable individual(s).
  - (+) You keep the call to a reasonable time period, and ask for that amount of time at the very beginning of the call. (Reshedule at a specific time to suit your customer's needs, if necessary.)
  - (+) You are courteous at all times without arguing, defending or selling your respondent. Your role is restricted to being a very accurate reporter.
  - (+) You provide your client at the end of the call the opportunity to add other comments, criticisms or suggestions not covered by the survey.
  - (+) You *always* thank your respondent for his cooperation and time in answering the questions. This is most important if hostile vibrations have been received!

Analyzing the Survey Results

After the survey has been completed, now what? First of all, tabulate all the replies to each question without any editing, expansion or editorializing! Then, proceed as follows:

- Are there certain areas where the vast majority of replies are consistent? If so, *believe the results,* whether painful or surprising.

• If there is considerable variance with replies to certain queries, do they fall into two or three distinct groups? If so, seek the "common denominator" to these groups, such as: industry application; manufacturing period; point of manufacture, etc.

- Be aware of strengths or weaknesses relating to "hardware" and "service" areas. Many companies do well with their manufactured products, but fall down in their "people oriented" contacts, or vice versa.
- Be honest in analyzing data comparing your company to your competitors.
  - (+) Each item with a reasonable consensus for your competition is an opportunity area in which improvement might be made.
  - (+) Each item with a reasonably favorable consensus for your company is an area where you might gain marketing advantage through emphasis.
- Evaluate your survey honestly in whether it achieved its intended information gathering. Recognize whether wording or construction of questions needs improvement to eliminate confusion, or to obtain important customer feelings. Should it be broadened, or reduced next time?
- Follow-up on any identified unresolved issues, and specific suggestions or

criticisms volunteered beyond the structured survey. Reply in a courteous but forthright manner. Do not infer that you might take additional actions, unless you actually intend to do so. And then, *make it happen!*

IMPROVING YOUR QUALITY POSTURE

Reacting to Survey Results

Regardless of the intent or wishes of top management, recognize that the survey results truly reflect your company quality posture, as perceived by your customers. This, in turn, is a fair approximation of how they think you are running your business. You may feel this scorecard is considerably lower than the lofty ideals incorporated into your company policies and procedures. Basically, your quality posture is an indication of the habits, attitudes, priorities and actions of your people in their daily work. It also shows the effectiveness of your management staff in accomplishing the truly important priority tasks (to the field), which are their responsibility. (And these particular tasks may not even be included in their personal or organizational objectives!)

There may not even be any apparent correlation between the current expense of testing, checking or inspection your products, and the survey results. If so, you may not be using the appropriate acceptance criteria, as far as your customers' needs are concerned, with resulting "quality by happenstance" from lot-to-lot.

Remember The Customer's Viewpoint

Recognize that your quality posture goes beyond the legal warranty limitations, and includes the professional, ethical and moral responsibilities towards your clients and the general public. Examples include:

- Identification of design or production situations which result in reduced useful product life *beyond* the normal warranty period.
- Identification of new applicational situations for products which are not compatible with certain product limitations.
- Identification of potential risks or hazards not previously known, attached to certain uses of well established products.
- Response of your company to unexpected emergencies of a customer, (explosion, fire, flood, tornado, etc.) concerning extraordinary replacement or service needs.

Accent The Positive

Your quality posture is influenced by your entire organization, not just the Quality Department.

- The environment for it is established by Executive Management, and without a visible, firm and consistent top level commitment, even extraordinary dedication by groups or individuals will be short lived and futile.
- Emphasis must be placed upon promoting the positive aspects of improving customer satisfaction, instead of reducing the negative aspects (such as scrap, rework, design and production cost reductions, etc.).

(+) Consider that you might be making certain expensive acceptance decisions based upon criteria which are immaterial to the field's functional needs, and ignoring other items with much greater priority needs for the applications served.

(+) Consider that emphasized cost reduction efforts lead to "corner cutting" without affecting present acceptance criteria, but could lead to degradation of field performance, reliability or serviceability.

- The influence of individuals with whom your customers have direct contact (for the initial sale, during production and shipment, installation, and subsequent handling of service needs) may be more meaningful to your future business with those customers, than whether your products fell somewhat short of optimal performance.

Significance Of Quality Posture

A good, positive quality posture contributes to your corporate health in a number of very tangible ways:

- It will increase your market share in future sales, displacing competitors whose quality posture is found to be wanting.
- It may make possible a price premium differential over the average competitor.
- It will prevent alienation of a customer who has suffered from objectionable field performance, if your handling of the problem was prompt, courteous and efficient. This can be even more true if temporizing measures are employed while permanent corrections are being worked out. (In short, if you show the customer "you care", and recognize his needs.)
- It will give your company a preferred position when you introduce new products or services, based purely upon the quality posture earned by your company on its existing lines.
- It helps your employees to have pride in their company, and in their personal performance. This, in turn, promotes constructive suggestions for still greater improvement of productivity and customer satisfaction.

SUMMARY

Look What Can Be Accomplished!

Consider the influence of quality posture, as perceived by the general U. S. public, towards Japanese products following World II, and during the current period:

- After W.W.II, the general feelings (and buying habits) of the average U.S. citizen were characterized by these terms: "cheap," "imitation", "second rate" and "non-reliable". Sales were therefore primarily based upon lowest cost only, where performance and reliability were down-graded for unimportant jobs or applications.
- But present day U.S. feelings towards many Japanese products rank them equal to, or superior to, many sophisticated products. In this regard, many people are willing to pay a price premium over competing U.S. goods, based mainly on present day quality posture comparisons.

This radical change in attitude and purchasing position is *not* the result of quality circles (for many large, successful Japanese companies have never used them, or only recently started some). Instead, Japanese companies have steadfastly applied basic Yankee Quality Assurance principles faithfully, always pointed towards long range improvements. Indeed, most every field failure (even those long past the warranty period) is analyzed for cause, since each one presents another possible opportunity for improvement

Mr. C.E.O., your company's quality posture is showing! How does it shape up for future competitive survival?

MEASURING QUALITY COSTS BY WORK SAMPLING

L. James Esterby
Director, Quality Assurance
Victor Equipment Company
Denton, Texas

ABSTRACT

Work sampling is a technique used, usually by the Industrial Engineer, to investigate the proportions of total time devoted to the various activities that comprise a job or work situation. As such, it may be employed to determine the various elements of quality costs. This is usually done as they occur at a particular segment of time, i.e., a snaphot of quality costs.

Such a snapshot can be a useful tool in two ways. First, when applied where no formal quality costs program exists, it becomes an excellent demonstration to management of the existence and relative magnitude of such costs. Secondly, it can be used as an audit of an existing quality costs program to evaluate the effectiveness and scope of the program.

When the study is conducted by the supervision and management of the cost producing departments under the direction of the Quality Department the method becomes particularly efficient and free of challenge.

MEASURING QUALITY COSTS BY WORK SAMPLING

Quality cost analysis is a method of determining costs associated with product quality and, more often, costs associated with the lack of product quality. Such costs are usually determined and collected through an accounting system which divides these costs into the components of appraisal, prevention, internal failure, and external failure. Such a process can be a very valuable management tool. However, two questions arise: First, whether a quality cost analysis system should be implemented in your firm; that is, are quality costs sufficiently large to be of a concern to management? Secondly, if a quality cost analysis system is in place, is the process capturing all cost associated with product quality? Both of these questions can be answered by a work sampling appraisal of the quality costs.

Work sampling is a technique used to investigate the proportion of total time devoted to various activities that comprise a job or work situation.[1] This process uses random observations of persons at work or machines in operation over a period of time and results in a proportion of total time applied to specific activities. Thus, work sampling can provide, in effect, a snapshot of quality costs at a given time. This snapshot provides in effect a stop-action picture of a time interval during the work operations. If this time interval is representative of the overall time frame, this stop action can then be extrapolated to provide the proportion of total time in a month, year or other time interval desired that is applied to a particular work element. When the work elements chosen are those associated with quality costs, the resultant is a proportion of the time spent on quality cost elements, and thus extrapolated presents the total time or total dollars associated with quality costs.

The first activity necessary for a work sampling study of quality cost is to identify the elements to be studied. We must enumerate all of those elements that we shall consider quality costs within our own firm. For a review of the elements of the quality cost I refer the reader to QUALITY COST — WHAT AND HOW.[2] This list will obviously vary from firm to firm and may not be met with total agreement even within a firm. Certain elements such as scrap and rework are relatively obvious quality cost elements. Other elements that may be less obvious and on which there is less consistent agreement could be development of prototypes, value analysis, preparation of repair literature, review of drawing change requests, investigation of line problems, etc. Each of these type elements may have a proportion that could truly be considered quality costs, and also have a portion that is considered a normal part of doing business. Each firm must, therefore, determine the list for itself. Exhibit 1 presents the elements considered for internal failure at the Victor Equipment Company. In order to save space in this manuscript, the complete list of elements is not presented but also includes prevention, appraisal, and external failure items. It is worth pointing out that this list of internal failure items included elements from quality engineering, inspection hourly, marketing, industrial engineering, design engineering, and production hourly. It is, therefore, not limited to a single department or even to the factory work force alone.

Once the list of elements has been determined, the next question to be asked is how many observations need be made for the study? Work sampling is based on the fundamental laws of probability if a given event can be either present or absent, i.e., good or bad, operating or not operating. Statisticians have derived the expression which shows the probability of x occurrences of an event in n observations to be :

$$(p + q)^n = 1$$

p = probability of a single occurrence

q = (1 − p) the probability of an absence of occurrence

n = number of observation

The expression $(p + q)^n = 1$ is expanded according to the binomial theorem. As n becomes large the binomial distribution approaches the normal distribution. The use of the normal distribution of a proportion provides a mean of p and a standard deviation of the $\sqrt{pq}$. From elementary sampling theory we know that $\hat{p}$ ($\hat{p}$ = proportion based on a sample) will not equal the true value of p. However, we can expect that $\hat{p}$ of any sample will fall within the range of $p \pm 2$ segma,

[1]Benjamin W. Niebel; *Motion And Time Study;* Richard D. Erwin, Inc.; Homewood, Illinois; 1971; pgs. 512-529

approximately 95 percent of the time. Utilizing these facts we can solve for n by the following:

$$n = \frac{\hat{p}\,(1 - \hat{p})}{\sigma_p^2}$$

For example, say if we wish to determine the percentage of time spent on rework and we assumed reworked to average 10 percent. Further we wish to be 95 percent confident that the true proportion fell within 25 percent of our estimated rework. Then according to sampling theory, ± 2 = 10 ± .25 (10) percent = 7.5 (12.5%–7.5%)/4=.05/4=.0125.

$$n = \frac{.10(1-.10)}{.0125^2} = 576$$

Thus, we need 576 observations to determine our true percentage of rework if we estimate rework to be 10 percent and we wish to be accurate within 25 percent of the true number.

Next we must determine the frequency of observations. First, one must consider any factor in the work year that could affect quality costs. Such a factor might be high absenteeism in mid-summer due to vacation time; high absenteeism in December due to holidays; a year-end or quarter-end rush to meet production schedules or a fiscal year-end house cleaning of scrap and rework. Such factors would determine both the length of time over which observation should be taken and the period within the year over which observation should be taken. Assuming there are no unusual or cyclic factors, one could take the observations over any part of the year. The duration of time then becomes merely one of convenience and availability of persons to take the observations. Oftentimes one calendar month will be an appropriate period for taking such observations. If, say, the above calculated 576 observations were to be taken in one calendar month of 20 working days we would then take 29 observations per day. It should be noted here that factors changing the number of work days in a month, that is, holidays or five-week accounting months, should not present a problem as the calculations will result in a proportion of total time spent during the month which can be readily extrapolated into proportion of time spent during a year or any other length of time.

Once we've determined the number of observations to be taken during the day the next question to be answered is, when must the observation be taken? To be statistically valid these observations need be taken randomly throughout the day so as not to be influenced by hourly fluctuations in activities such as a cleanup of rework during the later part of the day. I suggest the use of a random number table to select the hours of the day in which to take observations. The observer may make several observations at a time or one observation during each interval of time. An important point here is that a predetermined schedule be prepared

[2]ASQC; Quality Cost-Cost Effectiveness Technical Committee; *Quality Cost-What And How;* Second Edition; 1971

and adhered to in order to assure that no bias is introduced by the observer.

Once we have determined the items for investigations and the number of observations needed, we now need to turn our attention to who is to gather the information. Here I suggest the observer to be a person regularly within the department to be observed and familiar with the operations; that is, a supervisor, foreman, or manager of a department. This has two effects: (1) the supervisor of the department is very apt to understand and know well the jobs being performed and thus can readily make the observations and determination of what a person is doing at a particular time; (2) by having the observations made by a supervisor who is typically in the department, the work force is less apt to be concerned by being observed or less conscious of the fact that they are being observed. Regardless of who the observer will be it is important to bring the supervisors of the department on board at the outset of the plan for the work sampling study. They can assist in determining the elements that apply to their departments. They can also assist in determining random sampling for their daily work schedules. In this way you will establish a coordination and a cooperation with the people involved. It is important to stress that this information will be useful to them in being able to manage their departments better and will assist them better in obtaining aid from such support departments as tooling, engineering, etc., in correcting problems that may be discovered.

In order to simplify data gathering, a work sheet should be constructed, identifying each element to be observed and providing for simple check marks or hash marks to record the observed results (see Exhibit 2). Here I suggest that the work sheet not only include those elements associated with quality cost but also all general elements associated with the working group, as can be seen from Exhibit 2. Elements such as direct labor, waiting parts, setup machines, etc., are work elements included in the study. This is done for two reasons: First, it provides the department gathering the data with additional information which may be very useful to them. The study could provide such information as setup time, materials, handling, personal time, etc. In fact any item that could be a proportion of worker or machine time could be studied, whether a component of quality costs or not. Secondly, by providing for observation of all work elements we are not biased for, nor against the quality cost elements that we are seeking. It is often possible to accentuate an item being studied, either positively or negatively, by centering attention on that item. It is of major importance not to bias the results by the method of data gathering.

Once the data sheets have been completed we are now ready for the data reduction. This simply requires the summation of times or numbers of observations shown for each element. The proportion of time then becomes the number of observations for a given element divided by the total number of observations for all elements. If this data were taken over a one-month period time and rework showed 110 observations out of 1,000 total observations the resultant would be 11 percent of the time is being spent on rework. Given this percentage figure we can then determine the total number of hours or dollars spent on any elements, say rework, for a given month, quarter or year by merely multiplying by the appropriate number of total hours spent and the dollar per hour wage rate.

This, then, provides the mechanics for performing the work sampling study beginning with a definition of the elements of your quality cost system establishing the necessary sample size, providing for the data gathering, establishing the method of data gathering, and finally reducing the data into a result that provides total costs per element per unit time.

It was indicated above that work sampling may be used for two reasons: (1) To determine the relative magnitude of quality costs in a firm in which quality cost analysis program is not being utilized; (2) to perform periodic audit against an existing quality cost system to evaluate its efficiency and effectiveness. When used for the former, this method will provide an appraisal to management of the potential of a quality cost analysis program and will thus provide management support for such a system. At this point, it is particularly important to recognize that absolute accuracy of the numbers is not important. Rather the interest should be placed on the relative magnitudes. For example, in a firm of $10,000,000 sales it would not be important to determine the accuracy of existing rework costs of say, $50,000 plus or minus $5,000 a year but rather that management become aware that rework costs are in the order of $50,000 or $250,000 or $500,000.

Further, in this initial study, it may not be necessary to evaluate a great array of elements. One should consider here only the major elements and only those upon which there is consensus of opinion that they do, in fact, constitute quality costs. At this point, if the elements do not present significant consensus of opinion or if the elements are divided too finely, management may well take issue with the entire process and therefore the whole work sampling study may lose its effectiveness. A final point relative to this area and in fact is true of both the initial work sampling study and periodic audits, is that, if any errors exist in the estimates of cost, they should be errored to the conservative side. Thus, anyone wishing to take issue with the numbers would of necessity have to argue that the true numbers are larger, in fact, than those reported.

When using work sampling for a periodic audit of an existing quality costs analysis program, the emphasis becomes one of greater accuracy in the percentages determined, further definition of quality cost elements and perhaps more importantly the identification and recording of elements that heretofore had not been considered in the quality cost analysis. This process also provides a measure for the accuracy of the capture of total quality cost and cost by each element. That is, is the ongoing quality cost system actually obtaining accurate measures of the cost elements? This provides for reinforcement to management of the quality cost system and for an evaluation as a basis of improvement and refinement of the system.

This process was implemented at Victor. At the outset a list of cost elements was developed by each of the four major topics and subdivided by department affecting those particular quality costs, i.e., appraisal, prevention, internal and external failure. This list was reviewed with each department manager and consensus was obtained. The resulting overall list was presented to the president's staff for their concurrence. Once their participation was received the program became a company program (or division or plant as appropriated) rather than a Quality Department program.

Sample sizes were determined and in the case of the production departments, area managers were selected to perform the actual observations, each manager being responsible for approximately one-third of the operation department. Work sheets were prepared and random observation times were determined. Further, through the use of the work sheets and some up-front practice each manager was able to make observations of approximately one-third of the total work force in fifteen to twenty minutes. Four such reviews were made daily and the process was continued for a month.

The results were accumulated and reported to all participants and to the president and his staff. It is important that the results be reported to all participants. Remember it is a company program, their program, and they should participate in the results, not just the work. Some major results of this work sampling analysis were as follows: actual total quality costs according to the work sampling were 82 percent higher than that reported under the existing quality cost system. Internal failure costs were 128 percent higher than had been reported. A significant design engineering component of internal failure was discovered that had not previously been recognized or evaluated. And, finally, quality cost analysis became a real and understood management activity.

CONCLUSION

Work sampling as an evaluation of quality costs is not only a useful theory, but is in fact a very practical method of measuring quality cost. The total man hours involved in such a study are quite small relative to the information gained. The process is easy to establish, administer, yields very accurate results, and the process is adaptive to any form of organization. While the above examples have centered around a manufacturing firm, the process can be applied to any profit or not for profit organization whether dealing with a product or a service.

Some knowledge of work sampling and quality cost analysis techniques are required. However, this paper and its bibliography present sufficient background for a novice to establish the program. Consensus on major quality cost item definitions is readily obtainable and questionable items can be disregarded. Observers are easily trained and results are obtained with a very small expenditure of time. The data reduction involves only simple arithmetic while the results are statistically sound. And finally, the process can be applied to any organization whether for profit or not, dealing with either a product or service.

BIBLIOGRAPHY

1. ASQC; *Quality Cost-What And How;* Second Edition; 1971
2. Niebel, Benjamin; *Motion And Time Study;* Richard D. Erwin, Inc.; Homewood, Illinois; 1971; pgs 512-529

MINIMIZING THE COST OF INSPECTION
by

William J. Latzko
Irving Trust Company
New York, New York

ABSTRACT

This paper presents the proof that when a lot of size N is submitted for inspection by taking a sample of size n, and defectives are either replaced with effective items or discarded, the minimum cost is achieved with either of two strategies: $n=0$ or $n=N$ and no other. The determination of which strategy gives the minimum cost is based on a function relating the cost of inspection, k_1, the cost of a defective product that got into the system, k_2, and lot quality, p. Examples are presented and rules given for tracking the quality level, p, as well as what to do if the quality level of the lot is totally unknown.

INTRODUCTION

Acceptance sampling has been the premier tool of quality professionals for over four decades. With the availability of documented plans such as Dodge-Romig[1]and MIL-STD-105D[2], acceptance sampling is ubiquitous. Not only are the plans used for incoming inspection but also for production control. They are probably misused as often as not.

Most of the plans look towards minimizing the cost of inspection. They leave it up to the reader to take other costs into consideration. Specifically, these plans do not cover the cost of a defective item that is included in the accepted lot.

Some researchers have considered the downstream costs. Hald[3] and Smith[4] approached the problem from a prior probability view. They assumed underlying distributions and worked out formulas for determining n*, the optimum sample size and c*, the optimum acceptance number.

Martin[5] and Dayton[6] approached the problem from a more empiric side. They arrived at similar conclusions to Hald and Smith. However, the preoccupation of all these authors with the probability of acceptance for a given sampling plan and lot quality prevented them from seeing that in the case of non-destructive sampling the minimum cost is achieved by either accepting the lot as is or by 100% screening, depending on the lot quality, p. Deming demonstrated this remarkable conclusion in his recent work.[7] This paper will discuss the conclusion and suggest applications in both service industry and manufacturing.

THEORY

When a vendor submits a lot of material the receiver can take a sample and examine the sample. When the sample is found to contain less than a critical amount of defectives, it is accepted for further use. If the sample contains the critical amount or more defects, the lot is rejected. Defective items in the sample are either replaced or removed so that the sample contains only good items. If the lot is rejected it is fully screened and defective items replaced with good items or simply removed.

There are essentially two costs involved in this process:

1) the cost of inspection and

2) the cost of a defective which slipped into the process.

The following notation is used to develop the models:

- N the number of pieces in the lot
- n the number of pieces in the sample. Dr. Deming notes that the use of an acceptance sampling plan implies that the items are selected by use of random numbers from the lot.
- p the average incoming fraction defective
- q = 1-p
- P the proportion of lots set off for screening at initial inspection.
- Q = 1-P. This is the probability of acceptance, the ordinate of the OC curve.
- k_1 the cost to test one piece at the start.
- k_2 the incremental cost of operations due to a defective piece in the process. This includes the cost of removing and replacing the defective item total cost.
- Y total cost
- x = n/N the sampling fraction
- K = k_2/k_1 the cost ratio

Whatever is used as a plan of acceptance, the following conditions hold:

$$\text{if } n=0 \text{ then } P=0 \text{ and } Q=1$$
$$\text{if } n=N \text{ then } P=1 \text{ and } Q=0$$

As Dr. Deming points out, "We need to know nothing more about P and Q in order to perceive what to do about inspection of incoming materials."[7]

Front End Cost

The initial cost Y_1, is the cost of sampling n items at k_1/item plus the cost of replacing defectives. The number of defectives is given by np. However, since the replacements have a percentage p defective it takes, on the average, 1/q items to get a good one. In addition, there is a probabililty P that the remaining lot (N-n) be screened. Considering all of this, the initial cost can be expressed as:

$$(1)\ Y_1 = nk_1 + npk_1/q + (N-n)\,Pk_1 + (N-n)\,Ppk_1/q$$
$$= Nk_1/q\,(P + Qx)$$

Processing Cost

The cost of problems downstream in the production process is based on the number of what remains in the lot after sampling (N−n), times the probabillity of acceptance, Q. These lots contain p defectives at cost k_2 and replacement cost of k_1/q as explained above.

$$(2)\ Y_2 = (N-n)\,Qp\,(k_2 + k_1/q)$$
$$= Nk_1/q\,[(1-x)\,Qp\,(kq + 1)]$$

Total Cost

The total average cost of the operation is the sum of the above two costs.

$$(3)\ Y = Y_1 + Y_2$$
$$= Nk_1/q\ (P + Qx) + Nk_1/q\ [(1-x)\ QpKq + Qq]$$

This equation reduces down to the following form:

$$(4)\ Y = (Nk_1/q)\ [1 + Qq\ (Kp-1)\ (1-x)]$$

Where

$$(5)\ K = k_2/k_1$$

Effect Of Not Replacing Defectives

In many cases, the defective items are removed from both the sample and the screen rather than replaced. This will alter the total cost equation only slightly. The up-front cost of equation (1) now becomes:

$$(6)\ Y'_1 = nk_1 + (N-n)\ Pk_1$$
$$= Nk_1\ (P + Qx)$$

While equation (2), the downstream cost, becomes:

$$(7)\ Y'_2 = (N-n)\ Qpk_2$$
$$= NQpk_2\ (1-x)$$

While the total cost becomes:

$$(8)\ Y' = Nk_1\ (P + Qx) + NQpk_2\ (1-x)$$
$$= Nk_1\ [1 + Q(1-x)\ (Kp-1)]$$

From which can be seen the relationship of the cost is:

$$(9)\ Y = Y'/q$$

Economic Decision

The decision related to minimizing cost is the same whether we replace defectives with good items or not. Both Equation (4) and (8) have the same characteristics: the minimum cost is achieved by adjusting the sample ratio $x = n/N$ with respect to the cost/quality ratio $Kp = (K_2/k_1)p$. In both cases, minimum cost is achieved by minimizing the expression $Q(Kp-)\ (1-x)$.

$$(10)\ MIN(Y) = MIN\ (Q(Kp-1)(1-x))$$

It is obvious that the following conditions hold for Y to be minimum:

TABLE 1. CONDITIONS TO MINIMIZE COST

Kp	x	n	Q
>1	0	N	0
<1	1	0	negative
= 1	immaterial		0

It can also be seen that any alternative plan is less than minimum. These are only two realistic solutions based on the ratio Kp: n=0 or n=N, no inspection or 100% inspection.

DISCUSSION

Authors prior to Deming recognized the cost breakeven point Kp or in their terms:

(11) $p^* = k_1/k_2$

Where p* is the critical lot percent defective. Their concentration on the probability of acceptance for given plans leads them to miss the fact that no sampling plan other than n=0 or n=N can be optimum.

Martin comes closest when he says "In the segment where p>[p*] while the cost floor indicates the tightest plans — the loosest plan that still shows a [probability of acceptance] of O and the OC curve is a proper selection."[5] He missed the fact that this occurs when n=N at all times and indeed his figure 2 demonstrates the convergence.

Dayton on the other hand does not recognize the case of n=0 although he does recognize the need for "100% screening inspection (P=[p*] and unstable), (p>[p*])."[6]

A great deal of the decision is based on the lot quality p. Deming prepared a graph of Kp=1 (Figure 1) which can be used to determine for a given value of K the region in which p leads to a decision to sample or not. Equations (11) can also be used in this way. If the process average of the supplier is much less than p*, zero inspection will minimize costs. If it is much above p*, screening is indicated. If the value is close to p* screening is advisable as a conservative measure.

It is of course possible to work out the probability of different states of the world assuming a beta distribution and so obtain the expected value of each option choosing the best cost expected value. Or one could use the Bayesian approach of Smith[4] assuming a prior beta distribution with parameters alpha = beta = 1. From a practical side it does not appear worthwhile to go through these technical exercises since they are all based on estimates anyway and the use of n=N is not much more costly than n=0 near the point of indifference (Kp=1).

In the same way, a new supplier should state his process average. If he cannot, the question is raised, "why deal with this vendor since he does not have any control of his process?" If circumstances force you to reluctantly use such a vendor, the only safe plan is n=N until more is known about the vendor. Because of the uncertainty, it also turns out to be the most cost effective plan.

Most of the authors suggest the use of a surveillance sample when using the plan n=0. This is not necessary. Good production records can be used to estimate the actual lot quality and a Shewhart chart (or CUSUM Chart) can be maintained to track vendor performance and conformance to their process average.

APPLICATIONS

The most obvious application is receiving purchased materials. A sub-assembly is received from a manufacturer in lots of 1500 assemblies. It takes approximately 2 hours to test the assembly for an average cost (including burden) of $24 per assembly. The process average of the manufacturer is 2% and recent quality information confirmed this experience with lots received. To replace defective parts in final inspection requires $780 of fully costed labor. What sampling plan should be used?

$Kp = k_2/k_1 * p = \$780/24 * .02 = .65 < 1$

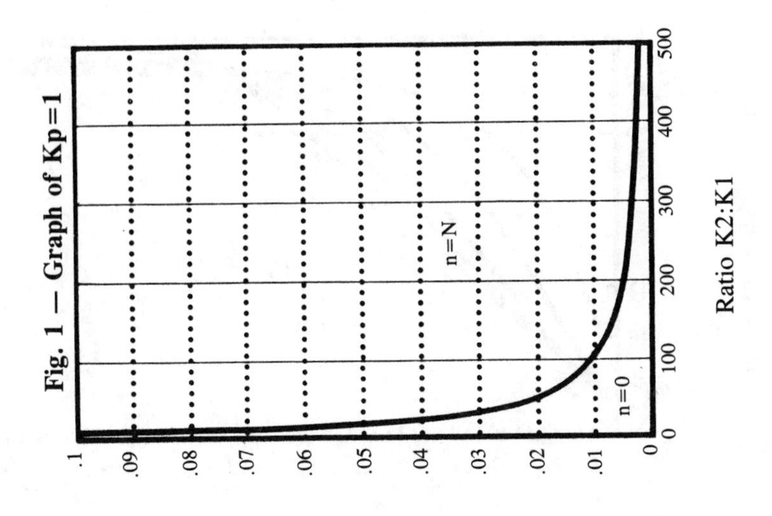

Since Kp is less than 1 do not inspect. Table 2 below is a numerical evaluation showing the policy of n=0 to be optimum. Figure three gives a graphic illustration showing for various values of p that n=0 or n=N is the minimum cost decision.

TABLE 2. COSTS UNDER VARYING SIZE SAMPLES

$k_1 = \$24,\ k_2 = \$780,\ p = .02$

Size of Sample* n	Acceptance Number* c	Sampling Fraction x = n/N	Prob of Reject P	Sample & Screen Cost Y_1	Downstream Cost Y_2	Total Cost $Y=Y_1+Y_2$
0	—	0	0	$ 0	$24,135	$24,135
125	5	1:12	.05	4,745	21,017	25,762
125	3	1:12	.25	11,480	16,593	28,072
125	2	1:12	.49	19,561	11,283	30,844
125	1	1:12	.78	29,327	4,867	34,194
125	0	1:12	.91	33,704	1,991	35,695
1500	—	1	1	36,735	0	36,735

*Sample Size 125 and acceptance number from MIL-STD-105D Level II (Plan K) single sampling.

Fig. 2 Comparative Cost Curves

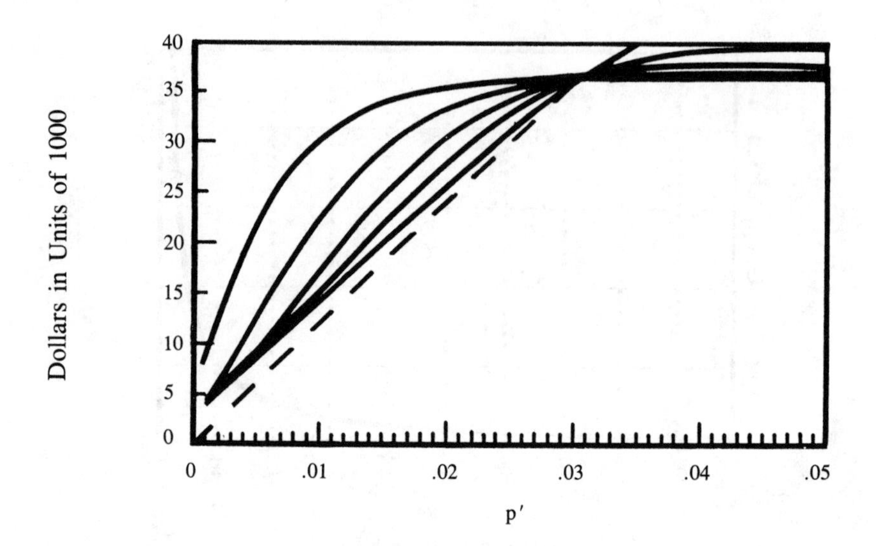

Another application, common to the service industry, is the decision to inspect or not when work moves from one department to the next or even to the customer. Consider a hypothetical but realistic situation in financial transactions. The cost of inspection is approximately \$0.25 per item (k_1) while the average loss is approximately \$500 (k_2). In this case,

$$p^* = .25/500 = .0005$$

or 5 defects per 10,000. Since very few operations are better than 10 defects per 10,000 or worse, it is obvious that 100% inspection is the least cost policy. ($Kp = 2 > 1$)

It should be noted that the presumption that 100% screening finds all defects is not always true, particularly in service industries. This topic has been covered elsewhere.[8] The method described here only makes the initial decision of 100% inspection or no inspection at all.

BIBLIOGRAPHY

1. Dodge, H. F. and Romig, H. G. *Sampling Inspection Tables,* New York, John Wiley 1944, 2nd Ed. 1959.
2. U. S. Department of Defense, Military Standard, *Sampling Procedures and Tables For Inspection By Attributes* (MIL-STD-105D), (Washington D. C.: U. S. Government Printing Office, 1963).
3. Hald, A. "The Compound Hypergeometric Distribution and a System of Single Sampling Inspection Plans Based on Prior Distributions and Costs", *Technometrics,* Vol 2, No. 3, August 1960, p 275-340.
4. Smith, Barnard E. "The Economics of Sampling Inspection", *Industrial Quality Control,* Vol 21, No. 9, March 1965, p 453-458.
5. Martin, Cyrus A. "The Cost Breakeven Point in Attribute Sampling" *Industrial Quality Control,* Vol 21, No.3, September 1964, p 137-144.
6. Dayton, Joseph F. "Fine Tuning Inspection For Minimum Cost", *Quality,* November 1977, p 46-49.
7. Deming, W. Edwards. *Management of Statistical Techniques For Quality and Productivity,* 1981, Chapter 8.
8. Chang, D.K.C. and Latzko, W. J. "Quality Improvement Program, An Application", *Thirtieth Annual Technical Conference* (Toronto: American Society For Quality Control).

"Q" COSTS/RESULTS — THE PETROLEUM INDUSTRY

Clyde W. Brewer
Otis Engineering Corporation
Dallas, Texas

ABSTRACT

Early 1978 brought another new quality requirement and the first to be imposed on many segments of the oil and gas industry. This document is ANSI/ASME SPPE with -1 applicable to manufacturers of equipment and -2 applicable to test laboratories. This introduced the total quality concept to this area for the first time. Management applications utilizing quality costs emerged as a result of this change.

This paper will describe the actual results of a quality program which was introduced during 1978 and which has already produced some dramatic reductions in the cost of quality. The paper will also describe the requirement in a generic review and a look at consensus standards regulation which has some contemporary management flexibility built in.

GENERAL

The first oil well was drilled in 1859 in Titusville, Pennsylvania by E. L. "Colonel" Drake. Today, there are over 670,000 producing wells in the United States. During 1980 alone total U. S. well completions topped out at 62,375. It has been estimated that in excess of 70,000 U. S. completions will be made annually during the late 1980's. The needs during this period will be 100,000 completions.

One standard barrel of oil contains 42 gallons. This measure was established back in Drake's days when oil was transported in 42 gallon whiskey barrels. From one barrel we get approximately 20 gallons of gasoline, 12 gallons of fuel oil and 10 gallons of other petro products such as kerosene, lubricating oils and greases. During 1979 the U. S. alone consumed 7.1 billion barrels of oil. At this consumption rate, it is easily understood why the major oil companies must continue to press onward in the search for additional petroleum supplies both inland and off-shore.

Many oil companies are drilling deeper in hopes of finding additional oil supplies. Drake's original oil well scarcely pierced the earth's surface at a mere 69 1/2 feet. During 1979, 621 of the wells completed dipped into the ultra-deep zone. At 29,622 feet, the deepest was Gulf Oil Corporation's 2 Emma Lou, Unit No. 1 in Pecos County, Texas. Within the next few years completions will bottom out around 50,000 feet or 9.5 miles below the earth's surface.

In 1969, a major blowout occurred in the Santa Barbara Channel. This event was the catalyst which has generated the quality involvement in the petroleum industry.

Although the Department of Interior had been granted jurisdiction over the Outer Continental Shelf 16 years prior to this blowout, the controls which existed to regulate offshore operations at that time were less than sufficient to prevent such catastrophies. In the Santa Barbara blowout it was estimated that up to 780,000 barrels of oil were lost. More critical than the loss of oil was the potential loss of human lives, wildlife and plants.

The Santa Barbara blowout was the turning point in the development of offshore safety and pollution controls. Shortly after the blowout, the people of California joined forces and pushed the government into action. The result of this public outbreak is visible today in the various laws, agencies, and committees that currently exist in the interest of offshore safety enhancement and pollution prevention.

The regulations developed for this purpose did not single out the oil companies nor the drillers as responsible for oil spills and blowouts. Instead, the regulations were developed to cover any and all phases of offshore operations. This included manufacturers, assemblers, and testers of safety equipment for offshore use. The United States Geological Survey, or USGS, was appointed by the Department of Interior as the agency responsible for developing and enforcing the laws necessary to ensure safety and the prevention of pollution in offshore operations.

The USGS developed a series of regulations or orders governing both drilling and production operations and the development and testing of equipment to be used in Outer Continental Shelf operations. These orders, referred to as OCS Orders, are divided into four geographical categories: the Atlantic, the Gulf of Alaska, the Gulf of Mexico, and the Pacific. The content of each order is basically the same for each geographical area but may vary slightly, such as to accomodate the arctic climatic conditions of the Alaskan Gulf or new frontier conditions.

Let's look first at the control lines. The U. S. Department of Interior, having jurisdiction over the Outer Continental Shelf, is the peak of the regulation control pyramid.

The Department of Interior granted authority to the USGS. The USGS developed the orders. Now we come to the next level in the control pyramid — the operators. When we say "operators" we are referring to the production companies. Compliance is the operator's role. Failure to comply with OCS orders can result in heavy fines and in the shut down of a well or wells, resulting in lost production income.

At the bottom is Otis and all other manufacturers, assemblers, and testers of safety and pollution prevention equipment in the control pyramid. Even though this industry is not directly subject to compliance with the USGS, our customers (the operators) are. Therefore, if we wish to remain in offshore equipment sales, we must design our equipment to comply with current safety and pollution control standards.

Before we get too wrapped up in the regulations set forth in the OCS orders, there are two thoughts you should keep in mind:

1. When we speak of SPPE, we are referring to safety and pollution prevention equipment, and
2. Of the 14 OCS Orders currently in existence, only one pertains directly to this industry — OCS Order #5 Production Safety Systems.

In 1977, the American National Standards Institute, or ANSI, approved and adopted two new generic quality standards, ANSI/ASME SPPE -1 and -2. These two standards deal directly with safety and pollution prevention equipment to be used in offshore oil and gas operations. The American Society of Mechanical Engineers, or ASME, and their representatives serve as auditors in the certification and accreditation of SPPE manufacturers and testing laboratories. They are joined in this effort by appointed representatives of other member organizations in the National SPPE Committee. SPPE -1 outlines Quality Assurance requirements for manufacturers and assemblers of SPPE. SPPE -2 defines the compliance requirements for laboratories who test SPPE.

To qualify as a manufacturer and assembler of SPPE for offshore oil and gas operations, industry must comply with SPPE -1. This caption was taken from OCS Order #5:

> "Safety and pollution prevention equipment *shall conform* to the following Quality Assurance standards:

a. ANSI/ASME's Standard 'Quality Assurance and Certification of Safety and Pollution Prevention Equipment Used in Offshore Oil and Gas Operations,'

which is ANSI/ASME SPPE -1 and

b. ANSI/ASME'S Standard 'Accreditation of Testing Laboratories for Safety and Pollution Prevention Equipment Used in Offshore Oil and Gas Operations,' "

which is ANSI/ASME SPPE -2.

Industry is further committed to OCS compliance as outlined in Sections 3.2 and 4.3 of OCS Order #5. Section 3.2 states that:

> "Surface controlled and subsurface controlled subsurface safety equipment . . . which are installed on new installations or replaced on old installations after February 1, 1980 *shall conform* to API Spec. 14A . . . "

API Specification 14A, entitled "Specification for Subsurface Safety Valves" covers all aspects of subsurface safety valves, safety valve locks and safety valve landing nipples. Included are minimum acceptable standards for materials, manufacturing and testing of both surface and subsurface controlled safety valves.

Section 4.3 of OCS Order #5 states that:

> "All wellhead surface safety valves . . . which are installed on new installations or replaced on old installations after February 1, 1980 *shall conform* to API Specification for Wellhead Surface Safety Valves for Offshore Service, API Spec. 14D."

We began the task of preparing to meet certification requirements both as a manufacturer and assembler in February, 1978. A task force was appointed to develop procedures and documentation necessary to achieve this certification.

To comply with the requirements of SPPE-1, our quality program is designed, organized and documented to provide controls over all activities which affect quality. Otis was the first company to obtain authorization to monogram subsurface safety valves with the OCS emblem. Having the certification of authorization existing and future SPPE as being OCS qualified.

Simultaneous to meeting the regulatory requirements a total quality program was designed using proven quality costs principles. No preventative quality had ever been introduced into this industry and the SPPE document has no mention of "cost of quality." Selling the concept became a matter of results rather than a forced input from a regulatory document.

The initial step was a management commitment to provide the freedom to accomplish the task. There had never been a quality policy established in the over 40 plus years the company had been in operation. The following policy was established in February 1978 and the program was underway:

> **"POLICY**
>
> Product quality and reliability at Otis Engineering is the responsibility of every employee involved in the design, manufacture, delivery or services of equipment to our customers. The equipment and services which Otis supplies must always adhere to the highest quality standards in order to protect human lives and the environment during operation.
>
> The Quality function exists at Otis Engineering to measure and report the level of quality in any area associated with designing, manufacturing, delivering, installing or servicing equipment for our customers.
>
> The Quality organization provides reports for management through monitoring and/or auditing the various operations within the Company. These reports include action items which must be corrected or improved by management in order to maintain compliance with customer, industry, and regulatory standards.
>
> The Director of Quality is assigned the responsibililty for definition and administration of the Quality policy and reports directly to the Senior Vice President and Technical Director. The Director of Quality is authorized to resolve all conflicts between Quality policy and other policies, procedures or operations within the Company. The Director of Quality will review the Quality organization on an annual basis to assess its adequacy and will take appropriate corrective actions, where necessary, to provide effective implementation; this assessment will be accomplished through an audit conducted by persons other than those within the Quality organization. A report of this audit is provided to the President. The Director of Quality is responsible for maintaining a written Quality policy which is distributed throughout the management of the Company. In addition, a Quality program, consisting of a summary of the various applicable functions in the Company, will be developed and maintained by the Director of Quality for each unique customer or regulatory specification which requires a program description."

The first major hurdle was to establish "conformance" as the only criterion for the appraisal process. This change alone introduced a 900% increase in rejection documents and obviously a major objection concerning deliveries which were at least 90 days late with many items two years behind schedule already. Management pressure was obviously intense to walk softly. Backed by the policy and a firm management commitment, we now had the vehicle to find out what the problems actually were in our operations. We had excellent field failure data to begin with.

Next was accumulation of the cost of problems. Our accounting data included exacting cost relative to scrap and rework. We also had very accurate data on the Quality Department cost which was also in accounting dollars. To establish a base, develop trend prediction capability and obviously a method to determine successful results, we captured cost beginning in 1976. With this we began the task of educating management in the quality costs principles.

When the surge of spoilage cast increased as a result of the 900% increase began to appear, the management pressure was in direct proportion to the curve. Again, the policy and top management support allowed continuation of a Quality program on all products, not just our safety equipment. This equipment, in fact, only accounted for 20% of our total business.

The next series of slides will dramatically reveal the program success. Building the program was a step-by-step process. First, we had to justify additional people. As you will see the overall population of the Quality Department was frozen to gain creditability. Fortunately, the manning had been all appraisal so that adding the prevention was a matter of selecting QA personnel from within or adding more QA people as attrition occurred in the QC process. This produced the stability in the cost of operating the department.

Results as you have seen have been dramatic but during this program another phenomenon occured. Look at the traditional cost curve we all are familiar with. Now, look at spoilage plotted over Quality Department cost and you see a nice bell curve. From this we have perceived a need for change in our historical chart and realistically acknowledge that (1) spoilage will increase in a direct proportion to the amount of quality effort wherever you begin the program. The success then becomes how quickly you can get the two to cross back over and find the saturation point or equilibrium point to maintain an optimum. This can also reveal when you need to reduce the intensity of the program in departmental costs. As a member of the Quality Cost Technical Committee, I intend to pursue the changing of our traditional curves. This type information is more palatable to management than infinite data. We should someday be able to quantify enough data to even recommend points on the curve as related to business.

CONCLUSION

In closing, I will share with you some of the tools we have used to accomplish this success. We have reports, not data, that furnish information which is sensitive to individuals and accumulated at first-line supervisor level. We now produce a report which Paretorizes defects. Soon, we will be able to dollarize this data. Our main thrust was use of quality cost as a measure of performance in our manufacturing areas. The slide now shows this formulation which was the primary contribution to success. Look what happened when this decision was made.

We are now in the finesse area of corrective action. We now have a fully qualified QA staff. Improvement will continue until we saturate this technique, then we have an incentive program designed to capture the group dynamics improvements possible through peer pressure, motivational approaches, etc. With our initial success established, our secondary thrust cannot possibly fail and justification for additional funding for a quality incentive will be guaranteed. Another paper will be proposed for our Denver conference in 1984 to share the successes of the two phases noted.

PRINCIPLES OF QUALITY COSTS

Jack Campanella
Fairchild Republic Company
Frank J. Corcoran
Singer Kearfott Division
Little Falls, New Jersey

ABSTRACT

This presentation is based on the forthcoming ASQC publication "Principles of Quality Costs", in which the complete field of Quality Costs will be explored and shaped into the current discipline. The first part explores the Quality Costs concept and deals with Quality Cost history, philosophy and theory. The second part discusses the implementation of Quality Cost programs, including how to start, definitions, cost collection, summary, analysis and reporting. This session is a tutorial, designed to provide the tools required to understand and implement a Quality Cost program.

INTRODUCTION

"Principles of Quality Costs" is a forthcoming publication of the American Society for Quality Control (ASQC) upon which this tutorial is based. The publication is being developed by the Quality Costs Technical Committee as the standard Quality Cost text and will establish principles, definitions, applications, examples, and methods of Quality Cost management. It will be a completely new guide, expanding on "Quality Costs — What and How," incorporating the latest techniques and establishing a set of teaching Quality Cost systems. The material herein has been obtained from many sources, including inputs from Quality Cost Technical Committee members and publications.

QUALITY COST CONCEPT

History

The concept of the economics of quality can be traced back to the early 1950s. Chapter I of Dr. J. Juran's first "Quality Control Handbook",[1] published in 1951, was titled "The Economics of Quality" and contained discussions of the "Cost of Quality" and his now famous analogy of "Gold in the Mine". However, most other papers and articles of the time were based on specific applications such as "The Economic Choice of Sampling Systems in Acceptance Sampling" by J. Sittig in 1951.[2]

Among the earliest articles on Quality Cost systems as we know them today were Harold Freeman's 1960 paper "How to Put Quality Costs to Use",[3] and Chapter 5 "Quality Costs", of Dr. A. V. Feigenbaum's famous text "Total Quality Control"[4] in 1961. They are among the earliest writings categorizing Quality Costs into the costs of Prevention, Appraisal, Internal and External Failure.

In December, 1963, the Department of Defense issued MIL-Q-9858A, "Quality Program Requirements",[5] making "Costs Related to Quality" a requirement on many Government contractors and subcontractors. From that point on, the subject boomed. In 1964 the Industrial Engineering Department of Stanford University did a research study for the Air Force Systems Command and published the QUICO (Quality Improvement Through Cost Optimization) System[6] ultimately leading to the May 1967 release of Quality and Reliability Assurance Technical Report TR8, "A Guide to Quality Cost Analysis",[7] by the Office of the Assistant Secretary of Defense for Installations and Logistics.

The ASQC Quality Costs Technical Committee was formed in 1961, and in 1967 published "Quality Costs — What and How",[8] the most popular document on the subject and the largest seller of any ASQC publication. Other popular committee publications are the 1977 "Guide to Reducing Quality Costs"[9] and more recently, the "Guide for Managing Vendor Quality Costs"[10] published in 1980.

The more than 300 excellent articles and papers listed in the current "Bibliography of Articles Relating to Quality Cost — Concepts and Improvement",[11] maintained by the Quality Costs Technical Committee, further attest to the popularity of the subject and to the recognition given to it throughout industry.

Applicable Specifications

As mentioned above, military specification MIL-Q-9858A makes Quality Costs a requirement on many Government contracts. Paragraph 3.6 of that document requires the contractor to "maintain and use quality cost data as a management element of the quality program". Except for requiring the identification of the costs of "prevention and correction of nonconforming supplies", this paragraph is not very definitive or restrictive. It permits a very wide interpretation and allows "the specific quality cost data to be maintained and used" to be "determined by the contractor".

A more recent Military Standard, MIL-STD-1520B[12] (original issue MIL-STD-1520, 1 May 1974), requires Air Force contractors and subcontractors who have MIL-Q-9858A on contract to collect "the costs associated with nonconforming material" with an objective of providing "current and trend data to be used for contractor and Government management review and appropriate action". This document is much more specific than MIL-Q-9858A and lists the actual costs that must be collected and summarized.

In addition to these documents, more and more contracts, both Govenment and commercial, are spelling out Quality Cost requirements . . . from the basic collection of scrap and rework costs, to full-scale Quality Cost programs.

Philosophy

The real value of a Quality Program is ultimately determined by its ability to contribute to improved customer satisfaction and to profits. That is the environment in which Quality Management exists, and is the principle reason why Quality Costs should be an integral part of an effective quality management system.

To develop the concept of Quality Costs, it is necessary to establish a clear picture of the difference between Quality Costs and the cost of the Quality Department. It is important that we don't view Quality Costs as the expenses of the quality function. Fundamentally, each time work must be redone, we are adding to the cost of quality. The most obvious examples are the reworking of a manufactured item, the retesting of an assembly, or the rebuilding of a tool because it was originally unacceptable. Other examples may be less obvious, such as the repurchasing of parts or the response to a customer complaint.

In short, any cost that would not have been expended if quality was perfect, contributes to the cost of quality. Unfortunately, many such costs are overlooked or unrecognized simply because most accounting systems are not designed to identify them. It is for this reason that the technology or system of Quality Costs was created. It was designed to recognize the costs of "doing things over" as a significant addition to the cost of quality management, and to show them collectively as an otherwise hidden opportunity for profit improvement.

As seen in Figure 1, the most costly condition exists when a customer finds defects. Had the manufacturer, through much inspection and testing, found the defects himself, a less costly condition would have resulted. However, had the manufacturer's quality program been geared towards prevention of defects, defects and their resulting costs would have been minimized; obviously the most desirable condition.

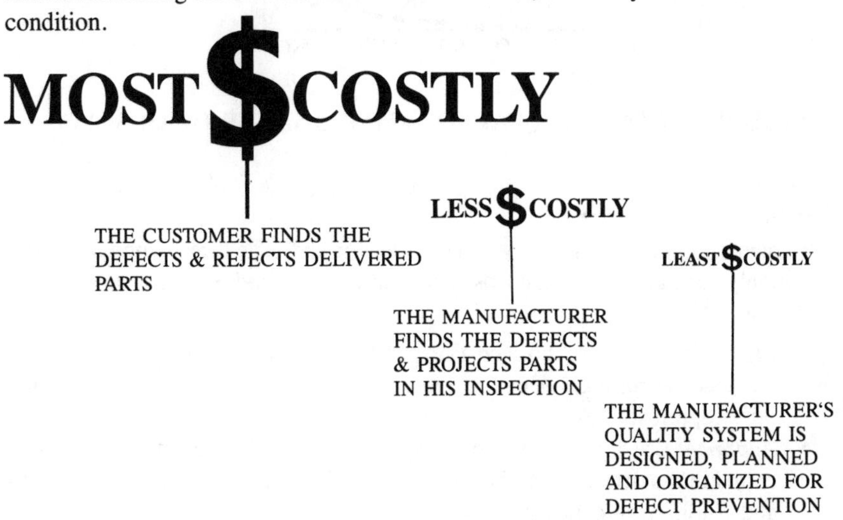

FIGURE 1. COMPARATIVE COST OF QUALITY

The basic problem is to strike the correct balance between the cost of quality and the value of quality. Figure 2 shows that total cost is at a minimum when the cost of quality is at an optimum point. It is at this optimum point that the correct balance between the cost and value of quality has been reached.

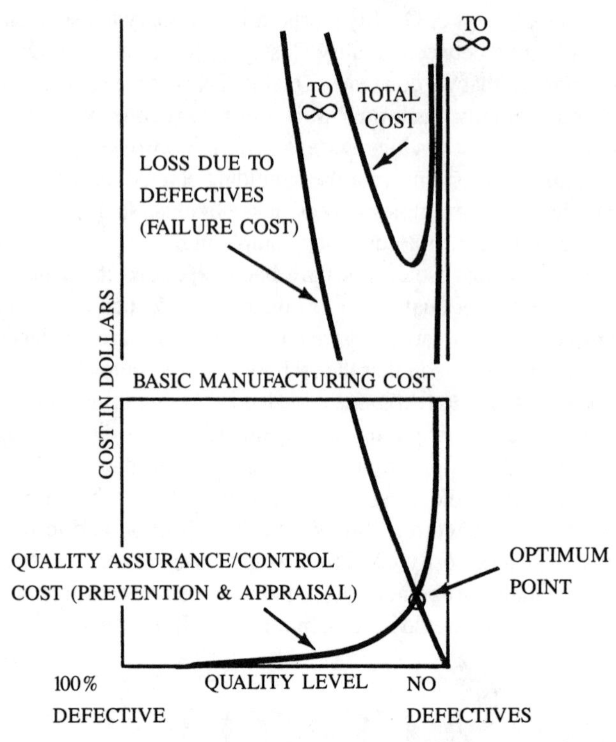

FIGURE 2. COST OF QUALITY VERSUS VALUE OF QUALITY

Theory

To manage Quality Costs, they must be categorized. The three major categories commonly used are Prevention Costs, Appraisal Costs and Failure Costs, defined as follows:

- *Prevention Costs* are those costs expended in an effort to prevent discrepancies, such as the costs of quality planning, supplier quality surveys, and training programs;
- *Appraisal Costs* are those costs expended in the evaluation of product quality and in the detection of discrepancies, such as the costs of Inspection, Test and Calibration control;
- *Failure Costs* are those costs expended as a result of discrepancies, and are usually divided into two types:
  - *Internal Failure Costs* are costs resulting from discrepancies found prior to delivery of the product to the customer, such as the costs of Rework, Scrap, and Material Review;
  - *External Failure Costs* are costs resulting from discrepancies found after delivery of the product to the customer, such as the costs associated with the processing of customer complaints, customer returns, field services and warranties.

Total Quality Cost is the sum of these costs, i. e. Prevention, Appraisal and Failure. It represents the difference between the actual cost of a product, and what the reduced cost would be if there were no possibility of failure of the products nor defects in their manufacture. It is, as described by Dr. J. Juran,[13] "gold in the mine" waiting to be extracted. (Quality Costs can run as high as 15 to 20 percent of sales in some companies and as low as 2.5 percent of sales in others.)

The objective is to bring the Total Quality Cost to a minimum while maintaining required quality levels. The basic concept is that an increase in the cost of prevention should result in a larger decrease in the cost of failure, thereby reducing the Total Quality Cost. (An ounce of prevention is worth a pound of failure.) When this no longer happens, Prevention Costs have been "saturated", and no further dollars should be invested in prevention until conditions change or a breakthrough[14] is achieved. It is at this point, the *saturation point,* that further costs of prevention become larger than the savings afforded.[15] It may be determined for example, that the cost of a quality training program (a Prevention Cost) was responsible for a large decrease in Failure Costs. Future programs of this type would therefore be indicated since the value of the program (the decrease in Failure Costs) exceeded the cost of the program. However, this too can reach a saturation point, where the program costs exceed the savings afforded. With proper analysis, this will be indicated by the Quality Cost program.

While an increase in the cost of prevention should result in a decrease in the cost of failures, it may also cause Appraisal Costs to decrease somewhat. The reduction in Failure Costs may justify an increase in sampling inspection due to the confidence gained in improved quality, thereby reducing the amount of inspection performed.

An increase in Appraisal Costs may also result in some decrease in Failure Costs, because of the resulting increase in the proportion of discrepancies found in-house. As we saw in Figure 1, external failures, those found by the customer, inevitably cost more. Returned material costs, retrofit costs, and possible loss of future business due to unhappy customers, are examples of expensive External Failure Costs. Appraisal Costs and Failure Costs should change in opposite directions, Appraisal Costs going up with Failure Costs going down, *until an optimum point is reached.* An investment in Appraisal Costs beyond the optimum point is uneconomical. The determination of the optimum point should be part of the Quality Cost analysis.

Trend Analysis

Total Quality Cost, compared to an applicable base, results in an index which may be plotted and periodically analyzed in relation to past indices. The base should be representative of and sensitive to fluctuations in business activity. Some bases commonly used are Manufacturing Direct Labor, Net Sales Billed, and Cost Input. Figure 3 is a sample plot of a Quality Cost Index (I) by month, indicating an improvement trend for the second half of the year.

It should be obvious that increases in the expenditures for prevention and appraisal will not show immediate reductions in Failure Costs because of the time lag between the cause and effect.[16] This lag can be observed on a Quality Cost Index trend chart. For this purpose, it may be desirable to indicate on the chart when major changes to the quality program were made.

Figure 4 is the same Quality Cost Index trend chart illustrated in Figure 3. It is marked to indicate a change to the quality program. We can now see the reason for the steady improvement over the last five months; back in April, operator training programs were initiated.

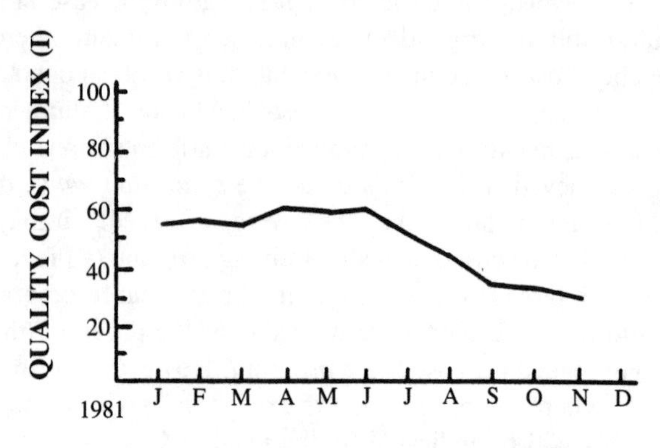

FIGURE 3. QUALITY COST TREND

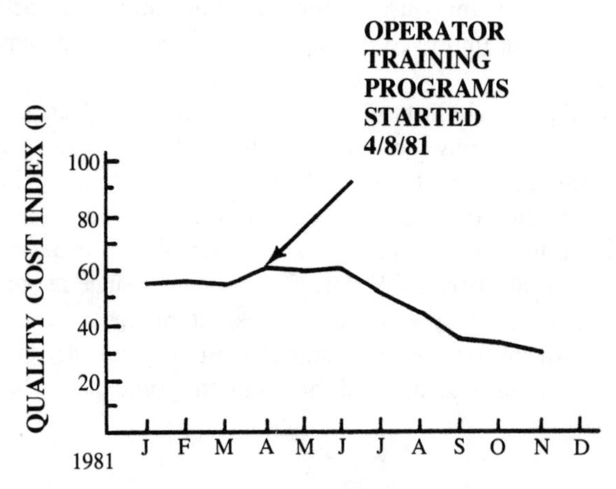

FIGURE 4. QUALITY COST TREND

The first effect was an increase in the Quality Cost Index due to the cost of the programs (Prevention Cost increased but Failure Costs remained the same). After a "cause and effect lag" of about two months, the value of the training began to become evident by a steady reduction in the Quality Cost Index (Failure Costs decreased while Prevention Costs remained the same). By November, a 45% reduction was indicated.

Obviously, the training programs were a worthwhile investment. Had no improvement been indicated after a reasonable amount of time, some action would be necessary. The programs would have to be re-evaluated and either revised or dropped in favor of some other course of action.

QUALITY COST IMPLEMENTATION

How to Start

The first step in starting a Quality Cost Program is to determine the need for such a program. This should be presented to management in a way which will justify the effort and will interest them in participating in the program. One way to do this is by establishing a trial program. The trial program can be simple . . . only the major costs need be gathered, and only data readily available need be included. You may find that much of the data required for the program is presently available. You may even estimate some of these costs, if necessary. Select a program, division, facility or area of particular interest to management. The results should be sufficient to sell them on the need for the program. Most trial runs will show eye-opening results . . . spectacular enough to make management sit up and take notice. They'll see that Quality Costs can run as much as 20% or more of net sales dollars and opportunities for significant savings will be identified (Juran's "Gold in the Mine"). Once top management is convinced, getting the all-important cooperation of the Accounting Department should be easy.

Defining Quality Cost Elements

With management sold, and with accounting "ready to go", the specific Quality Costs to be collected must be determined. To determine the Prevention Costs to be collected, the tasks performed in your company in an effort to prevent discrepancies, should be listed together with responsible departments. Remember that Quality Costs are not only incurred by the Quality Department. In a like manner, Appraisal Cost elements are determined by listing those tasks associated with the inspection and test of product for the detection of discrepancies.

For Failure Cost elements, you need to determine those costs which would not have been expended if quality were perfect. If quality were perfect, you would not have rework, nor would you have to respond to customer complaints, or take corrective action. Remember to divide Failure Costs into Internal and External categories.

Quality Cost elements may be different from company to company, especially so from industry to industry; however, the overall categories of Prevention, Appraisal and Failure Costs are always the same.

Cost Collection

Now that the specific costs to be collected have been determined, a method must be developed for their collection. Collection of Quality Costs should be the responsibility of the Controller. If top management was properly sold on the program, the Controller will have been charged with the task of heading this effort. The Controller with the help of the Quality Manager, should review the list of costs needing collection, determine which of these are available under the present accounting system, and decide where additions to the existing system need to be made. Occasionally, the mere addition of new cost element codes to the present charging system is sufficient. However, if necessary, the present system may be supplemented by separate forms designed especially for this purpose. Ideally, a complete system of cost element codes could be generated; coded in such a way that the costs of Prevention, Appraisal, Internal and External Failures could be easily distingushed and sorted. These codes would be entered on a labor distribution, charge, or time card together with the hours expended against the cost element represented by that code. The labor hours would be converted to dollars by data processing. An exception might be scrap. In many companies, the existing scrap reporting documents are forwarded to the Estimating Department, where the labor and material costs expended to the stage of completion of the scrapped items, are estimated. The Accounting Department then provides all collected Quality Costs to the Quality Department in a format suitable for analysis and reporting. Of course, training programs will be necessary to inform all personnel as to the method to be used to report their Quality Cost expenditures. The training should be repeated periodically and the collection system should be audited on a regular basis.

Summary and Analysis

Quality Costs may be summarized by company, by division, by facility, by department or by shop. They may be summarized by program, by type of program, or by the total of all programs. The decision must be predicated on the individual needs of *your* company.

Analysis can include comparison of Total Quality Cost to an appropriate measurement base such as Net Sales, Cost Input, or Direct Labor. This will relate the cost of quality to the amount of work performed. An increase in Quality Cost with a proportionate increase in the base is normal. It is the nonproportionate change that should be of interest. The index [Total Quality Cost/Measurement Base] is the factor to be analyzed. The goal is to bring this index to the point of most economical operation — the optimum point. The index may be plotted so that trends representing present status in relation to past performance and future goals may be analyzed.

Other methods of analysis include study of the effect that changes in one category (Prevention, for example) have on the other categories, and on the Total Quality Cost. This technique can provide insight into where the quality dollar can most wisely be spent. Increases in Failure Costs must be investigated to determine where

Prevention Costs must be expended to reverse these trends and reduce the Total Quality Cost. Losses must be defined, their causes identified, and preventive action taken to preclude recurrence.

Other existing quality systems, such as that for defect reporting and analysis, may be used to identify significant problems. While the losses are distributed among numerous causes they are not uniformly distributed. A small percentage of the causes will account for a high percentage of the losses. This is an adaptation of "Pareto's Principle."[17] These causes are the "vital few", as opposed to the "trivial many." Concentration on prevention of the "vital few" causes will achieve maximum improvement at a minimum of cost. The goal is to determine and attain an optimum level, where return for the effort expended is greatest. By its very nature, this will have the effect of improving quality, while reducing costs.

Reporting

There are almost as many ways to report Quality Costs as there are companies reporting them. That is because how they are reported depends to a large extent on who they are being reported to and what the report is trying to say. The best way for you is the way that's best for your company and purpose. The examples provided as Figures 5 through 9 are samples of reporting techniques that have worked in companies with established Quality Cost programs[18],[19].

The amount of detail included in the Quality Cost report generally depends upon the level of management the report is geared to. To top management, the report may be a scoreboard, depicting through a few carefully selected trend charts, the status of the Quality Program . . . where it's been and the direction it's heading. Savings afforded over the report period and opportunities for future savings might be identified. To middle management, the report might provide Quality Cost trends by department or shop to enable identification of areas in need of improvement. Reports to line management might provide detailed cost information, perhaps the results of a Pareto analysis, identifying those specific areas where corrective action would afford the greatest improvement. Scrap and rework costs by shop are also effective charts, when included in reports to line management.

Charts are used in presenting data and trends pictorially. Properly done, they can bring home a point or tell a story simpler, better, and more interestingly than the raw data they represent. However, to keep them simple, they should rarely be designed to portray more than one idea. Charts with more than one idea usually do not present any of them effectively. Charts can be presented in various ways depending on their purpose. Line graphs are best for depicting trends, while bar charts are useful for showing proportions and comparisons. Circle or "pie" charts are also used for showing proportions, and are particularly effective in illustrating the slice of each silver dollar expended for Prevention, Appraisal, Internal and External Failures. However, circle charts are limited in that they cannot be plotted over time as can bar charts.

Use by Management

Once the Quality Cost program is implemented, it should be used by management to justify and support improvement in each major area of product activity. Quality Costs should be reviewed for each major product line, manufacturing area, or cost center. The improvement potential that exists in each individual area can then be looked at and meaningful goals can be established. The Quality Cost system then becomes an integral part of quality measurement. The proper balance is to establish improvement efforts at the level necessary to effectively reduce the total cost of quality and then, as progress is achieved, adjust it to where Total Quality Costs are at the lowest attainable level. This prevents unheeded growth in Quality Costs and creates improved overall quality performance, reputation and profits.

A Quality Cost program, based on the concept and methods of implementation endorsed by the ASQC's Quality Cost Technical Committee[20] and broadly presented herein, can be used by management as an aid towards achieving its goal of an optimum quality program at a minimum Quality Cost. The program will measure the value of the quality effort, identify the strong and weak points of the quality program, indicate how the quality dollar can be spent most effectively, and provide quality improvement while reducing costs.

QUALITY COSTS — By Program

FARMINGDALE

Month August 1978

CODE	ELEMENT DESCRIPTION	PROGRAM (In Thousands)					TOTAL ALL PROGRAMS
		A-14	SSVT	F-4	747	Misc	
K	MATERIAL REVIEW ACTIVITY	71.8	—	.7	7.7	—	80.2
L	CORRECTIVE ACTION	249.8	—	.9	2.7	—	253.4
X	TROUBLESHOOTING/FAILURE ANALYSIS	47.2	—	—	.8	—	48.0
R	REWORK/REPAIR	128.6	.4	6.3	26.1	—	161.4
P	SCRAP	19.2	—	—	0.5	—	19.7
V	RWK/RPR/SCRAP — VENDOR RESP.	27.1	—	.1	2.6	—	29.8
U	PROCESSING OF CUSTOMER COMPLAINTS	5.4	—	—	1.4	—	6.8
I	PROCESSING OF CUSTOMER RET'N'D MAT'L	23.3	—	—	—	—	23.3
J	FIELD SERVICES	—	—	—	—	—	—
Y	WARRANTY COSTS	.1	—	—	—	—	.1
TOTAL "UNQUALITY" COSTS		572.5	.4	8.0	41.8	—	622.7
QUALITY PREVENTION AND APPRAISAL		718.2	.9	8.3	69.8	38.6	835.8
TOTAL QUALITY COSTS		1290.7	1.3	16.3	111.6	38.6	1458.5
MFG DIRECT LABOR COSTS		6973.8	7.5	116.5	882.8	396.5	8377.1
SCRAP/REWORK/REPAIR AS % OF MFG D/L		2.1	5.3	5.4	3.0	—	2.2
COST INPUT		20943.4	66.1	978.4	1946.6	104.1	24038.6
TOTAL QUALITY COSTS AS % OF COST INPUT		6.2	2.0	1.7	5.7	37.1	6.1

FIGURE 5. QUALITY COSTS BY PROGRAM

TOTAL QUALITY COSTS
(FARMINGDALE — A-10 PROGRAM)

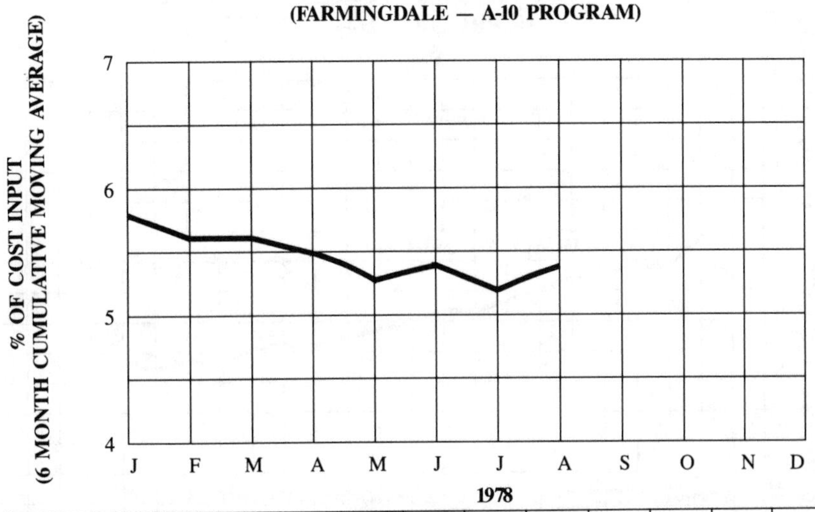

% OF COST INPUT
(6 MONTH CUMULATIVE MOVING AVERAGE)

7
6
5
4

J F M A M J J A S O N D

1978

Tot. Qual. Cost*	937.9	921.8	1380.5	1217.2	888.9	1385.5	1157.5	1290.7				
Cost Input*	16429	18984	25028	21051	18562	25363	24546	20943				
% of Cost Input	5.7	4.9	5.5	5.8	4.8	5.5	4.7	6.2				
6 Mon. Mov. Avg.	5.8	5.6	5.6	5.5	5.3	5.4	5.2	5.4				

*In Thousands

F-A10-1

FIGURE 6. TOTAL QUALITY COSTS BY MONTH

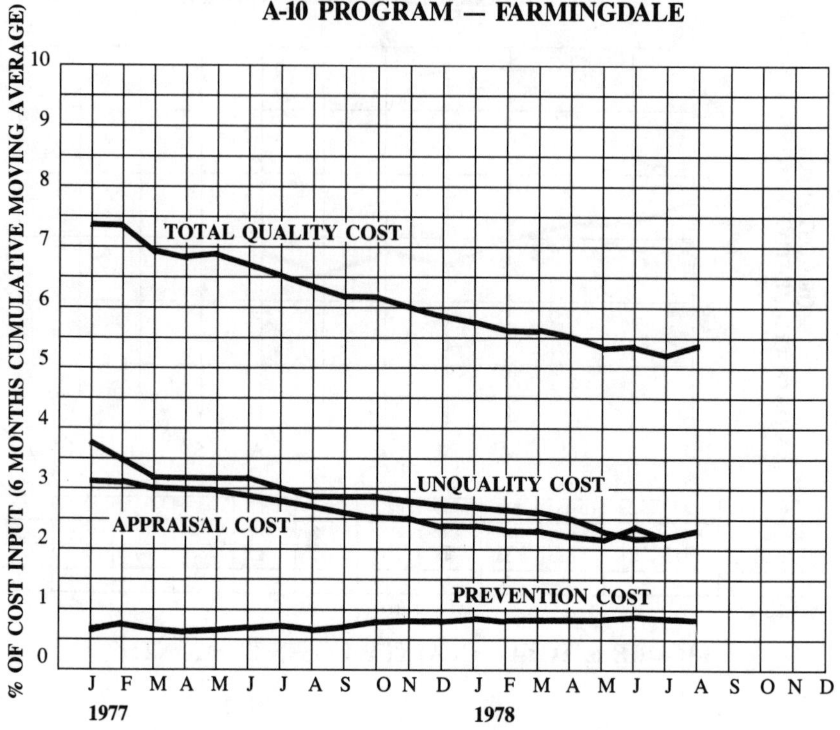

FIGURE 7. QUALITY COSTS BY CATEGORY

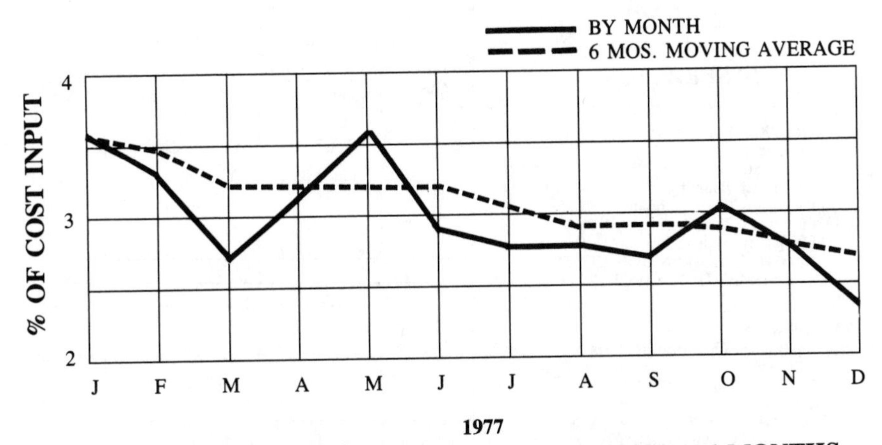

FIGURE 8. FAILURE COSTS BY MONTH WITH A 6 MONTHS MOVING AVERAGE FOR TRENDING

QUALITY COST REPORT
FOR THE MONTH ENDING ________
(In Thousands of U. S. Dollars)

DESCRIPTION	Current Month			Year to Date		
	Quality Costs	As a Percent of		Quality Costs	As a Percent of	
		Sales	Other		Sales	Other
1 PREVENTION COSTS						
1.1 Product Design						
1.2 Purchasing						
1.3 Quality Planning						
1.4 Quality Administration						
1.5 Quality Training						
1.6 Quality Audits						
TOTAL PREVENTION COSTS						
PREVENTION TARGETS						
2 APPRAISAL COSTS						
2.1 Product Qualification Tests						
2.2 Supplier Production Inspection and Test						
2.3 In Process and Final Inspection and Test						
2.4 Maintenance and Calibration						
TOTAL APPRAISAL COSTS						
APPRAISAL TARGETS						
3 FAILURE COSTS						
3.1 Design Failure Costs						
3.2 Supplier Product Rejects						
3.3 Material Review and Corrective Action						
3.4 Rework						
3.5 Scrap						
3.6 External Failure Costs						
TOTAL FAILURE COSTS						
FAILURE TARGETS						
TOTAL QUALITY COSTS						
TOTAL QUALITY TARGETS						

MEMO DATA	Current Month		Year to Date		Full Year	
	Budget	Actual	Budget	Actual	Budget	Actual
Net Sales						
Other Base (Specify)						

FIGURE 9. QUALITY COSTS BY ELEMENT

Quality Costs as a Budgeting Tool

An additional benefit to be gained from a Quality Cost program is its ability to be used as a budgeting tool. Once Quality Cost elements have been established and costs are being collected against them, a history is generated which can be used to determine the average cost per element. These averages can be used as the basis for future quotes and "estimates to complete". Budgets can be established for each element. Then, going full circle, the actuals collected against these elements can be used to determine budget variances and, as with any good system of budget control, action can be taken to bring over or under running elements into line. Figure 10 illustrates a report providing actual-versus-budget figures for each Quality Cost element.

Reducing Quality Costs

"An ounce of prevention is worth a pound of cure". That old maxim is as true as ever when it comes to reducing Quality Costs. An increase in the cost of prevention should result in a larger decrease in the cost of failures, thereby reducing Total Quality Costs.

DATE ____________
MONTH ____________

QUALITY COST REPORT

	MONTH		YEAR TO DATE	
PREVENTION	**ACTUAL**	**BUDGET**	**ACTUAL**	**BUDGET**
P1.1 ENGINEERING EFFORT	20.0	21.0	200.0	200.0
P1.2 QUALITY ADM.,PLNG. & SERV.	5.0	5.0	50.0	50.0
P1. TOTAL	25.0	26.0	250.0	250.0
APPRAISAL				
A1.1 INCOMING INSPECTION	20.0	25.0	200.0	250.0
A1.2 IN-PROCESS & FINAL INSPEC.	50.0	55.0	600.0	700.0
A1.3 TESTING	50.0	50.0	600.0	600.0
A1.4 PRODUCT AUDIT	10.0	12.0	100.0	150.0
A1.5 OTHER TESTS-RELIAB. ETC.	10.0	11.0	100.0	100.0
A1.6 EQUIP. CALIBRATION	10.0	12.0	50.0	50.0
A1. TOTAL	150.0	165.0	1650.0	1850.0
INTERNAL FAILURES				
N1.1 DEFECTIVE WORK/JUNK	30.0	35.0	350.0	400.0
N1.2 REMAKES, REWORK & RETEST	30.0	20.0	400.0	300.0
N1.3 INVESTIGATION	10.0	5.0	100.0	150.0
N1. TOTAL	70.0	60.0	850.0	850.0
EXTERNAL FAILURES				
E1.1 WARRANTY — COMPLAINTS	10.0	20.0	100.0	200.0
E1.2 FIELD REPAIR — (IN WARRANTY)	2.0	5.0	50.0	50.0
E1.3 FIELD REP. — (OUT OF WARRANTY)	1.0	5.0	50.0	75.0
E1.4 OTHER	2.0	5.0	50.0	25.0
E1. TOTAL	15.0	35.0	250.0	350.0
Q1. TOTAL	260.0	286.0	3000.0	3300.0
- RATIO TO TOTAL SALES	3.2%	5.0%	3.0%	3.2%
- RATIO TO TOTAL STANDARD COST OF OUTPUT	5.1%	4.6%	4.6%	4.7%

FIGURE 10. QUALITY COST ACTUALS VERSUS BUDGET

Quality improvement results in cost improvement. Designing and building a product right the first time always costs less. Solving problems with existing products by finding their causes and eliminating them results in measurable savings. To cash in on these savings requires that the quality performance of the past be improved. Ways and methods to do this are described in the ASQC publication, "Guide for Reducing Quality Costs"[21]. The purpose of the publication is to provide guidance to general management and professionals engaged in Quality Program Management, to enable them to structure and manage programs for Quality Cost reduction. It describes techniques for using Quality Cost data in programs to reduce costs and improve profits.

Figure 11 is extracted from the ASQC "Guide", and shows how Quality Cost analysis bridges the gap between the elements of a prevention-oriented quality program and the means used by general management to measure performance — the Profit and Loss statement. It shows the flow of Quality Cost information from the working quality assurance level to the Total Cost of Quality level and ultimately to the Profit and Loss statement. Every dollar saved because of improved quality has a direct impact on profit!

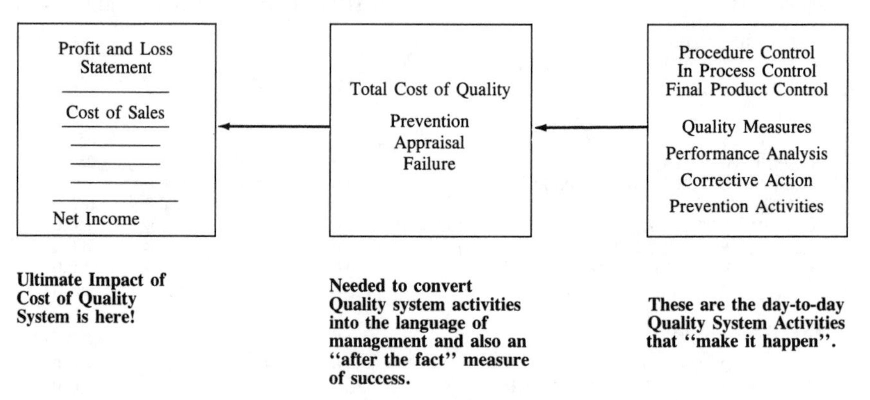

FIGURE 11. QUALITY COST SYSTEM

Corrective Action

A key factor in the reduction of Quality Costs is corrective action. Quality Costs do not reduce themselves, they are merely the scorecard. They can tell you where you are, and where your corrective action dollar will afford the greatest return.

Once a target for corrective action is identified, through Pareto and/or other methods of Quality Cost analysis, the action necessary must be carefully determined. It must be individually justified on the basis of an equitable cost trade-off (e. g., a $500 rework problem versus a $5000 solution). At this point, experience in measuring Quality Costs will be invaluable for estimating the payback for individual corrective action investments. Cost benefit justification of corrective action is a continuing part of the program.

Some problems have fairly obvious solutions. They can usually be fixed immediately (e. g., replacement of a worn bearing or a worn tool). Others are not so obvious (such as a marginal condition in design or processing) and are almost never discovered and corrected without the benefit of a well-organized and formal approach. Marginal conditions usually result in problems that can easily become lost in the accepted cost of doing business. Having an organized corrective action system justified by Quality Costs will surface such problems for management's visibility and action. The true value of corrective action is that you only have to pay for it once; whereas failure to take corrective action may be paid for over and over again.[22]

Auditing Quality Costs

Implemention of a Quality Cost program will be effective only as long as it continues to accurately measure true Quality Costs within the organization.[23] The financial establishment has long recognized that adequate initial implementation of sound and reasonable procedures is, in itself, insufficient to maintain an accurate reporting system. Periodic audits are required to determine if the system is functioning as it was designed to and if it is still conceptually adequate. Some companies use their financial auditors to review the cost collection system, and their Quality Auditors to review the balance of the program; however, there is no method applicable to all companies. Each company knows best how to audit its own cost systems.

Often, however, the major emphasis is on auditing to determine whether the system is functionally adequate and too little, if any, emphasis is on reviewing to determine whether the system is still conceptually adequate. An annual ''conceptual'' review of the total Quality Cost system with more frequent monthly or quarterly ''functional audits'' of major portions of the system is usually sufficient.

CONCLUSION

The concept of Quality Costs has expanded to where it has become a principal management tool. Definitions and standards have been developed and refined along with techniques and methods for implementation. The Quality Cost Program is the bridge between line and executive management. It provides a common language, measurement and evaluation system which proves that quality pays in increased profits, productivity and customer acceptance.

This paper was an attempt to summarize the material to be included in the planned ASQC publication "Principles of Quality Costs", but is in no way all inclusive. As indicated earlier, the material used was obtained from many sources. References were provided wherever possible. Special thanks to the many members of the Quality Costs Technical Committee who have submitted material for this effort.

References

1. Juran, J. M., *Quality-Control Handbook,* First Edition, McGraw-Hill Inc., 1951.
2. Sittig, J., *The Economic Choice of Sampling Systems in Acceptance Sampling,* Bulletin of International Statistics Institute, 1951.
3. Freemen, H. L., *How to Put Quality Costs to Use,* Transactions of the Metropolitan Conference, ASQC, 1960.

4. Feigenbaum, A. V., *Total Quality Control,* McGraw-Hill, Inc., 1961.
5. Department of Defense, *MIL-Q-9858A, Quality Program Requirements,* 16 December 1963.
6. Morgan, E. D. and Ireson, W. G., *Quality Cost Analysis Implementation Handbook,* Stanford University, Department of Industrial Engineering, 27 March 1964.
7. Office of the Assistant Secretary of Defense (Installations and Logistics), Quality and Reliability Assurance Technical Report TR8, *A Guide to Quality Cost Analysis,* 31 May 1967.
8. Quality Cost Effectiveness Technical Committee, ASQC, *Quality Costs — What and How,* American Society for Quality Control, 1977.
9. Quality Costs Technical Committee, ASQC, *Guide for Reducing Quality Costs,* American Society for Quality Control, 1977.
10. Quality Costs Technical Committee, ASQC, *Guide for Managing Vendor Quality Costs,* American Society for Quality Control, 1980.
11. Quality Costs Technical Committee, ASQC, *Bibliography of Articles Relating to Quality Cost Concepts and Improvement,* 1 January 1981.
12. Department of Defense, *MIL-STD-1520B, Corrective Action and Disposition System for Nonconforming Material,* 3 July 1980.
13. Juran, J. M., *Quality Control Handbook,* Third Edition, McGraw-HIll, Inc., 1974.
14. Juran, J. M., *Managerial Breakthrough,* McGraw-Hill, Inc., 1964.
15. Campanella, J., *A Simplified Approach to the Use of Costs Related to Quality,* Transactions of the 13th Annual All Day Conference on Quality Control, L. I. Section, ASQC, 1975.
16. Office of the Assistant Secretary of Defense (Installations and Logistics), op. cit.
17. Juran, J. M., op. cit., (Third Edition).
18. Campanella, J., *The Fairchild Republic Company Quality Cost Program,* transactions of the 33rd Annual Technical Conference, ASQC, 1979.
19. Hagan, J. T., *Quality Costs,* ITT, N. Y., 1981.
20. Quality Cost Effectiveness Technical Committee, ASQC, *Quality Costs — What and How,* Second Edition, American Society for Quality Control, 1971.
21. Quality Cost Technical Committee, ASQC, *Guide for Reducing Quality Costs,* American Society for Quality Control, 1977.
22. Hagan, J. T. op. cit.
23. Quality Cost Effectiveness Technical Committee, ASQC, op. cit., Second Edition.

QUALITY COSTS — PAY

James Demetriou, Manager of Quality Administration
ITT Avionics Division
Clifton, New Jersey

ABSTRACT

A quality cost system, like accounting, is an effective management tool for measuring and controlling cost, with the objective of improving quality, productivity and profitability. In initiating a quality cost system, the greatest difficulty can be convincing management of the benefits. To do this, it becomes necessary to sell management people on the advantages. This presentation will highlight the major features one must know and use to both sell management and initiate a quality cost system and will use examples and results from a system presently operating.

INTRODUCTION

Can anyone envision a modern business being run without a financial accounting system? In today's business the financial accounting system is used as a standard of measure for performance of the enterprise and forecasting. The unit of measure used is a common one understood by all — the dollar. How many times in our daily lives, in and out of business, do we ask the question; how much does it cost?

Should not management ask how much does quality cost; and, more importantly — how much does the lack of quality cost? These questions are best answered in the language all management understands — dollars. A quality cost system will accomplish this, and, like an accounting system, is an effective management tool for measuring and controlling quality costs. Once this measure is available, it can be utilized to improve the quality system with resultant increases in productivity and bottom-line profit.

Managers in a modern business are usually aware of the cost of the Quality Control Department through normal accounting procedures and if asked what the cost of quality is, would give that figure. They would, that is, if they had not been educated by their Quality Department and were completely unaware of the significant difference between "the cost of quality" and "quality costs."

For those of you who are not familiar with this difference between the two statements or have used them interchangeably, I will explain.

"Quality costs" are invariably assumed to be the cost incurred by Quality Department, for the obvious reason that they are described as Quality, not Manufacturing, Accounting, Procurement, etc.; therefore, they have to be those of the Quality function?!!? However, all the expenditures of a modern business on activities and actions which are directly involved with producing a "quality" item are expenditures which may be fairly categorized as the "cost of quality." These costs will include not only those of the Quality function but also some of the costs incurred by or caused by Engineering (design changes), Manufacturing (rework, repair, design changes, scrap) and Procurement (wrong parts, etc.). All of these costs must be recognized as "costs of quality" — the costs involved in making the product meet the specifications, physical and performance-wise — also known as "quality of product." It remains then for those Quality functions which have not educated their top management about the difference between quality cost/cost of quality to establish a dialogue with them. This must be done by first getting their attention as in the proverbial donkey and the 2″ × 4″. Instead of a club, however, Quality must use a very attractive lure — money. Top management must be shown that a properly established and implemented cost of quality program can highlight problems in their manufacturing operations, in terms of cost of quality — in other words, bottom-line dollars.

The system and procedures necessary for a cost of quality program are very often already available in a reasonably well-organized company with an effective Quality Department and Accounting system. This presentation will cover the major activities and actions necessary to organize and implement a cost of quality system.

Organization

To have a cost of quality system, there must be a quality system and an accounting system which retrieves the necessary data. The existence of an adequate quality system depends on organization of the company into clearly defined functions and goals as evidenced by organization and functional organization charts.

Procedures

Definition of functions and their goals must be described in procedures which will set the standard(s) for accomplishing the goals. Thus, adequate planning must be made to set the standards — see FIGURE 3, Management Feedback Cycle diagram, for sequence of events once standards are set.

As stated earlier, a cost of quality system can only exist if other ingredients are present — the system depends on the developed data of other functions and cost of quality reports will inevitably include actions by other departments.

The Finance Department is responsible for establishing the procedures for collecting cost of quality data and reporting it. At this point, there is no problem, since everyone believes the numbers from Accounting. However, since there are no generally accepted accounting principles defining cost of quality as such, the system for obtaining this latter data and reporting it must be developed and integrated into the existing, approved accounting system. This defining of quality costs must be done by the department most familiar with this data, namely, the Quality Department. This also becomes an opportunity to educate the Accounting Department on cost of quality and the savings potential that can be realized.

The data to be represented by the Cost of Quality Report will be in three general categories with detailed subgroups. FIGURE 1 is an example of a typical format for reporting, representing these categories. There may be more or less subgroups than the examples shown depending on your industry and the detail you desire. The important thing is to place the costs in what you determine is the proper category.

Reporting

The method/format used in reporting the cost of quality in any given company must be that which is best for that company or management style. FIGURES 1 and 2 are examples of how cost of quality may be reported.

Cost of quality as a percent of sales is typically considered to be made up of three categories, as shown earlier. It is also typical that if more is expended on Prevention Activity, then Appraisal and Failure should reduce in some fairly equal proportion. Equally, if more is expended on Appraisal, then Field Expenditures should reduce. The ideal is to spend most on Prevention and to reduce overall cost of quality expenditures.

Typically, a target is set each year for the cumulative end of year cost of quality as a percent of total sales. Each month, the data from Accounting is plotted by each category and sub-category, on a cumulative basis. The data is then entered into the formats developed and plotted on a cumulative curve on which the previously calculated cumulative target has been plotted for the year. In this way, an easily read, visual presentation of performance is available. Other data is entered into a matrix in the form of dollar targets and actuals, both monthly and year to date, again providing detailed reference data. The matrices and chart, plus a summary analysis of the performance for the reporting month is put together as a monthly report to senior management, PMO, Quality Engineering for review and action on major contributors to the total cost of quality.

COST OF QUALITY
YEAR TO DATE AND CURRENT MONTH
FOR

CATEGORY	YEAR END ( ) % COQ	YEAR TO DATE			CURRENT MONTH		
		FORECAST ($000's)	ACTUAL ($000's)	% OF ACTUAL COQ	FORECAST ($000's)	ACTUAL ($000's)	% OF ACTUAL COQ
PREVENTION Quality Engineering	24.4	622.0	586.0	19.1	61.9	57.5	14.9
Quality Administration	6.0	129.9	129.7	4.2	12.7	13.3	3.4
Total Prevention	30.4	751.9	715.7	23.3	74.6	70.8	18.3
APPRAISAL Receiving Inspection	6.5	242.4	237.4	7.8	25.5	23.6	6.0
All Other Inspection	6.8	203.0	228.7	7.5	25.9	27.2	7.0
Test	6.2	420.9	183.6	6.0	91.6	28.8	7.4
Standards & Calibration	7.2	176.7	162.5	5.3	18.6	18.2	4.7
Vendor Quality	6.8	168.7	175.5	5.7	16.1	16.7	4.3
Travel Expense	1.6	76.7	59.6	1.9	9.7	9.3	2.4
Mat'ls. & Evaluation Lab.	5.0	107.0	101.2	3.3	11.6	9.2	2.4
Product Acceptance	13.9	355.8	314.0	10.3	35.2	36.4	9.4
Total Appraisal	54.0	1751.2	1462.5	47.8	234.2	169.4	43.6
FAILURE Rework Labor	8.7	240.8	271.3	8.9	22.4	47.1	12.2
ECR/ECN Labor	5.7	481.5	522.3	17.0	44.8	86.3	22.3
Scrap	0.4	43.3	85.0	2.8	4.2	11.9	3.1
Warranty	0.5	35.7	5.5	0.2	3.3	1.6	0.5
Total Failure	15.3	801.3	884.1	28.9	74.7	146.9	38.1
Total Quality Costs	2430.7	3304.7	3062.3		383.5	387.1	
Sales	71570.9	71654.0	71991.3		7222.0	8091.1	
COQ as a % OF Sales	3.4	4.6	4.3		5.3	4.8	

FIGURE 1

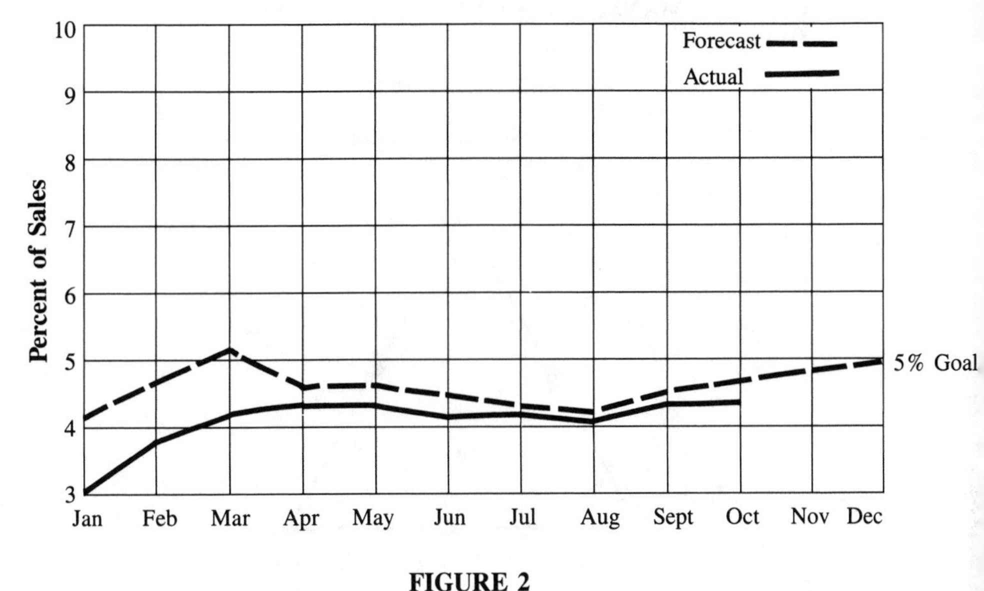

FIGURE 2

Analysis

Here again we must analyze the cost of quality as we would our monthly budget statements. We must look at the largest variances and/or dollar amounts being expended and concentrate on those areas. In other words, don't spend your time trying to save money on pencils and paper clips.

FIGURE 3, Management Feedback Cycle, can be used to implement and control your budget or the cost of quality reporting system. The top of the triangle is where you set the standard of performance you expect by procedures and/or dollar amounts. This can also be considered your goal(s).

Your cost of quality reporting is the lower right of the triangle or the measurement of performance to the established standard. The difference between the standard and measured value on budget reporting is known as variance and the same would apply for reporting cost of quality.

MANAGEMENT FEEDBACK CYCLE

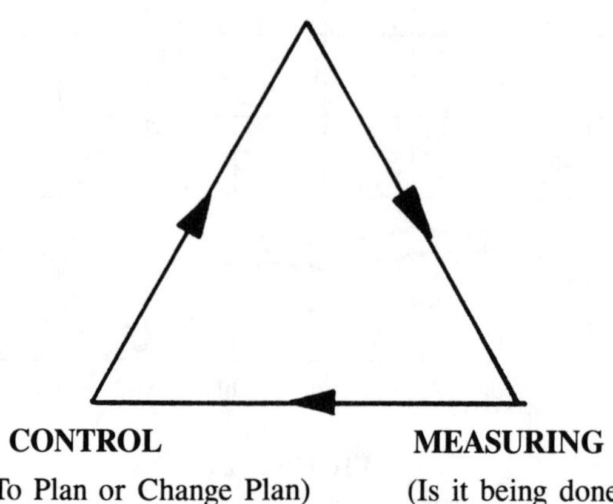

FIGURE 3

Corrective Action

Corrective action for those areas identified as high cost of quality contributors must be initiated. Experience shows that when managers and other concerned individuals receive copies of their report they immediately note out of control categories within their respective areas of responsibility. FIGURE 4 is a typical form that can be used to show cost of quality by program and/or functional area. When all else fails the existing company quality and corrective action system can be utilized.

FAILURE COSTS ($)

Program/Responsibility		Test		Fabrication		Assembly		Inspection			FIELD WARRANTY	
Name	Quality Engineer	ECR/ECN	REWORK	ECR/ECN	REWORK	ECR/ECN	REWORK	ECR/ECN	REWORK	SCRAP		TOTALS
		—	—	—	—	—	188	212	—	—	—	400
		—	—	—	1304	142	598	—	—	—	—	2044
		—	—	1554	1249	921	1962	—	238	2820	—	8744
		—	948	1970	3912	1558	4935	106	1455	—	—	14884
		18115	1422	2358	3495	47864	5463	344	556	7000	—	86617
		—	223	111	666	945	141	—	159	450	—	2695
		—	725	8905	11662	779	920	—	159	867	—	24017
		—	1003	222	915	71	378	—	661	—	—	3250
		—	—	56	195	—	—	—	—	—	—	251
		—	—	—	—	—	—	—	78	960	1564	2602
		—	—	—	168	—	1322	—	—	—	—	1490
	Sub-totals											
	ECR/ECN	18115		15176		52280		662				86233
	Rework		4321		23566		15907		3306			47100
	Scrap									12097		12097
	Warranty										1564	1564
	Totals	22436		38742		68187		3968		12097	1564	146994

FIGURE 4

Results

Results to be obtained from the establishment of a cost of quality program will be proportionate to the effort and care with which the program is thought out, set up and implemented, including top management involvement. As stated earlier, top management's attention is easily obtained when bottom line dollars are associated with a proposed program. The cost of quality program can be very fruitful in this respect. An example of the cost avoidances that have been obtained by an operating system in a large company is shown in TABLE 1.

TABLE 1

COST AVOIDANCE HISTORY

Year	C of Q as % Sales	$/Year Saved	Cum. $ Saved
1969	13.5		
1970	12.4	518,958	518,958
1971	10.1	791,476	1,310,434
1972	8.5	560,080	1,870,514
1973	7.96	260,982	2,131,496
1974	7.32	392,537	2,524,033
1975	6.71	384,562	2,908,595
1976	5.83	626,639	3,535,234
1977	5.86	(22,500)*	3,512,734
1978	5.95	(48,600)*	3,464,134
1979	5.50	282,000	3,746,134
1980	5.25	310,647	4,056,781

*Negative cumulative savings resulted from new programs and higher than anticipated level of engineering changes.

Summary

In summation, a cost of quality system is a valuable tool that may be utilized by Management for many purposes, some of which are to:

- Establish quality cost reduction goals
- Determine if maximum benefit is being derived from the cost of quality expenditures
- Determine the department budget for specific areas
- Identify areas needing improvement
- Identify areas in control

Most important, costs of quality are a management score card for measuring quality's contribution to profit margins.

AUTHOR INDEX

Agnone, Anthony 79

Breeze, John D. 467, 475
Brewer, Clayton C. 327
Brewer, Clyde W. 315, 557
Brown, F. X. 215, 297

Campanella, Jack 563
Cartin, T. J. 5
Cawsey, R. A. 179
Cerzosimo, Ronald R. 61
Corcoran, Frank J. 321

Dawes, Edgar W. 71, 149, 239
Dobbins, Richard K. 127, 139, 201, 535
Demetrious, James 581
Dwyer, Michael J. 15

Esterby, L. James 413, 543

Goeller, W. D. 437
Goetz, Victor J. 251
Grimm, Andrew F. 101, 207
Gryna, Frank M. 223
Gunneson, Alvin 375, 527
Gurunatha, T. 251

Hagan, John T. 89
Harrington, H. James 163, 397

Kennedy, William J. 1
Kiang, T. David 169
Kimble, David L. 271
Koga, Yohei 19
Kolacek. O. G. 157
Kroeger, Robert C. 259

Latzko, William J. 95, 549
Lee, Dennis D. 483
Liebesman, Burton S. 425

Mayben, J. E. 359
Moore, William N. 53, 233

Oak, A.D. 165
Ortwein, William J. 517

Patel, Narendra S. 285
Pollard, Ronald L. 191

Raouf, A. 29
Rogers, C. B. 35

Sadowski, Joseph 11
Scanlon, Frank 303, 391, 499
Shainin, Dorain 257
Stalcup, R. W. 387
Stenecker, Robert G. 117

Whitton, A. Q. 43
Wilhelm, W. C. 337
Williams, Ronald J. 491
Winchell, William O. 185, 245, 309, 447

Zaludova, A. H. 455
Zerfas, James F. 279

TITLE INDEX

Activities for Reduction of User's Costs 19
Attacking Quality Costs 117

Bayesian Cost Analysis of Rectifying Rejected Lots 271
A Business Performance Measure of Quality Management 179

Can Quality Cost Principles Be Applied to Product Liability? 257
Computer Isolation of Significant Quality Costs 359
A Cost-Determined Quality Control Plan for Adjustable Process 1
Cost-Effective Quality 15
Cost Effectiveness of Corrective Action — Quality Costs — A Place for Decision Making and Corrective Action 139
Cost Improvement Through Quality Improvement 391
The Cost of Software Quality Assurance 437
A Cost Oriented Quality Control System 11
Cost Reduction Through Quality Management 303
Cost Quality Control Chart for Variables 165

Developing a Cost Effective Program — How to Start 251

Extending Effectiveness of Quality Cost Programs 201

Guide for Managing Vendor Quality Costs 309
Guide for Managing Vendor Quality Costs 447
Guide for Reducing Quality Costs 491
Guide for Reducing Quality Costs 279

The Hidden Aspect of Vendor Quality Cost 185
Honeywell's Cost Effective Defect Control Through Quality Information Systems 61
How to Effectively Implement a Quality Cost System 375

Improved Productivity Through Quality Management 499
Innovation in Quality Costs in the New Decade 327

Life Cycle Costing — A New Dimension for Reliability Engineering Challenge 169

Management Budget Control: Quality Labor Standards 191
Measuring Quality Costs by Work Sampling 543
A Method for Predicting Warranty Costs 207
Methods for Selling Total Quality Cost Systems 43

Minimized Cost Sampling Technique 483
Minimizing the Cost of Inspection 549
Mr. CEO, Your Company's Quality Posture Is Showing! 535

Optimizing Attribute Sampling Costs — A Case Study 71
Our Only Output Is Information 387

A Participatory Approach to Quality 527
The Philosophy and Usefulness of Quality Costs 233
Prediction of Human Performance in a Quality Control System: A Step Towards Cost Control 29
Principles of Quality Costs 563

"Q" Costs Results: The Petroleum Industry 557
Quality Cost — A Key to Productivity 397
Quality Cost Analysis: A Productivity Measure 413
Quality Cost and Profit Performance 215
Quality Cost — A Place for Evaluation of Customer Satisfaction 163
Quality Cost — A Place for Financial Impact 157
Quality Cost at Work 89
Quality Cost Management for Profit 127
Quality Cost Measurement and Control 79
Quality Costs and Strategic Planning 297
Quality Cost — A New Perspective 259
Quality Costs — A Place On The Shop Floor 149
Quality Costs — A Review and Preview 315
Quality Costs Can Be Sold, Part I 467
Quality Costs Can Be Sold, Part II 475
Quality Costs — Pay 581
Quality Costs Principles — A Preview 321
Quality Costs — What Does Management Expect 223
Quality Costs: Where Are They in the Accounting Process? 101
Quality Optimization via Total Quality Costs 455
Quality Program Modeling for Cost Effective Tailoring 337
Quality/Reliability Challenges for the 1980's 351

Reducing Appraisal Costs 239
Reducing Clerical Quality Costs 95
Reducing Failure Costs and Measuring Improvement 245
Reducing Manufacturing Costs While Maintaining Reliability and Quality 5
Reducing Quality Costs 53

Selection of MIL-STD-105D Plans Based on Costs 425
Study Costs and Improve Productivity 517
Source Surveillance and Vendor Evaluation Plan 285

Uncovering the Hidden Costs of Defective Material 35